Selected Topics in Power, RF, and Mixed-Signal ICs

EDITORS

Yan Lu
University of Macau, Macao, China

Chi-Seng Lam
University of Macau, Macao, China

Tutorials in Circuits and Systems

For a list of other books in this series, visit www.riverpublishers.com

Series Editor

Franco Maloberti

President IEEE CAS Society
University of Pavia, Italy

Published 2017 by River Publishers
River Publishers
Alsbjergvej 10, 9260 Gistrup, Denmark
www.riverpublishers.com

Distributed exclusively by Routledge
4 Park Square, Milton Park, Abingdon, Oxon OX14 4RN
605 Third Avenue, New York, NY 10017, USA

Selected Topics in Power, RF, and Mixed-Signal ICs / by Yan Lu, Chi-Seng Lam, Franco Maloberti.

ISBN 978-87-93609-40-2 (print)

Table of contents

Introduction

Integrated circuits and systems are the backbone of modern information society, and play a vital role on enabling the internet-of-things (IoT), big data, autonomous vehicles, drones, and many emerging applications. For example, numerous wireless sensing devices are going to be deployed for environment or health monitoring. For such systems, the hardware consists of three major parts: communication, sensing, and power management. Due to the amount of transmitted/processed data and the number of devices are enormous, low-power and energy-efficiency have become crucial requirements for these devices.

This book covers several selected advanced topics in the areas of power, radio frequency, and mixed-signal integrated circuits and systems. Both their fundamental operation principles and the state-of-the-art advancements will be addressed. The book includes eight chapters, and will introduce very-high-frequency DC-DC switching converters, analog and digital low-dropout (LDO) regulators, sub-sampling frequency synthesizers, successive approximation (SAR) and hybrid analog-to-digital converters (ADCs), CMOS image sensors, CMOS temperature sensors, CMOS millimeter-wave (mm-Wave) power amplifiers, and ultra-low power Zigbee/BLE transmitters.

Chapter 1 focuses on the analysis and design of fast-transient and very-high-frequency switching converters. Switching loss and conduction loss are considered concurrently and an optimal switching frequency could be inferred for a particular application. Time-domain parameters such as line transient, load transient and reference tracking will be considered. Frequency-domain parameters such as maximum unity-gain bandwidths of various compensation schemes are discussed. Integrated circuit techniques of implementing fast-transient and very-high-frequency (VHF, 30MHz to 300MHz) switching converters are reviewed.

Chapter 2 discusses the LDO regulators, which are indispensable in system-on-a-chip (SoC) designs because they provide fast-transient response and good power supply ripple rejection. Conventional analog LDO regulators can perform excellent in normal supply voltages, while the digital LDO regulators recently become very popular for its low-voltage operation, good large-signal performance and process-scaling features. The design considerations, circuit techniques, and recent advancements on both analog and digital LDO regulators, will be addressed in this chapter.

Chapter 3 analyzes the sub-sampling PLLs, which is a low-power analog building block and enables low-noise frequency synthesis because of its high-loop-gain characteristics. In this chapter, the principle of sub-sampling PLLs will be introduced. Then, the concept is extended to fractional-N synthesis. Detail implementation and analysis are covered for GHz frequency synthesis.

Chapter 4 introduces a high-performance hybrid ADC architecture with digitally-assisted techniques. SAR ADCs achieve excellent power efficiency due to its simple architecture and dynamic operation, while its conversion speed is limited by its sequential conversion. Hybrid ADC takes the design advantages of multi conventional ADC architecture to optimize the conversion speed, resolution and power dissipation. A hybrid ADC architecture, combining the flash, SAR and pipelined ADC with interleaving and op-amp sharing schemes to achieve near GHz sampling rate and 11-bit resolution, is introduced in this chapter. Moreover, some digitally-assisted solutions that fix the conversion errors from offset and DAC mismatches are also introduced.

Chapter 5 introduces recent advancement of CMOS image sensors. The fundamental operation principle is mentioned in detail and then some recent topics including shrinking pixel pitch, high sensitivity, high speed, etc. Finally, biomedical applications with CMOS image sensor technology are demonstrated such as retinal prosthesis, fluorescence detection for ELISA, and implantable micro imaging devices.

Chapter 6 talks about the ultra-low-power and energy-efficient high accuracy CMOS temperature sensors, which play an important role in the upcoming IoT era, with emerging applications including medical, surveillance, and environmental monitoring and many more. Being one of the most common measurement parameters, high performance CMOS temperature sensors become an important design element of such microsystems. This chapter covers the fundamental knowledge and operating principles of CMOS temperature sensors. The main sources of inaccuracy will be identified, followed by possible circuit/system level solutions to achieve high power/energy efficiencies. Case studies of state-of-the-art CMOS temperature sensors will also be introduced.

Chapter 7 focuses on realizing CMOS mm-Wave power amplifiers (PAs) towards more output power, higher efficiency and broader bandwidth. The mm-Wave transceivers in advanced CMOS still poses many challenges at device, circuit and architecture levels. In addition to generic difficulties, such as high-frequency operation and low active gain, mm-Wave designers must deal with issues like low breakdown voltage, high interconnect loss, unwanted mutual coupling, poor device matching, inaccurate PDK high-frequency models, strict design rules, long EM-simulation time, etc. In this chapter, all the aforementioned design challenges will be described. Novel design techniques at mm-Wave will be presented, followed by prior-art PA and TX designs in advanced CMOS technologies.

Chapter 8 deals with ultra-low power Zigbee/BLE transmitters for IoT applications, such as wearable or implantable healthcare monitoring. The small physical dimension of those products has significantly limited the battery size, pressuring the power budgets of the radios at both architectural and circuit levels. The transceiver efficiency has been significantly reduced in the past few years, thanks to both CMOS technology down scaling and the development of the new circuit techniques for power reduction. In this tutorial, the design of ultra-low power transmitters for Zigbee/BLE standard will be discussed from system architecture to circuit techniques. In addition, a 2.4-GHz transmitter using a function-reuse Class-F DCO-PA achieving 22.6% system efficiency at 6-dBm Pout will be presented as a case study.

CHAPTER 1

Very-High-Frequency and Fast-Transient DC-DC Switching Converters

Wing-Hung Ki and Lin Cheng

The Hong Kong University of Science and Technology
Hong Kong, China

This chapter focuses on the analysis and design of fast-transient and very-high-frequency switching converters. Switching loss and conduction loss are considered concurrently and an optimal switching frequency could be inferred for a particular application. Time-domain parameters such as line transient, load transient and reference tracking will be considered. Frequency-domain parameters such as maximum unity-gain bandwidths of various compensation schemes are discussed. Integrated circuit techniques of implementing fast-transient and very-high-frequency (VHF, 30MHz to 300MHz) switching converters are reviewed.

1  Outline

- Why VHF converters?
- Switching frequency on efficiency
- Converter speed in frequency and time domains
- Frequency domain: bandwidth and UGF extension
- Time domain: fast transient techniques

VHF is the short-form of very-high-frequency, that is, from 30MHz to 300MHz. We will see how switching loss would affect the efficiency of a switching converter. Now, the speed of a converter can be considered in either the frequency domain or in the time domain. More will be said on this later.

2  Why VHF Switching Converters?

Why switching converters?
- High efficiency ($\eta_{ideal} = 1$)

Switching frequency (f_s) of switching converters:
- Fixed: PWM control
- Variable: Hysteretic, constant on-time, constant off-time, etc.

Control methods of PWM converters:
- Voltage-mode control
- Current-mode control

Fast transient responses need high unity-gain bandwidth (UGF)
High UGF needs high f_s (e.g. UGF = $f_s/5$, $f_s/10$)
Small form factor needs high f_s

$\Rightarrow$ High f_s: VHF (very high frequency, 30 – 300 MHz)

A switching converter achieves the ideal efficiency of unity if no parasitic exists. The switching frequency can be fixed or varied, but fixed frequency is preferred as the spectrum is then predictable. Fixed-frequency converters are also known as PWM converters, as the pulse-width is modulated to control the output voltage. Both voltage-mode control and current-mode control can be employed. Now, fast transient responses need high unity-gain bandwidth, and high UGF requires a high switching frequency.

3 Voltage-Mode Buck Converter

A switching converter consists of the power stage and the feedback network.

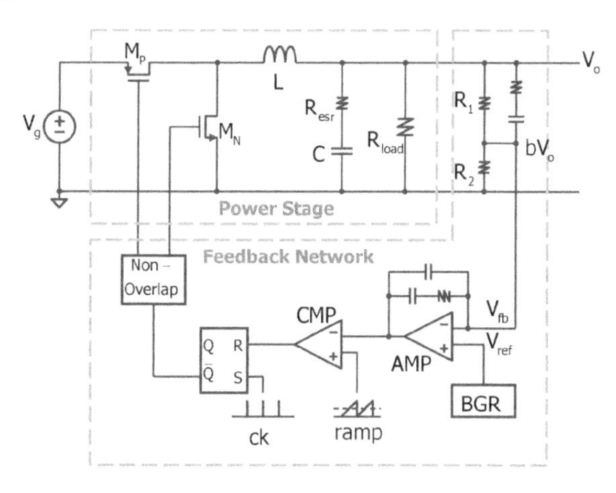

This slide shows a voltage-mode buck converter. It consists of the power stage and the voltage-mode control feedback network.

4 Conduction Loss of Buck Converter

Conduction loss can be computed by considering the power stage.

Employ volt-second balance, the voltage conversion ratio (M) of the buck converter working in continuous conduction mode (CCM) is

$$\frac{V_o}{V_g} = M = \frac{D}{1 + D\frac{V_s(D) + V_{R\ell}(D)}{V_o} + D'\frac{V_d(D') + V_{R\ell}(D')}{V_o}}$$

$$= \frac{D}{1 + \frac{R_\ell}{R_L} + D\frac{R_s}{R_L} + D'\frac{R_d}{R_L}}$$

Let the converter works in continuous conduction mode (CCM). By using volt-second balance, the voltage conversion ratio can be computed as shown. It is equal to D if the parasitics are zero.

5 Switching Loss and Efficiency

Switching loss of the power transistors M_P and M_N is proportional to the switching frequency f_s:

$$P_{sw-loss} = C_{gT} V_g^2 f_s \qquad C_{gT} = C_{gP} + C_{gN}$$

In accounting for $P_{sw-loss}$, the efficiency of the CCM buck converter is

$$\eta = \frac{P_o}{P_o + P_{cond-loss} + P_{sw-loss}} = \frac{V_o I_o}{V_g I_g + C_{gT} V_g^2 f_s}$$

$$= \frac{1}{1 + \frac{R_\ell}{R_L} + D\frac{R_s}{R_L} + D'\frac{R_d}{R_L} + \frac{C_{gT} R_L f_s}{D^2}}$$

With no parasitic,

$$\eta_{ideal} = 1$$

Switching loss is due to switching the gate capacitances of the power transistors at the switching frequency. By adding it to the conduction loss, the efficiency of the converter can be computed.

6 Switching Converter Efficiency: Example

Technology: TSMC 45nm
V_g=3V, D=0.6, R_L=10Ω (V_o simulated = 1.71V)
R_ℓ=0.2Ω, R_s=0.35Ω, R_d=0.29Ω, C_{gP}=26.2pF, C_{gN}=16.5pF

$$\eta = \frac{1}{1 + 0.02 + 0.021 + 0.0116 + 0.001168 \times f_s / 1M}$$

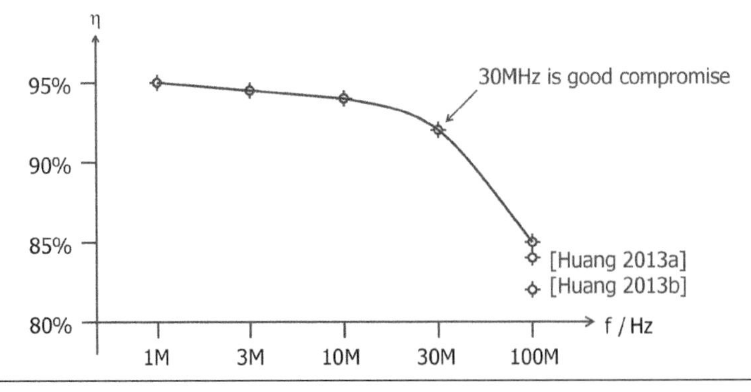

Using the design parameters stated in this slide, the efficiency starts to drop sharply when the switching frequency is higher than 30MHz, and 30MHz is a good compromise between relatively low switching loss and small inductor and capacitor values.

7  Switching Converter Speed

The speed of a switching converter can be characterized either in the frequency domain or the time domain.

Frequency domain:
    Loop bandwidth (= unity-gain bandwidth UGF)

Time-domain:
    Line transient
    Load transient
    Reference tracking

For time-domain waveforms, there are techniques developed to limit the overshoot and the undershoot.

The speed of a switching converter can be characterized in either the frequency domain or the time domain. In the frequency domain, we would like to extend the unity-gain bandwidth of the loop-gain function. In the time domain, circuit techniques on how to achieve fast transient responses while reducing overshoot and undershoot will be considered.

8  Frequency Domain: Loop-Gain Functions

A regulated switching converter can be decomposed into the power stage and the feedback network. The loop-gain function (whether using voltage-mode control or current-mode control) is given by

$$T(s) = A(s) \times P(s)$$

A(s) is the compensator function

P(s) is the power-stage function

N.B. P(s) the frequency response of the power stage plus non-compensator components (for example, the ramp voltage in voltage-mode control).

The loop-gain function of a switching converter can be decomposed into the compensator function and the power-stage function.

9  T(s) of Voltage-Mode CCM Converters

Loop-gain functions of voltage-mode PWM CCM converters with trailing-edge modulation are compiled. Parasitic resistances except ESR are excluded [Ki 1998].

Buck: $$T(s) = A(s) \times \frac{b_o V_o}{D V_m} \cdot \frac{1 + sCR_{esr}}{1 + \frac{sL}{R_L} + s^2LC} \qquad b_o = \begin{cases} b & \text{OTA} \\ 1 & \text{opamp} \end{cases}$$

Boost: $$T(s) = A(s) \times \frac{b_o V_o}{D' V_m} \cdot \frac{[1 - sL/(D'^2 R_L)]}{1 + \frac{sL}{D'^2 R_L} + \frac{s^2LC}{D'^2}}$$

Buck-boost: $$T(s) = A(s) \times \frac{b_o |V_o|}{DD' V_m} \cdot \frac{[1 - sDL/(D'^2 R_L)]}{1 + \frac{sL}{D'^2 R_L} + \frac{s^2LC}{D'^2}}$$

The resistor divider $b=R_2/(R_1+R_2)$ is part of the power-stage function P(s). If the compensator A(s) is realized by an OTA, the ratio b shows up in P(s); however, if an op-amp is used, b does not show up in P(s) (it becomes 1).

10  Voltage-Mode Converter: Loop Gain Function

In discussing fast-transient converters, one important parameter is the unity-gain bandwidth (UGF).

The loop gain function of the buck converter with voltage mode control operating in CCM ignoring ESR is given by [Ki 1998]

$$T(s) = A(s) \times \frac{bV_o}{DV_m} \cdot \frac{1}{1 + \frac{sL}{R_L} + s^2LC} = A(s) \times \frac{P_o}{1 + \frac{1}{Q}\frac{s}{\omega_o} + \frac{s^2}{\omega_o^2}}$$

The resonance frequency ω_o and the pole-Q are

$$\omega_o = \frac{1}{\sqrt{LC}} \qquad Q = R_L\sqrt{\frac{C}{L}}$$

The converter enters DCM at

$$R_{L(BCM)} = \frac{2L}{D'T} \Rightarrow Q_{BCM} = \frac{2}{D'}\frac{1}{\omega_o T}$$

P(s) of CCM buck consists of a pair of complex poles, and the denominator can be expressed in terms of the (angular) resonant frequency and its pole-Q. Q attains the maximum value when the converter is operating at the boundary of CCM and discontinuous conduction mode (DCM), which is called the boundary conduction mode (BCM).

11  Voltage-Mode Converter: Bandwidth Limitation

For voltage mode buck, the output voltage ripple is given by

$$\frac{\Delta V_o}{V_o} = \frac{D'}{8}\frac{1}{LCf_s^2}$$

If $\Delta V_o/V_o$=0.01 and D=0.5, then the complex pole-pair is at

$$\omega_o = 0.4f_s \quad \Rightarrow \quad f_o = \frac{\omega_o}{2\pi} \approx \frac{f_s}{16}$$

$$Q_{BCM} = \frac{2}{D'}\frac{1}{\omega_o T} = 10$$

In using dominant-pole compensation, to have adequate gain margin GM of 6dB (=2), UGF has to be reduced by 10×2=20 times:

$$UGF = \frac{1}{20}\times\frac{f_s}{16} = \frac{f_s}{320}$$

If f_s=1MHz, then UGF is at around $f_s/320$ = 3.125kHz.

The output voltage ripple can be derived using charge balance analysis. If we assume it is 1% of the output voltage and the duty ratio is 0.5, then the resonant frequency (in Hz) is 1/16 of f_s, and the maximum pole-Q is 10. If dominant-pole compensation is used, to make sure that the resonant peak is 6dB lower than unity-gain, the unity-gain frequency of the control loop has to be $f_s/320$.

12  Voltage-Mode Buck: T(s) with R_δ

UGF of $f_s/320$ is too low. Fortunately (or unfortunately), the converter inevitably has parasitic resistors such as R_{ESR}, R_ℓ (inductor series resistor), R_s (switch resistance) and R_d (diode resistance), and the loop gain function is modified as [Ki 1998]

$$T(s) \approx A(s)\times\frac{bV_o}{DV_m}\cdot\frac{1}{1+s\left(\frac{L}{R_L}+CR_\delta\right)+s^2LC}$$

where

$$R_\delta \approx R_{ESR}+R_\ell+DR_s+D'R_d$$

This R_δ is at least 200mΩ, thus reducing Q_{BCM} to around 3. With GM to be 6dB, UGF is reduced by 3×2=6 times, and

$$UGF = \frac{1}{6}\times\frac{f_s}{16} \approx \frac{f_s}{100}$$

If f_s=1MHz, then UGF is at around $f_s/100$ = 10kHz.

The low UGF is due to the high pole-Q. Yet, parasitic resistors lower the pole-Q (by a factor of 3~4, for example) with only minor effect on the resonant frequency, and hence, the UGF can be extended 3 to 4 times, and UGF is now around $f_s/100$.

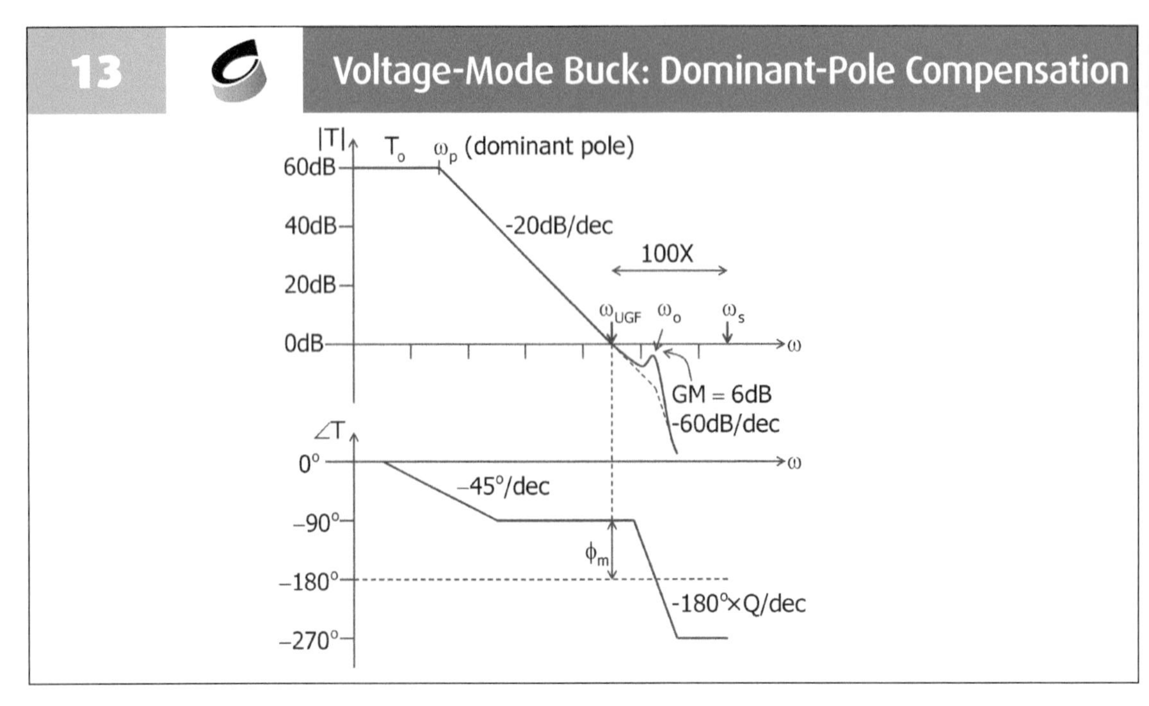

This slide shows an example of a dominant-pole compensated loop-gain function for a voltage-mode buck converter. The phase margin approach 90° but the phase is dropping fast.

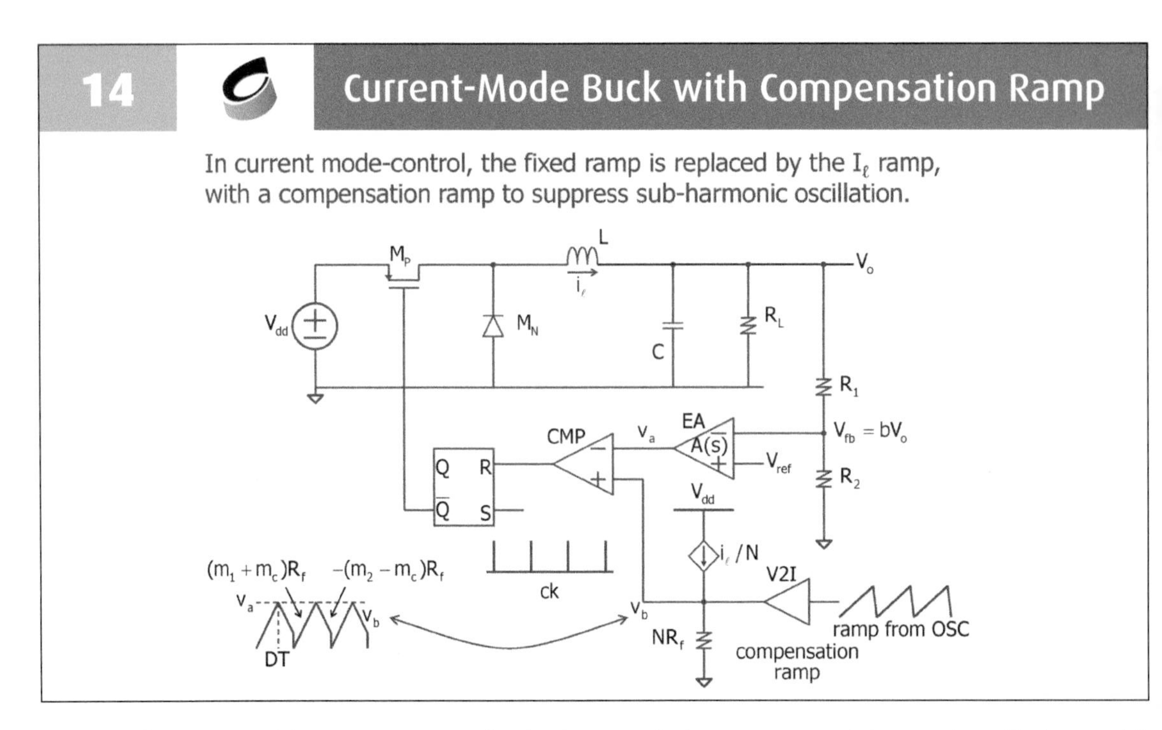

Next, let us consider a current-mode buck converter. Note that a compensation ramp is needed to suppress sub-harmonic oscillation, and it is added to the inductor current ramp as shown.

15  Current-Mode Buck: T(s) (1)

The loop-gain function of the buck converter with current-mode control operating in CCM ignoring ESR is given by [Ki 1998]

$$T(s) = \frac{A(s)b \times \frac{1}{CR_f}\frac{1}{n_1 D'T}}{s^2 + s\left(\frac{1}{CR_L} + \frac{1}{n_1 D'T}\right) + \frac{1}{n_1 D'T}\frac{1}{C}\left(\frac{1}{R_L} + \frac{(n_1 D' - D)T}{L}\right)}$$

with

$$n_1 = 1 + \frac{m_c}{m_1},\ m_c > \frac{m_2}{2} \Rightarrow n_1 > \frac{2-D}{2D'}$$

and

$$R_{L(BCM)} = \frac{2L}{D'T}$$

In general, the two poles are real, as discussed next.

The compensation ramp m_C is accounted for by the slope factor n_1. The current-mode buck converter working in CCM has two real poles.

16  Current-Mode Buck: T(s) (2)

If the poles are real and far apart, the denominator could be simplified as shown below.

$$\text{Buck:}\quad T(s) = A(s) \times b_o \cdot \frac{R_L || R_a}{R_f} \cdot \frac{1 + s/\omega_z}{\left(1 + \frac{s}{\omega_a}\right)\left(1 + \frac{s}{\omega_{t1}}\right)}$$

$$R_a = \frac{L}{(n_1 D' - D)T} \qquad n_1 = 1 + \frac{m_c}{m_1}$$

$$\omega_z = \frac{1}{CR_c} \qquad \omega_a = \frac{1}{C(R_L || R_a)} \qquad \omega_{t1} = \frac{1}{n_1 D'T}$$

For two real poles that are farther apart, pole-zero compensation could be used to extend the bandwidth.

With two real poles, pole-zero compensation can be used to extend the UGF. The two real poles are relatively far apart, and their locations are approximated as shown. This T(s) also includes the ESR zero at $-1/CR_C$.

17 Current-Mode Buck: Bandwidth Limitation

To compute the upper limit of UGF w.r.t. f_s, we simplify the current mode case as follows. Let D=0.5 and choose n_1=2 such that sub-harmonic oscillation could be suppressed even for D=0.667. The loop-gain function at $R_{L(BCM)}$ = 2L/D'T is then

$$T_{BCM}(s) = A(s)b_o \frac{25}{3}\frac{T}{CR_f}\frac{1}{(1+s25T/3)(1+sT)}$$

$$= \frac{T_{o(BCM)}}{\left(1+\dfrac{s}{\omega_{1(BCM)}}\right)\left(1+\dfrac{s}{\omega_2}\right)}$$

Note that at BCM, the equivalent resistance R_a cannot be ignored and hence, the simplify equation would give a larger error.

By choosing D=0.5, n_1=2, output voltage ripple is 1% of output voltage, and the converter operates in BCM, the two poles are approximately 8 times apart, with the high-frequency pole, in Hz, at $f_s/(2\pi)$, or $f_s/6$, and the low-frequency pole (in Hz) at approximately $f_s/50$.

18 Current-Mode Buck: Dominate-Pole Compensation (1)

Dominant Pole Compensation: The UGF with a phase margin of 70° should be $\omega_{UGF} = \omega_{1(BCM)}/3 = f_s/25$, that is

$$\text{UGF} = \frac{1}{2\pi}\times\frac{f_s}{25} \approx \frac{f_s}{160}$$

Clearly, there is little advantage ($f_s/320$ for R_δ=0) to even worse ($f_s/100$ for $R_\delta \neq 0$) when compared to the equivalent voltage-mode converter.

If dominant-pole compensation is used, and to achieve a phase margin of 70°, the UGF should be 3 times lower than the low-frequency pole. Hence, the UGF is ~$f_s/160$.

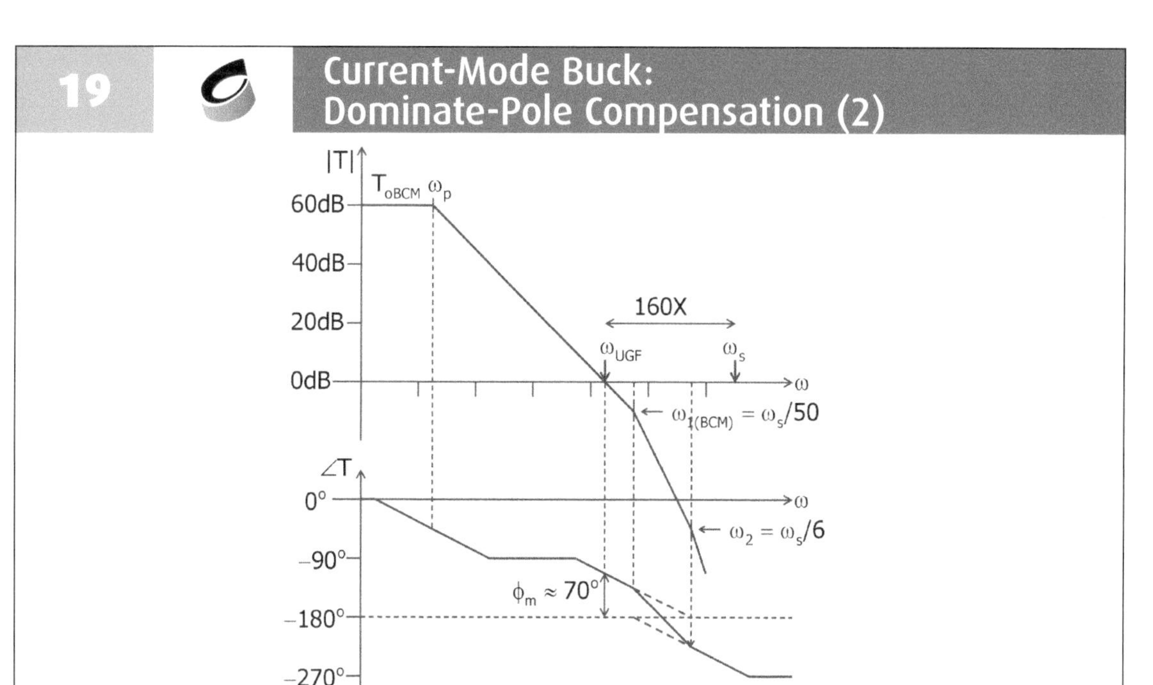

This slide shows the Bode plots of the loop-gain function T(s), with UGF ~ $f_s/160$.

20  Current-Mode Buck: Pole-Zero Cancellation (1)

Pole-zero cancellation compensation may be performed at ω_1, and should be done at the highest load current I_{omax} (smallest load resistance). Take $I_{omax} = 5I_{o(BCM)}$, then

$$\omega_{1max} = \frac{\omega_2}{3} \approx \frac{f_s}{20}$$

By adding a zero at ω_{1max} (using Type II compensator with an OTA) such that UGF = $f_2/3$, a phase margin of 70° is achieved, and

$$UGF = \frac{1}{3} \times \frac{f_s}{2\pi} \approx \frac{f_s}{20}$$

Hence, a current mode converter could have a unity-gain frequency 5 times higher than its voltage-mode counterpart.

By adding a zero to nullify the effect of the low-frequency pole, the high-frequency pole becomes the first pole, and the UGF should be 1/3 of the first pole. Hence, UGF=$f_2/3$~$f_s/20$, and this is 5 times higher than the voltage-mode counterpart.

21 Current-Mode Buck: Pole-Zero Cancellation (2)

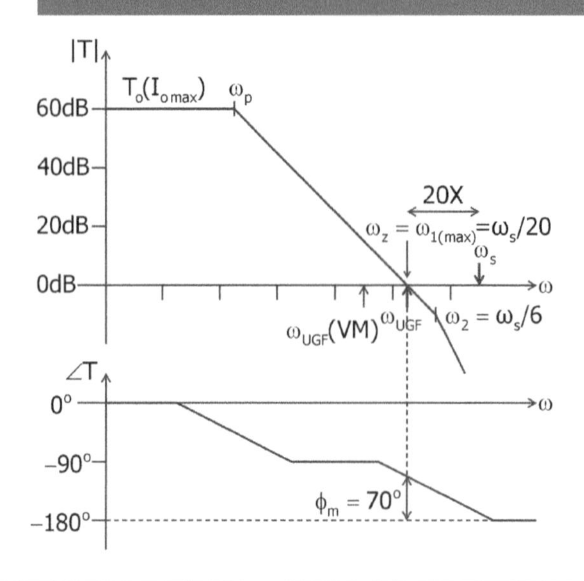

This slide shows the resultant loop-gain function of the current-mode buck converter operating in CCM.

22 VM Buck UGF Extension: Highpass Notch (1)

[Lee 2000] used a notch filter to nullify the effect of the complex pole at ω_0 of the buck converter, which depends only on L and C, and is fixed after design is done.
A highpass notch (HPN) filter was eventually used.

HPN compensator:

V_a — HPN — $A_1(s)$ — V_{fb} (−), V_{ref} (+)

$$HPN(s) = \frac{1 + \frac{1}{Q_L}\frac{s}{\omega_L} + \frac{s^2}{\omega_L^2}}{1 + \frac{1}{Q_H}\frac{s}{\omega_H} + \frac{s^2}{\omega_H^2}} \qquad \omega_L \approx \omega_o = \frac{1}{\sqrt{LC}}$$

N.B. $A_1(s)$ is the compensator with dominant pole compensation.

Pole-zero cancellation is applicable to current-mode converters. For voltage-mode converters, [Lee 2000] suggests to use a high-pass notch filter with complex zeros to nullify the phase drop (180°) due to the complex poles.

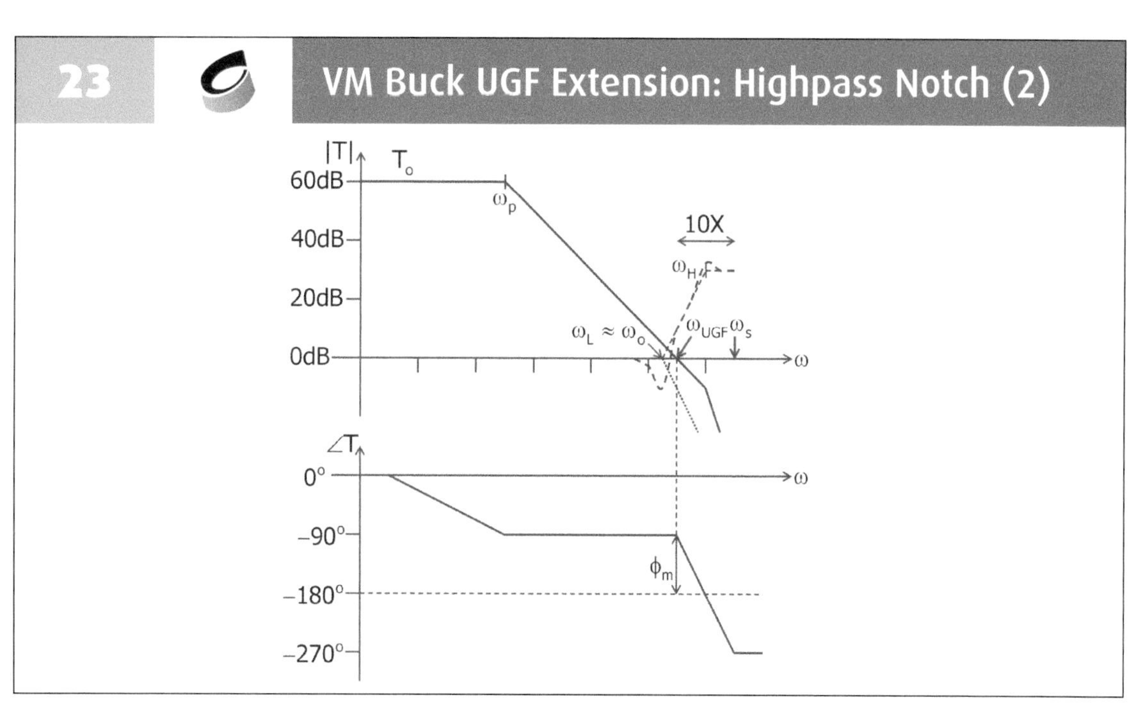

In [Lee 2000], the UGF was designed to be $f_S/10$.

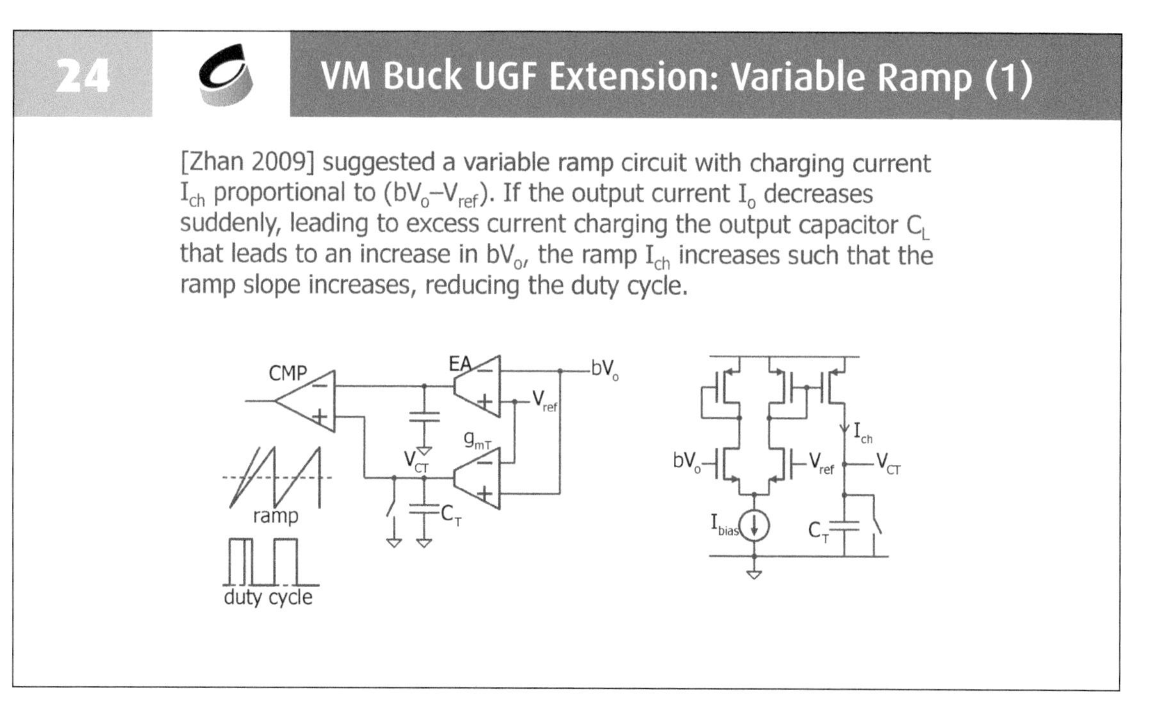

Consider a voltage-mode converter. Observe that if the load current decreases, the duty ratio should be decreased temporarily to avoid excessive overshoot. If the charging current of the ramp capacitor C_T is made inversely proportional to the load current, the ramp slope increases, and the duty ratio is reduced.

25  VM Buck UGF Extension: Variable Ramp (2)

Using the method in [Ki 1998], the loop-gain function is modified to:

$$T'(s) = A'(s) \times \frac{bV_o}{DV_m k^2} \cdot \frac{1}{1 + s\left(\frac{L}{k^2 R_L} + CR_\delta\right) + s^2 \frac{LC}{k^2}}$$

$$k = \sqrt{1 + \frac{bV_o g_m T}{I_{ch}}}$$

$$\omega_o' = \frac{k}{\sqrt{LC}} = k\omega_o$$

$$Q' = \frac{k(1+a)}{(1+k^2 a)} Q \qquad a = \frac{CR_\delta R_L}{L}$$

It can be shown that for a>1, Q'<Q, which is true at $R_{L(BCM)}$ = 2L/D'T.

Mathematically, the loop-gain function is modified by the parameter k that is related to the charging current through $g_m T$. As k>1, the resonant frequency increases while the pole-Q decreases, and the UGF can then be extended.

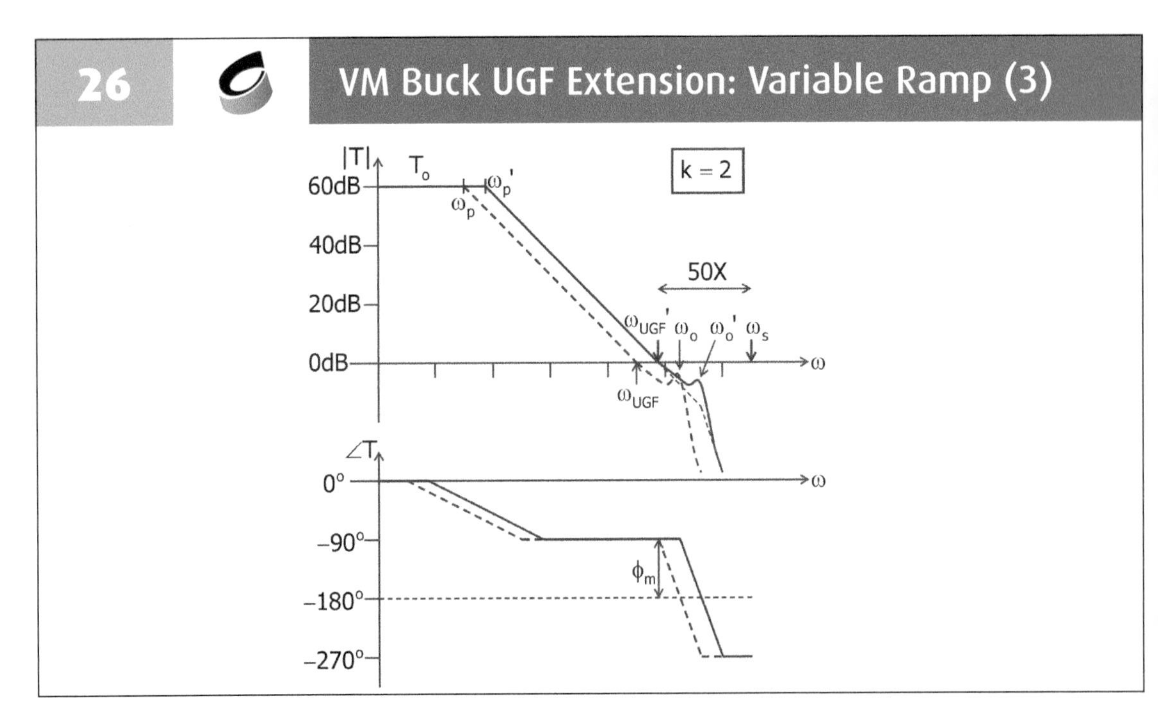

However, as $g_m T$ cannot be made too large and so does k, the extension of the UGF is limited.

27 Speed Considerations in Time Domain

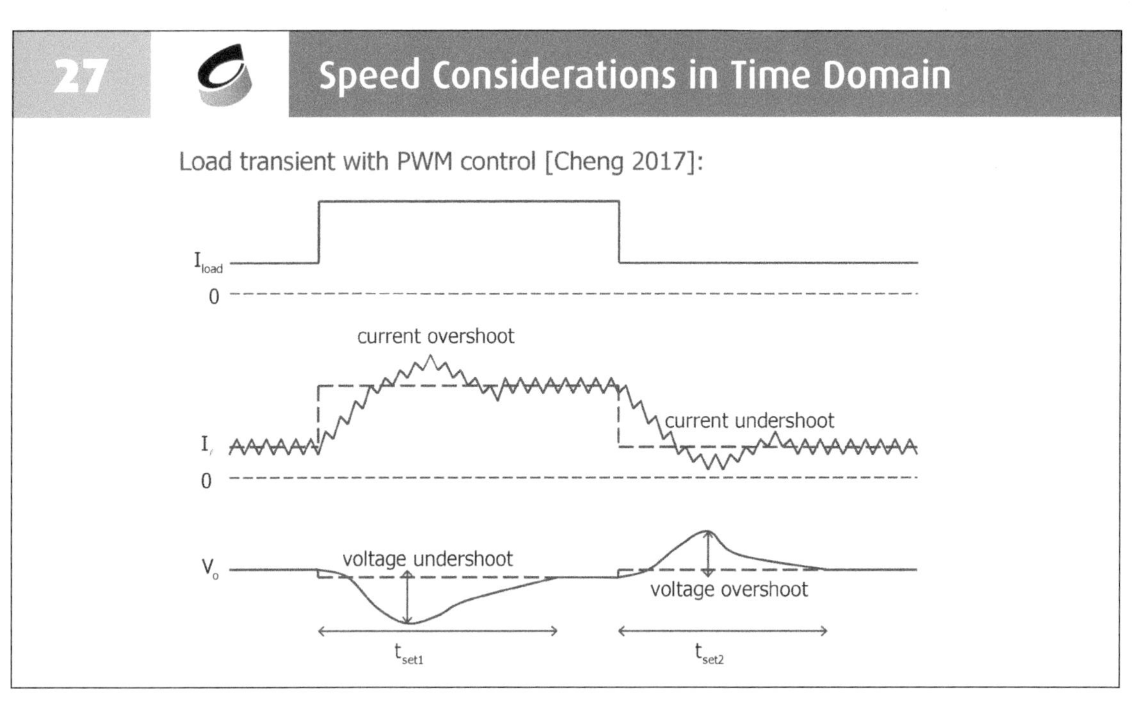

Let us turn our attention to time-domain speed consideration. When the load current increases abruptly, the inductor current cannot change as fast, and the delay leads to output voltage undershoot, which commands the inductor current to overshoot for correction. An analogous "counter"-mechanism occurs when the load current decreases abruptly.

28 Load Change Under Hysteretic Control

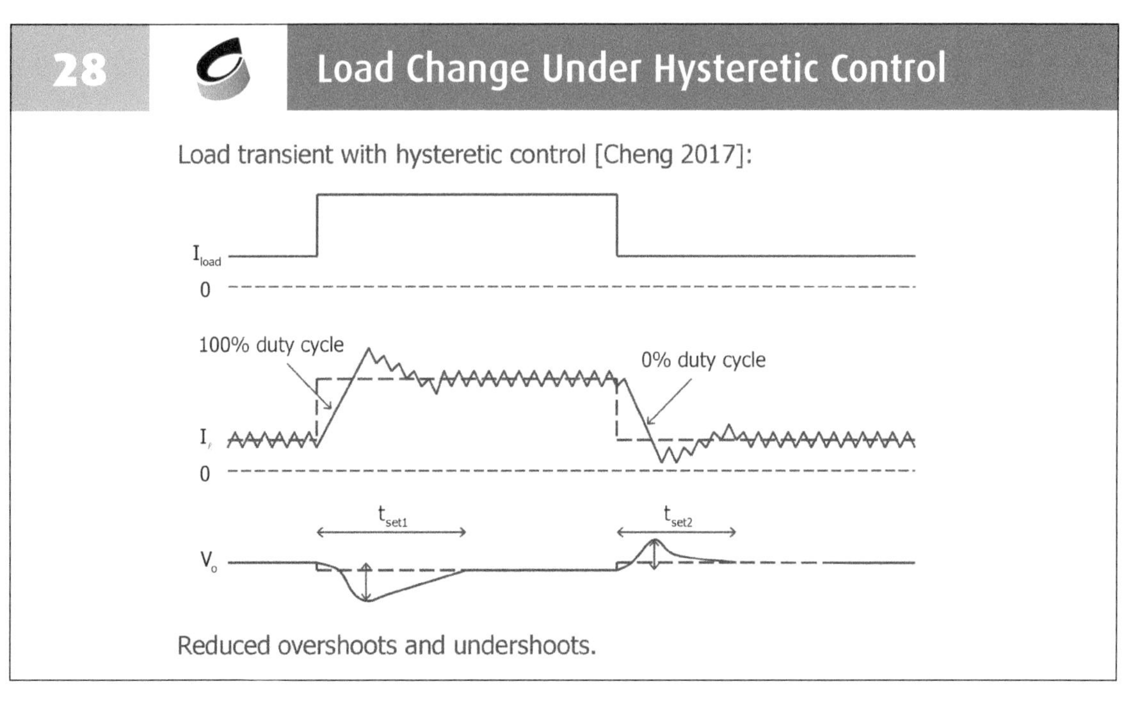

The advantage of hysteretic control is as follows. When the load current increases abruptly, the control loop allows the duty ratio to go to 1, and the inductor current increases with full speed that helps in reducing the output voltage undershoot. When the load current decreases abruptly, the duty ratio goes to 0, and the output capacitor is drained by the load current.

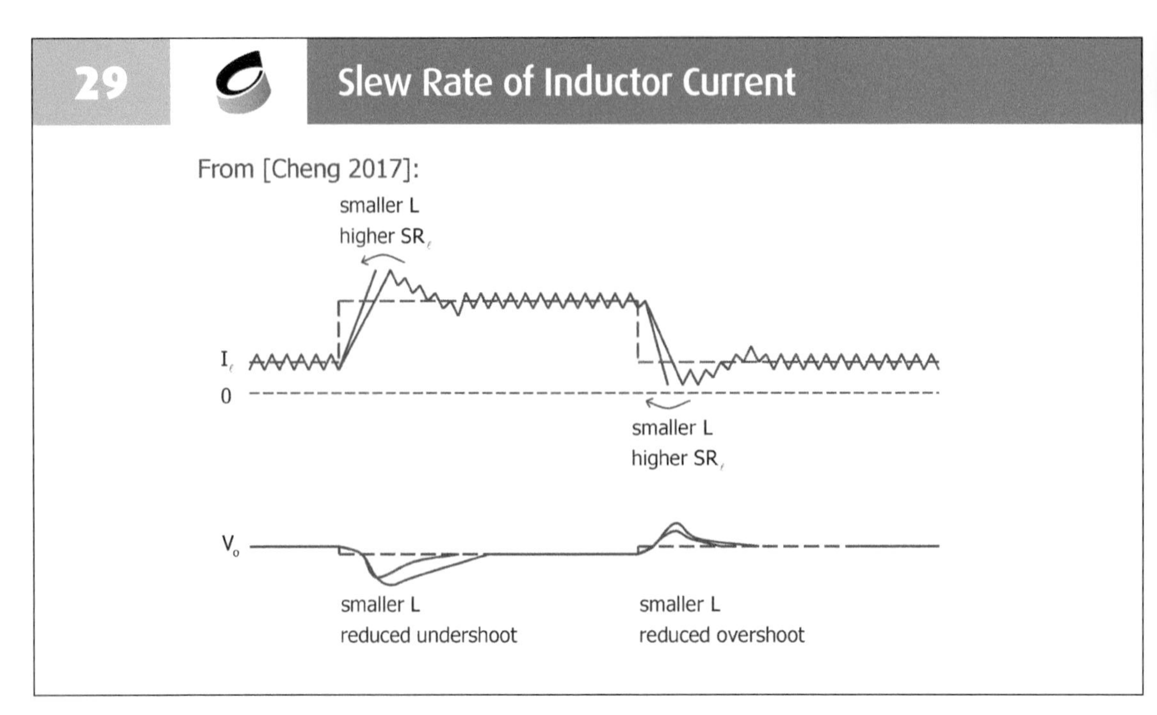

If a smaller inductor is used, the slew rate is higher. When the load current increases/decreases abruptly, the inductor current approaches the steady-state level faster, leading to reduced undershoot/overshoot.

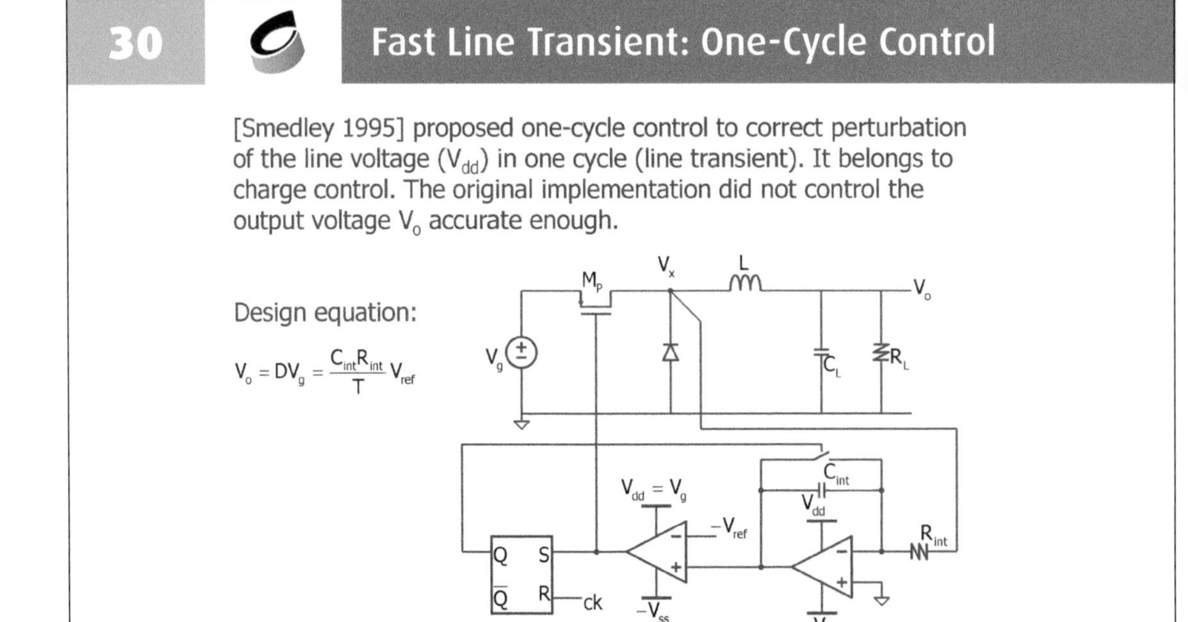

Consider a CCM buck converter. To tackle line transients, the duty ratio has to change as the input voltage changes. Both V_g and D can be obtained from the switching node V_X. By integrating V_X and comparing with the reference voltage $-V_{ref}$, V_O can be controlled by the equation shown in the slide. In principle, the change in V_g can be dealt with in one cycle, and the method is called one-cycle control.

31  Integrated Buck Converter with One-Cycle Control

The implementation in [Smedley 1995] needs both a positive and a negative power supply. [Ma 2004] modifies the scheme such that only one positive supply is needed. More importantly, the improved scheme could work in both CCM and DCM.

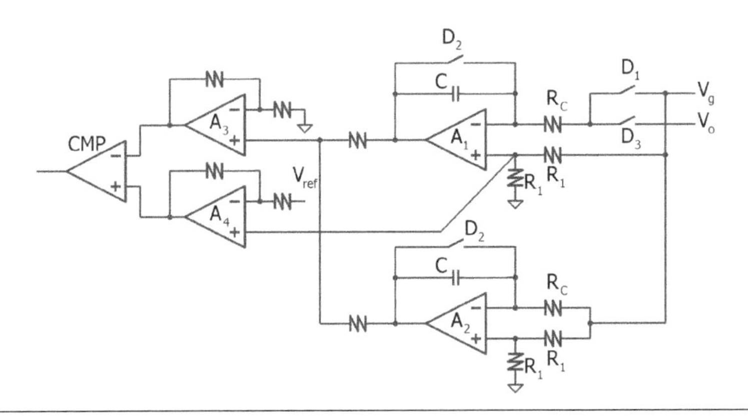

The implementation in [Smedley 1995] needs both positive and negative power supplies, and only CCM is considered, while [Ma 2004] proposed an integrated-circuit implementation that only needs one positive power supply, and both CCM and DCM are taken care of.

32  Reference Tracking: End Point Prediction (1)

Besides fast line and load transients, fast reference tracking is also important for dynamic voltage scheduling (DVS) applications.

[Siu 2006] introduced the concept of end point prediction (EPP). It is a feedforward scheme, bypassing the slow error amplifier EA, and predicts the correct duty ratio for CCM operation as the reference voltage V_{ref} changes.

$$\frac{V_o}{V_g} = \frac{V_{ref}/b}{V_g} = D = \frac{V_a - V_L}{V_H - V_L}$$

$$V_a = (V_H - V_L)\frac{V_{ref}}{bV_g} + V_L$$

Design

$$V_H = bV_g + V_L$$

$$\Rightarrow \quad V_a = V_{ref} + V_L$$

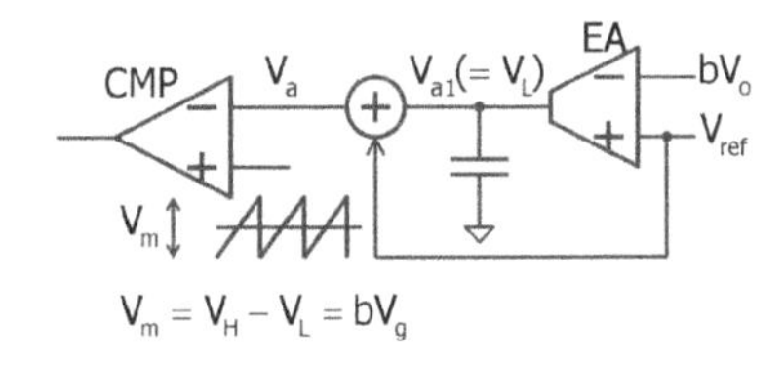

The duty ratio is a function of the compensator output voltage and the ramp amplitude as shown. For reference tracking, when V_{ref} changes, the final compensator output voltage, that is, the end point, is known in advance. Hence, a circuit can be designed to change the duty ratio, and the concept of end-point prediction was introduced in [Siu 2006].

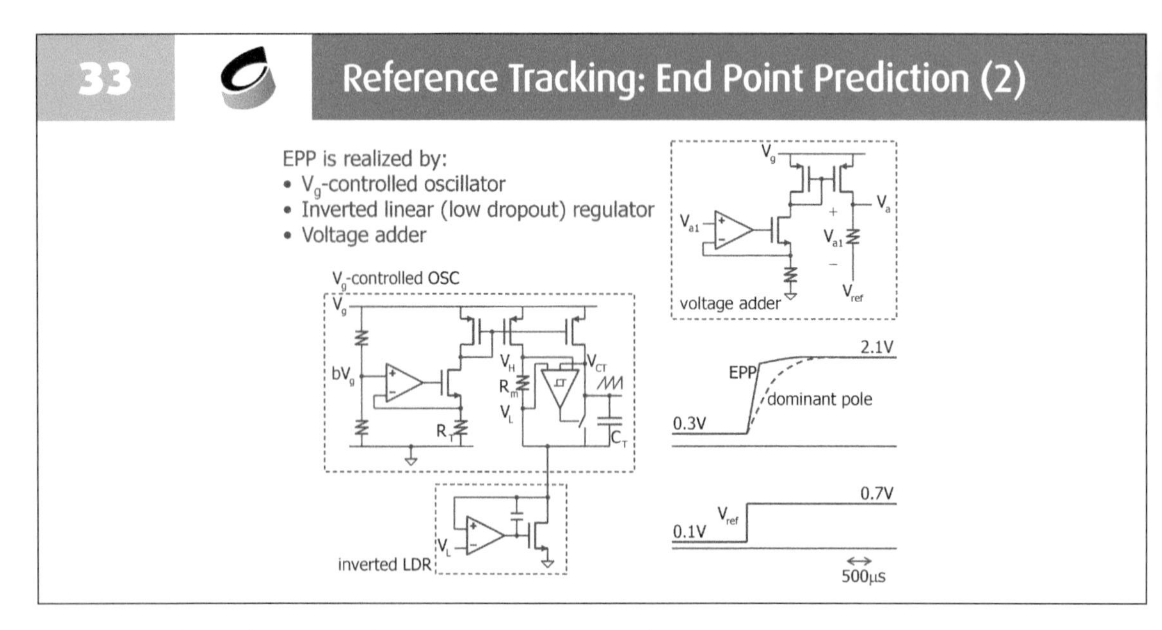

The inverted low dropout regulator generates the low-threshold voltage V_L that can sink current, and the V_g-controlled oscillator generates the voltage ramp signal V_{CT}. The voltage adder adds the output of the compensator V_{a1} to V_{ref} to generate V_a that is to be compared with V_{CT} by the PWM comparator. Both simulation and measurement results showed that tracking speed is greatly improved.

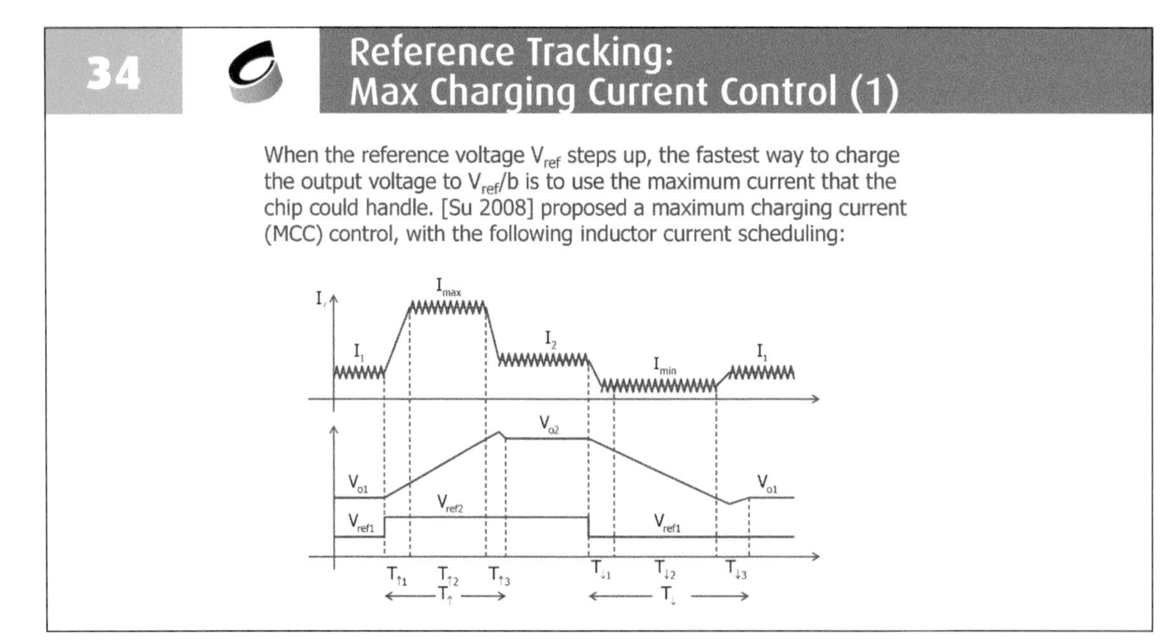

When V_{ref} steps up, the fastest way to charge the output capacitor is to have unity duty ratio and use the maximum current that the chip can handle. Consider that V_{ref} changes from V_{ref1} to V_{ref2} suddenly, and V_o should change from V_{o1} to V_{o2}, and the load current should change from I_1 to I_2. Initially, the inductor current ramps up with unity duty ratio until I_{max} is reached, and switching resumes. V_o ramps up from V_{o1} and overshoots past V_{o2}. To lower V_o, the inductor current is reduced (again with unity duty ratio) until reaching I_2. An analogous action occurs when V_{ref} switches from V_{ref2} down to, say, V_{ref1}.

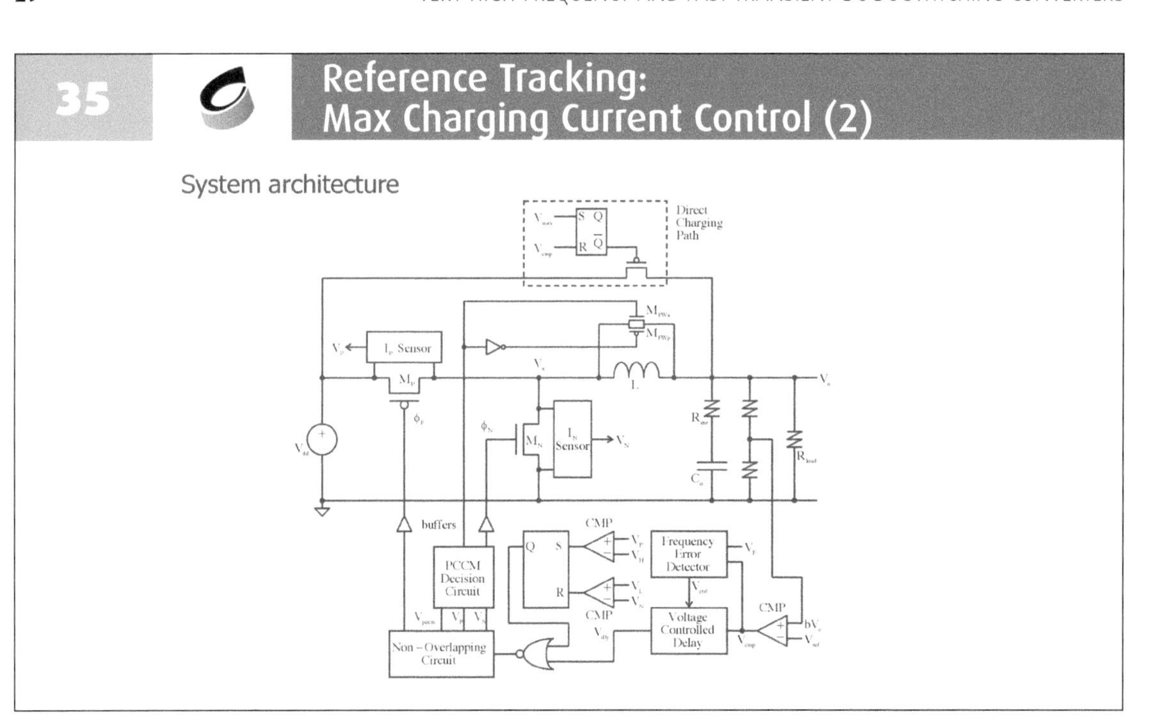

The above maximum charging current control was proposed in [Su 2008]. Readers may refer to [Su 2008] for detail circuit implementation.

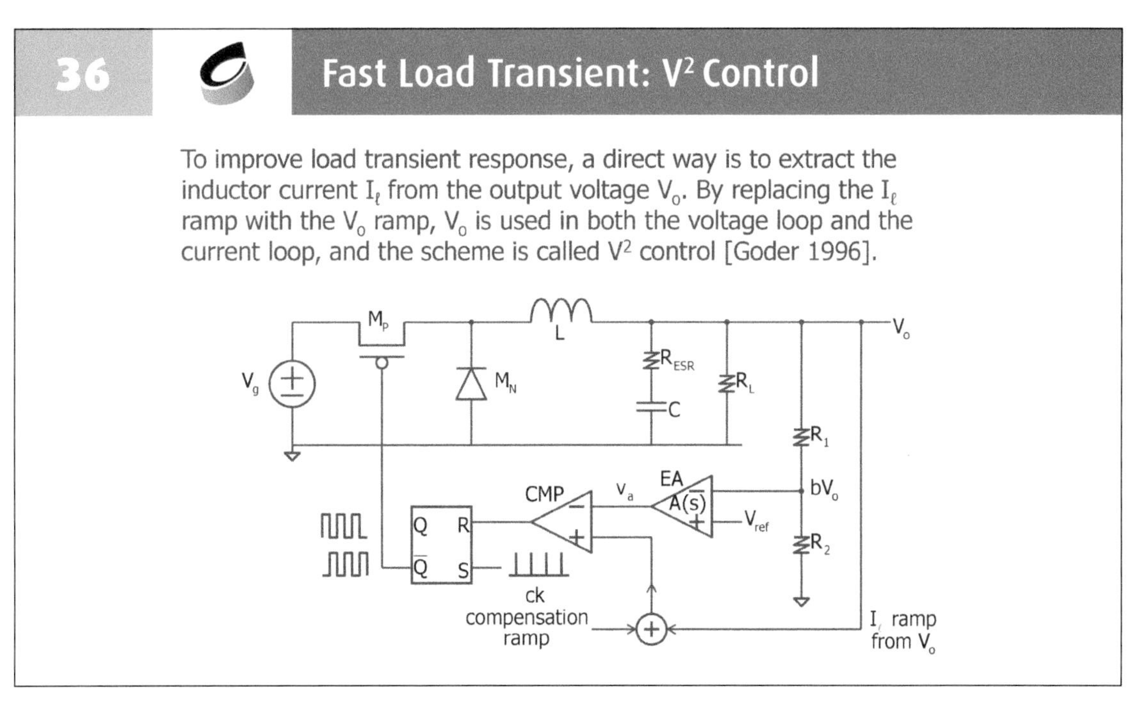

One way to improve load transient responses is to extract the inductor current directly from the output voltage. By replacing the inductor current ramp with the output voltage ramp, V_O is used in both the voltage loop and the current loop, and the scheme is called V^2 control.

37 Basic Principle of V² Control

Inductor current is extracted from ESR of C:

$$V_{o(ac)} = V_{c(ac)} + V_{ESR} = V_{c(ac)} + I_{\ell(ac)} \times R_{ESR}$$

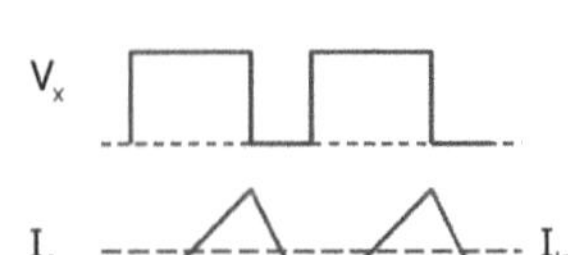
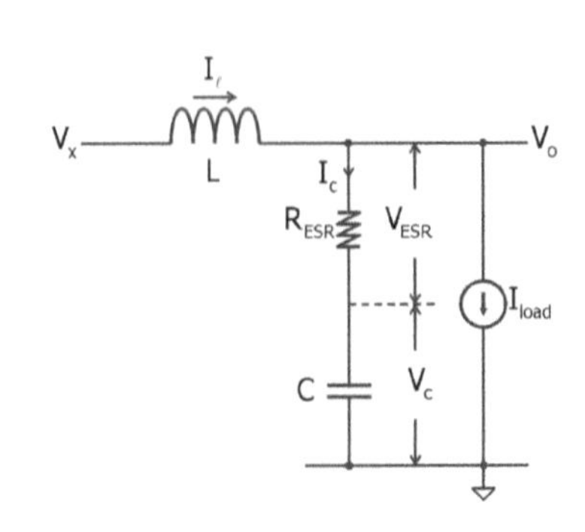

The output voltage consists of the capacitor ripple voltage plus the inductor current times the ESR. Hence, a large ESR is preferred, but the efficiency will then be bad.

38 Implementation of V² Control

The second voltage loop bypasses the slow compensator, and feeds forward to the control loop.

To guarantee that $V_{ESR} >> V_{c(ac)}$, a large ESR is needed.

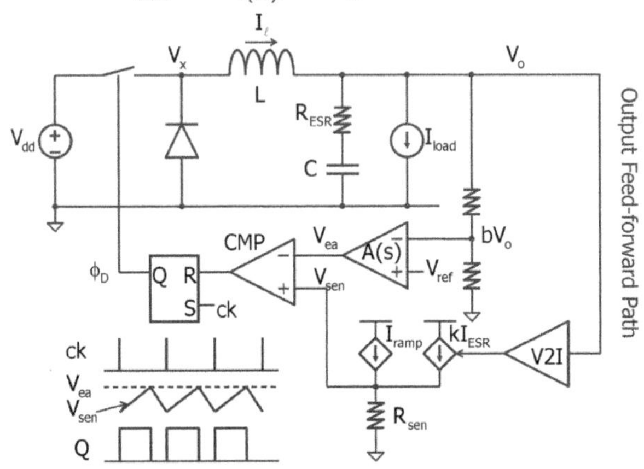

V^2 control is fast because the second voltage loop does not go through a lowpass filter such as the compensator.

V^2 Control with Derivative of Output Voltage

[Mai 2008] suggested extracting I_ℓ by differentiating V_O:

$$V_{o(ac)} = V_{c(ac)} + V_{ESR} \Rightarrow \frac{dV_o}{dt} = \frac{dV_{o(ac)}}{dt} = \frac{I_{\ell(ac)}}{C} + \frac{dI_c}{dt} \times R_{ESR}$$

what is needed   offset with jumps

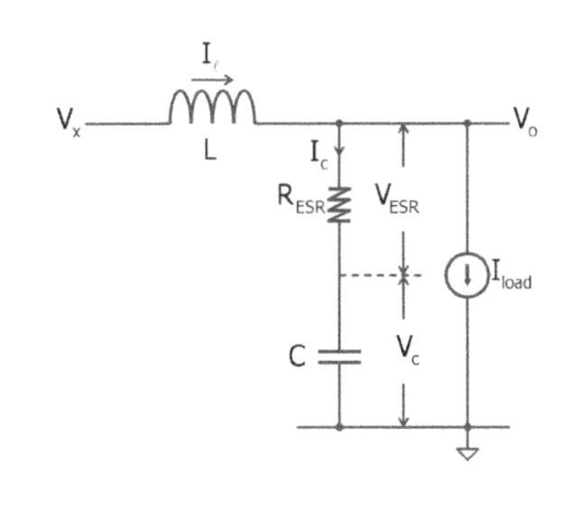

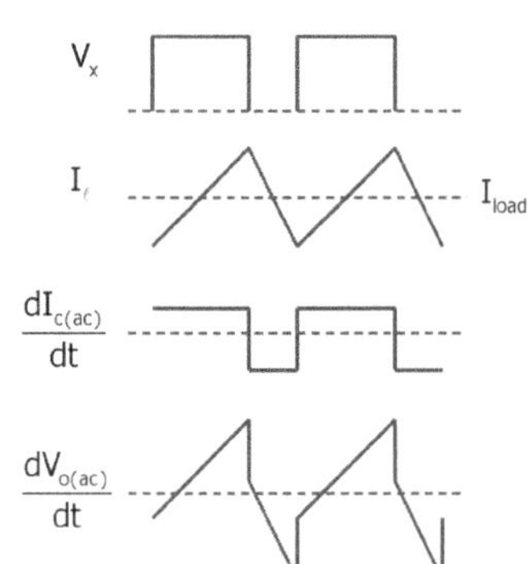

The inductor current cannot be effectively extracted when the ESR is small. [Mai 2008] suggested to extract the inductor current by differentiating the output voltage instead.

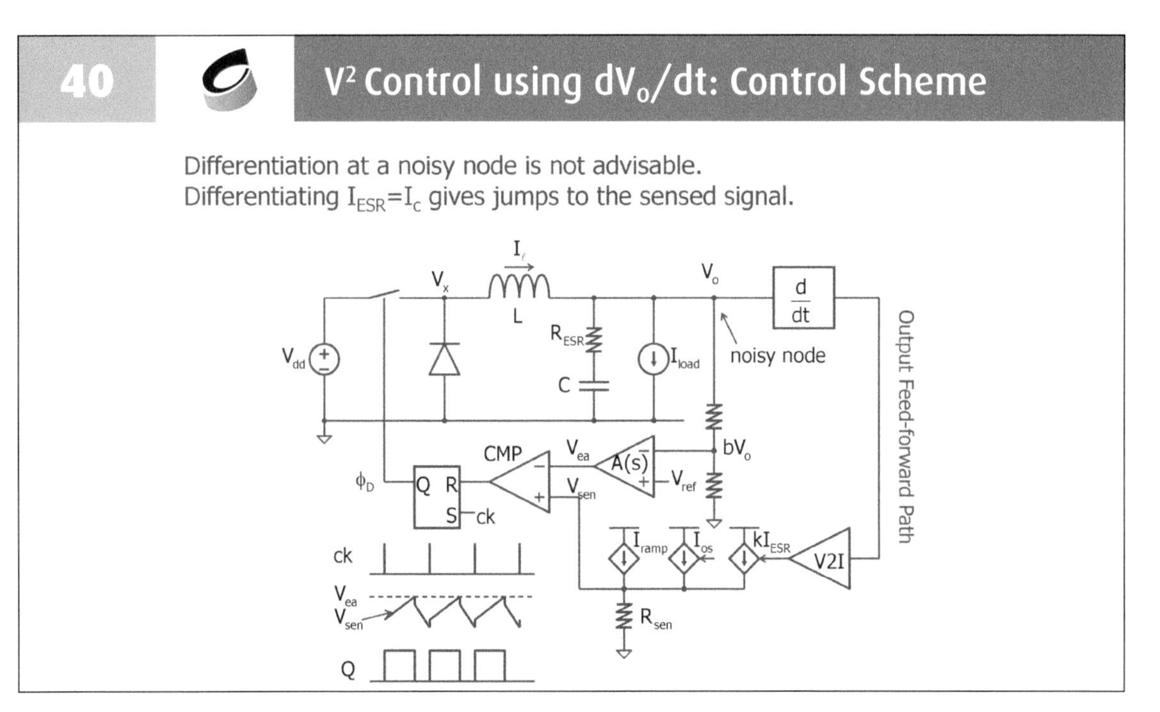

However, the output voltage is very noisy due to switching glitches and unrelenting changes in the load current. Differentiating the noisy node exacerbates the noise. Differentiating the capacitor current also results in jumps.

V^2 Control with RC Filter (1)

[Su 2009] noted that the node V_f gives the information of I_ℓ.
Bottom plate of C_f is connected to V_o (AC ground).

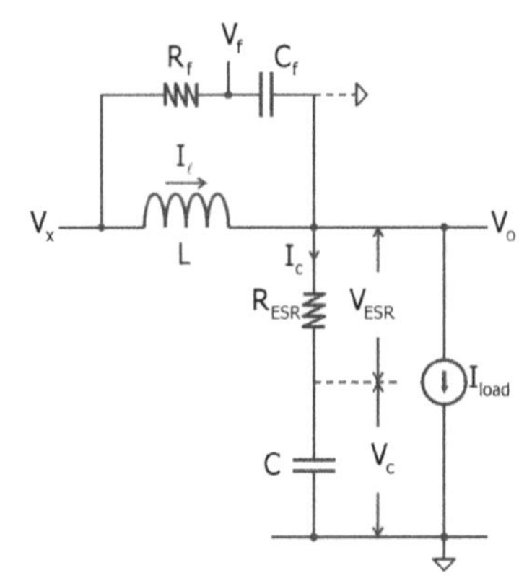

V_o is the output of a 2nd order RLC filter

↓

$$V_o = \int\left(\int V_x dt\right)dt$$

↓

$$\frac{dV_o}{dt} = \int V_x dt$$

↓

$\frac{dV_o}{dt}$ is the output of an integrator

↓

1st order RC filter is an integrator

Instead, the inductor current may be obtained by integrating the switching node V_X.

42 V^2 Control with RC Filter (2)

For $0 < t < DT_s$: $V_f(t) = V_f(0) + (V_{dd} - V_o)(1 - e^{-t/R_fC_f})$

$$\approx V_f(0) + (V_{dd} - V_o)\frac{t}{R_fC_f}$$

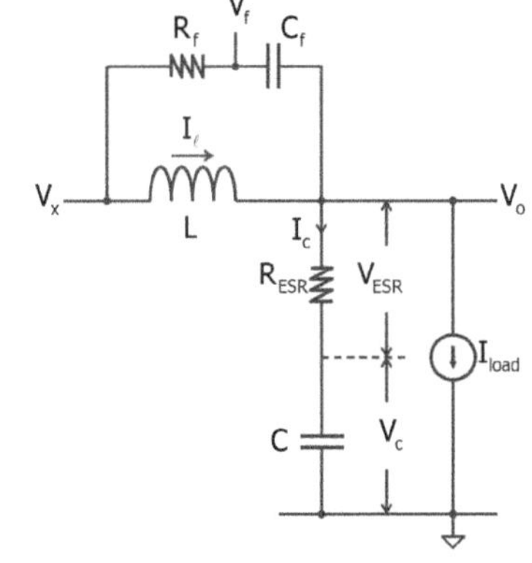

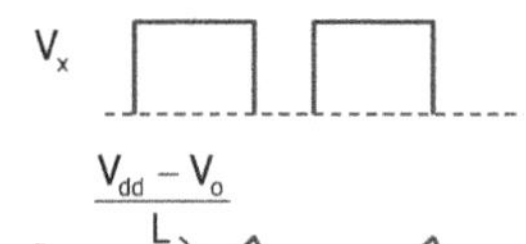

In [Su 2009], R_fC_f is designed to have a large time constant, and hence V_f essentially ramps up and down linearly. In State 1, V_f ramps up with a slope proportional to $(V_{dd}-V_o)$.

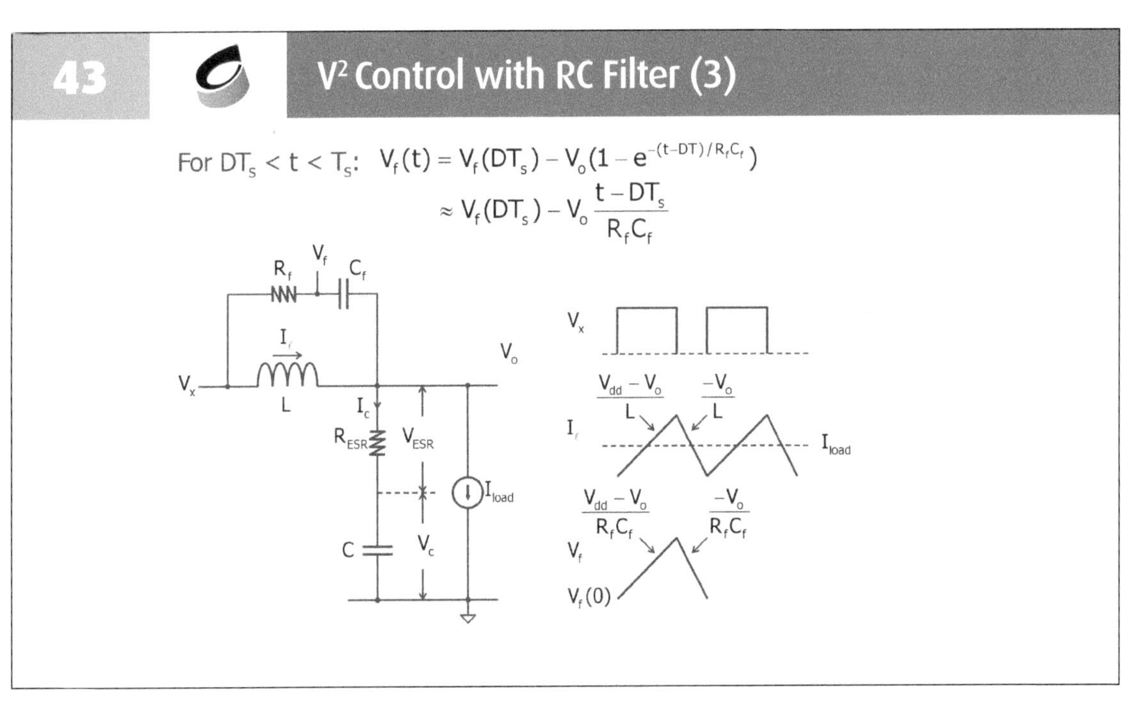

In State 2, V_f ramps down with a slope proportional t$_O$ $-V_O$. Hence, the inductor current can be obtained by sensing V_f.

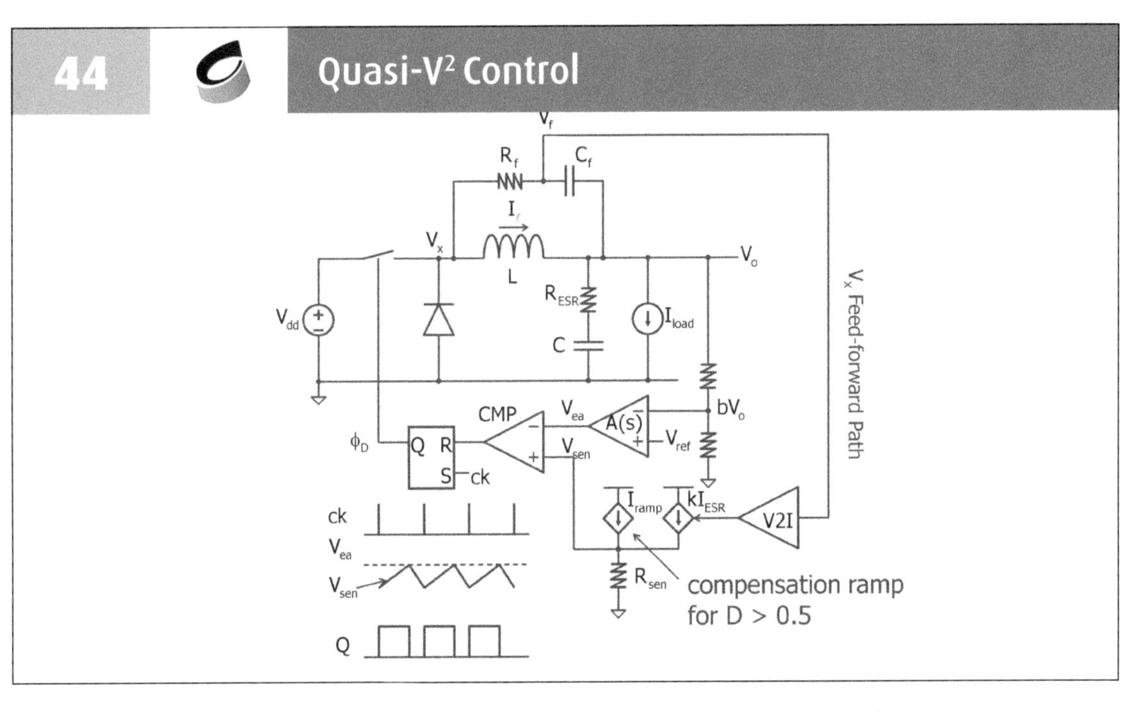

It is called a quasi-V² control because the output voltage is not sensed directly. Remember that it is a current-mode control, and a compensation ramp (through I_{ramp}) has to be added.

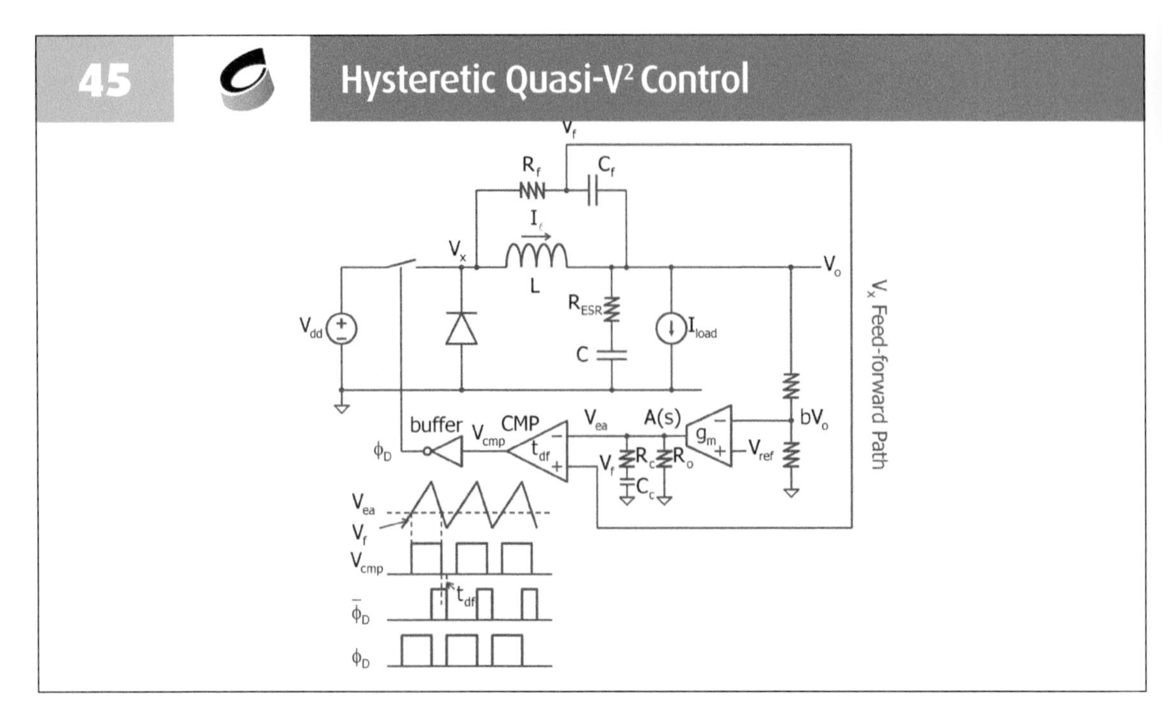

Instead of using a fixed-frequency clock, switching can be achieved by the delays due to the rise time and the fall time of the comparator and the logic. Whenever the sensed voltage (related to the inductor current) V_f passes V_{ea} from below or from above, the comparator will trip later.

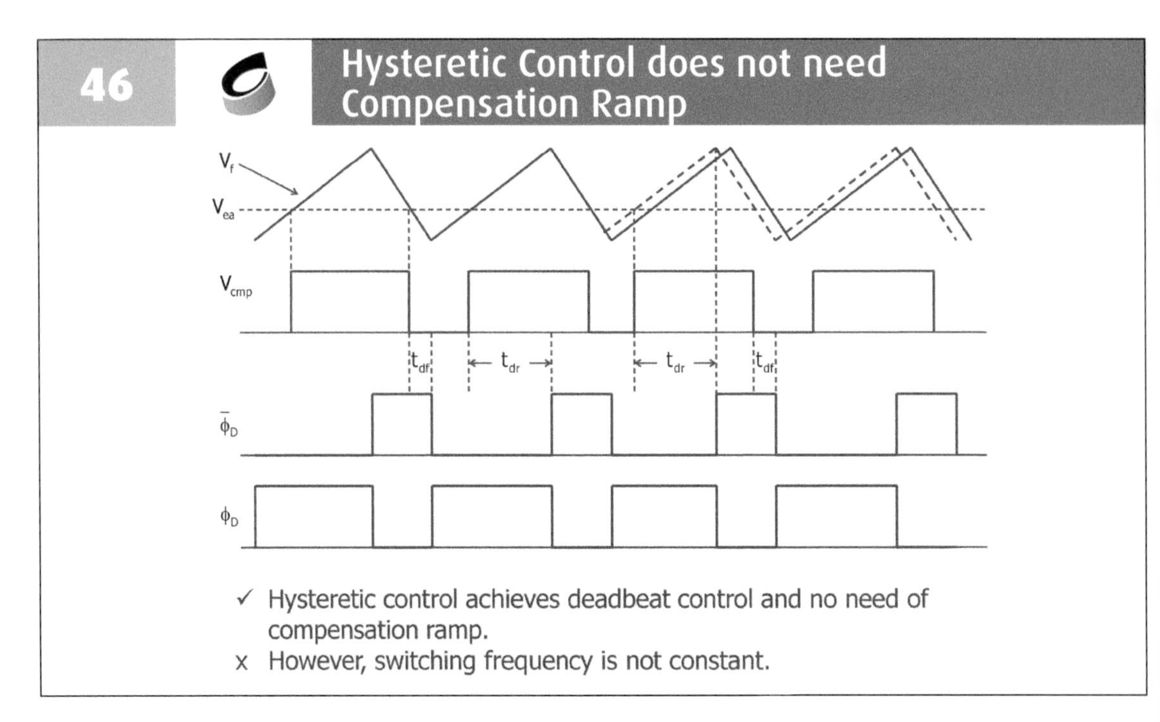

For example, when the inductor current switches to ramp up from the valley value, it will rise above V_{ea}, and the comparator will trip after the delay due to finite rise time, which then turns off the switch, and the inductor current starts to ramp down. The opposite actions then follow.

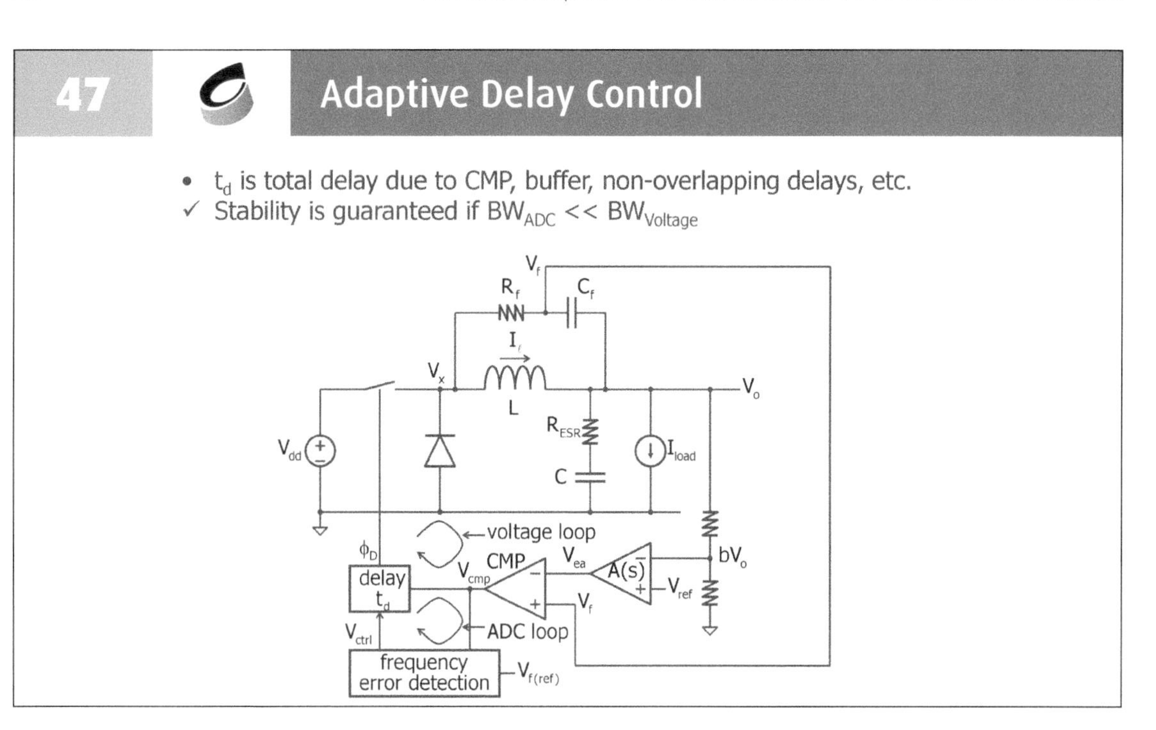

The variable frequency can be made to be approximately constant by using a frequency error detector to monitor and control the delays. Stability is guaranteed if the bandwidth of the adaptive delay control loop is much slower than the voltage feedback loop.

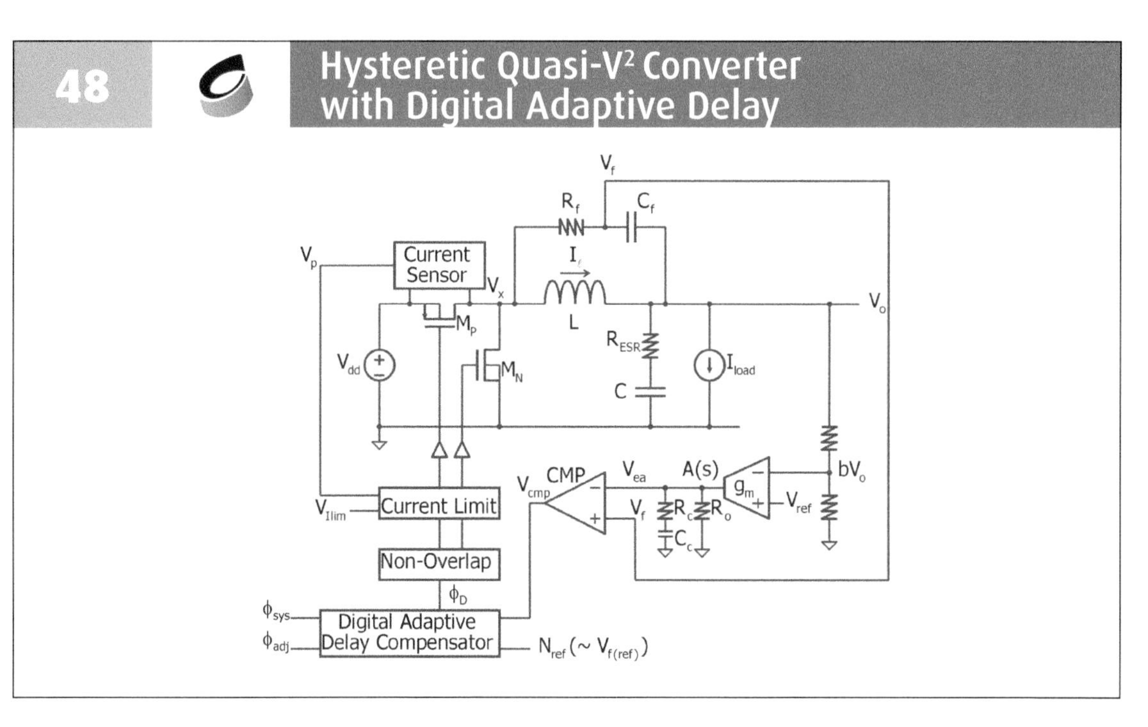

The adaptive delay control can be implemented digitally.

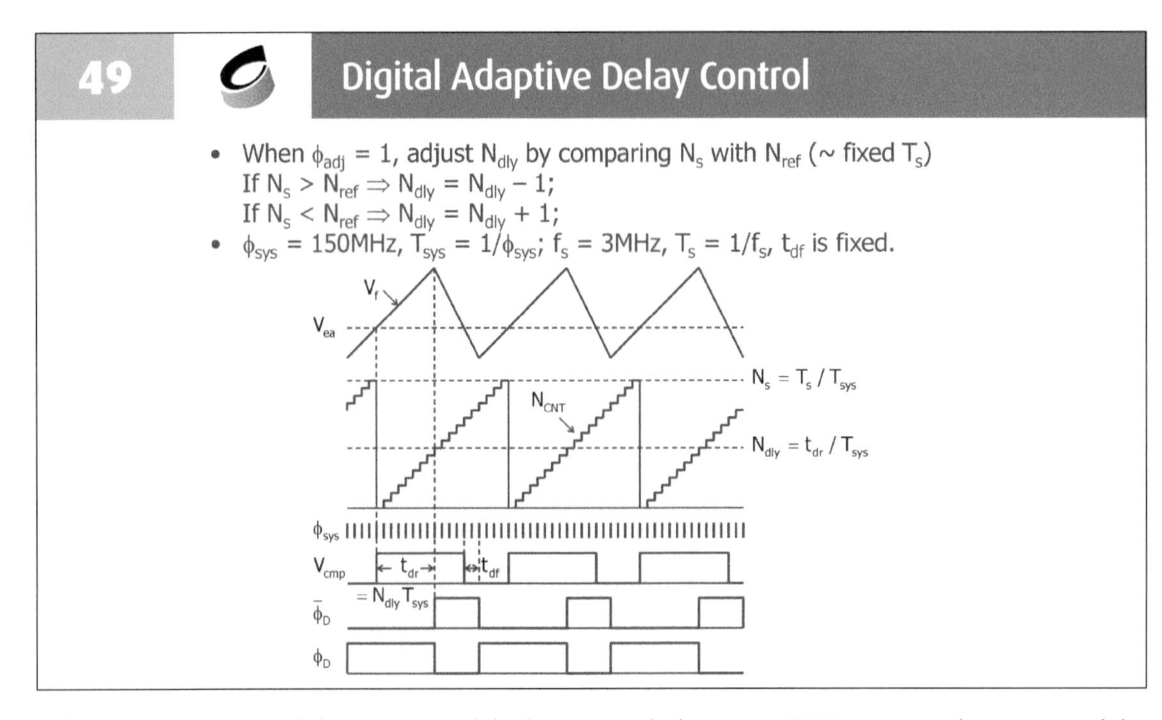

The DC-DC converter switches at 3MHz while the system clock runs at 150MHz. Hence, the accuracy of the duty-ratio is approximately 8 bits.

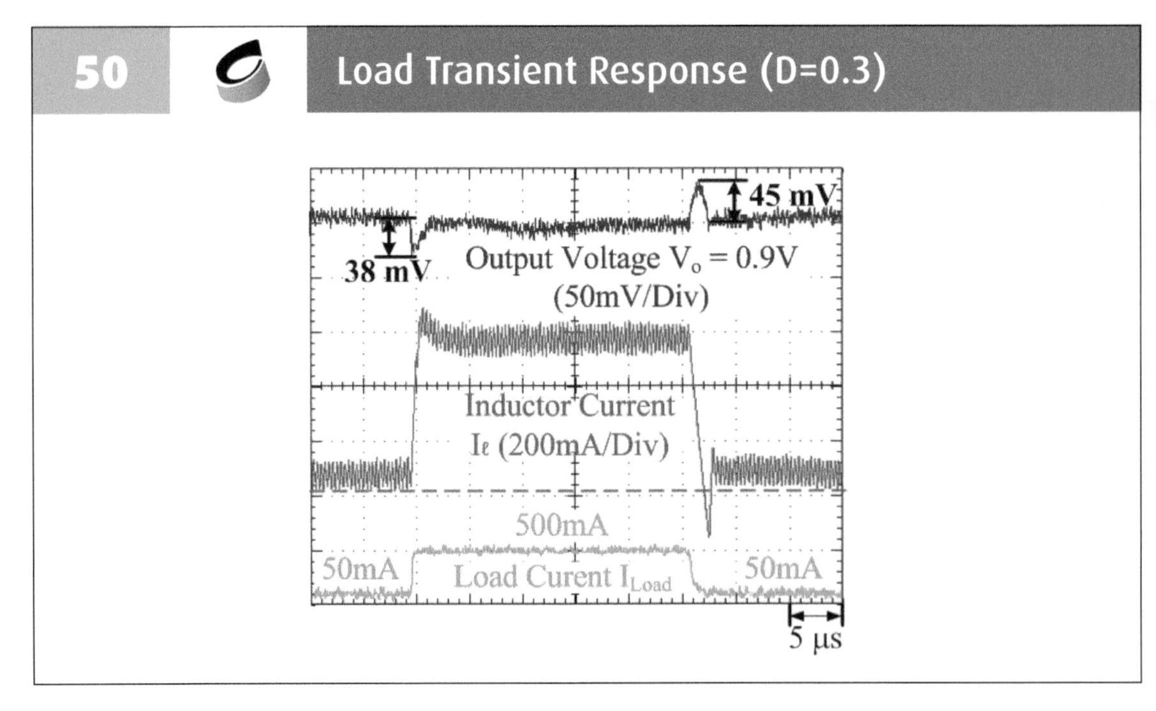

When the load current increases from 50mA to 500mA, the inductor current increases continually (duty ratio is 1). When the load current decreases from 500mA to 50mA, the inductor current decreases continually (duty ratio is 0).

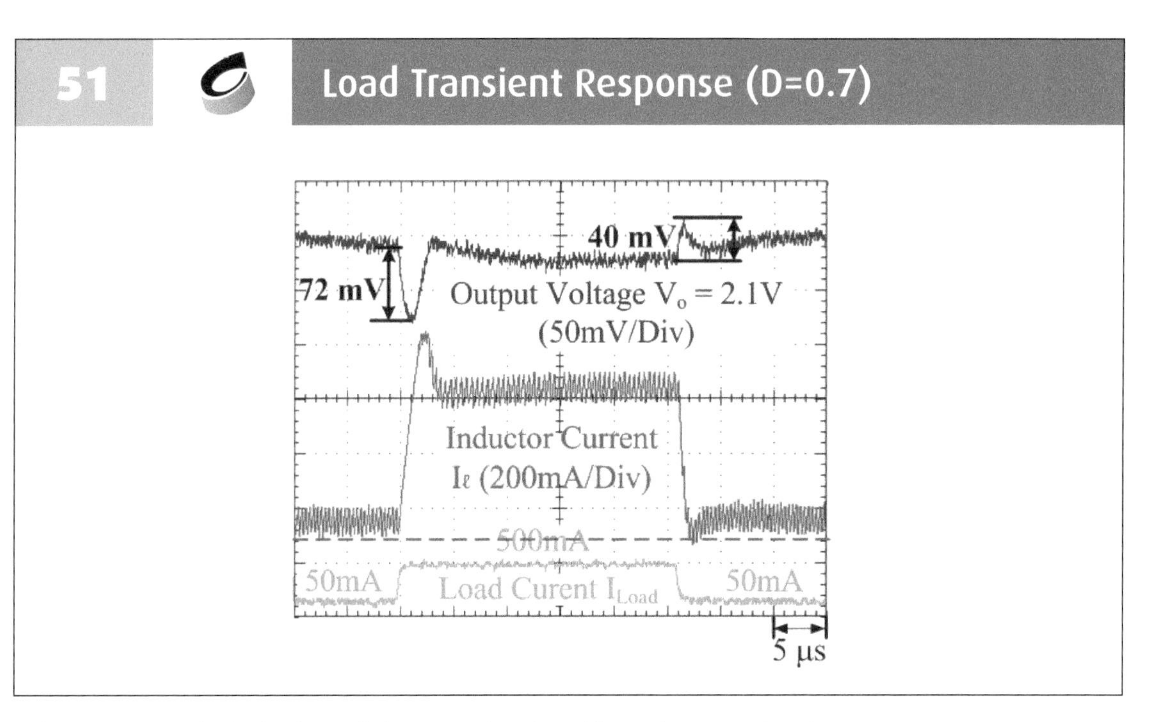

The scheme works for both small duty ratio (=0.3) and large duty ratio (=0.7).

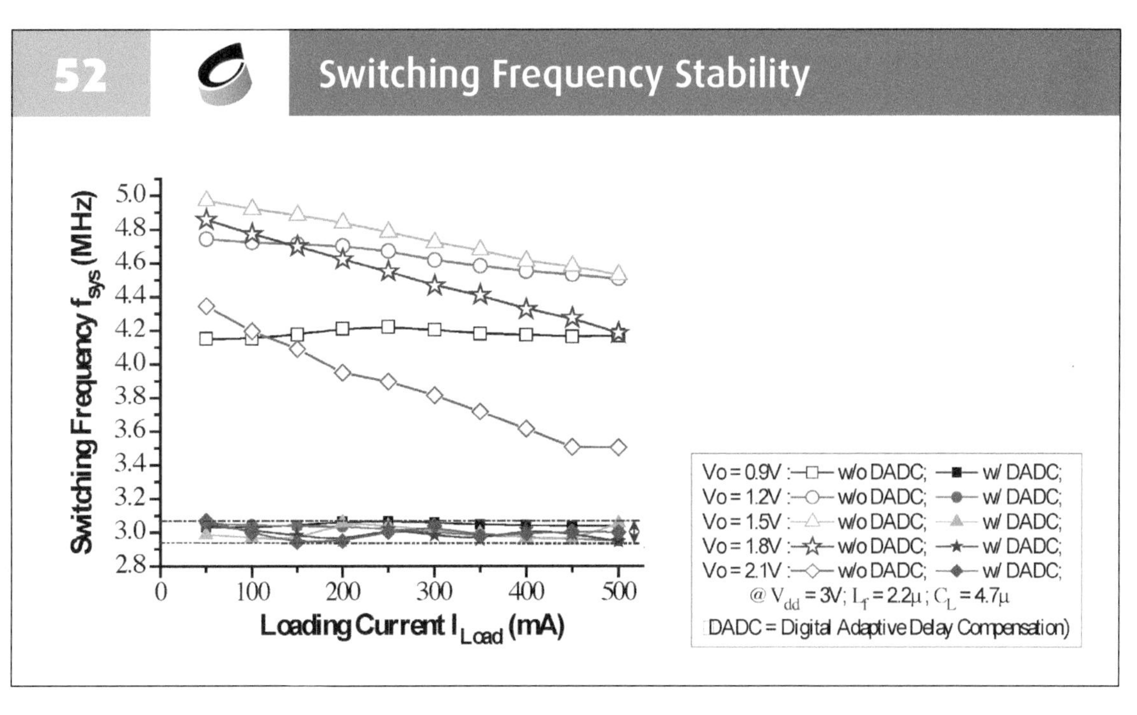

By activating the function of digital adaptive delay compensation, the switching frequency is well controlled to around 3MHz.

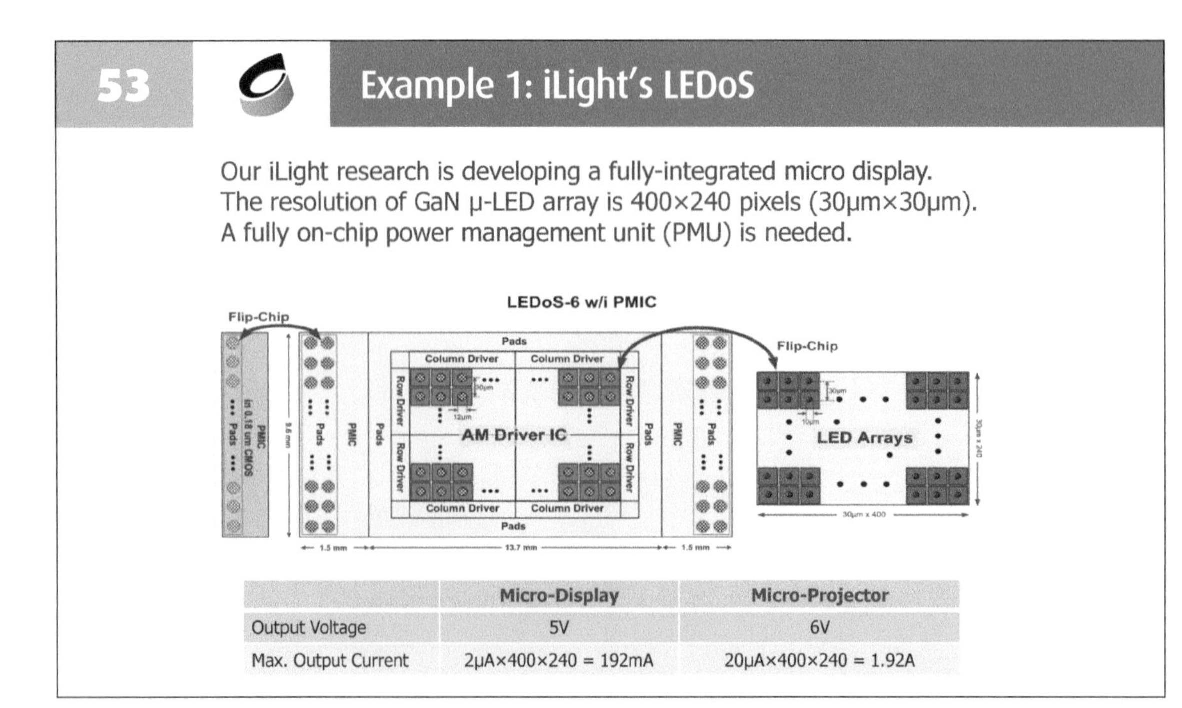

	Micro-Display	Micro-Projector
Output Voltage	5V	6V
Max. Output Current	2μA×400×240 = 192mA	20μA×400×240 = 1.92A

One application of VHF switching converters is to power up an LED micro-display. 400x240 LED pixels are to be flip-chip bonded on the active matrix (AM) driver IC, while powered up by distributed micro switchers with interleaving phases around the perimeter. These switchers have to be very tiny, and so they have to switch at very high frequency (say 30MHz).

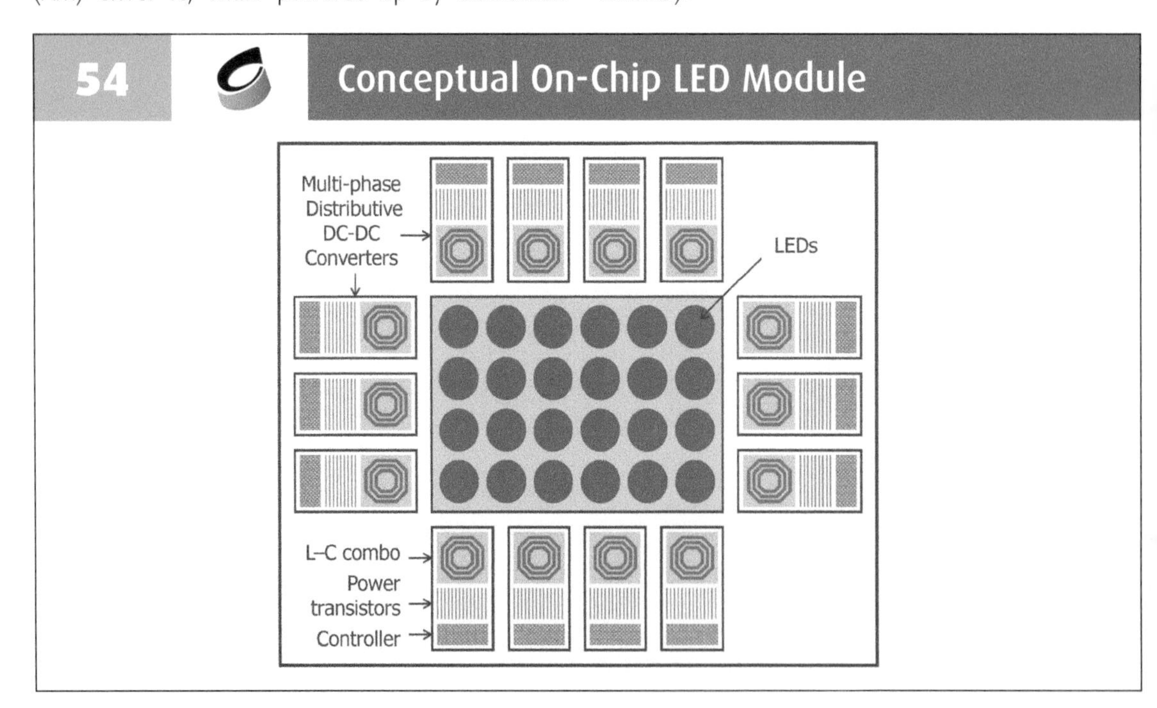

The eventual goal is to integrate both the inductor (~100nH) and capacitor (~100nF) on-chip, with f_S ~ 30MHz. The values can be estimated as follows.

55  L – f_s Consideration

The output voltage ripple of a CCM buck converter is

$$\frac{\Delta V_o}{V_o} = \frac{1}{8}\frac{(1-D)}{LCf_s^2}$$

By assuming $\Delta V_o/V_o$=0.01 (1% ripple) and D=0.5, the values of L and C can be estimated by

$$L = C = \frac{2.5}{f_s}$$

Consider operating the converter in the VHF (30 – 300 MHz) range. In particular, for f_s = 30MHz,

$$L = \frac{2.5}{30M} = 83nH$$

We may pick L = 100nH as a convenient value.

Let us consider the buck converter for simplicity. If we assume the output voltage ripple to be 1% of the output voltage, with a duty ratio of 0.5, and a switching frequency of 30MHz, then L and C could be 83nH and 83nF respectively.

56  L – f_s Consideration: Example

The μ-display PMU needs a boost converter. Just for argument's sake, consider using a buck converter instead, with V_g=5V, V_o=4V (which gives D=0.8). The inductor is

$$L = C = \sqrt{\frac{100 \times 0.2}{8f_s^2}} \approx \frac{1.6}{f_s}$$

If f_s=30MHz, then L≈50nH. Let us pick L=100nH.

For the buck converter to work in CCM, the output current I_o has to be

$$I_o > \frac{\Delta I_\ell}{2} = \frac{m_2 D' T}{2} = \frac{V_o}{2L} D' \frac{1}{f_s} \approx \frac{V_o}{32} = 125mA$$

Note that when the μ-display is fully turned on, the output current is 192mA.

If the output current is 192mA, then the converter works in continuous conduction mode (CCM).

57 On-Chip Passive Components

From previous discussion, at f_s=30MHz, L~50nH and C~50nF. For good input and output filtering, two filtering capacitors are needed.

On-Chip Capacitors: limited to ~10fF/μm² [Huang 2013b], and 100nF needs 10mm² of silicon area ⇒ interim solution

On-Chip Inductors: recent achievements at HKUST [Fang 2014, Wu 2015]:

	L	f_s	Q	R_ℓ	Area
[Fang 2014]	131nH	800kHz	7.4	60mΩ	1mm²
[Wu 2015]	88nH	10MHz	5.9	450mΩ	2mm²

To build a 50nF capacitor and a 50nH inductor on-chip is quite impractical, even in 2017, as they will occupy too huge a silicon area. One interim solution is to consider system-in-package (SiP) instead of system-on-chip (SoC).

58 Interim Implementation

Surface Mount Capacitors

muRata offers 100nF to 220nF chip ceramic capacitors with L×W×T = 0.5×0.5×0.3 mm³.

Surface Mount Inductors

CoilCraft offers 100nH chip inductors that work up to 2.2GHz with DC resistance < 75mΩ, and L×W×T = 1.14×0.635×0.71 mm³.

For example, surface-mount capacitors and inductors could be very tiny, with footprints on the order of 0.5mm x 1mm, the sizes of an IC.

59 PMU Layout

Technology: CSMC 0.18μm

M_P: W/L = 60,000/0.5
R_P = 140mΩ
area = 280μm × 320μm

M_N: W/L = 20,000/0.6
R_N = 110mΩ
area = 180μm × 220μm

Pad: area = 70μm × 70μm

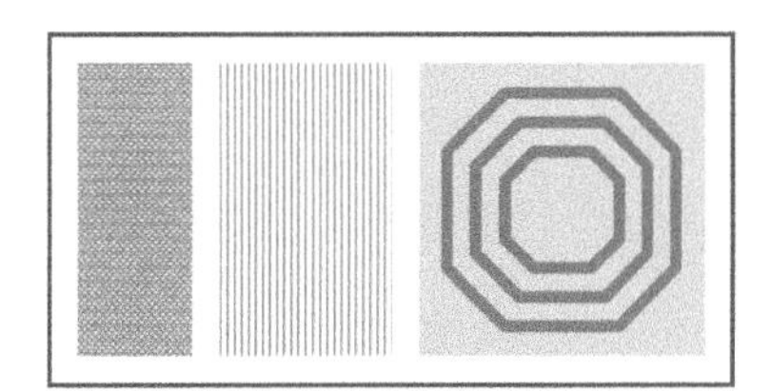

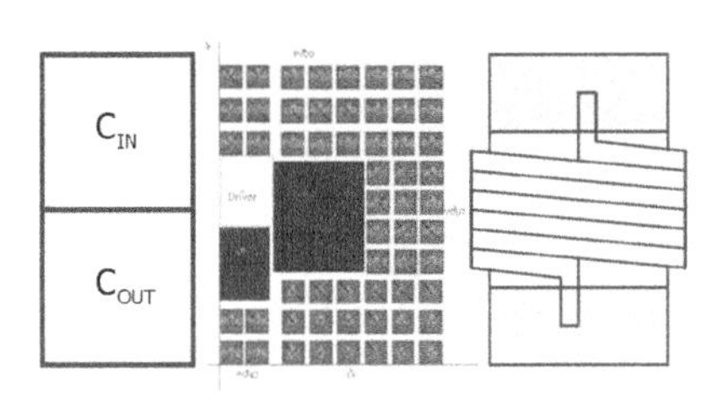

Consider a 0.18μm process with power transistors that can handle 5V, R_{ds}(on) of 0.15 ohm can be achieved with a reasonable silicon area (0.3mm x 0.3mm, say). Therefore, all components can fit in a footprint of 1.25mm x 2.5mm.

60 LEDoS PMU

PMU with SiP (system in package) implementation is realizable, and is our immediate interim goal.

PMU with SoC (system on chip) implementation is doable but efficiency is a problem. With further improvements in on-chip inductors and on-chip capacitors, it is expected to be realizable in two years' time with good performance.

For the LED micro-display, system-in-package could be an interim solution to the eventual system-on-chip design.

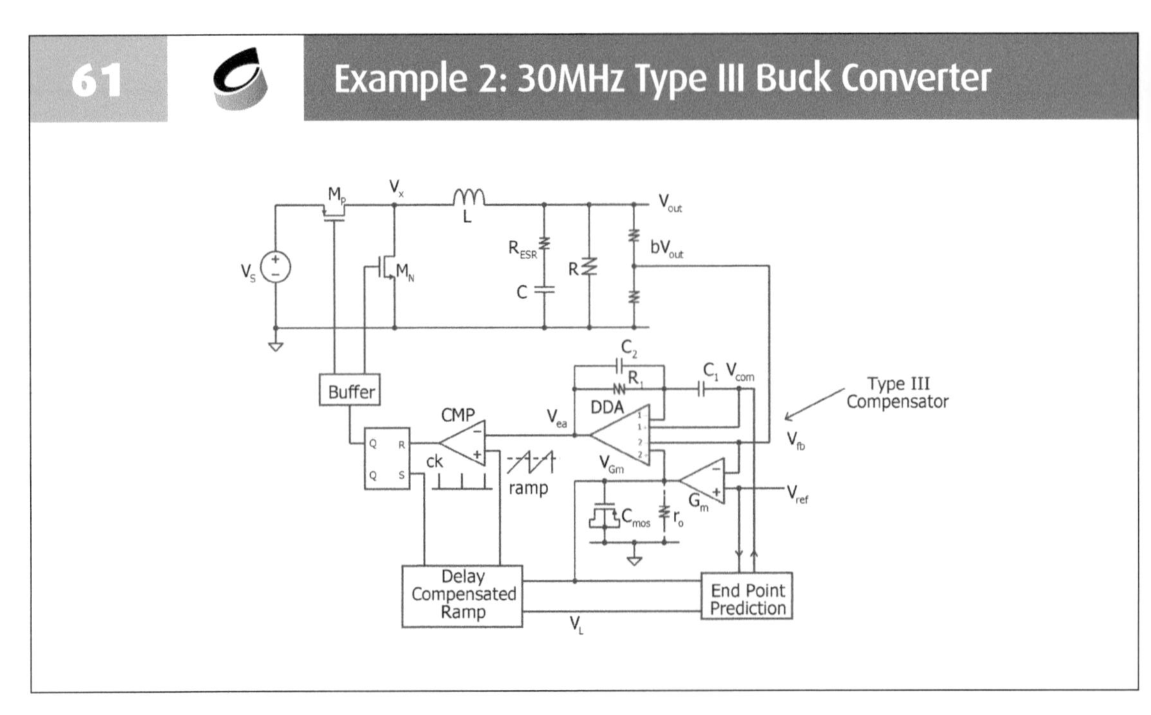

A second example is to design a 30MHz buck converter with three distinct features. The first is a ramp generator that works accurately up to 70MHz. The second is an area-efficient compensation scheme. The third is an end-point prediction scheme.

62 

Controller Considerations

Problems with VHF or even HF (3 – 30 MHz) operation:

(1) Current-mode control needs high-speed current sensor that is difficult to design and consumes much power

Solution: use voltage-mode control

(2) Circuit/comparator delays affect accuracy of ramp

Solution: delay-compensated ramp

(3) Traditional Type-III compensator needs large ratios of R and C

Solution: DDA-based Type-III compensator with reduced area

A current-mode switching converter that works at 30MHz needs a very fast current sensor that consumes too much power. Voltage-mode control should be considered. The generic R_TC_T oscillator has large overshoots and undershoots that needs to be improved. And the generic Type-III compensator takes up too much area and has to be improved.

63 

R_TC_T Oscillator

The R_TC_T oscillator generates a ramp that synchronized with the clock, which fits the requirement of a PWM switching converter.

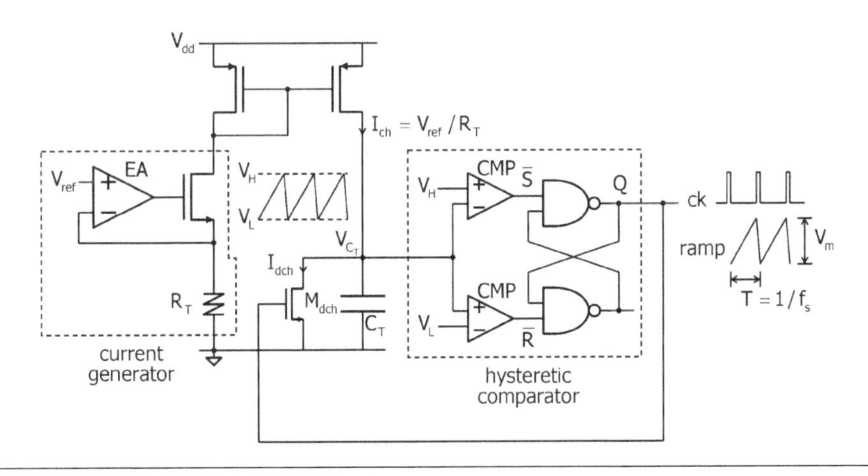

This figure shows a generic R_TC_T oscillator.

64 

R_TC_T Oscillator Operation Principle

The charging current I_{ch} is well-controlled by a bandgap derived voltage V_{ref} and an accurate 1% (external) resistor R_T (1% rule).

I_{ch} charges an accurate 1% (external) capacitor C_T slowly from the lower bound V_L to the upper bound V_H. The ramp excursion is V_m.

The hysteretic comparator trips when $V_{CT} > V_H$, and ck = 1.

When ck = 1, the NMOS M_{dch} is turned on, and discharges C_T with a large current I_{dch} that is around 10 times of I_{ch}.

When $V_{CT} < V_L$, the hysteretic comparator trips again, and ck = 0.

I_{dch} is not well-controlled, but the accuracy of the oscillation (switching) frequency f_s is well-controlled because it is dominated by the accurate I_{ch}.

The operation of the R_TC_T oscillator is discussed in this slide.

65 Delays in Ramp Generator

Due to comparator delay and logic delay (t_d), the C_T ramp is not bounded exactly by V_H and V_L, but overshoots above V_H and undershoots below V_L, causing shift in the oscillator frequency.

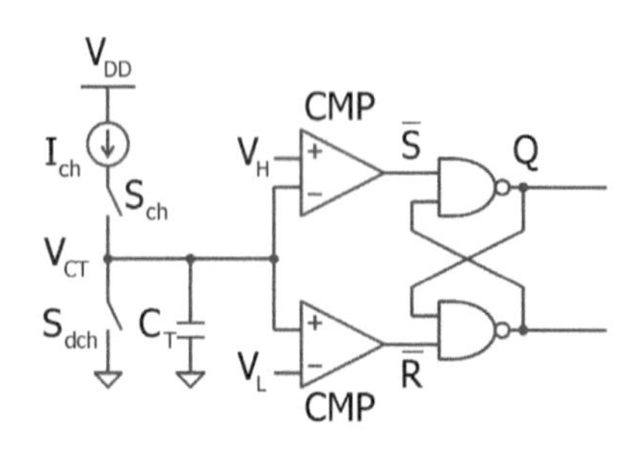

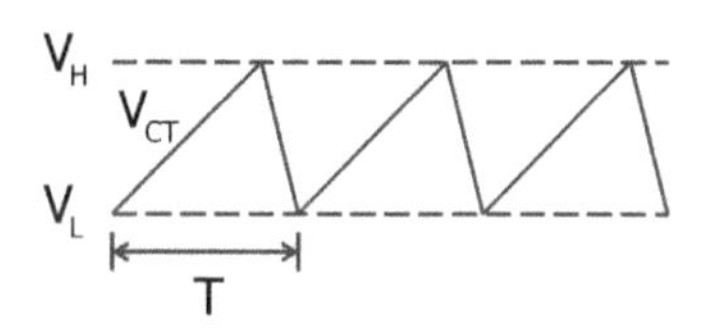

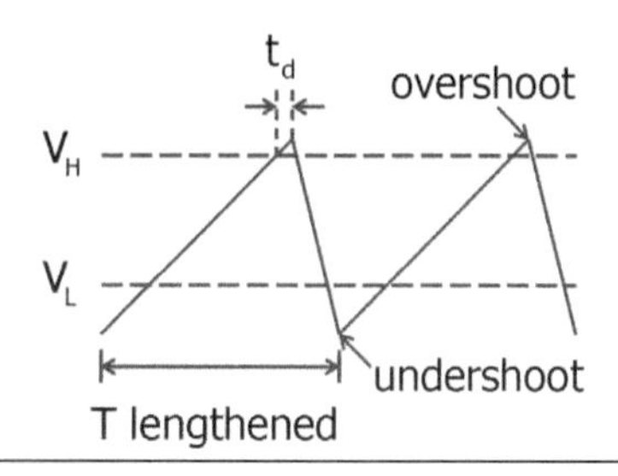

The switching frequency is affected by overshoots and undershoots that are caused by comparator delay and logic delay. To achieve accurate switching at the thresholds of V_L and V_H, the comparators should be triggered earlier appropriately, that is, the delays have to be compensated.

66 Delay Compensated Ramp Generator

The sampling capacitors C_{H1} and C_{L1} are very small compared to C_T. S_{H1} tracks with S_{ch} and samples $V_{CT,pk}$ in C_{H1}, and then stored in C_{H2} to be compared to V_H. The feedback loop with the G_{mH} OTA adjusts V_H' such that $V_{CT,pk} = V_H$ [Cheng 2014].

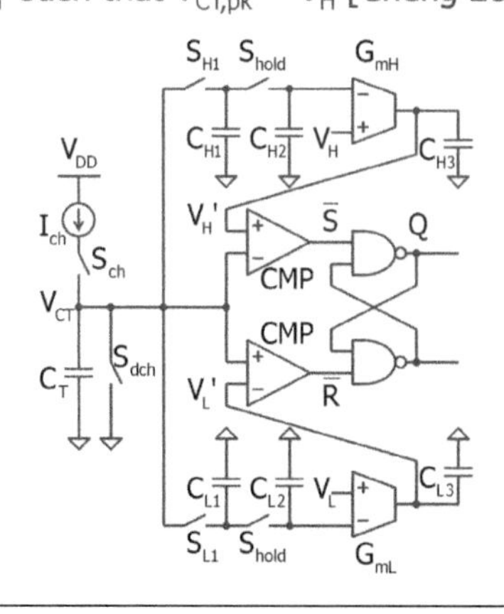

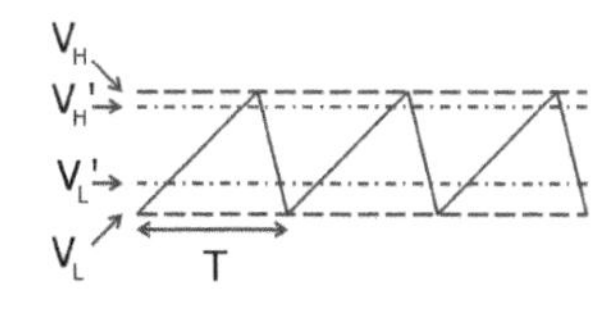

Delay compensation can be achieved by adding individual feedback loops for V_L and V_H. Consider the V_L branch. If C_{L1} (and C_{L2}) samples V_{CT} to be lower than V_L, then G_{mL} will raise its output such that the next time V_{CT} will trip at a higher V_L' (that is, it trips earlier). The effect is that the feedback loop adjusts V_L' so that the valley voltage is exactly V_L.

67 Accuracy Comparison of Ramp Signals

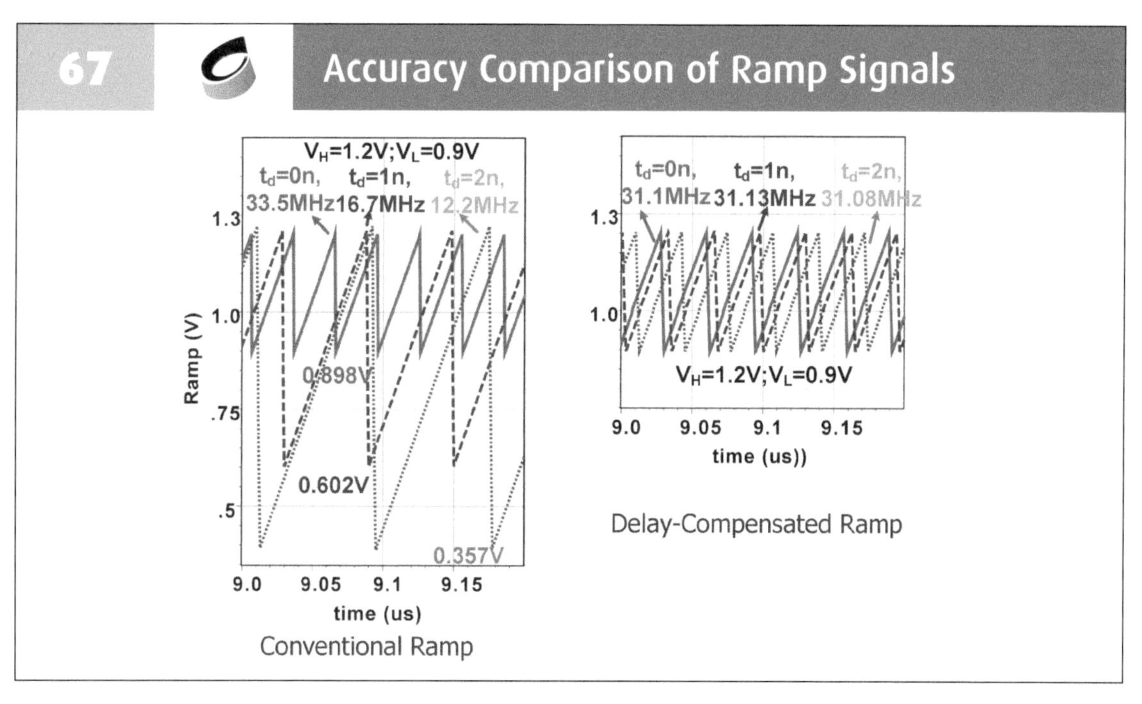

With no delay, the generic oscillator switches at 33.5MHz; but with t_d of 1ns and 2ns, f_S goes down to 16.7MHz and 12.2MHz. For the delay-compensated oscillator, the switching frequency maintains at around 31MHz for 0 to 2 ns delay.

68 Buck Converter with Type-III Compensation

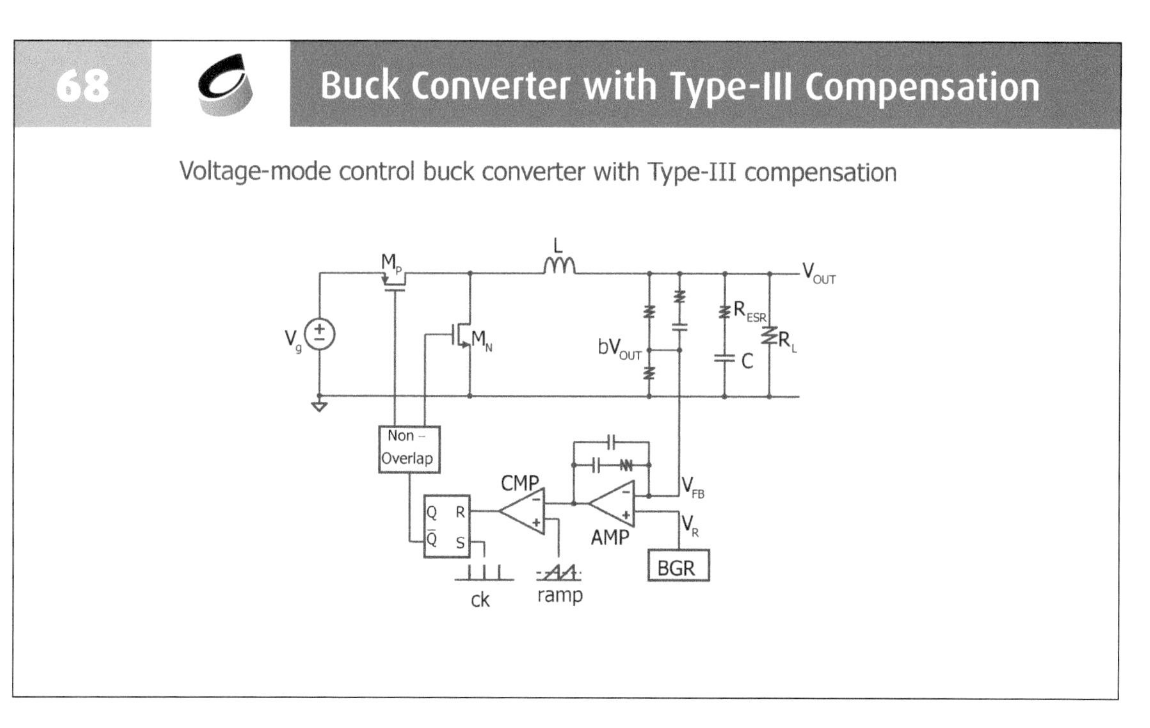

The bandwidth of the control loop is maximized by using Type-III compensation that has two zeros to compensate for the phase drop due to the complex poles.

69 Type III Compensator

Type III compensator consists of two pole-zero pairs, and phase boosting of 180° is possible to compensate for complex poles.

$$\frac{V_{out}}{V_{in}} = -\frac{(1+sC_2R_2)[1+sC_3(R_1+R_3)]}{s(C_1+C_2)R_1(1+s(C_1||C_2)R_2)(1+sC_3R_3)}$$

$$A(s) \approx \frac{(1+sC_2R_2)(1+sC_3R_1)}{sC_2R_1(1+sC_1R_2)(1+sC_3R_3)}$$

$$(C_1 << C_2,\ R_1 >> R_3)$$

The traditional Type-III compensator that built around an op amp has interdependent poles and zeros. For example, changing R_2 affects one pole and one zero. Moreover, R's and C's differ by about two orders, leading to large R's and C's.

70 System Loop Gain with Type-III Compensation

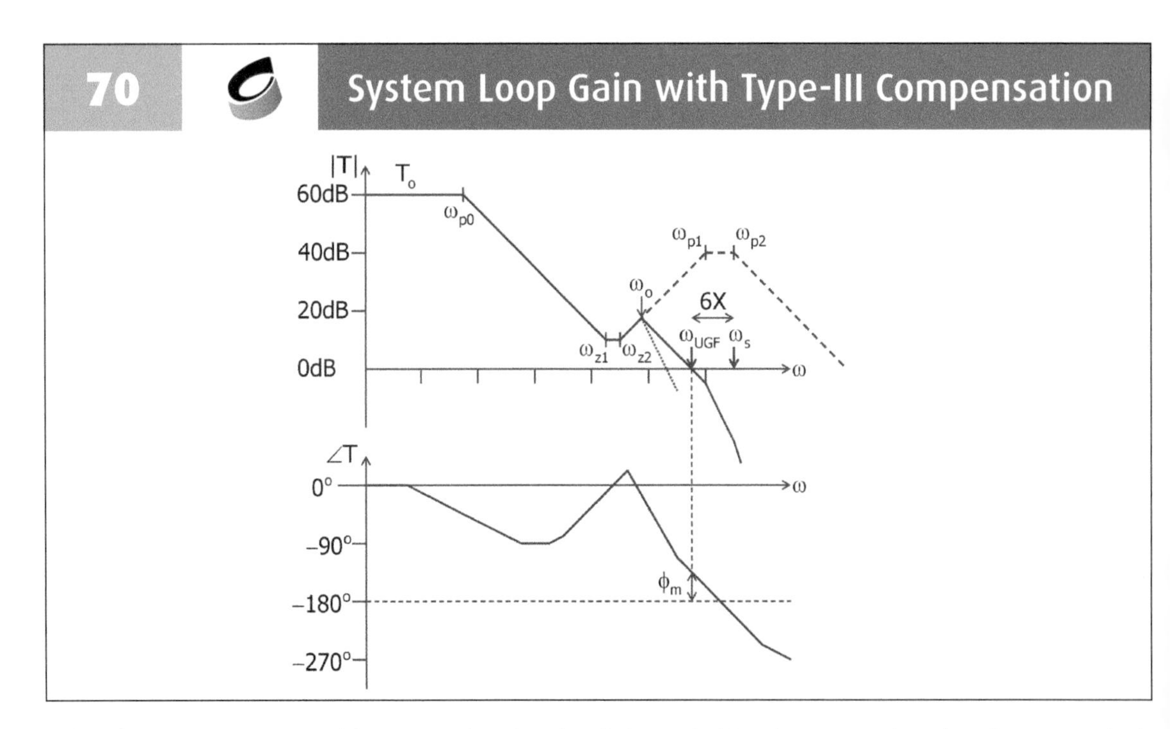

This design gives a control loop UGF that is 1/6 of the switching frequency, but the phase margin is only 45°.

71  Pole-Zero Cancellation in Type III Compensator

The system loop-gain function of the VM buck converter is

$$T(s) = A(s) \times P(s)$$

$$\approx \frac{(1+s/z_1)(1+s/z_2)}{(s/\omega_{t1})(1+s/p_1)(1+s/p_2)} \times \frac{P_o(1+s/z_{ESR})}{\left(1+\frac{1}{Q}\frac{s}{\omega_o}+\frac{s^2}{\omega_o^2}\right)}$$

By employing pole-zero cancellation between p_1 and z_{ESR}, that is

$$\frac{(1+s/z_{ESR})}{(1+s/p_1)} = 1$$

We have

$$T(s) \approx \frac{P_o}{\left(\frac{s}{\omega_{t1}}\right)} \times \frac{(1+s/z_1)(1+s/z_2)}{\left(1+\frac{1}{Q}\frac{s}{\omega_o}+\frac{s^2}{\omega_o^2}\right)} \times \frac{1}{\left(1+\frac{s}{p_2}\right)}$$

This slide shows the analytic system loop-gain function. Note that the ESR zero is used to cancel one of the low-frequency poles.

72  Type III Compensation Schemes

	[Mattingly 2003] Intersil	[Meeks 2008] TI	[Rahimi 2010] IR (3A)	[Rahimi 2010] IR (3B)
z_1	$0.5\omega_o$	$0.9\omega_o$	$0.75\omega_o$	$0.5\omega_{UGF}\sqrt{\frac{1-\sin\phi_m}{1+\sin\phi_m}}$
z_2	ω_o	ω_o	ω_o	$\omega_{UGF}\sqrt{\frac{1-\sin\phi_m}{1+\sin\phi_m}}$
p_1	z_{ESR}	ω_{UGF}	z_{ESR}	$\omega_{UGF}\sqrt{\frac{1-\sin\phi_m}{1+\sin\phi_m}}$
p_2	$\omega_s/2$	$10\omega_{UGF}$	$\omega_s/2$	$\omega_s/2$
ω_{UGF}	$\omega_s/10$		$\omega_s/8$	

Type 3A: $\omega_o < \omega_{UGF} < z_{ESR} < \omega_s/2$

Type 3B: $\omega_o < \omega_{UGF} < \omega_s/2 < z_{ESR}$

Different researchers argue for different poles-zeros assignments. The one suggested by Intersil could be a good and conservative choice.

73 Intersil Type-III Compensation Scheme

The Bode plots show both $|T|$ and $\angle T$ of the Intersil assignment. The phase margin is around 70°.

74  Large Ratios of C and R

The Bode plots of T(s) shows that the zeros and poles are far apart:

$$A(s) \approx \frac{T_o(1+s/z_1)(1+s/z_2)}{(1+s/p_o)(1+s/p_1)(1+s/p_2)} \qquad \omega_{UGF} \approx \omega_s/10$$

$$\approx \frac{(1+s/z_1)(1+s/z_2)}{(s/T_o p_0)(1+s/p_1)(1+s/p_2)} \qquad \omega_t \approx 50p_2$$

$$A(s) \approx \frac{(1+sC_2R_2)(1+sC_3R_1)}{sC_2R_1(1+sC_1R_2)(1+sC_3R_3)}$$

The ratios of C and R are then very large:

$$\frac{p_2}{z_2} \approx \frac{1/C_3R_3}{1/C_3R_1} = \frac{R_1}{R_3} \approx 100$$

$$\frac{\omega_t}{p_1} \approx \frac{1/C_1R_3}{1/C_1R_2} = \frac{R_2}{R_3} \approx 100 \qquad \Rightarrow R_1 \approx R_2$$

$$\frac{p_1}{z_1} \approx \frac{1/C_1R_2}{1/C_2R_2} = \frac{C_2}{C_1} \approx 30$$

Using the Intersil assignment, the ratios of poles and zeros, and thus the ratios of R's and C's, can be computed. The ratios of resistors are as large as 100, and the ratio of capacitors is 30. The ratios pose big problem for IC implementation.

75  VM Buck UGF Extension: Pseudo Type III (1)

[Wu 2010] suggested a pseudo Type III (PT3) compensator that is composed of a lowpass filter (LPF) and a bandpass filter (BPF) added together. A simplified (but fairly accurate) computation is shown below:

$$LPF(s) = \frac{G_1}{(1+s/\omega_p)} \approx \frac{G_{vy}}{s/\omega_{t1}} \qquad BPF(s) = \frac{G_{vy}(1+s/\omega_z)}{(1+s/\omega_1)(1+s/\omega_2)}$$

$$PT3(s) \approx \frac{G_{vy}\left[1+s\left(\frac{1}{\omega_{t1}}+\frac{1}{\omega_1}+\frac{1}{\omega_2}\right)+s^2\left(\frac{1}{\omega_{t1}\omega_z}+\frac{1}{\omega_1\omega_2}\right)\right]}{\left(\frac{s}{\omega_{t1}}\right)\left(1+\frac{s}{\omega_1}\right)\left(1+\frac{s}{\omega_2}\right)}$$

$$\approx \frac{G_{vy}(1+s/\omega_{t1})(1+s/\omega_z)}{s/\omega_{t1}(1+s/\omega_1)(1+s/\omega_2)}$$

To reduce the component ratios, the Type-III function was realized by adding a lowpass function to a bandpass function, and it is called a pseudo Type-III compensator.

76 VM Buck UGF Extension: Pseudo Type III (2)

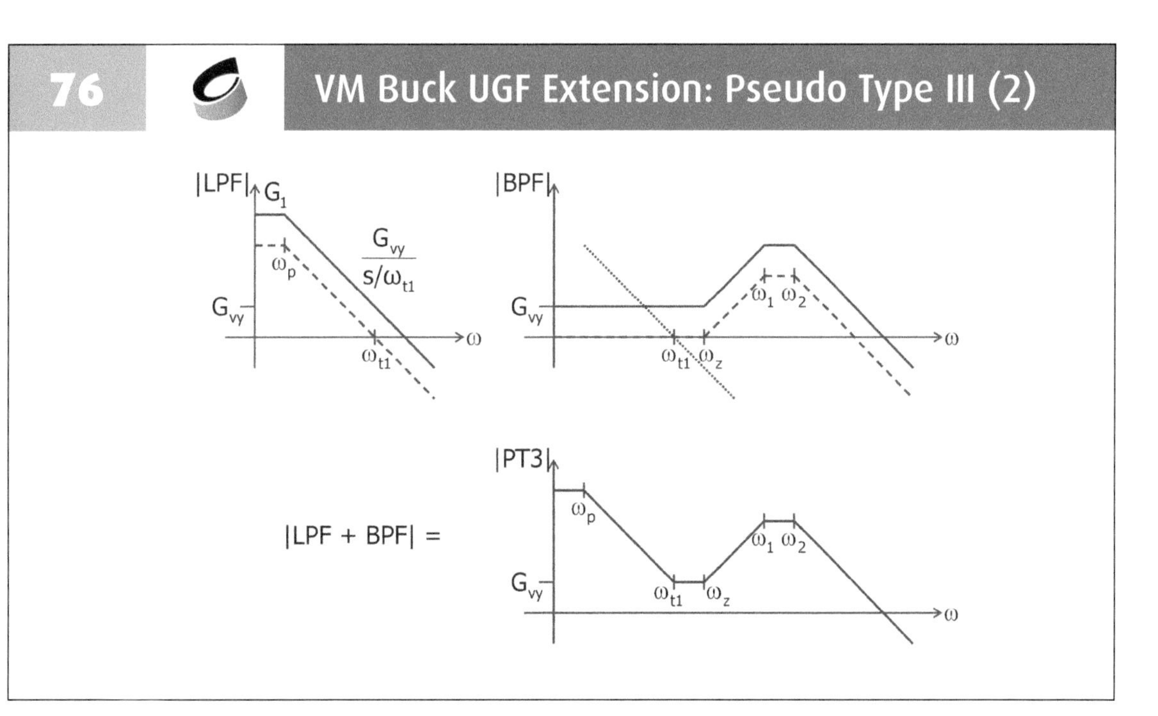

Note how the functions are added on the Bode plots.

77 VM Buck UGF Extension: DDA Type III Compensator

By using a differential difference amplifier (DDA) [Cheng 2014], the compensator can be more area-efficient.

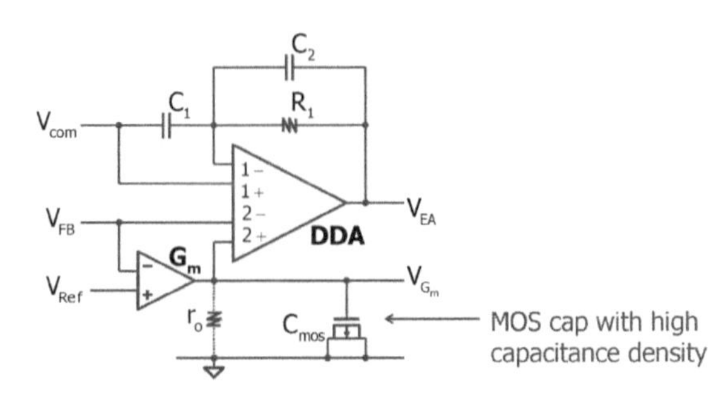

$$A(s) = -\frac{V_{EA}}{V_{FB}} = (1 + G_m r_o)\frac{(1 + sC_{mos}/G_m)(1 + s(C_1 + C_2)R_1)}{(1 + sC_{mos} r_o)(1 + sC_2 R_1)}$$

[Cheng 2014] proposed to build the Type-III compensator around a differential difference amplifier instead of an op amp. One advantage is that the low-frequency pole can be realized using a MOS capacitor that is referenced to ground. A second advantage is the flexibility in assigning component values to realize the poles and zeros.

78 Differential Amplifier

The differential amplifier (DA) has one pair of differential inputs. Working with proper feedback, the steady-state equation is

$$V_{1+} = V_{1-}$$

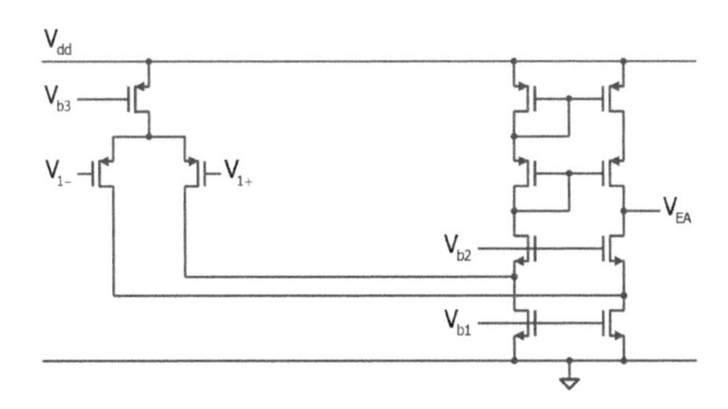

The working of the DDA is similar to the working of a differential amplifier. For a generic amplifier with feedback, $V_{1+} = V_{1-}$.

79 Differential Difference Amplifier

The differential difference amplifier (DDA) has two pairs of differential inputs. With proper feedback, the equation is

$$V_{1+} + V_{2+} = V_{1-} + V_{2-}$$

$$\Rightarrow \quad (V_{1+} - V_{1-}) = -(V_{2+} - V_{2-})$$

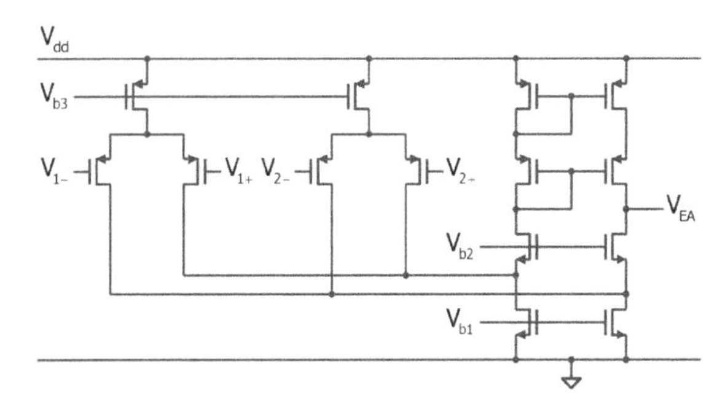

Similarly, when the DDA is working as an amplifier, $V_{1+} + V_{2+} = V_{1-} + V_{2-}$.

80 Matching Compensator Frequency Response

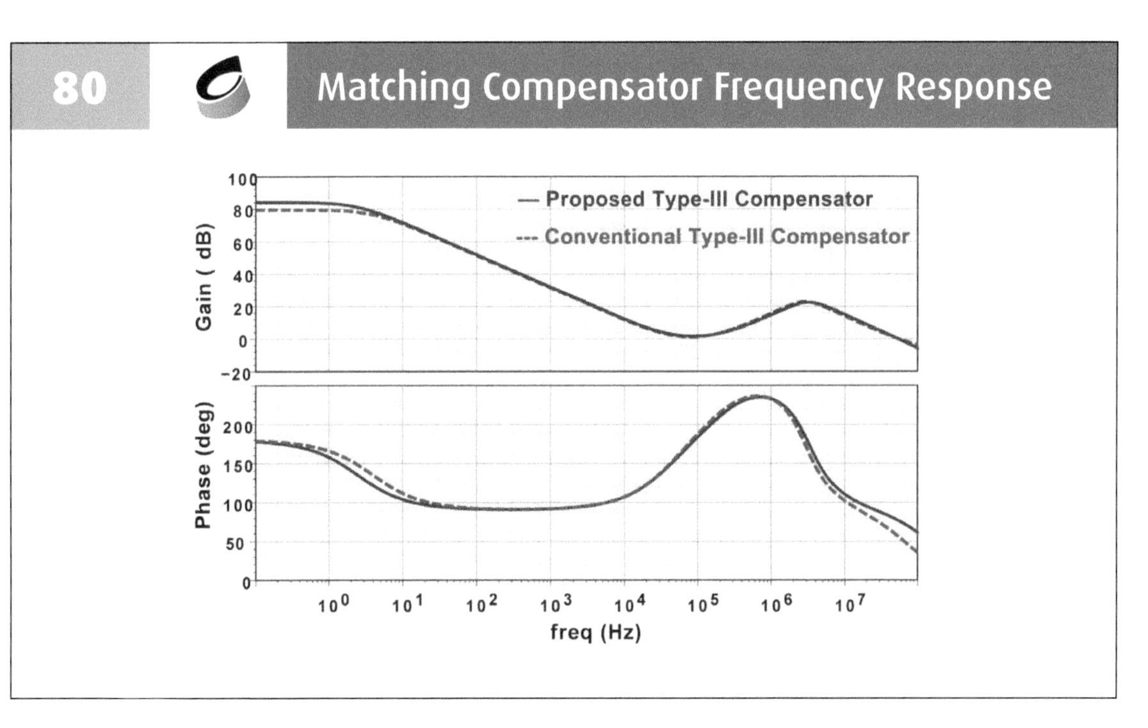

The DDA Type-III compensator is made to be approximately the same as the conventional Type-III compensator, so that their transient responses can be fairly compared.

81  Comparison of Silicon Area

Design	Conventional	Proposed
Capacitor	25pF MIM	5pF MIM 10pF MOS
Resistor	400kΩ	200kΩ
Amplifier	1	2
Area	$0.048mm^2$	$0.019mm^2$

60% area-reduction

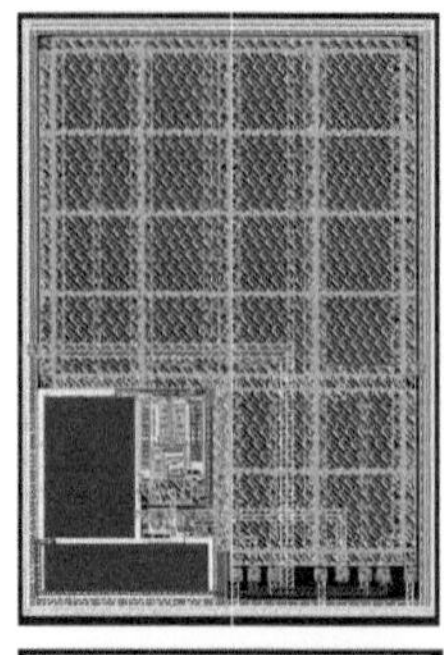

By using a CMOS capacitor referenced to ground and by using smaller resistors, the area of the DDA Type-III compensator is reduced by 60% as compared to the conventional Type-III compensator.

82  End-Point Prediction using DDA Compensator

End-point prediction requires: $V_{ea} = \frac{V_H - V_L}{bV_g} V_{ref} + V_L$

R_1 results in: $V_{ea} = V_{1-}$

If we assign: $V_{com} = V_{1+} = V_{ea}$

Then: $V_{1+} = V_{1-}$

From: $(V_{1+} - V_{1-}) = -(V_{2+} - V_{2-})$

$\Rightarrow \quad V_{2+} = V_{2-}$

Hence: $V_{fb} = V_{ref} = V_{G_m}$

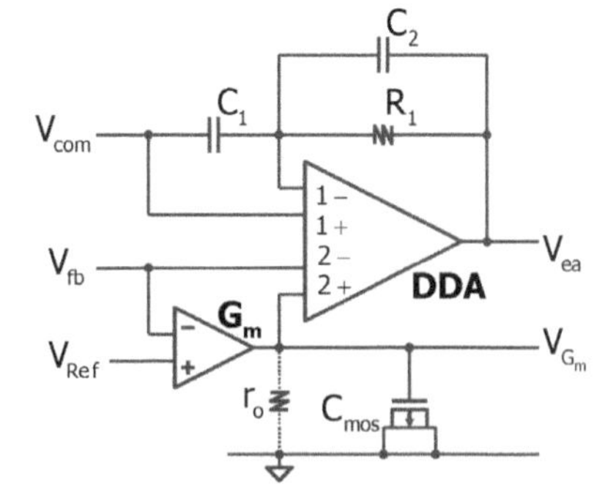

End-point prediction can be achieved by designing the output of the compensator (V_{ea}) as shown (refer to pp. 33 also). Now, the action of the DDA gives $V_{fb} = V_{ref} = V_{Gm}$.

83 EEP Implementation using DDA Compensator

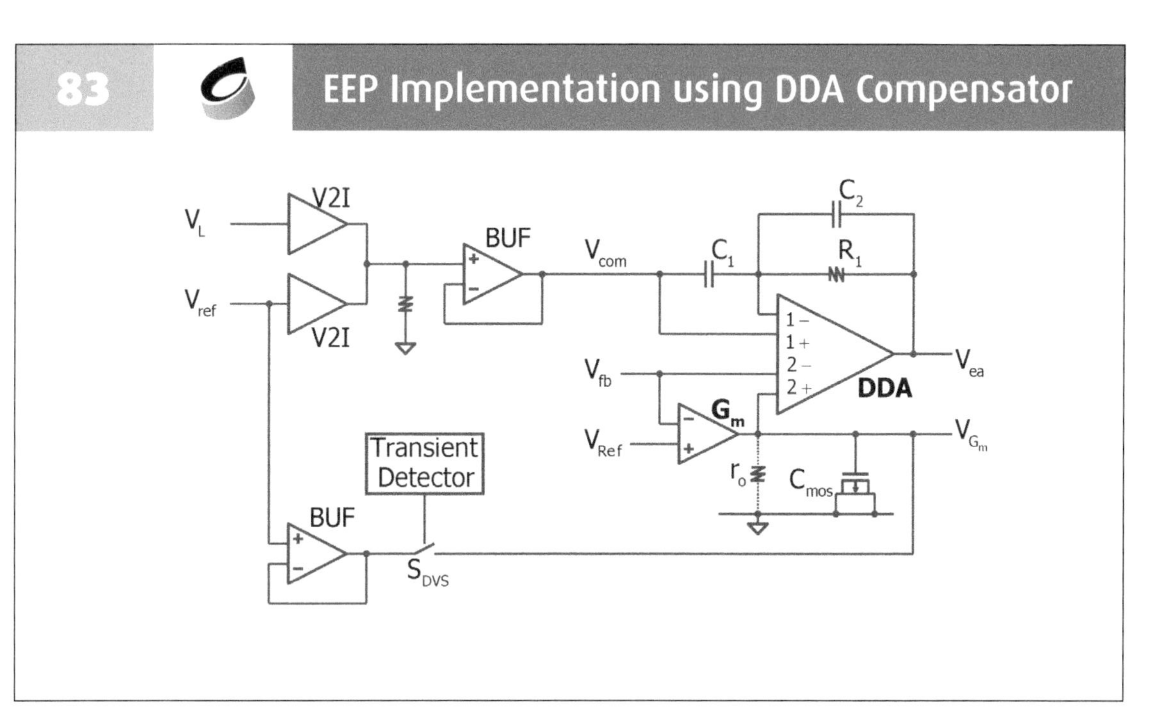

Therefore, end-point prediction is achieved by using the circuit shown.

84 Fast-Transient Switching Converters: Comparison

Publication	Zheng [JSSC 2011]	Kudva [JSSC 2011]	Lee [TPE 2012]	This Work [Cheng JSSC 2014]	
Technology	0.13μm	0.13μm	0.25μm	0.13μm	
Topology	Buck-Boost	Buck	Buck-Boost	Buck	
Control Scheme	Hysteresis	Voltage-Mode	Current-Mode	Voltage-Mode	
Max. P_{out}	0.4W	0.27W	1.4W	3.6W	
f_{sw}	10MHz	300MHz	5MHz	10MHz	30MHz
L	1μH	2nH	1μH	0.33μH	0.33μH
C_{out}	1μF	5nF	0.88μF	3.3μF	1μF
Nominal V_g	1.5V	1.2V	2.5~4.5V	3.3V	3.3V
V_{out} Range	0.9~2.2V	0.3~0.88V	3V	0.37~2.85V	0.45~2.4V
Peak Eff.	92.1%	74.5%	91%	91.8%	86.6%
Up-tracking	93.3μs/V	2.6μs/V	20μs/V	1.67μs/V	0.67μs/V
Down-tracking	26.7μs/V	3.6μs/V	15μs/V	4.44μs/V	1.56μs/V

Loosely speaking, the higher the switching frequency of the converter, the faster the transient responses can be achieved. However, by adding the circuitry for end-point prediction, [Cheng 2014] attained the best up-tracking and down-tracking speeds, especially when it was operating at 30MHz.

85  Example 3: 30MHz Type III Buck Converter

It is not convenient to use a voltage bias circuit to bias V_{com}.

We may instead set V_{com} equal to V_{ea} (through a buffer):

$$V_{com} = V_{1+} = V_{1-} = V_{ea}$$

$$\Rightarrow \quad V_{2+} = V_{2-}$$

Requirement of input range is relaxed.

OTA can be removed [Cheng 2017].

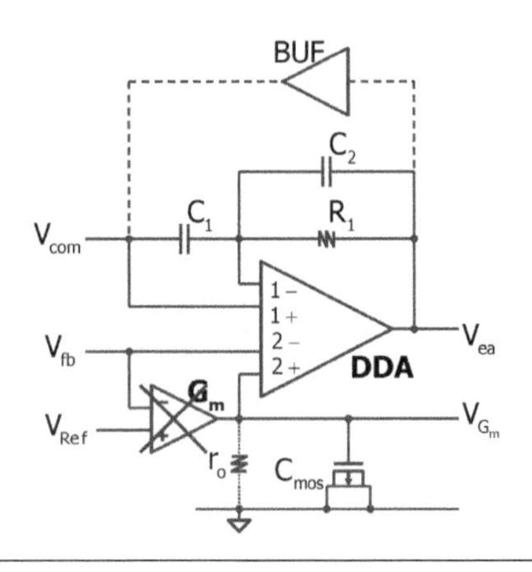

In the [Cheng 2014] design, V_{com} of the DDA compensator has to be biased by a DC voltage, and this is very inconvenient. If a buffer is added at V_{ea}, V_{com} may then be biased by V_{ea} instead. We may turn the buffer into a G_m stage, and move the original G_m stage to the new position [Cheng 2017].

86 New DDA Type III Compensator

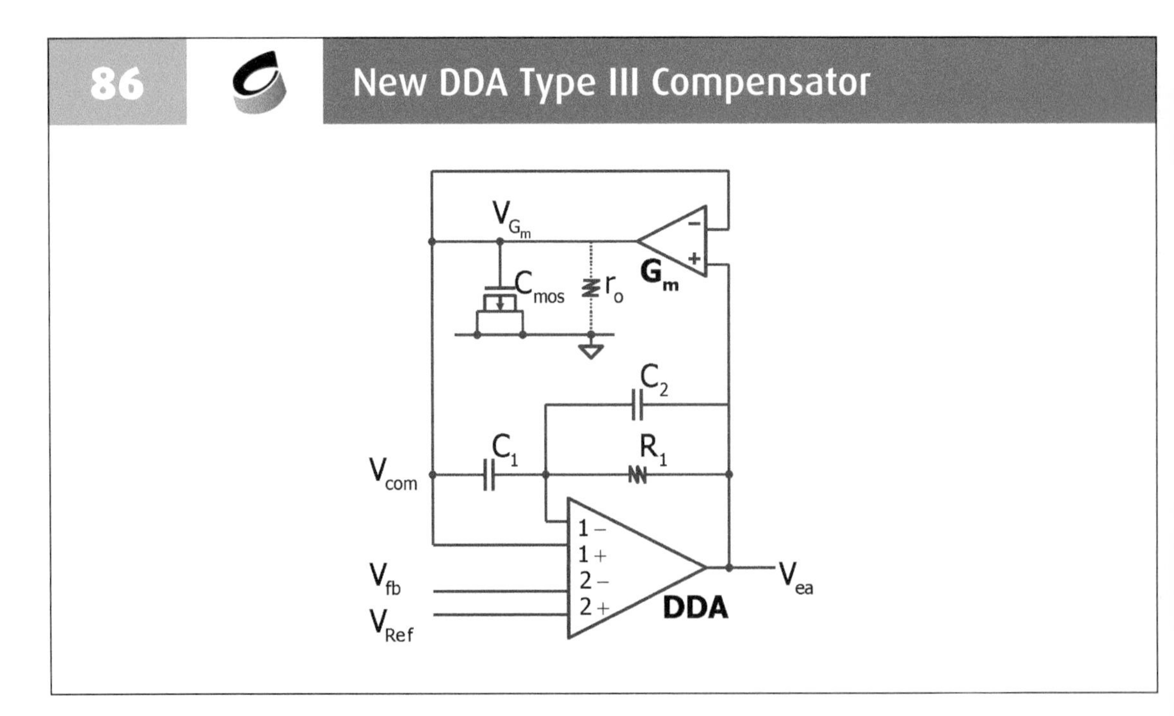

A new DDA Type-III compensator is thus obtained after the reconfiguration. It is important to re-derive the compensator function to make sure the correct function is obtained.

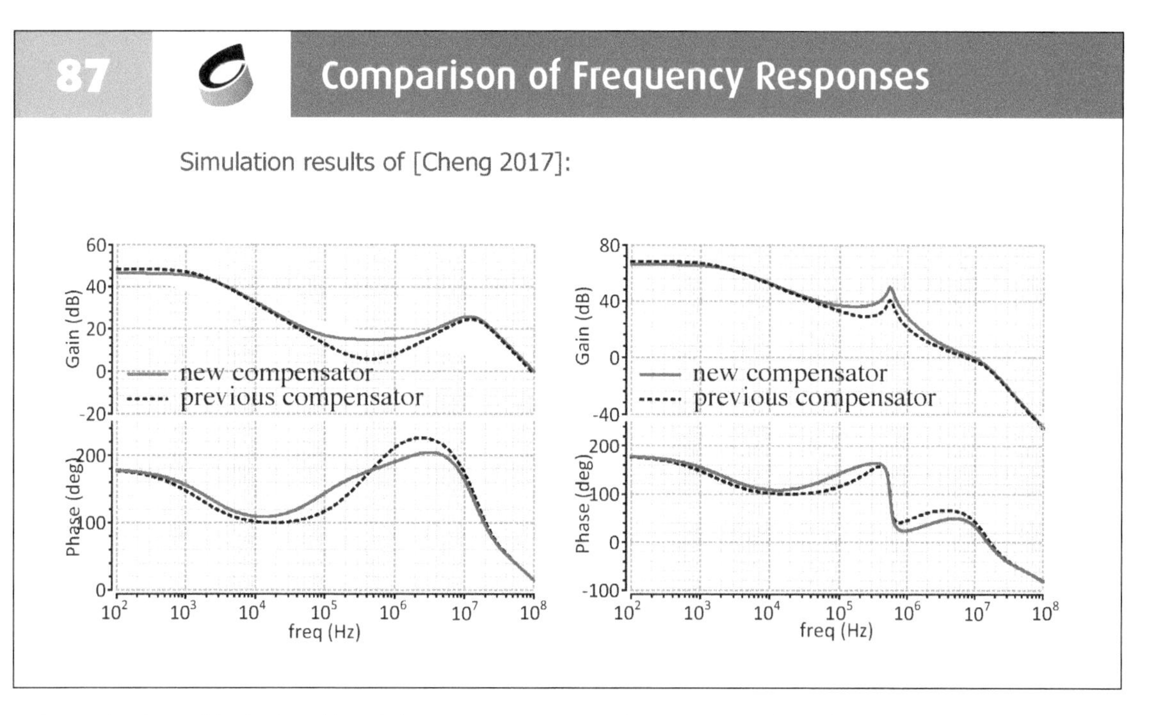

The Bode plots of both the old and the new compensator functions are plotted for fair comparison.

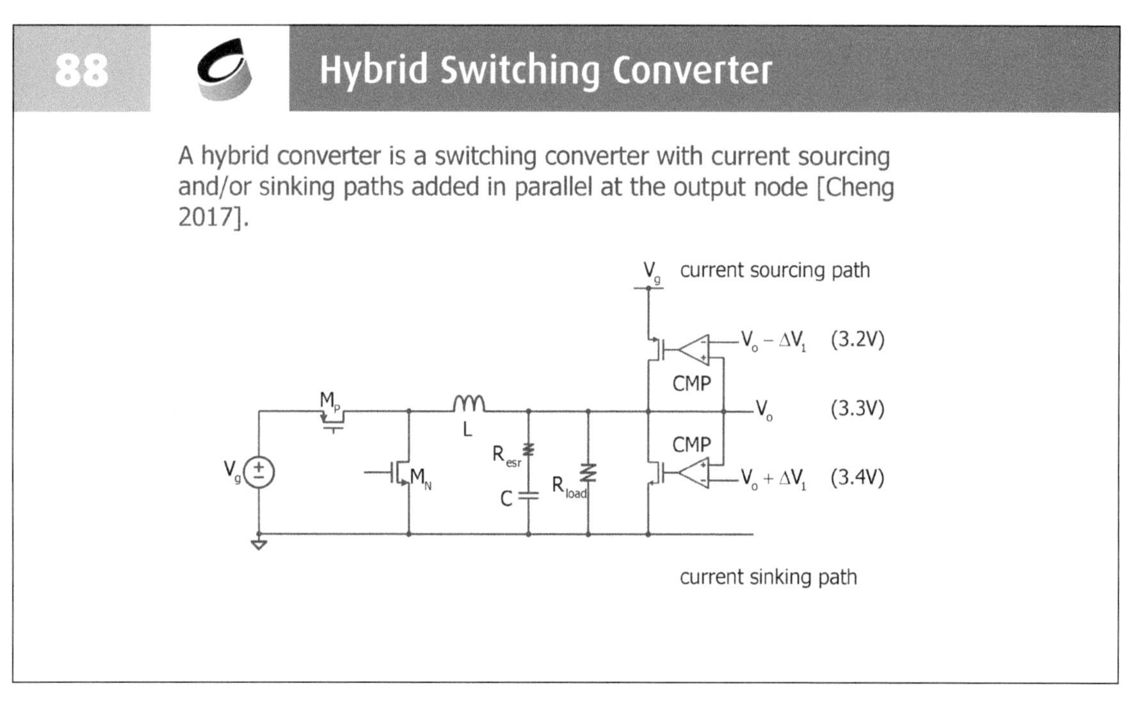

To further enhance the load transient responses, current source and current sink can be added to the output node. If V_O is too low, a current source can be activated to charge the output node faster. Similarly, if V_O is too high, a current sink can be activated to discharge the output node faster.

89  Hybrid Switching Converter: Design Issues

Design issues of a hybrid switching converter are:

- Current sourcing/sinking circuits consume minimum I_Q
- Fast turn-on and fast turn-off of current sourcing/sinking circuits
- Current sourcing/sinking comparators
- Current sourcing/sinking linear regulators
- Seamless mode transition between PWM and sourcing/sinking modes

The addition of the current source and the current sink in parallel to the output node is called a hybrid scheme. The design issues of a hybrid switching converter is stated in the slide.

90  V_{ea}-Based Hybrid Converter

The output of the compensator V_{ea}:
- stays within the range of ramp signal in the steady state
- swings out of the range during a large load transient

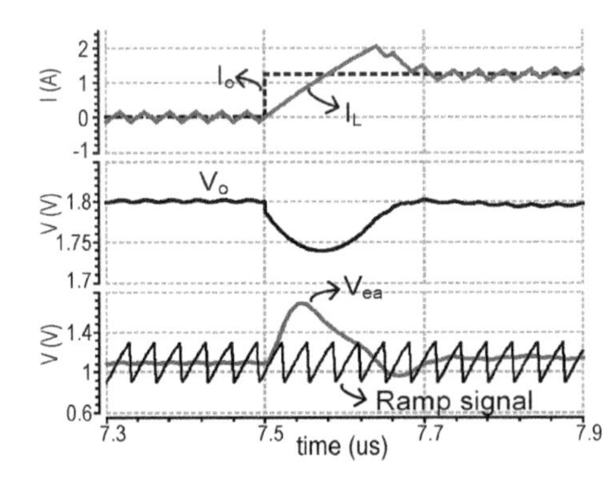

The output voltage is very noisy, while V_{ea} is a constant in the steady state. For small load change, the output voltage will have small overshoot and undershoot, and V_{ea} will change moderately. However, if the load current suddenly increase a lot, V_0 will have a large undershoot, and to respond, V_{ea} will jump high to correct it, and may even jump out of the range of the ramp signal.

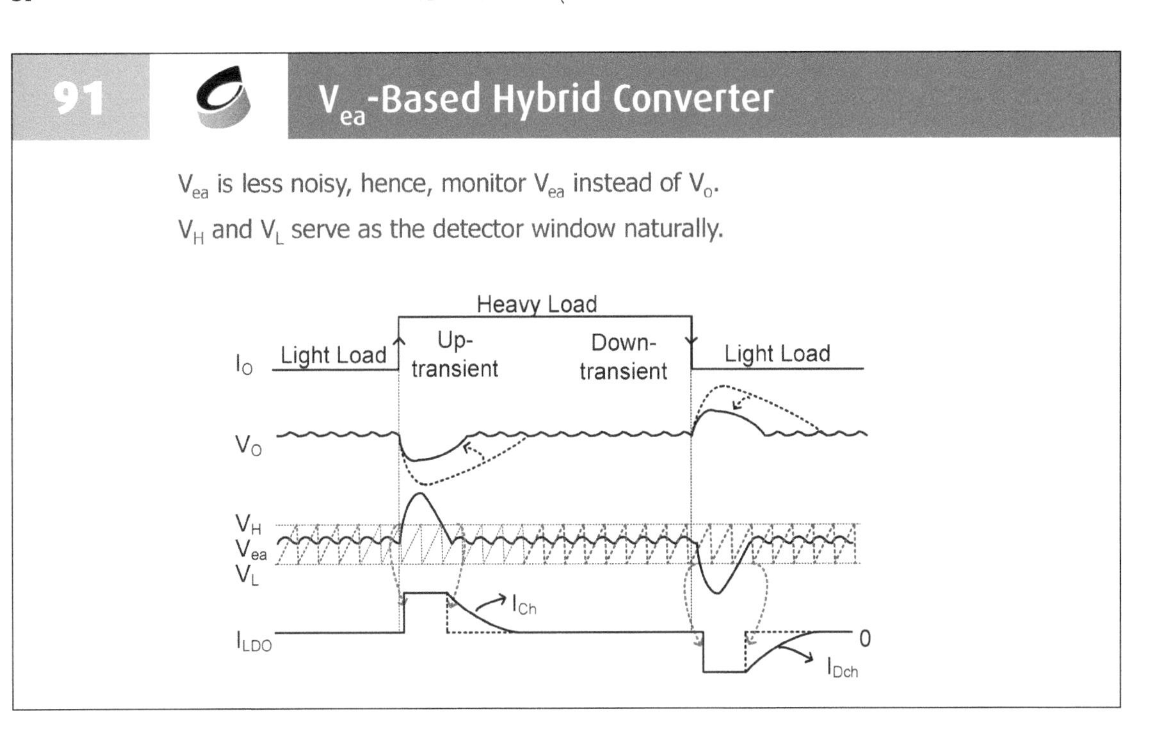

Hence, the high and low thresholds (V_H and V_L) of the ramp may serve as the detector window. When V_{ea} jumps higher than V_H, the current sink is activated; and when V_{ea} jumps lower than V_L, the current source is activated.

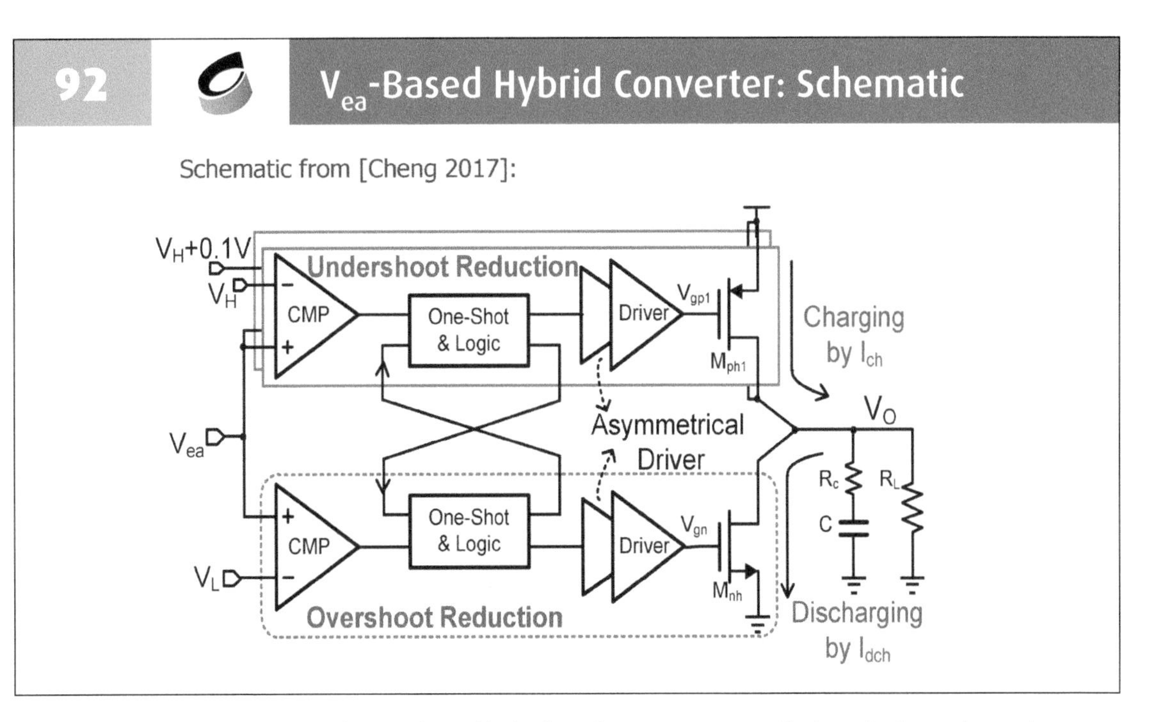

The circuit implementation of the V_{ea}-based hybrid converter is shown in the slide. To control the amount of charging and discharging current, a one-shot timer is installed, such that a limited amount of charge is injected to or drained from the load capacitor.

93  Fast-Transient Switching Converters: Comparison

Publication	ESSCIRC[5]	ISSCC [2]	ISSCC [4]	ISSCC [1]	This Work	
Year	2012	2014	2015	2016	**2017**	
Technology	0.13μm	0.18μm	65nm	0.18μm	**0.13μm**	
Switching frequency	10MHz	40MHz	30MHz	30MHz	**30MHz**	
Inductor	1μH	78nH×4	90nH×4	220nH×4	**90nH**	
Capacitor	1μF	0.47μF×2	0.47μF	0.62μF	**0.47μF×2**	
Nominal V_g/V_o	3.7V/1.2V	3.3V/1.2V	1.8V/1.5V	3.3V/1.8V	**3.3V/1.8V**	
Peak efficiency	84.5% (@V_o=1.2V)	86.1% (@V_o=1.6V)	87% (@V_o=1V)	86.5% (@V_o=2.5V)	**90.7%/88%/83.6% (@V_o=2.4V/1.8V/1.2V)**	
Control scheme	Voltage-mode +Hybrid	ZDS Hysteretic	Voltage-mode	Current-mode +CCS/LTO	**Voltage-mode only**	**Voltage-mode +Hybrid**
Up-Transient: I_o step (rise time)	0.3A(20ns)	5A(5ns) [4 phases]	0.4A (10ns) [4 phases]	1.8A(5ns) [4 phases]	**1.25A(2ns)/0.62A(2ns)**	
Up-Transient: V_o droop (% of V_o)	55mV (4.6%)	118mV (9.8%)	80mV (5.33%)	100mV (5.6%)	**72mV/33mV (4%/1.83%)**	**36mV/12mV (2%/0.67%)**
Up-Transient: 1% settling time	1500ns	230ns	600ns	133ns	**220ns/150ns**	**125ns/0ns***

By using the improved DDA Type-III compensator and the hybrid scheme, both overshoot and undershoot are maintained to be small; and the 1% settling times are on the order of 100ns. If the overshoot or undershoot is lower than 1% of the output voltage, one may even consider the output to be immediately settled (1% settling time is then considered to be 0).

94  Conclusions

- **Switching frequency f_s of 30MHz is a good comprise of UGF and efficiency**
- **Voltage-mode control with no high-speed current sensor is preferred**
- **Area-efficient Type III compensators are useful**
- **Hybrid switching converters could attains fastest transient responses**

To conclude, switching frequency of 30MHz, for now, is a good compromise of transient responses and efficiency. Voltage-mode control should be used, and Type-III compensator could push the UGF of the control loop to 3MHz. With an accurate hybrid scheme, the output settling times can be reduced to around 100ns to 200ns.

95 

References

1. [Smedley 1995] K. M. Smedley and S. Cuk, "One-cycle control of switching converters," *IEEE Tran. on Power Elec.*, pp. 625-633, Nov. 1995.
2. [Goder 1996] D. Goder and W. R. Pelletier, "V2 architecture provides ultra-fast transient responses in switch mode power supplies," *High Freq. Power Conv.*, pp. 19-23, Sept. 1996.
3. [Ki 1998]W. H. Ki, "Signal flow graph in loop gain analysis of DC–DC PWM CCM switching converters", *IEEE Tran. Circ. Sys. I*, pp. 644-655, June 1998.
4. [Lee 2000] H. Lee, P. Mok and W. H. Ki, "A novel voltage-control scheme for low-voltage DC-DC converters with fast transient recovery," *IEEE Int'l Symp. on Circ. and Syst.*, pp. I-256-I259, May 2000.
5. [Mattingly 2003] D. Mattingly, "Designing stable compensation networks for single phase voltage mode buck regulators," *Tech. Brief TB417*, Intersil, 2003.
6. [Ma 2004] D. Ma, W. H. Ki and C. Y. Tsui, "An integrated one-cycle control buck converter with adaptive output and dual loops for outer error correction," *IEEE J. on Solid-State Circ.*, pp. 140-149, Jan. 2004.
7. [Siu2006] M. Siu, P. Mok, K. N. Leung, Y. H. Lam, and W. H. Ki, "Avoltage-mode PWM buck regulator with end-point prediction", *IEEE Trans. on Circ. and Syst. II*, pp.294-298, April 2006.
8. [Mai2008] Y. Y. Mai and P. Mok, "A constant frequency output-ripple-voltage-based buck converter without using large ESR capacitor," *IEEE Trans. on Circ. and Syst. II*, pp. 748-752, Aug. 2008.
9. [Meeks2008] D. Meeks, "Loop stability analysis of voltage mode buck regulator with different output capacitor types – continuous and discontinuous modes," *Appl. Report SLVA301*, TI, April 2008.
10. [Su 2008] F. Su, W. H. Ki and C. Y. Tsui, "Ultra fast fixed frequency hysteretic buck converter with maximum charging current control and adaptive delay compensation for DVS applications," *IEEE J. of Solid-State Circ.*, pp. 815-822, April 2008.
11. [Su2009] F. Su and W. H. Ki, "Digitally assisted quasi V2 hysteretic buck converter with fixed frequency and without using large-ESR capacitor," *IEEE Int'l Solid-State Circ. Conf.*, pp. 446-447, Feb. 2009.
12. [Zhan2009] C. Zhan and W. H. Ki, "Loop bandwidth extension technique for PWM voltage mode switching converters," *IEEE Asian Solid-State Circ. Conf.*, pp. 325-328, Nov.2009.
13. [Rahimi2010] A. M. Rahimi, P. Parto, P. Asadi, "Compensator design procedure for buck converter with voltage-mode error-amplifier," *Appl. Notes AN1162*, IR, 2010.

14. [Wu2010] Y. Wu, S.Tsui and P. Mok, "Area- and power-efficient monolithic buck converters with pseudo Type-III compensators," *IEEEJ. of Solid-State Circ.*, pp. 1446-1455, Aug.2010.
15. [Huang 2013a] C. Huang and P. Mok, "An 84.7% efficiency 100-MHz package bondwire-based fully integrated buck converter with precise DCM operation and enhanced light-load efficiency," *IEEE J. Solid-State Circ.*, pp. 2595-2607, Nov. 2013.
16. [Huang 2013b] C. Huang and P. Mok, "A 100 MHz 82.4% efficiency package-bondwire based four-phase fully-integrated buck converter with flying capacitor for area reduction," *IEEE J. Solid-State Circ.*, pp. 2977-2988, Dec. 2013.
17. [Cheng 2014] L. Cheng, Y. Liu and W. H. Ki, "A 10/30 MHz fast reference-tracking buck converter with DDA-based Type-III compensator," *IEEE J. Solid-State Circ.*, pp. 2788-2799, Dec. 2014.
18. [Fang 2014] X. Fang, R. Wu, L. Peng and J. Sin, "A novel integrated power inductor with vertical laminated core for improved L/R ratios." *IEEE Elec. Device Lett.*, pp. 1287-1289, Dec. 2014.
19. [Wu 2015] R. Wu, N. Liao, X. Fang, and J. Sin, "A silicon-embedded transformer for high-efficiency, high-isolation, and low-frequency on-chip power transfer," *IEEE Trans. Elec. Devices*, pp. 220–223, Jan. 2015.
20. [Cheng 2017] L. Cheng, W. H. Ki, "A 30MHz Hybrid Buck Converter with 36mV Droop and 125ns 1% Settling Time for a 1.25A/2ns Load Transient," *IEEE Int. Solid-State Circ. Conf.*, pp. 188-189, Feb. 2017.

CHAPTER 2

Design of Low-Power Fast-Transient Analog and Digital Low-Dropout Regulators

Yan Lu

University of Macau
Macao, China

Low-dropout (LDO) regulators, which provide fast-transient response and good power supply ripple rejection, are indispensable in system-on-a-chip (SoC) designs. Conventional analog LDO regulators can perform excellent in normal supply voltages, while the digital LDO regulators recently become very popular for its low-voltage operation, good large-signal performance and process-scaling characteristics. This chapter addresses the design considerations, circuit techniques of LDO regulator, and will discuss the recent advancements on both analog and digital LDO regulators.

1 

Outline

- **Introduction of Low Dropout Regulator (LDO)**
- Analog LDO Circuit Techniques and Design Examples
- Digital LDO Circuit Techniques and Design Examples

2 

What does an LDO do?

- All we want is a clean DC supply.
- But... this is what we get.

- The LDO needs to resist the load/line transients and the input ripples.

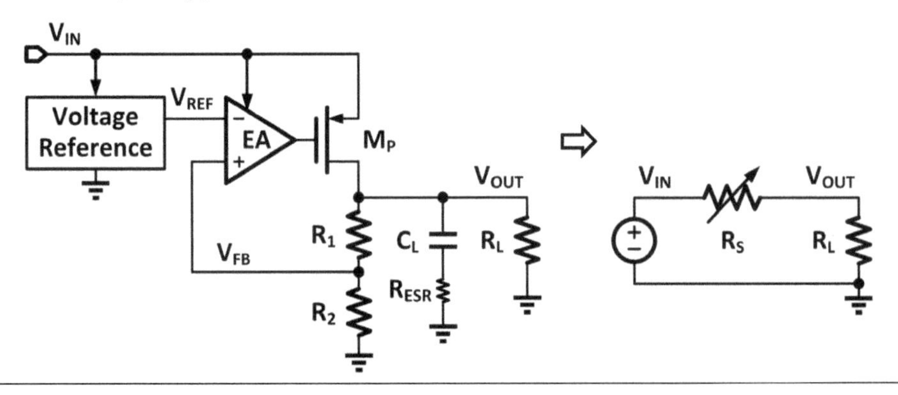

Firstly, let's start with some basics. What does an LDO do? For a power supply, all we want is a clean DC voltage. However, the LDO needs to resist the load transient, line transients, and input ripples.

An LDO is basically a linear amplifier that has large output current capability. The error amplifier (EA) compares the reference voltage and the feedback output voltage, adjusts the gate voltage of the power transistor M_P, and therefore adjust the output current according to external variations. In the equivalent circuit model, the LDO acts like an adjustable resistor, making sure a correct output voltage when the input voltage or the load resistor changes.

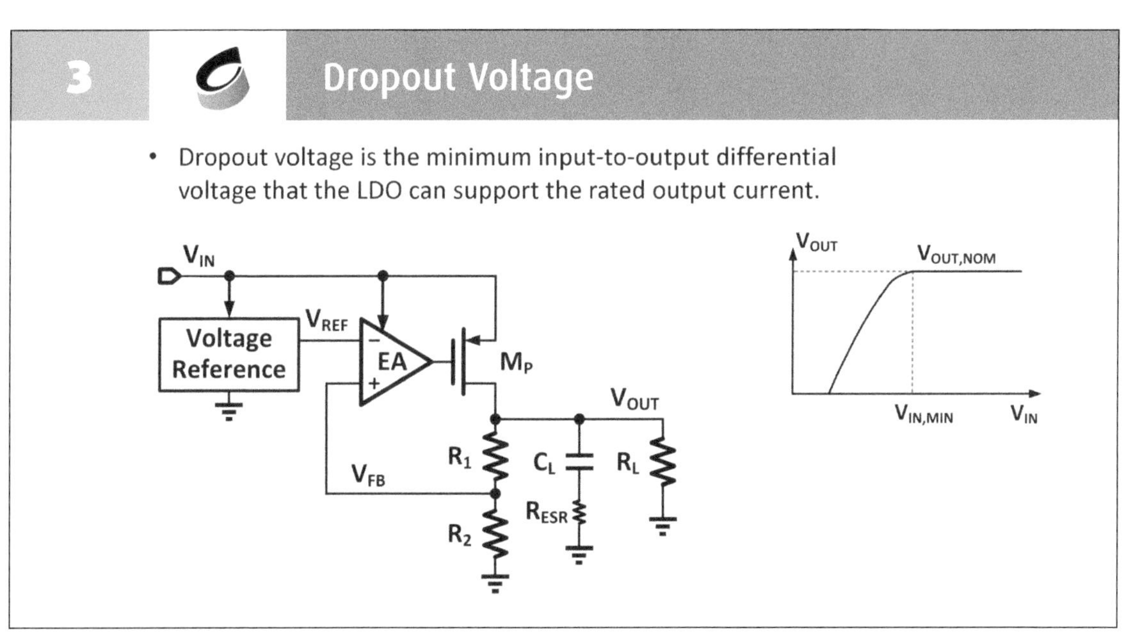

The dropout voltage V_{DO} is defined as the minimum input-to-output differential voltage that the LDO can sustain the rated maximum output current. As shown in the voltage relationship, $V_{DO} = V_{IN,MIN} - V_{OUT,NOM}$.

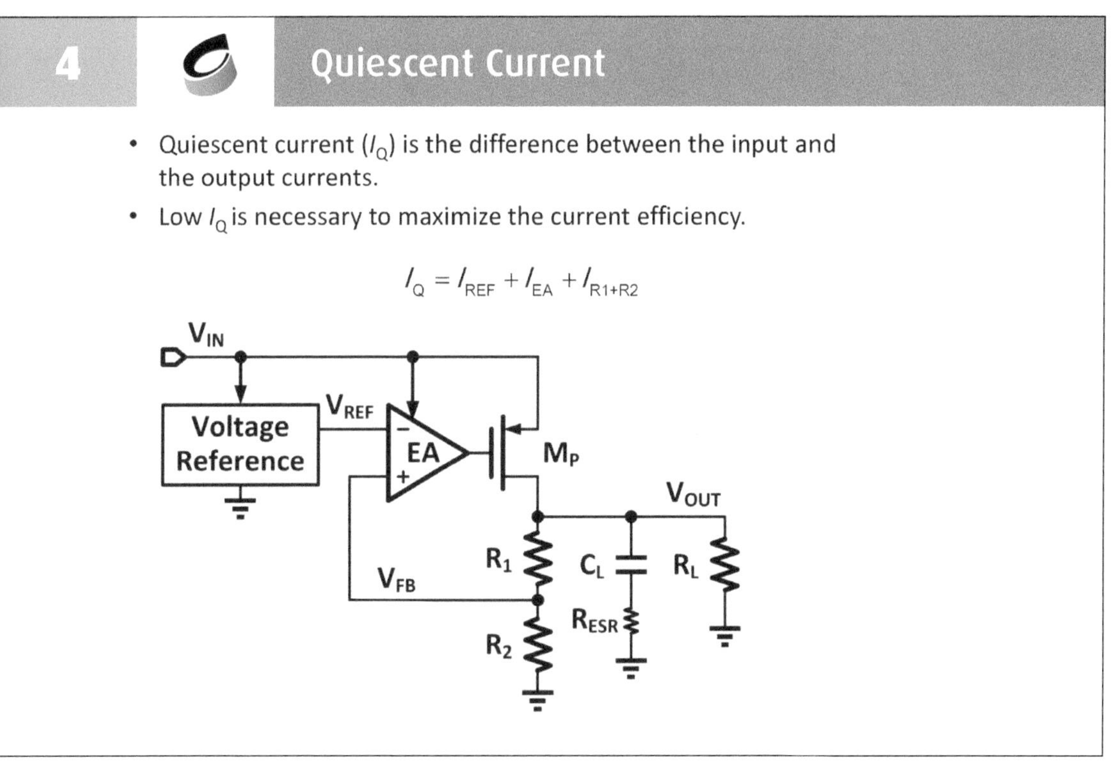

Quiescent current is the difference between the input and the output currents. Obviously, a low I_Q is necessary to maximize the current efficiency. It mainly consists of three parts: current consumption of the reference voltage generator, current of the error amplifier, and the current of the resistor divider.

5  Efficiency

- The efficiency of LDO regulators is limited by the quiescent current and input/output voltages.
- Usually, I_Q is much smaller than the load current I_L.

$$\text{Efficiency} = \frac{I_L V_{OUT}}{(I_L + I_Q) V_{IN}} \times 100$$
$$\approx \frac{V_{OUT}}{V_{IN}} \times 100$$

Examples:

1) V_{IN} = 1.2V, V_{OUT} = 1.0V, I_L = 100mA, I_Q = 0.1mA, Efficiency = 83.3%

2) V_{IN} = 3.7V, V_{OUT} = 1.0V, I_L = 100mA, I_Q = 0.1mA, Efficiency = 27%

Let's look at the equation for the efficiency calculation. It equals to the output power divided by the input power. Usually, I_Q is much smaller than the load current I_L, therefore, can be neglected. Then, the efficiency simply equals to V_{OUT}/V_{IN}. Therefore, efficiency is the main drawback of an LDO regulator, as it drops linearly when the input and output voltage difference increases.

Let's consider the following two examples, if we have a pre-stage DC-DC converter gives us a 1.2V input, and our targeted output is 1.0V, and I_Q is 0.1mA while delivering 100mA output current, the efficiency basically equals to 1/1.2 = 83.3%. If we directly connect the LDO to a Li-ion battery which has a 3.7V output voltage, the efficiency will be only 27% when the other conditions are the same. Therefore, LDO is usually used as a post-stage regulator at the point-of-load to clean up supply.

6  Power Supply Rejection (PSR)

- In the battery-powered applications, supply ripple rejection in the frequency band between 1 MHz and 100 MHz is especially important, because the output of a DC-DC switch mode power supply (SMPS) is commonly used to supply the LDO.
- For 4G baseband signal, its bandwidth can be up to 20MHz.

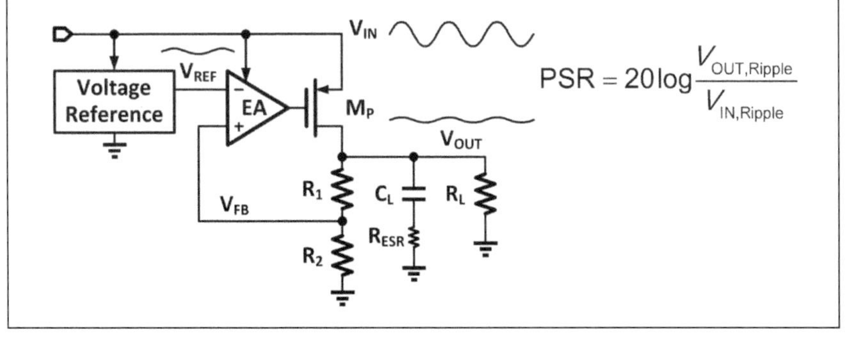

So, now let's look at the power supply rejection (PSR). It is defined as the ratio between the output voltage ripple and the input voltage ripple at certain frequencies. To test the PSR, we can apply sine waves at different frequencies to V_{IN} and observe the attenuated output ripples.

In the battery-powered applications, PSR in the frequency band between 1 MHz and 100 MHz is especially important. Because the pre-stage DC-DC converter is switching at such frequencies, and it will generate supply ripples in the range of mV to tens of mV. In the pervious chapter, Prof. Ki from HKUST introduced how to operate DC-DC converters at tens of MHz or even hundreds of MHz.

From the application point of view, PSR in MHz range is also important. Like in the 4G communication system, the bandwidth of the baseband signal could be up to 20MHz, and for 5G, its bandwidth may be over 100MHz.

7

Load Transient Response

- The transient response is a function of the LDO response time, the load capacitance (C_L), the equivalent series resistance (ESR) of the load capacitor, and the maximum load-current ($I_{O,MAX}$).
- The worst case of the output voltage variations occurs when the load current transitions from zero to its maximum rated value or vice versa.

$$\Delta V_{Tran,MAX} = \frac{I_{O,MAX}}{C_L}\Delta t + \Delta V_{ESR}$$

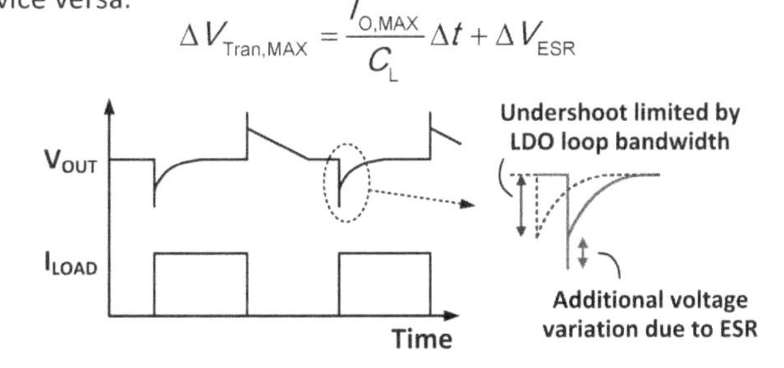

For the load transient response, a output capacitor C_L is commonly used to provide some buffer current when the regulation loop is too slow to response. An capacitor unavoidably has an equivalent series resistor (ESR). In some conventional designs, additional ESR is added to generate a left-half-plane (LHP) zero for compensating the non-dominant pole and boosting the phase margin. Or, the ESR could come from unwanted parasitic routing resistance. Anyhow, the ESR will contribute additional voltage variation during the load transient events. During the load transient from low to high, the capacitor will supply most of output current before the LDO loop can react. Then, the capacitor current will result in an opposite voltage on the ESR. Thus, additional voltage undershoot can be observed.

8

LDO Requirements

- To make a comparison, a figure-of-merit (FOM) is defined as

$$\mathrm{FOM} = T_R\frac{I_Q}{I_{MAX}} = \frac{C_L \times \Delta V_{OUT}}{I_{MAX}} \times \frac{I_Q}{I_{MAX}} \quad \text{[Hazucha JSSC 05]}$$

  - Fast transient response
  - Small quiescent current (I_Q)
- Besides...
  - Good power supply rejection (PSR)
  - Low dropout
  - Area-efficient
  - Performance scalable with process

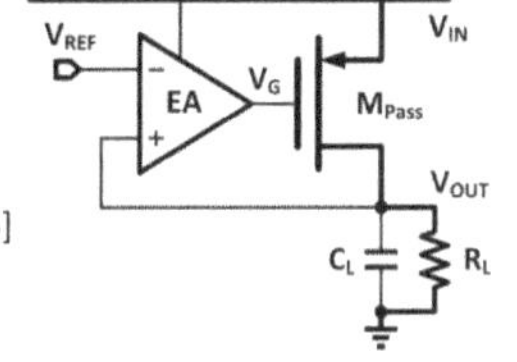

To make a comparison between different LDO designs, a figure-of-merit (FoM) is proposed by Peter Hazucha from Intel. The response time T_R is calculated by load capacitor C_L times ΔV_{OUT} divided by the maximum output current I_{MAX}. For better FoM, we need to use less I_Q to sustain a higher maximum current. To achieve that, a fast response, in the meanwhile low power, LDO is required. Therefore, it's a power and speed tradeoff.

Besides, good power supply rejection (PSR) is a very important specification for noise-sensitive analog and RF loadings. Low dropout voltage means high efficiency. Also, the fully-integrated LDO should be area-efficient. And, its performance should be scalable with process, which means the performances should automatically become better when we migrate the same circuit structure to a more advanced process. Note that, the last feature does not necessarily to be true as to be discussed in the following slides.

9 Loop Stability

- There are at least two LF poles in an LDO:
  - p_{Gate}: Pole at the Gate of the Power MOS
  - p_{Out}: Pole at the Output

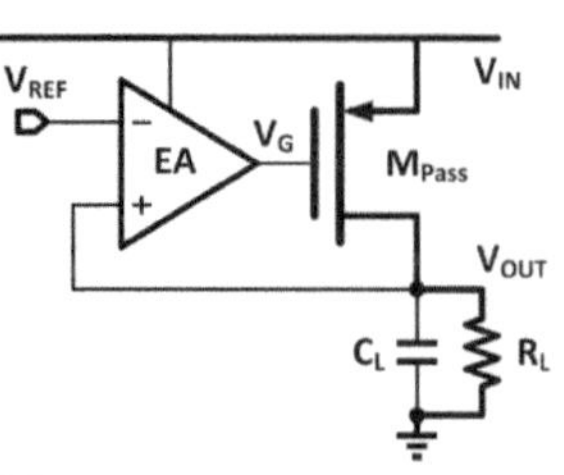

***p_{Out}* location would shift with load conditions**

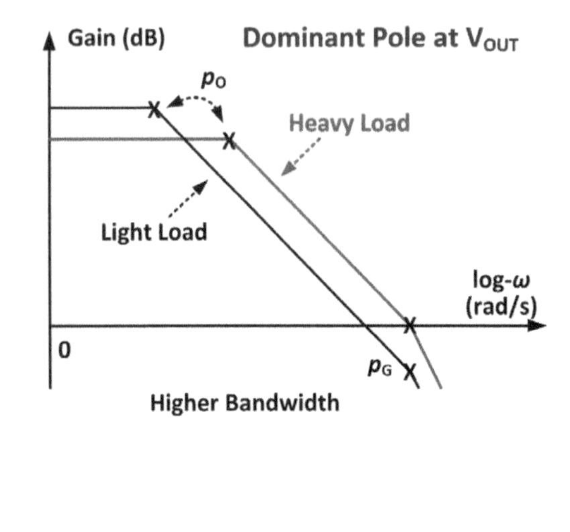

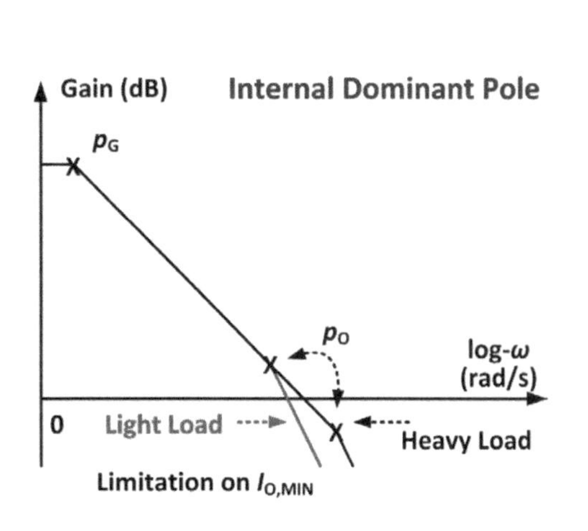

In a typical LDO design, there are at least two low frequency (LF) poles exist in the LDO. The gate of the power transistor has a large RC time constant because the EA has a high output resistance and the gate has a large parasitic capacitor. Meanwhile, the output node also has large capacitance which may come from the filtering capacitor C_L or from the load itself. Obviously, the output pole will shift with the load conditions as both the load resistance and the on-resistance of the power transistor vary with the output current.

Making the LDO loop stable is the most basic requirement for the LDO design. Therefore, the question we need ask ourselves when we start to design an LDO is where should I design the LDO dominant pole located at?

If we design the dominant pole at V_{OUT}, the unity-gain frequency (UGF) will be extended at heavy-load condition because the dominant pole p_{OUT} shifts to higher frequency. When the UGF approaches the first non-dominant pole, which could be p_G, the loop stability degrades. On the other hand, if we design the dominant pole to the internal node of the LDO, a quite low frequency internal pole is required. Because it is limited by p_O in the light-load or no-load conditions. For the stability consideration, there might be a minimum output current limitation on the internal dominant pole case.

10

PSR Considerations

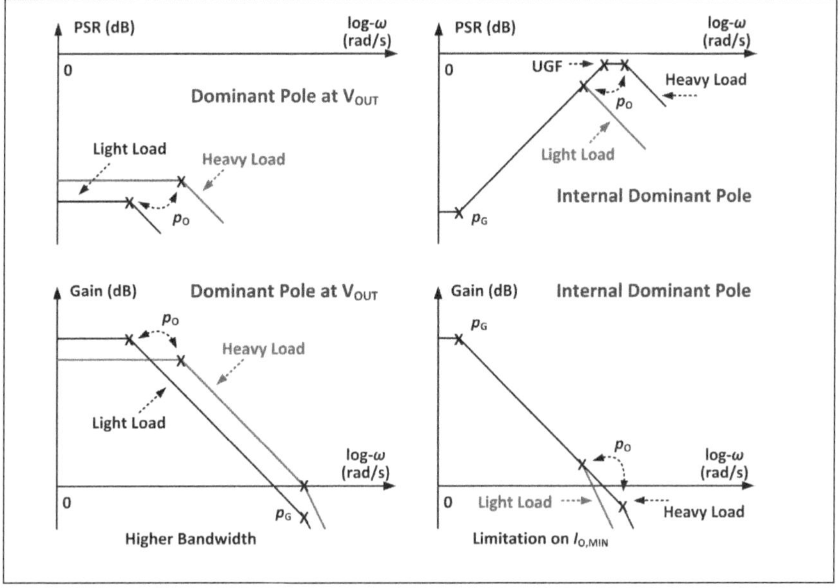

Also, for the PSR consideration, it is good to put the dominant pole at the output node. Because then we can allocate a large capacitor on the output node for filtering the supply ripples. At the frequencies higher than p_O, the high frequency ripples start to be filtered by the load capacitor.

For the internal dominant pole case, the PSR is mainly depends on the LDO regulation loop. When the loop gain decreases, the PSR becomes worse. At the frequency range around the UGF, its PSR drops to about 0dB which means no PSR.

11

Comparison for Dominant Pole Locations

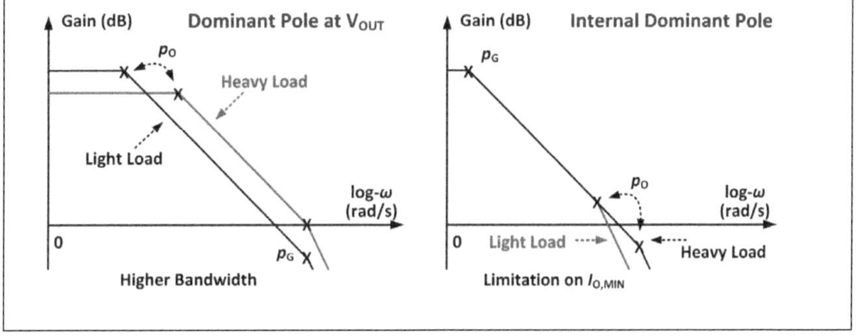

The table summarizes the pros and cons of designing the dominant pole at the output node or the internal node. We say the p_O dominant case can enjoy the process scaling because it would be easier to push the internal poles to higher frequencies with a better process. For the internal dominant case, the UGF can hardly be increased even with a better process, because it is basically limited by p_O at light load. The only drawback of the p_O dominant case is that it needs relatively more current to push the internal poles to higher frequencies, which will increase the LDO quiescent current.

Dominant pole	Output	Internal
Process Scaling	Yes	No
Limit on $I_{o,min}$	No	Yes
Unity Gain Freq.	√	X
Transient ΔV_{OUT}	√	X
PSR	√	X
I_Q	Larger	Smaller

12 Outline

- Introduction of Low Dropout Regulator (LDO)
- **Analog LDO Circuit Techniques and Design Examples**
- Digital LDO Circuit Techniques and Design Examples

13 Output Impedance of the PMOS and NMOS LDOs

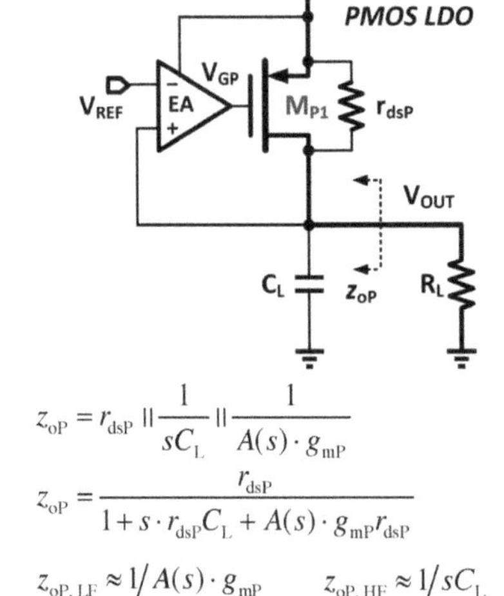

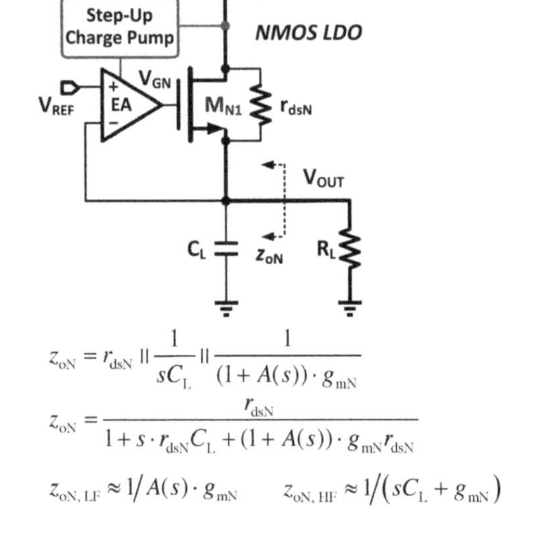

$$z_{oP} = r_{dsP} \,||\, \frac{1}{sC_L} \,||\, \frac{1}{A(s)\cdot g_{mP}}$$

$$z_{oP} = \frac{r_{dsP}}{1 + s\cdot r_{dsP}C_L + A(s)\cdot g_{mP}r_{dsP}}$$

$$z_{oP,LF} \approx 1/A(s)\cdot g_{mP} \qquad z_{oP,HF} \approx 1/sC_L$$

$$z_{oN} = r_{dsN} \,||\, \frac{1}{sC_L} \,||\, \frac{1}{(1+A(s))\cdot g_{mN}}$$

$$z_{oN} = \frac{r_{dsN}}{1 + s\cdot r_{dsN}C_L + (1+A(s))\cdot g_{mN}r_{dsN}}$$

$$z_{oN,LF} \approx 1/A(s)\cdot g_{mN} \qquad z_{oN,HF} \approx 1/(sC_L + g_{mN})$$

Y. Lu, W. Ki, and C. Patrick Yue, "An NMOS-LDO Regulated Switched-Capacitor DC-DC Converter With Fast-Response Adaptive-Phase Digital Control," *IEEE Transactions on Power Electronics*, vol. 31, no. 2, pp. 1294–1303, Feb. 2016.

We can choose either PMOS or NMOS transistor as the power transistor. Obviously, the PMOS is easy to be driven by low voltages. But the NMOS needs to be driven by a high voltage. To maintain a low drop-out voltage, a step-up charge pump circuit is usually required to generate a high voltage supplying the EA which drives the NMOS.

In terms of output impedance, since the NMOS power stage is configured as a source follower, it can provide an intrinsic transient response to the load transient. Because if the V_{OUT} drops during the load transient before the LDO loop can response, V_{GS} of the power NMOS will be automatically increased. Then, larger output current will be provided. Therefore, in the output impedance z_{oN} calculation, we get $1/g_{mN}$ in parallel with the other factors, where g_{mN} is the transconductance of the power NMOS.

14 

Replica LDO

- Uses the NMOS output stage for its *source-follower* property.
- Requires a voltage doubler to supply a higher gate-drive voltage.
- Sacrificing DC accuracy, should be acceptable for digital loads.

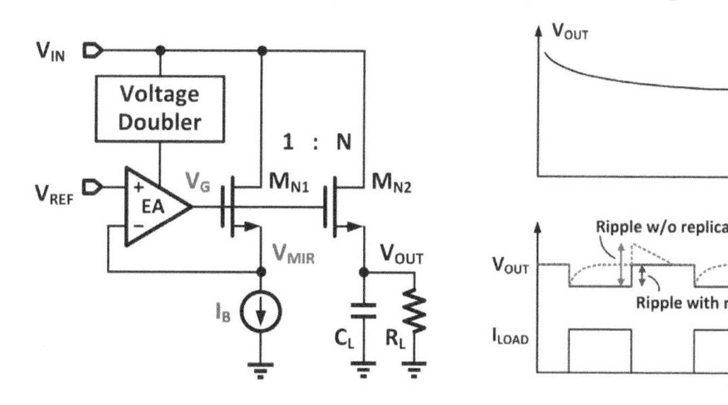

Based on the NMOS LDO characteristics, the replica LDO is a simple and good choice for digital loads. In the replica LDO, M_{N1} and M_{N2} are matched transistors with a size ratio of 1:N. If the bias current I_B and the output current also have the ratio of 1:N, V_{OUT} is approximately equal to V_{MIR}, while V_{MIR} is a mirrored voltage of the reference voltage V_{REF}. Of course, I_B and the output current can hardly matched. So, V_{OUT} changes with the load current. In heavy load, V_{OUT} drops, and in light load, V_{OUT} increases. If the load can tolerant the supply variation, this is not bad property. Because the absolute transient variation on V_{OUT} would be smaller as illustrated in the figure. An LDO with tight V_{OUT} regulation (the red dot line) has voltage undershoots and overshoots which increase the peak-to-peak transient variation.

15 

Flipped Voltage Follower (FVF)

- In simple source follower, the current through M_1 depends on the output current.
- The FVF has large sourcing capability, smaller output impedance.

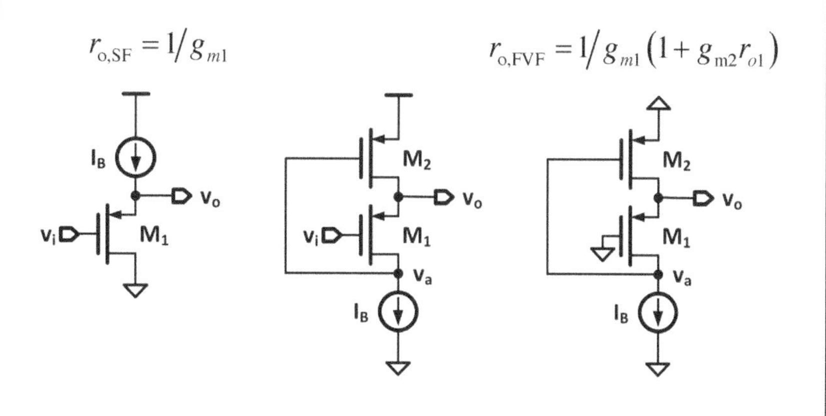

R. G. Carvajal, et al., "The flipped voltage follower: a useful cell for low-voltage low-power circuit design," *IEEE Trans. Circuits Syst. I: Regular Papers*, vol. 52, no. 7, pp. 1276–1291, Jul. 2005.

From the previous slides, we know that the simple source follower has an output impedance of $1/g_m$. In [Carvajal, TCAS-I 2005], a new type of source follower, flipped-voltage follower (FVF), was proposed. Compared to the simple source follower, the FVF has larger output current capability with a smaller output impedance. In the FVF topology, M_1 serves as a common-gate amplifier stage. If there is a small variation of vo, it will be amplified to v_a by $g_{m1}r_{o1}$ times, which is controlling the transconductance g_{m2} of M_2. Therefore, the output impedance of the FVF is $1/g_{m1}r_{o1}g_{m2}$.

16  Flipped Voltage Follower (FVF) based LDO

- Apply the FVF technique as a single-transistor controlled LDO.
- FVF is a single-ended structure.

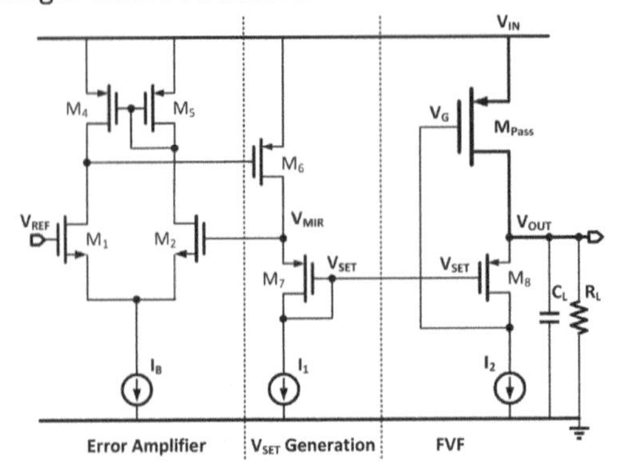

T. Y. Man, K. N. Leung, C. Y. Leung, P. K. T. Mok, and M. Chan, "Development of Single-Transistor-Control LDO Based on Flipped Voltage Follower for SoC," *IEEE TCAS-I*, vol. 55, no. 5, pp. 1392–1401, Jun. 2008.

The good sourcing current capability makes the FVF itself already a simple LDO, while V_{SET} only provides a DC bias for M_8. When I_1&I_2 and M_7&M_8 are matched, V_{OUT} should be approximately equal to V_{MIR} which is close to V_{REF}. The error amplifier and the V_{SET} generator only consume a tiny current as they only deal with DC voltages.

Compared to the conventional LDO with differential input pair, the FVF is a single-ended structure, which can achieve the same bandwidth with much smaller quiescent current.

17  Cascoded-FVF LDO

- Cascoded-FVF for larger DC gain.
- One more pole is introduced, Miller compensation may be required.

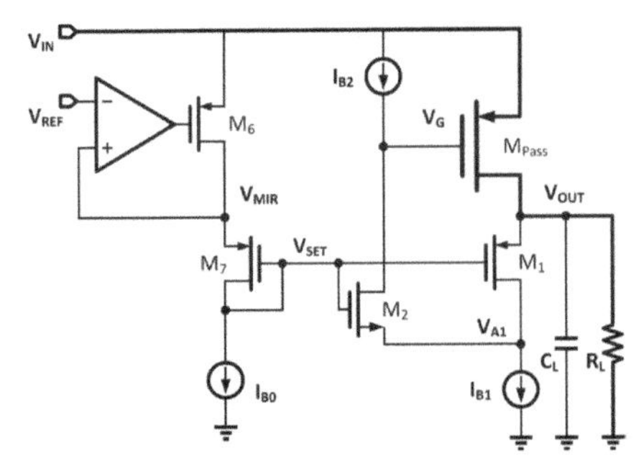

P. Hazucha, T. Karnik, B. A. Bloechel, C. Parsons, D. Finan, and S. Borkar, "Area-efficient linear regulator with ultra-fast load regulation," *IEEE Journal of Solid-State Circuits*, vol. 40, no. 4, pp. 933–940, Apr. 2005.

However, a simple topology only gives us small loop gain which results in poor DC regulation. For a larger DC gain, the casoded-FVF can be used. It employs M_2 as one more common-gate amplifier stage. Now, there are three poles from V_{OUT}, V_A, and VG in the loop. Therefore, Miller compensation circuit technique may be required.

18  A Typical LDO Regulator with Off-Chip Capacitor

- With impedance attenuation buffer
- Split one low-frequency pole into two relatively high frequency poles.

$$p_o = 1/(R_o C_L)$$
$$p_1 = 1/(r_{o,ea} C_{iB})$$
$$p_2 = 1/(r_{oB} C_P)$$

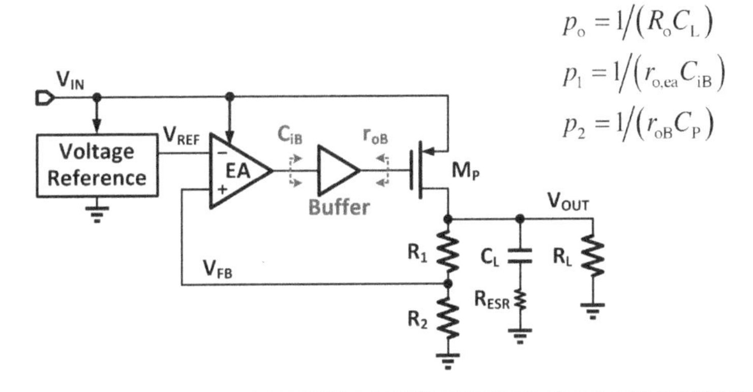

Here shows a typical LDO with off-chip capacitor and an inserted impedance attenuation buffer. With a large off-chip capacitor, the output pole can easily be the dominant pole. The buffer exhibits a low input capacitance to the EA and low output impedance to drive the power MOS. Therefore, the buffer splits one low-frequency pole on the gate of the power MOS into two relatively high-frequency poles.

19  Source Follower

- A simple source follower exhibits low input capacitance and low output resistance.

$$r_{o,SF} = 1/g_{m1}$$

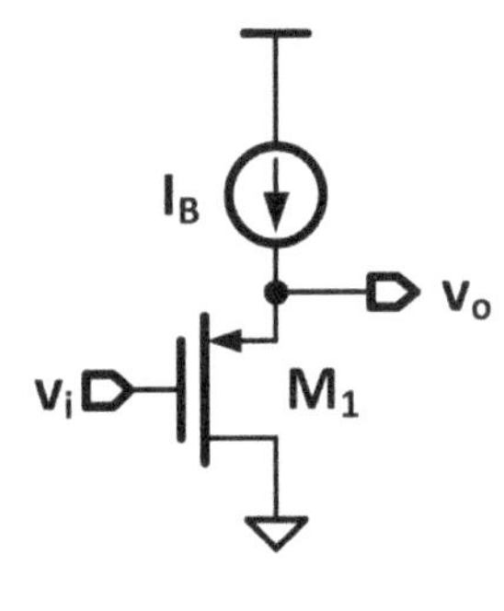

Again, the simplest way to implement a buffer is the source follower which gives an output impedance of $1/g_m$. However, if we want to further lower the output impedance, a large current is required by this simple structure.

20 Further Reduce the Output Impedance

- Reduce the output impedance by negative feedback.
- R_o reduced by $1+\beta A_0$ times.

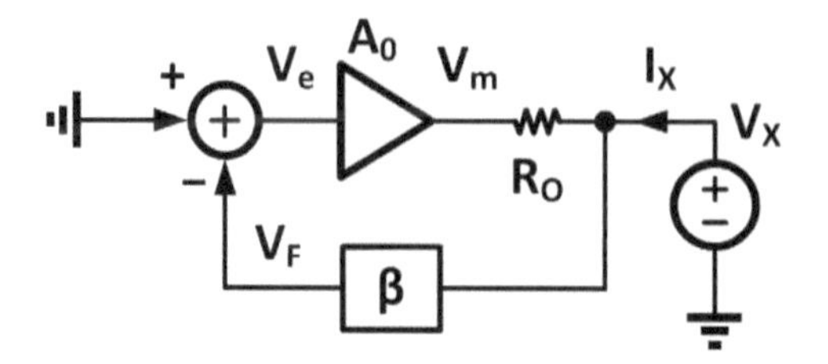

$$V_F = \beta V_X$$
$$V_e = -\beta V_X$$
$$V_M = -\beta A_0 V_X$$
$$I_X = [V_X - (-\beta A_0 V_X)] / R_O$$
$$\frac{V_X}{I_X} = \frac{R_O}{1+\beta A_0}$$

Instead of simply increasing the bias current of the source follower, we can use the favorable negative feedback property to further reduce the output impedance. As calculated in the slide, the output impedance can be reduced by $1+\beta A_0$ times if we have a negative feedback with a loop gain of βA_0.

21 Super Source Follower

- Reduce the output impedance by negative feedback.
- R_o reduced by $1+\beta A_0$ times.

$$r_{o,SF} = 1/g_{m1}$$

$$r_{o,SSF} = 1/g_{m1}(1+g_{m2}r_{o1})$$

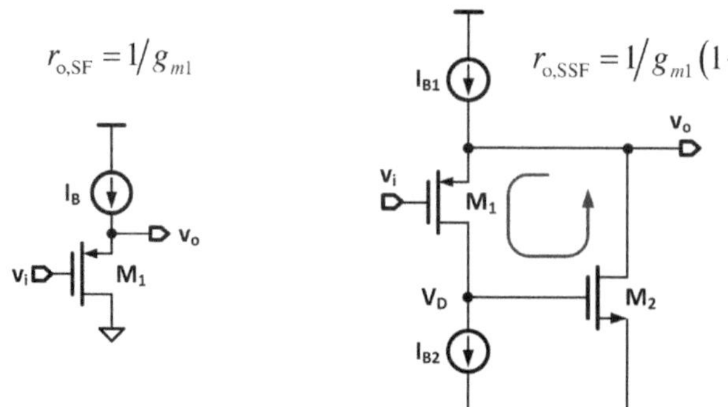

Paul R. Gray, et al. "Analysis and design of analog integrated circuits." John Wiley & Sons, 5th Edition, 2009.

Then, here comes the super source follower (SSF) which has one more transistor M_2 for output impedance reduction. Although there is one more branch consuming power in the SSF, the current is more effectively used. There will be more dynamic current to pull down the buffer output when its input goes down.

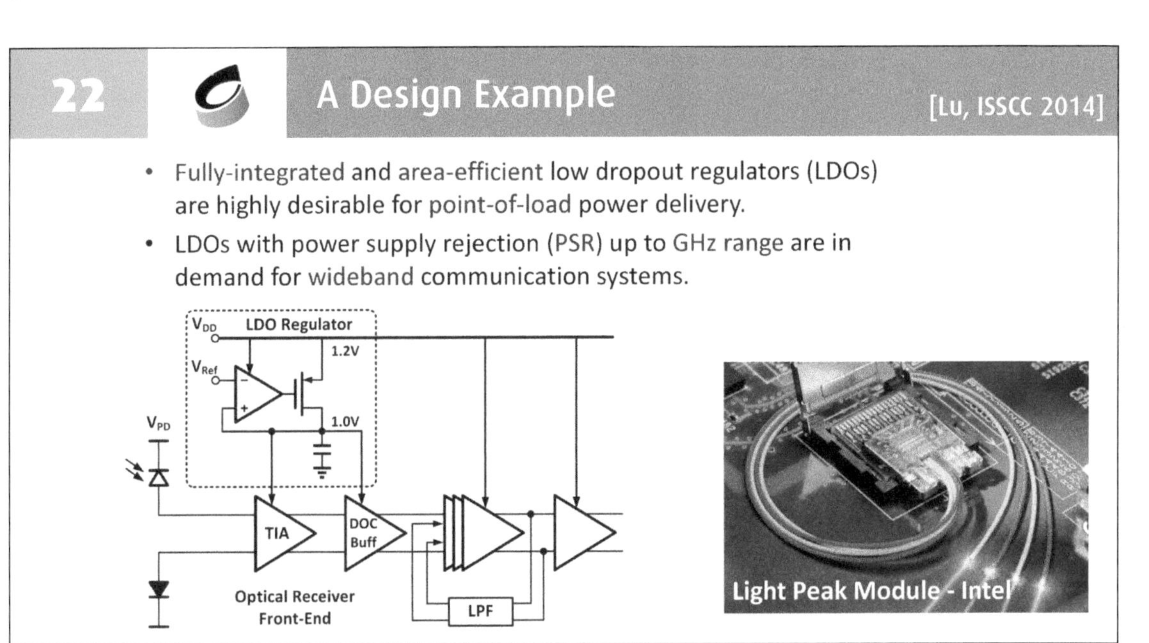

In the following several slides, we will discussed two design examples for the analog LDOs. The first example, which was designed for supplying the transimpedance amplifier (TIA) in an optical receiver, was presented on ISSCC 2014. In fiber communication systems, only one photodiode is used to sense on/off of the light. Therefore, the TIA would be either single-ended or pseudo-differential which is relatively more vulnerable to supply noise compared to the differential circuits. In this case, an LDO with PSR up to GHz range is in high demand for the fully-integrated wideband communication systems.

23 Comparison of the State-of-the-Art Non-Fully-Integrated LDOs

Publication	M. Al-Shyoukh JSSC 2007	H. Lam ISSCC 2008	M. El-Nozahi JSSC 2009	M. Ho JSSC 2010	M. Harwood ISSCC 2012
Technology	0.35μm	0.35um	130nm	90nm	40nm
V_{OUT}	1.8V	0.9V	1V	0.9V	N/A
Drop out	200mV	150mV	150mV	100mV	150mV
I_Q	20μA	4 to 164μA	50μA	9.3μA	N/A
Io,max	200mA	50mA	25mA	50mA	N/A
Capacitance	1μF	1μF	4μF	1μF	N/A
PSR	N/A	-50dB @1MHz	-56dB @10MHz	-35dB @10MHz	-30dB @1.5GHz
ΔV_{OUT} @T_{edge}	54mV @100ns	6.6mV @10ns	10mV @10ns	10mV @10ns	N/A
Load Reg.	34mV	3mV	N/A	4.1mV	N/A
T_R	270ns	132ns	1600ns	200ns	N/A
FOM	27ps	10.6ps	3200ps	37.2ps	N/A

*The **advantages** are highlighted with **GREEN** color, while the **disadvantages** are in **RED**.

Before we go into the design, let's have a literature review first. This table summarizes some LDOs that have off-chip capacitors. With a output capacitor in μF range, the LDO output variations due to load transient and input ripple can be very small, filtered by the output capacitor. However, in our case, we need a fully-integrated LDO for the point-of-load application without off-chip capacitor.

24 Comparison of the State-of-the-Art Fully-Integrated LDOs

Publication	K.N. Leung JSSC 2003	P. Hazucha JSSC 2005	V. Gupta ISSCC 2007	T.Y. Man TCAS-I 2008	J. Guo JSSC 2010	J. Bulzacchelli JSSC 2012
Technology	0.6μm	90nm	0.6μm	0.35μm	90nm	45nm SOI
V_{OUT}	1.3V	0.9V	1.2V	1V	0.5 to 1V	0.9 to 1.1V
Dropout	200mV	300mV	600mV	200mV	200mV	85mV
I_Q	38μA	6mA	50μA	95μA	8μA	12mA
Io,max	100mA	100mA	5mA	50mA	100mA	42mA
Capacitance	12pF	600pF	60pF	20pF**	50pF	1.46nF
PSR	-30dB @1MHz	N/A	-27dB @50MHz	N/A	0dB @1MHz	N/A
ΔV_{OUT} @T_{edge}	180mV @500ns	90mV @100ps	937mV	160mV @200ns	114mV @100ns	N/A
Load Reg.	±0.25%	90mV	N/A	14mV	10mV	3.5mV
T_R	2000ns	0.54ns	N/A	~300ns	N/A	0.288ns*
FOM	760ps	32ps	N/A	N/A	N/A	62.4ps*

* Simulated result ** Parasitic capacitance of the voltage probe

This table summarizes some fully-integrated LDOs. To achieve the fully-integrated feature, some specifications have been sacrificed in these works. High quiescent currents are used in [P. Hazucha, JSSC 2005] and [J. Bulzacchelli, JSSC 2012] for ultra-fast transient response, as their application is for high-performance CPUs. Cascoded power transistor with large dropout voltage is employed in [V. Gupta, ISSCC 2007] for superior PSR performance for analog loads. Meanwhile, some cap-less LDOs can only response to slow varying load transients.

25 LDOs with Off-/On-Chip Capacitor

- LDOs with **Off-Chip** μF range Capacitor
  - x Off-chip cap is conventionally added for filtering
  - – Dominant pole located at the output node
  - ✓ Small load transient glitches
  - ✓ High PSR

- LDOs with **On-Chip** sub-nF range Capacitor
  - – Dominant pole at the internal node (in previous designs)
  - x Large undershoot and overshoot during load transient
  - x Poor PSR

So, improving the PSR and transient performances of the fully-integrated LDO is our key design goal.

From the literature review in those two tables, we can found that the LDOs with off-chip capacitor only have small load transient glitches and can achieve high PSR, enabled by designing the dominant pole at the output node with a large capacitor. The performances of the fully-integrated LDOs still have plenty of room for improvements.

26  Schematic of the FVF Topology

- **Basic idea**: Use the p_{Out} dominant case with advanced process, push the internal poles to be higher than the UGF.
- **Technique**: Flipped Voltage Follower (FVF)

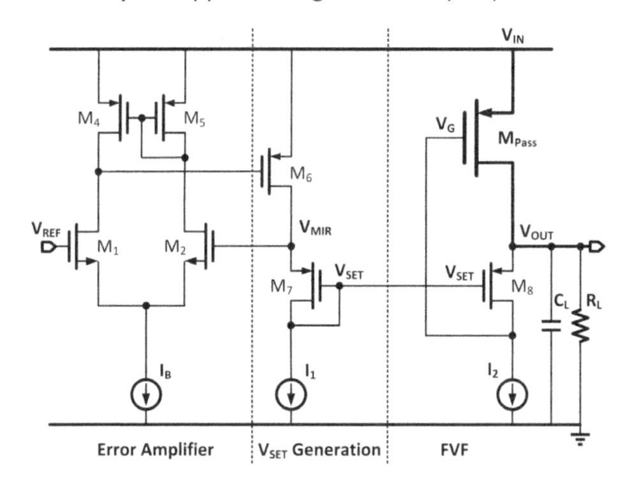

[Man TCAS-I 08]
0.35μm CMOS
p_{Gate} Dominant
Poor DC Regulation

In this design example, we started with the basic FVF topology and the design methodology of the non-fully-integrated LDO design. When we changed the 0.35μm process into 65nm process a few years ago, more design options were enabled by the advanced process.

27  Schematic of the BIA Topology

- **Basic idea**: Use the p_{Out} dominant case with advanced process, push the internal poles to be higher than the UGF.
- **Technique**: Buffer Impedance Attenuation (BIA)

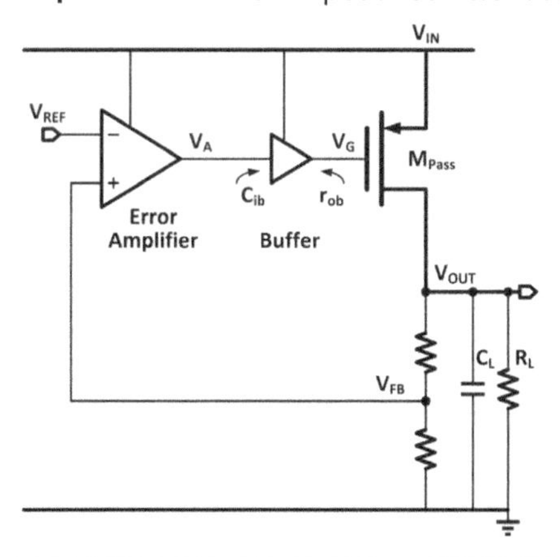

[Al-Shyoukh JSSC 07]
0.35μm CMOS
With 1μF Capacitor

The second step is to employ the popular impedance attenuation buffer.

28  Schematic of the FVF + BIA Topologies

- **Basic idea**: Use the p_{Out} dominant case with advanced process, push the internal poles to be higher than the UGF.
- **Techniques**: Flipped Voltage Follower (FVF)
  Buffer Impedance Attenuation (BIA)

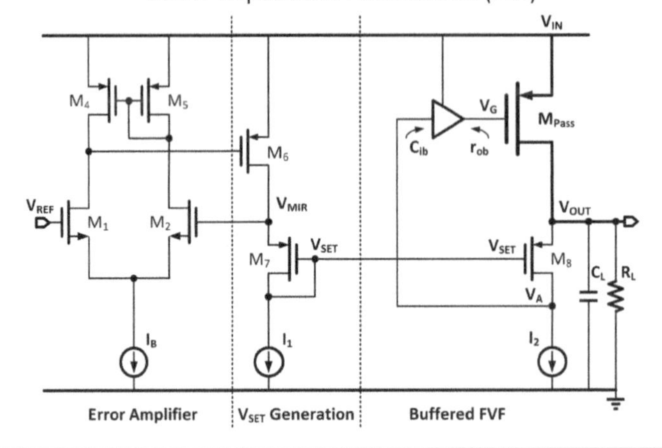

65nm CMOS

p_{Out} Dominant

Poor DC Regulation

Then, these two techniques are combined for designing the dominant pole at the output node. However, since the buffer only provides a gain less than one. The DC regulation problem of the FVF LDO becomes worse with the inserted buffer.

29  Schematic of the Proposed Tri-Loop LDO

- **Techniques**: Flipped Voltage Follower (FVF) and
  Buffer Impedance Attenuation (BIA)
- **Proposed 3-Input Error Amplifier (EA), and Tri-Loop Topology**

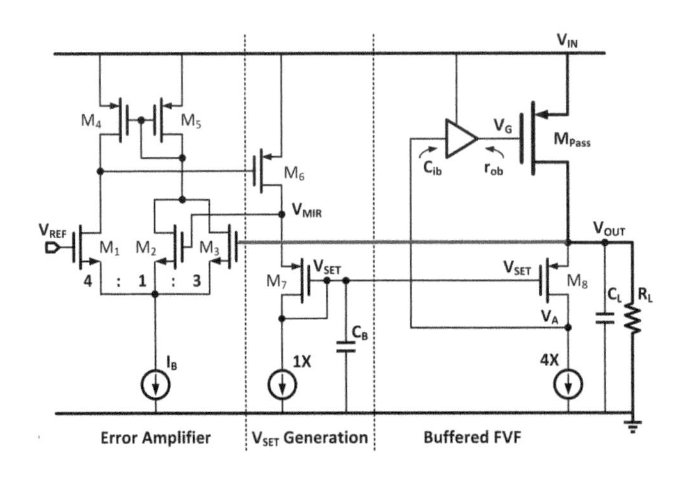

65nm CMOS

p_{Out} Dominant

Improved DC Reg.

140pF On-Chip Cap Including

7pF by-pass Cap C_B

50μA I_Q

Therefore, we proposed to use a three-input error amplifier to feed V_{OUT} back to the EA, for improving the DC regulation.

30  Tri-Loop in the Proposed LDO

- **Loop-1**: an ultra-fast low-gain loop with p_{Out} being its dominant pole and p_{Gate} be pushed to GHz range by BIA technique;
- **Loop-2**: a slow loop that generates V_{MIR} and V_{SET};
- **Loop-3**: feed V_{OUT} back to the EA to improve the DC accuracy.

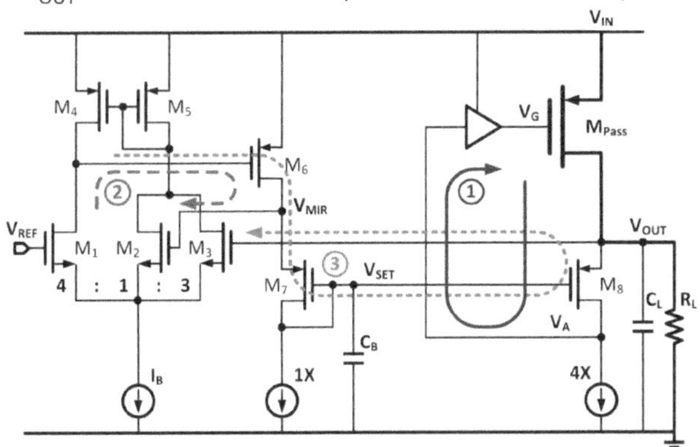

There are three loops exist in our proposed LDO. Loop-1 is a ultra-fast but low-gain loop with p_{OUT} being its dominant pole. It deals with fast load transient and high frequency PSR. Loop-2 is a slow local loop that generate V_{MIR} and V_{SET} for DC biasing. Loop-3 feed V_{OUT} back to the EA for improving the DC regulation.

31  Schematic of the Proposed Tri-Loop LDO

- **Techniques**: Flipped Voltage Follower (FVF) and Buffer Impedance Attenuation (BIA)
- **Proposed 3-Input Error Amplifier (EA), and Tri-Loop Topology**

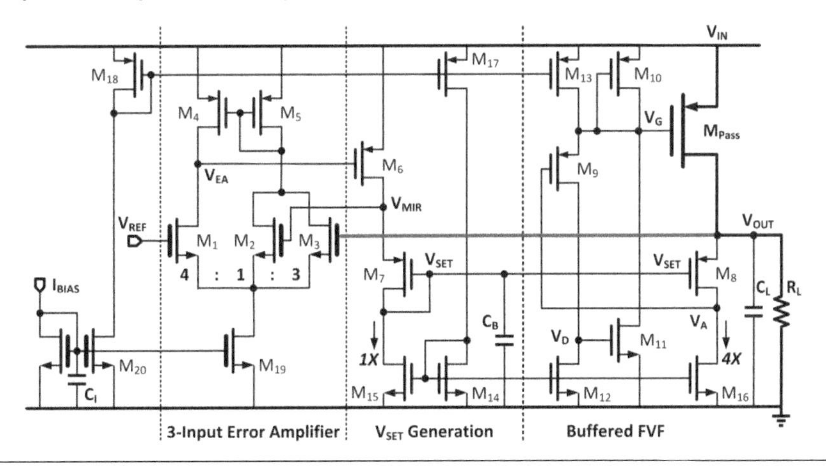

Here shows the full schematic of the proposed LDO. The buffer is implemented with a super source follower. A small by-pass capacitor C_B 7pF is used for preventing the noise on the left side goes to the right side.

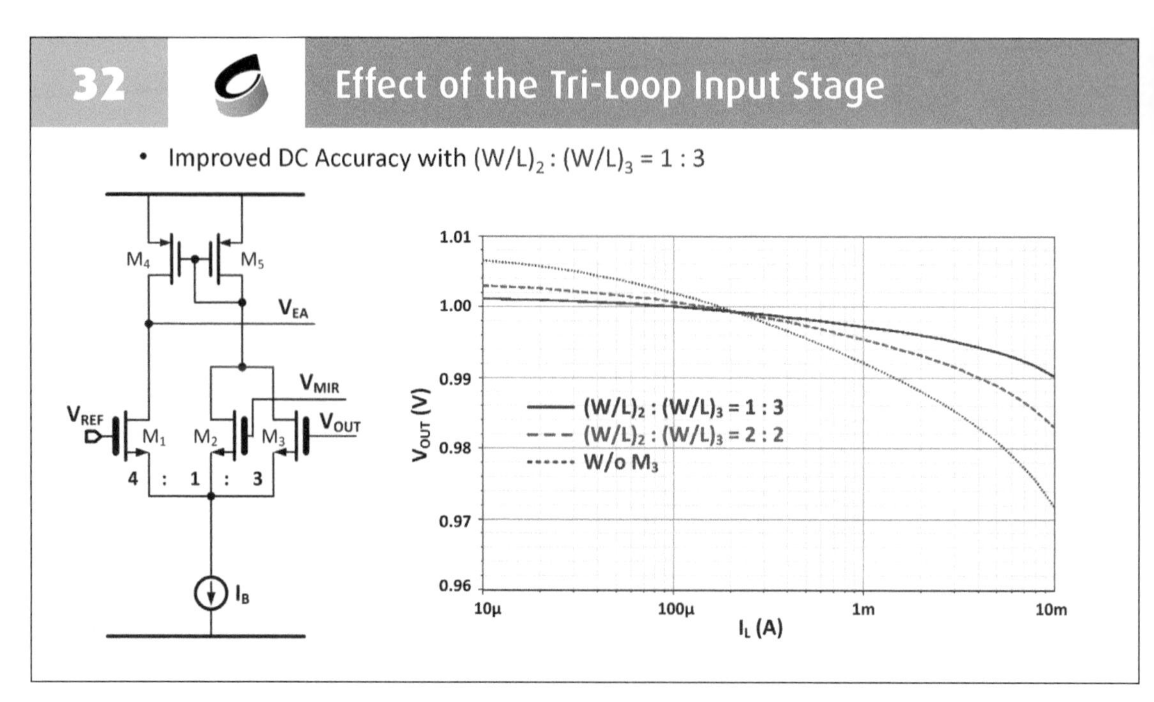

We can set different size (strength) to the three inputs of the EA. Of course, if the V_{OUT} terminal has larger strength, V_{OUT} should be closer to V_{REF}. The simulated curves show that the DC regulation performance has been improved by more than three times when comparing the cases with and without Loop-3.

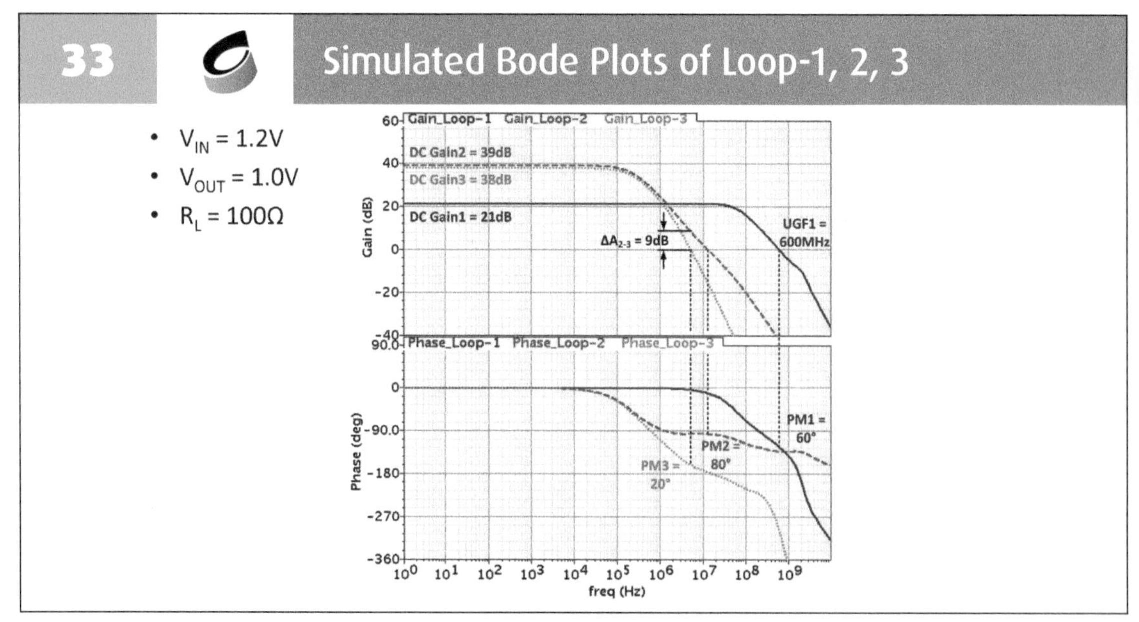

The Bode plots of the three individual loops can be simulated by breaking all the loops and inserting a small signal AC source to only one loop. From the simulated bode plots, we can see the UGF of Loop-1 is as high as 600MHz, which enables a ultra-fast transient response. Loop-2 and Loop-3 have certain part of overlapped circuit. Therefore, although the phase margin (PM) of the long Loop-3 is not quite enough, Loop-2 will help the entire LDO to be stable.

34 Stability Analysis of the Entire LDO

- Break the loop at V_{EA}
- All three loops are included in this analysis.

Or, we can investigate the stability of the entire LDO by only break one-point in the circuit for the AC simulation.

35 Simulated Bode Plots of the Entire LDO

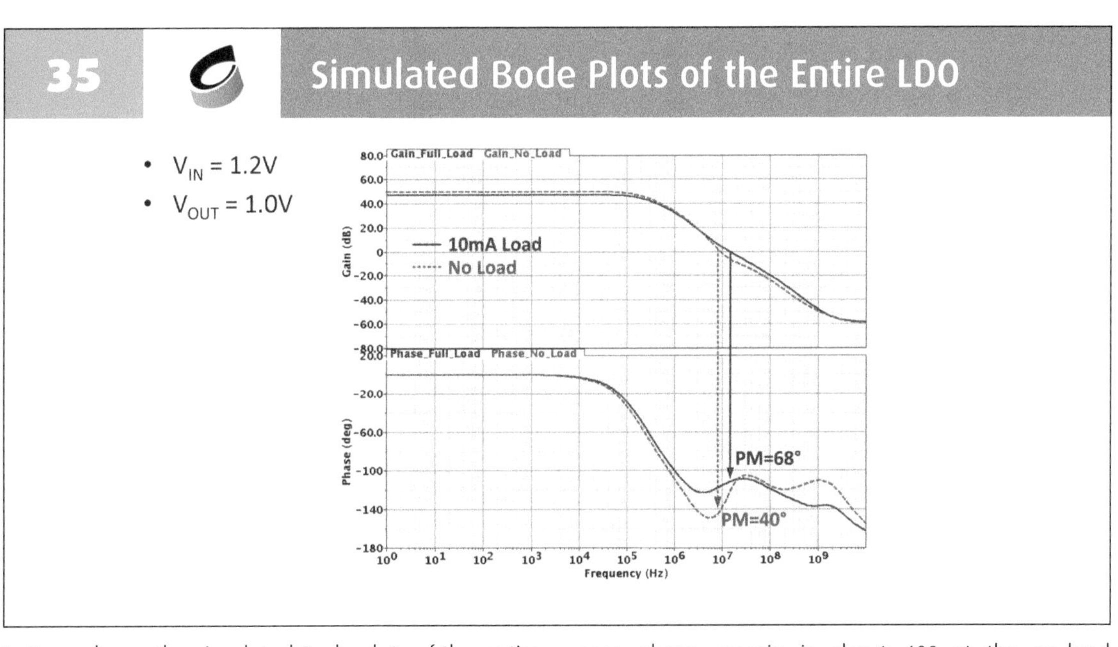

Here shows the simulated Bode plots of the entire LDO at the full-load and no-load conditions. The Loop-2, which is a shorter loop compared to Loop-3, provides a zero in the transfer function. The worst case phase margin is about 40° at the no-load condition when the output pole approaches the pole on V_{SET}.

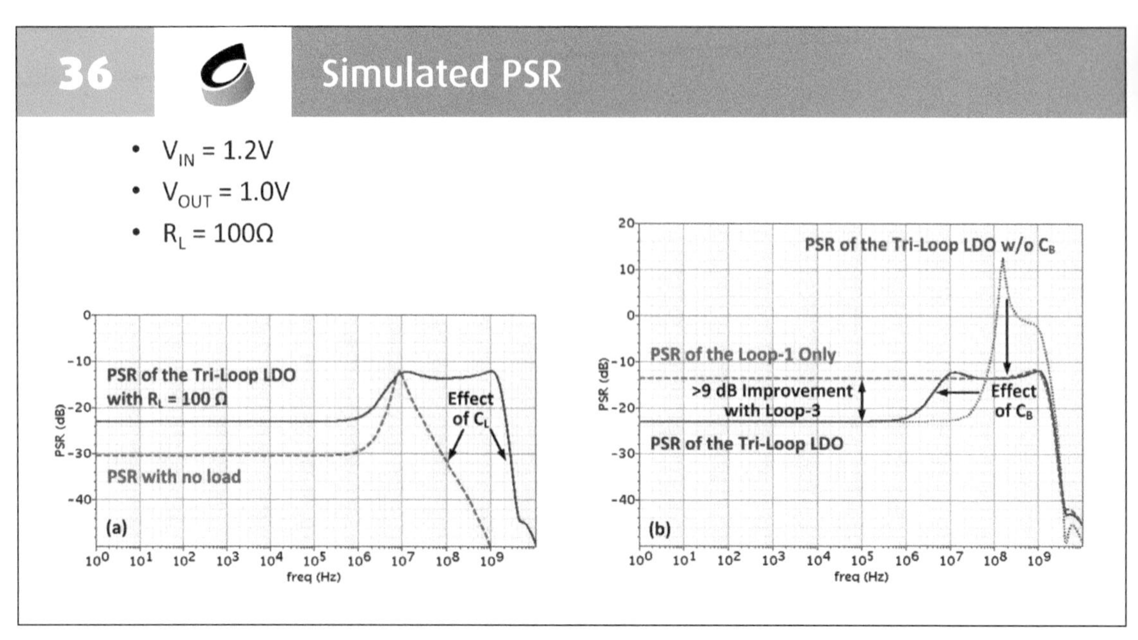

As we discussed before, the low frequency PSR is mainly contributed by the loop gain, while the high frequency PSR is provided by the output capacitor. Therefore, we can achieve a full-spectrum PSR by designing the dominant pole at the output node. With the help of Loop-3, the low frequency PSR has been improved by 9dB. And the filtering capacitor C_B helps to attenuate the medium frequency ripple from the left side of the circuit.

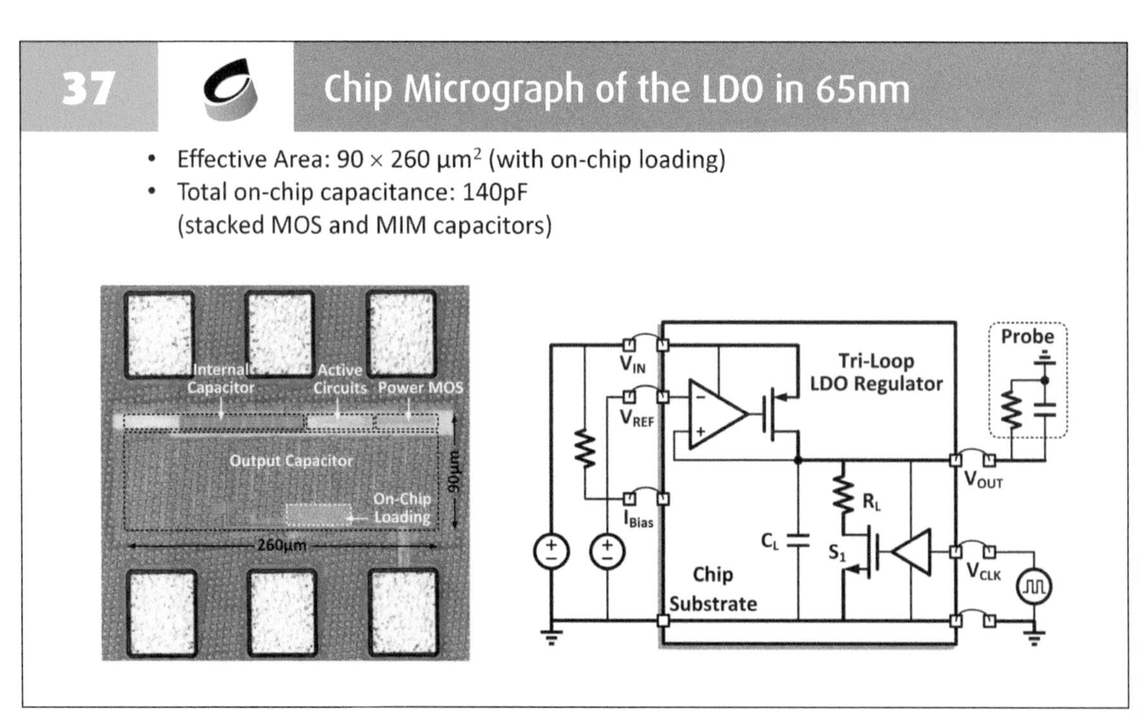

This design example is fabricated in a 65nm general purpose CMOS process. Most of the area is occupied by the output capacitor, which uses the stacked MOS and MIM capacitors for higher capacitor density. An on-chip loading that can provide ultra-fast load transient is also included for testing.

38 Ultra Fast Transient Response

- Rising/Falling edges of the load : < 200ps (0μA to 10mA)
- I_Q: 50μA; C_{Total}: 140pF; ΔV_{OUT}: 43mV (undershoot)
- Response time T_R: 0.6ns
- FOM: 3.01fs

$$FOM = T_R \frac{I_Q}{I_{MAX}} = \frac{C \times \Delta V_{OUT}}{I_{MAX}} \times \frac{I_Q}{I_{MAX}}$$

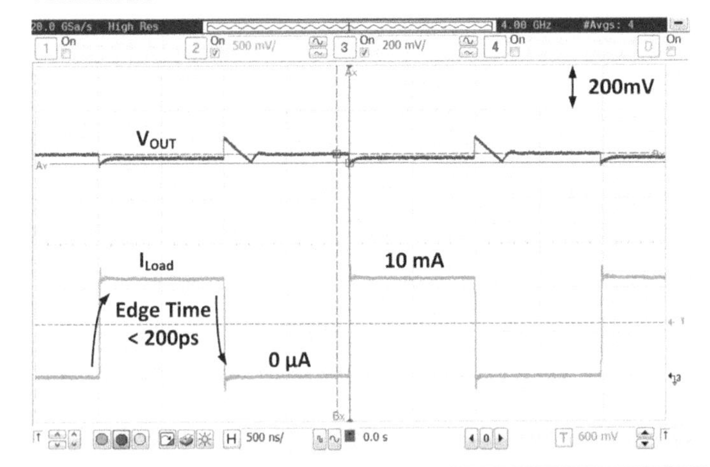

When the load current changes between 0μA and 10mA with <200ps edge time, the measured voltage undershoot is only 43mV which results in a calculated response time of 0.6ns and FoM of 3.01fs.

39 Measured PSR @ 1GHz and 2.5GHz

- V_{IN} = 1.2V, V_{OUT} = 1.0V, R_L = 100Ω

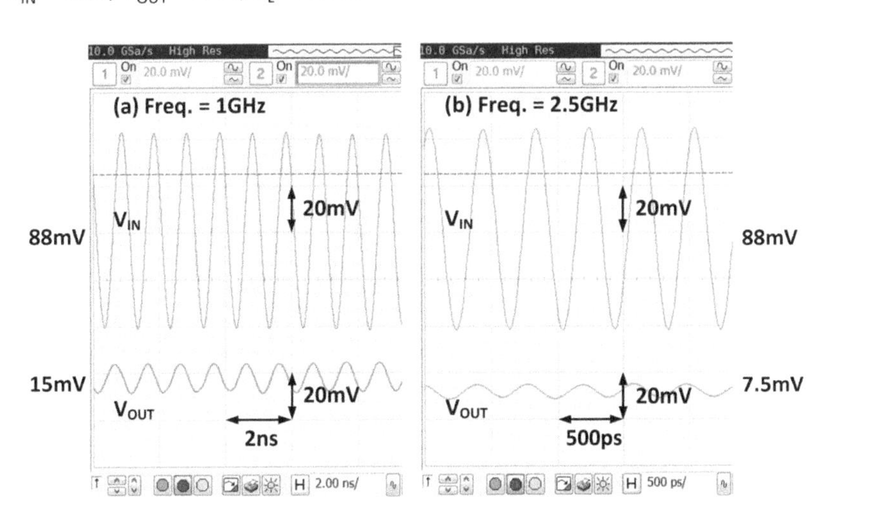

Here shows the measured waveforms for the PSR calculation. When the input ripples are 88mV at 1GHz and 2.5GHz, the output ripples are 15mV and 7.5mV, respectively.

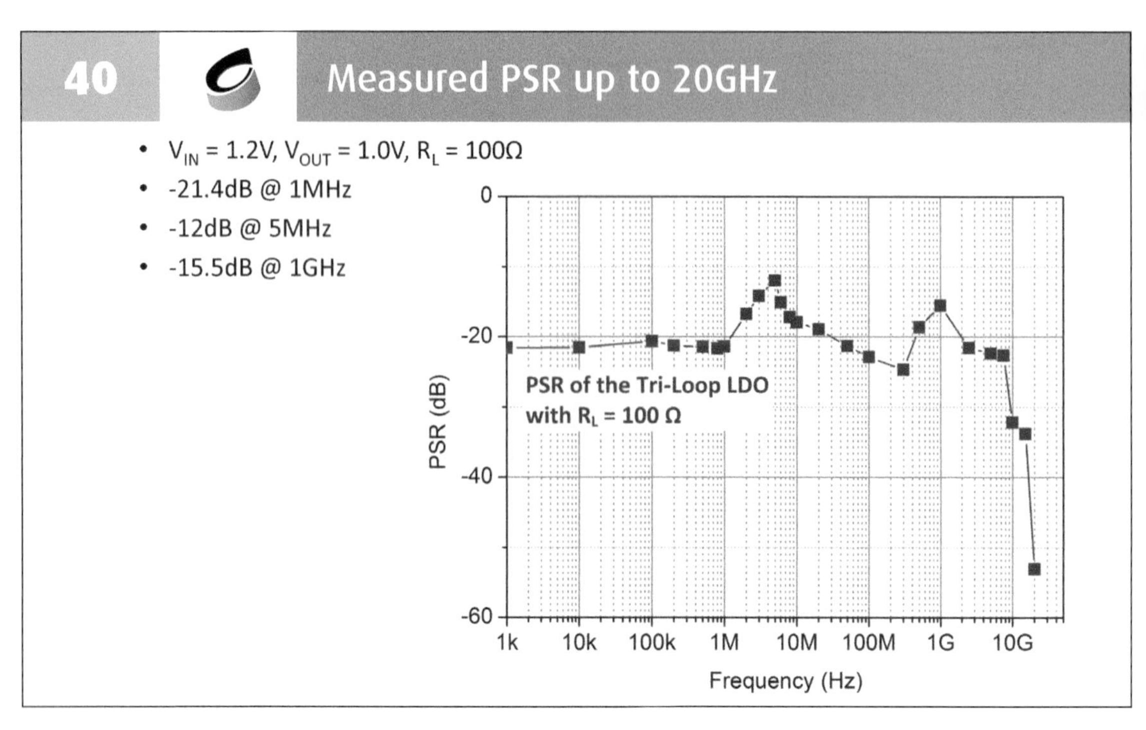

This plot summarized the PSR performance up to 20GHz, measured. For PSR at frequencies higher than 2.5GHz, a spectrum analyzer is used for the measurement. The worst case PSR of -12dB happens at 5MHz.

41 Comparison with the State-of-the-Art LDOs

Publication	P. Hazucha JSSC 2005	J. Guo JSSC 2010	J. Bulzacchelli JSSC 2012	This Work ISSCC 2014
Output Cap.	On-chip			
Technology	90nm	90nm	45nm SOI	**65nm**
V_{OUT}	0.9V	0.5 to 1V	0.9 to 1.1V	**1V**
Drop out	300mV	200mV	85mV	**150mV**
I_Q	6mA	8μA	12mA	**50μA**
Io,max	100mA	100mA	42mA	**10mA**
Capacitance	600pF	50pF	1.46nF	**140pF**
PSR	N/A	0dB @1MHz	N/A	**-15.5dB @ 1GHz**
ΔV_{OUT} @T_{edge}	90mV @100ps	114mV @100ns	N/A	**43mV @200ps**
Load Reg.	90mV	10mV	3.5mV	**11mV**
T_R	0.54ns	N/A	0.288ns*	**0.6ns**
FOM	32ps	N/A	62.4ps*	**3.01fs**

* Simulated result

From the comparison table, we can see that our design achieves ultra-fast transient response with only 50μA quiescent current and 140pF total capacitors, thanks to our different design methodology which applies to advance processes.

42  **Cascoded-FVF LDO with Two Buffer**

- If we use better process, we should have better performances.

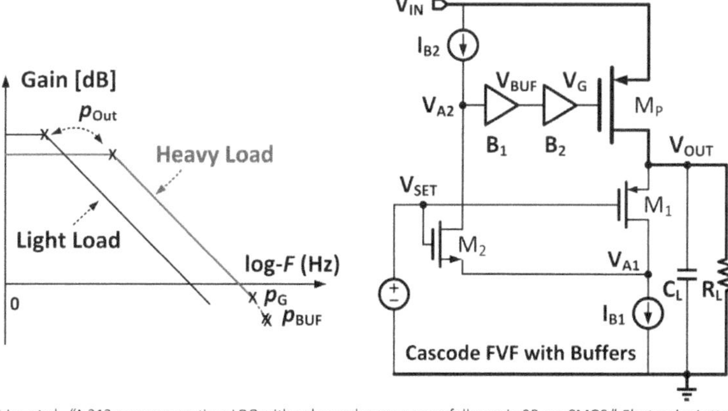

Y. Lu, et al., "A 312 ps response-time LDO with enhanced super source follower in 28 nm CMOS," *Electronics Letters*, vol. 52, no. 16, pp. 1368–1370, 2016.

As we have mentioned, a good LDO design should enjoy the process scaling. If we use better process, we should have better performances. Therefore, we applied the same LDO design methodology in a 28nm bulk CMOS process. In this second, design example, we keep the dominant pole at the output node. In the meantime, we would like to increase the loop with the cascoded-FVF structure. Since the loop bandwidth will be higher than our first design, two buffers are inserted to further push the internal poles to higher frequencies.

43  **An Enhanced Super Source Follower (E-SSF)**

- The inserted common-gate amplification stage formed by M_5 reduces the output impedance of the SSF buffer by a factor of $g_{m5}r_{o5}$, and thus provides a larger driving capability.

$$r_{oB,SSF} \approx \frac{1}{g_{m4} \cdot g_{m6} r_{o4}}$$

$$r_{oB,E\text{-}SSF} \approx \frac{1}{g_{m4} \cdot g_{m6} r_{o4} \cdot g_{m5} r_{o5}}$$

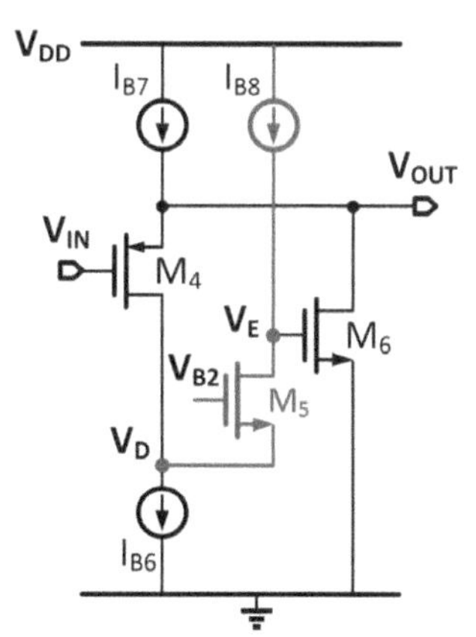

Also, in this design, we proposed an enhanced super source follower (E-SSF) for further output impedance attenuation. A common-gate amplifier stage, consists of M_5 and I_{B8}, is inserted to the SSF feedback loop to increase the local loop gain.

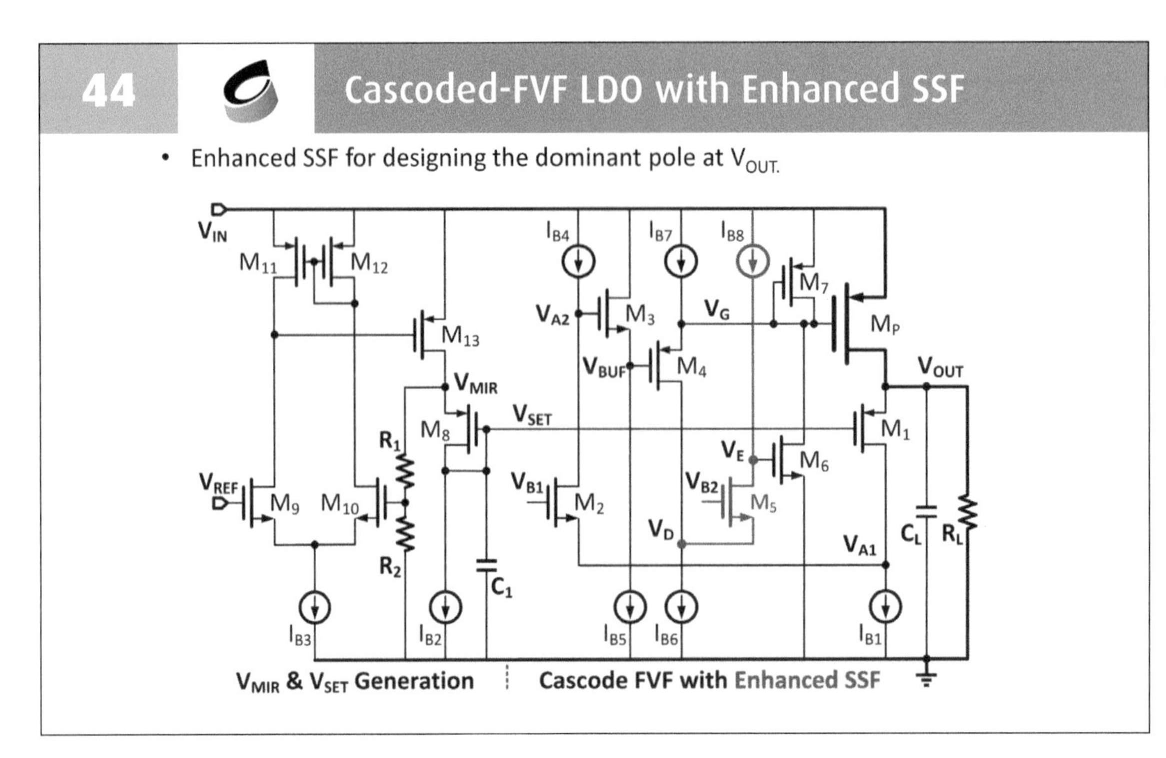

This slide shows the full schematic of the cascoded-FVF with enhanced SSF. M_1 and M_2 are the two common-gate amplifier stages in the cascoded-FVF loop. M_3 is the first buffer, while M_4 together with M_5 and M6 form the E-SSF as the second buffer.

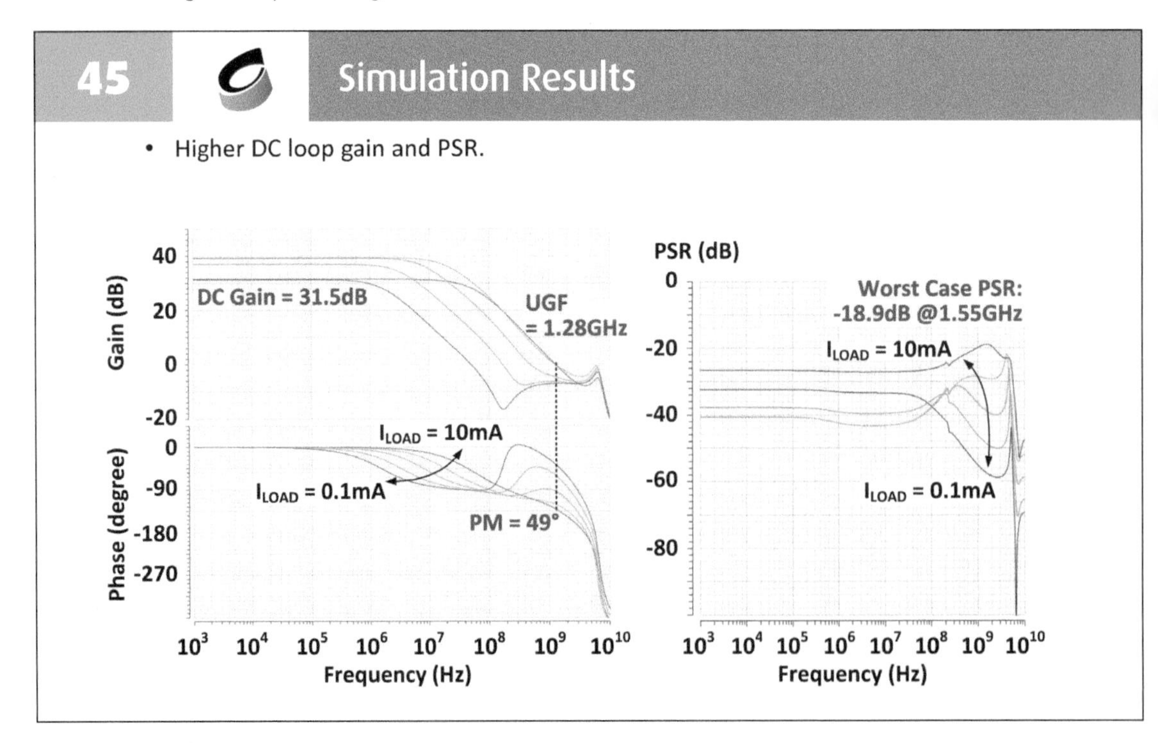

From the simulated AC responses, we can see that the loop bandwidth is as high as 1.28GHz in the full-load condition. The worst case PSR is -18.9dB at 1.55GHz.

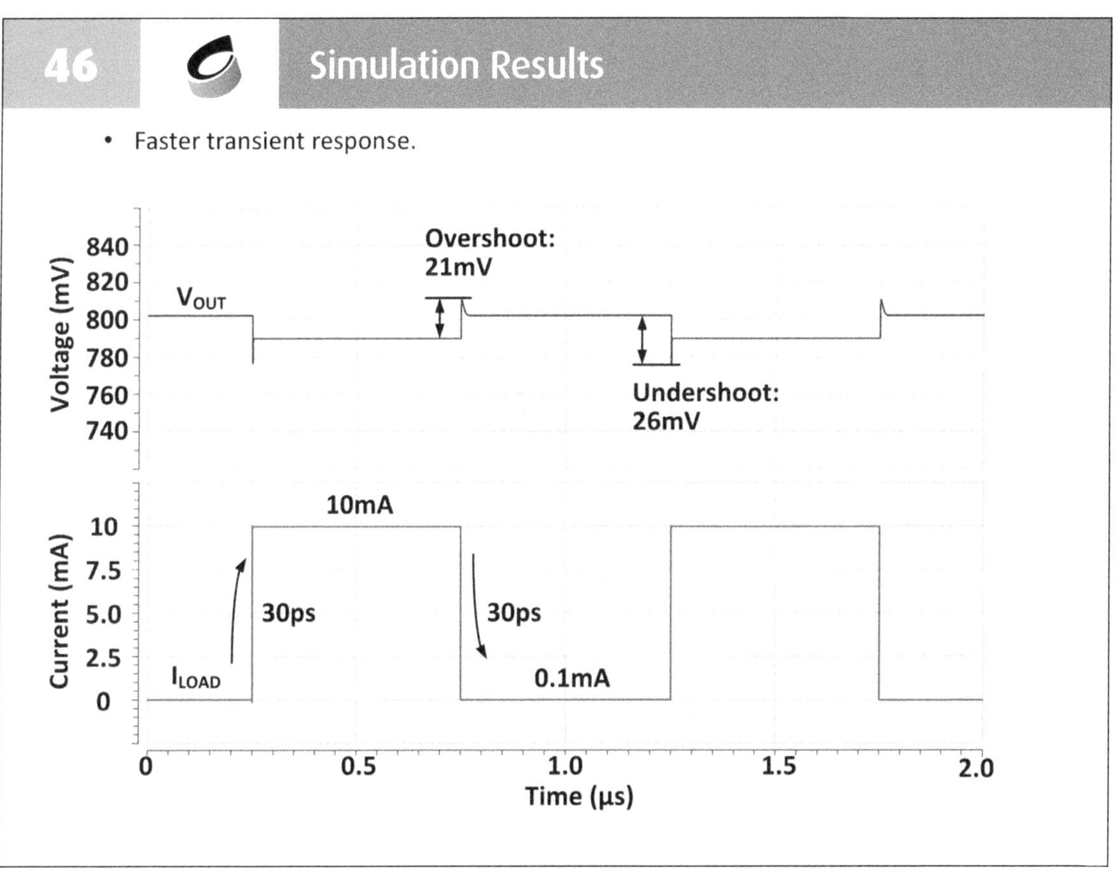

The voltage undershoot is only 26mV when the load changes from 0.1mA to 10mA. Therefore, we can conclude that the fully-integrated analog LDO can be scalable with process using our proposed design method.

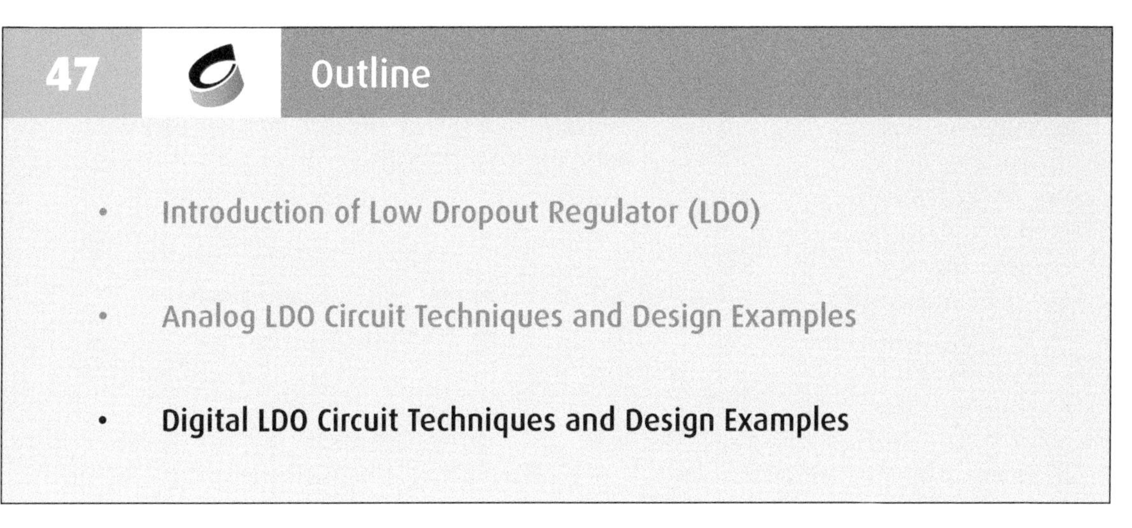

Next, we discuss the digital LDO circuits and design examples, which are becoming more popular in recent years.

48  Digital Low Dropout Regulator (Digital LDO)

✓ Low voltage operation (<0.6V)
✓ Benefit from process scaling
✓ Easy cooperation with digital loads
x May not be energy-efficient
x Limit cycle oscillation (LCO)

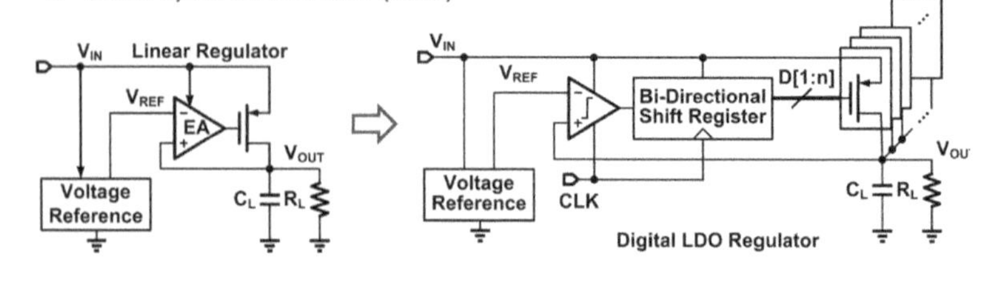

Analog LDO   Digital LDO [Okuma, CICC 2010]

The basic idea of the digital LDO is very simple. It divides the large power transistor in the analog LDO into many small slices (power switches) controlled by a bi-directional shift register. The bi-directional shift register gets its input signal from the comparator output. The clocked comparator compares V_{OUT} with V_{REF} at the beginning of every clock cycle, indicates the shift register to turn-on or turn-off one more power switch.

Comparing to the analog counterpart, the digital LDO can operate at low input voltages, like <0.6V, where the analog LDO can hardly have sufficient loop gain. And of course, the digital LDO can naturally benefit from process scaling. Also, the digital LDO can easily cooperate with the digital load to react to the load variations in advance. However, this conventional digital LDO is not energy-efficient, as it only changes one switch per clock cycle. For higher output accuracy, hundreds or even thousands of switches are required, resulting a slow response. Also, due to the finite output resolution, the digital control code may oscillate between adjacent codes even with a constant load current. This phenomenon is so-called limit cycle oscillation (LCO), which widely appears in digitally controlled systems.

49  Why Thermometer Code, not Binary?

- Thermometer code:
  - Simple, relatively more energy-efficient, little glitches, slow.

    00001111 → 00000111

- Binary code:
  - Fast, more circuit blocks, not efficient, large glitches.

    01111111 → 10000000

Immediately, people may ask why don't we use the binary code instead of the unary (thermometer) code for a fast response? Indeed, the response would be much faster if we employ the binary code. However, there are some major drawbacks of using binary code. Firstly, more comparators are needed for a multi-level analog-to-digital conversion which requires more conversion energy. Secondly, during the most-significant bit (MSB) transition, for example from 0111 to 1000, all the switches will be switched once, consuming huge energy but only changed one step (from 7 to 8) in terms of digital code. Meanwhile, the MSB transition will cause a large output glitch when the binary bits are not changed at exactly the same time. Therefore, the thermometer code is more suitable for the digital LDOs.

50

Loop Stability – Modeling of D-LDO

- Involves s-domain and z-domain
- A_{CMP} is the comparator gain
- SR is an integrator
- D/A is a zero-order hold (ZOH)

The loop stability analysis of a digital LDO is similar to an analog LDO, but a little bit more complicated. Because it involves both s-domain and z-domain. The clock comparator is one-bit analog-to-digital converter (ADC), which can be modeled as input-dependent-gain amplifier and a sample & hold switch. The shift register (SR) integrates the information from all the previous clock cycles, therefore, it is a DC pole in the small-signal model. The shift register can be set to shift one bit per cycle or multi bit per cycle, so there is a factor a in the model. The digital code is converted back to the analog domain by the power switch array, which can be modeled as a zero-order hold (ZOH). Then, the output node contributes a pole the same as the analog LDO.

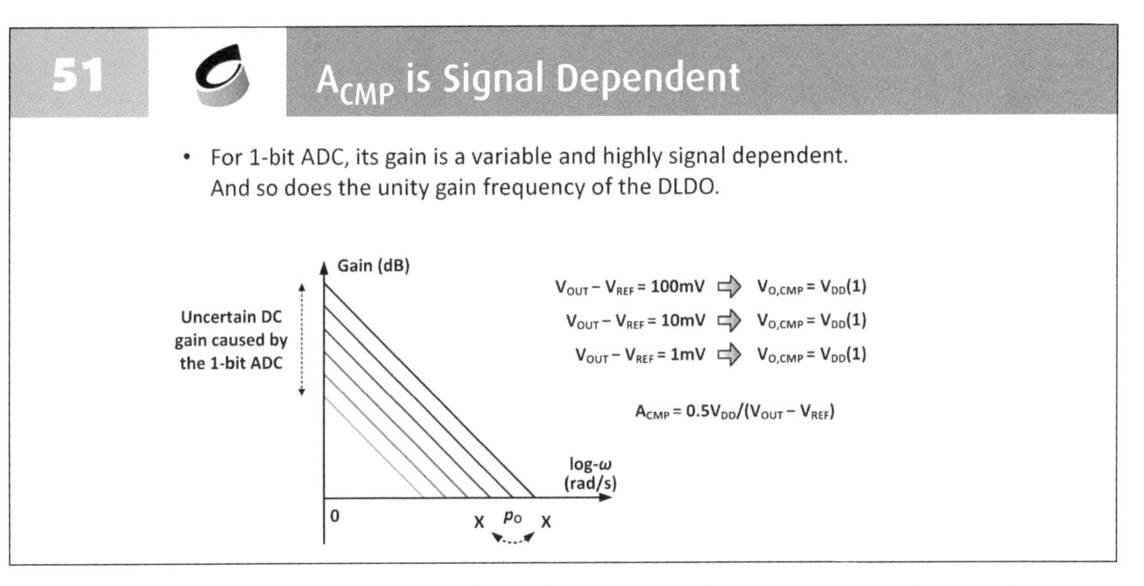

One property of the digital LDO small signal model is that the gain of the ADC is a variable and highly signal dependent. Because the ADC always outputs one certain digital output when the analog signal is within a quantization range, no matter the analog input signal is in the middle of the range or close to the boundary of the range. Therefore, the unity gain frequency of the DLDO also shifts with the varying input signal. When A_{CMP} is very large (V_{OUT} is very close to V_{REF}), the loop intends to oscillate because the UGF may exceed the output pole frequency. This is another explanation of the limit cycle oscillation phenomenon.

52 Limit Cycle Oscillation of D-LDO

- Limit cycle oscillation (LCO) happens in all the digitally controlled voltage regulators introduced by the quantization error of the ADC (comparator).

As we just mentioned, LCO comes from the quantization error of the ADC (comparator) with finite resolution. People defined the LCO modes to relate the LCO frequency to the clock frequency. If the digital codes oscillate between two adjacent codes, we say it's mode 1 LCO. If the codes oscillate between three codes, it's mode 2 LCO. The LCO could have mode 3, 4, 5, or more. Obviously, higher LCO mode means larger oscillation amplitude, if all the other conditions are the same.

53 LCO Amplitude Considerations

- Higher F_S results in larger **mode** (before V_{OUT} settles).
- Lower F_S allows longer charging/discharging time on C_L.
- Light load current results in larger R_L, and larger LCO **amplitude**.

The LCO mode and amplitude depend on a couple of factors. A higher clock frequency FS results in larger mode, because the comparator will sample V_{OUT} multiple times before V_{OUT} has been settled from the last D(t) change. On the other hand, if the LCO mode is the same, a slower FS will allow a longer charge/discharge time for the V_{OUT} node, which will result in a larger LCO amplitude. If the LCO mode is the same, LCO amplitude also depends on the loading condition. Light load current results in a larger equivalent load resistance RL, and therefore a larger LCO amplitude.

54  Reducing the LCO Amplitude

- **Three methods:**
  - Multi-bit ADC: more power, more complexity
  - Dead-zone (DZ) comparators: lose accuracy
  - Feed-forward (FF) path: simple and effective

M. Huang, *et al.*, "Limit cycle oscillation reduction for digital low dropout regulators", *IEEE Trans. Circuits Syst. II: Exp. Briefs*, Sep. 2016.

To reduce the LCO amplitude, there are some possible solutions here. By using a multi-bit ADC, the quantization error will be smaller, the digital LDO loop gain can be better defined, and therefore the låoop stability can be better designed. But it requires more power and more complicated circuit blocks. Instead of using one comparator and one boundary, dead-zone (DZ) comparators use two comparators and two boundaries. When VOUT is within the DZ, no switching activity for the digital controller. But the problem is how to define a proper DZ size. If the DZ is too large, we loss DC accuracy, if the DZ is too small, LCO happens. The third way of reducing the LCO amplitude was introduced in [Huang, TCAS-II, 2016]. A feed-forward (FF) path can be added in parallel with the main loop to guarantee the LCO mode equals to 1.

55 LCO Reduction with Feed-Forward Path

- An additional simple feed-forward (FF) path is added to maintain the LCO mode = 1.

M. Huang, *et al.*, "Limit cycle oscillation reduction for digital low dropout regulators", *IEEE Trans. Circuits Syst. II: Exp. Briefs*, Sep. 2016.

This slide shows the schematic of the digital LDO with an additional simple feed-forward (FF) path in parallel with the main loop to maintain the LCO mode = 1. This FF path can instantly response to the comparator output, performs like a zero in the frequency domain. By setting the strength β of the auxiliary power switches to 2 times of the main power switch, LCO mode=1 with the smallest LCO amplitude can be obtained.

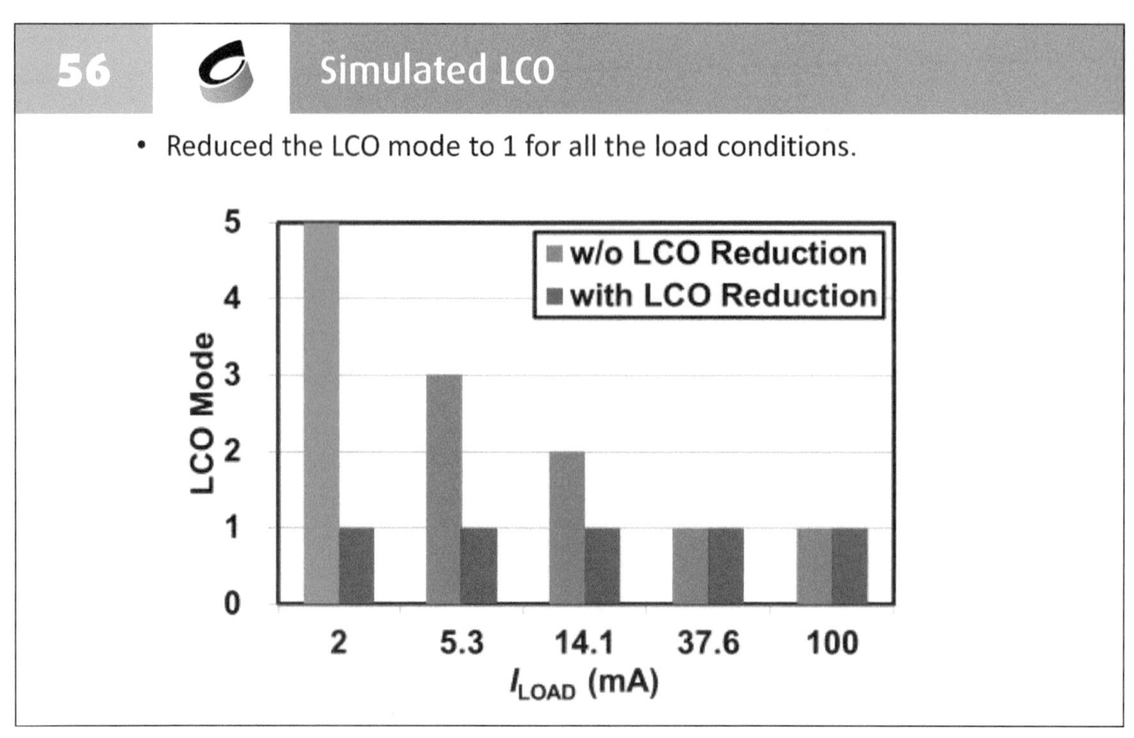

This slide shows the simulated LCO mode with and without the FF auxiliary path. The LCO mode maintains 1 with the proposed LCO reduction technique across a wide load current range.

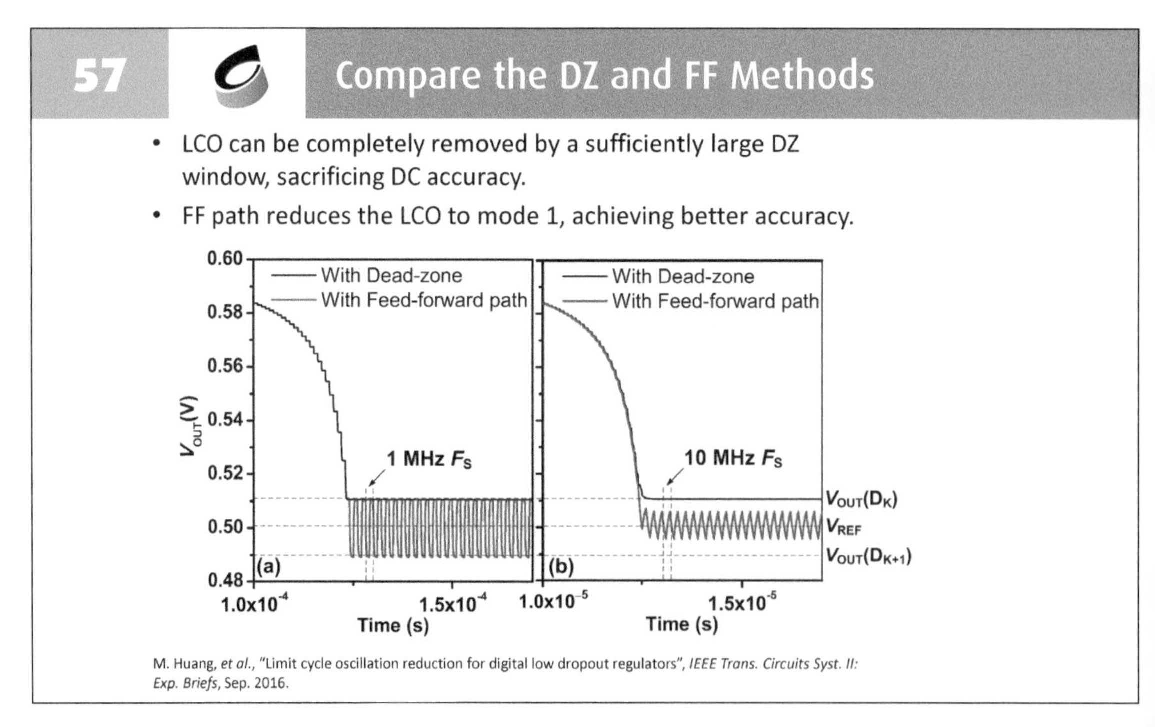

Here, we compare the DZ method and the FF method. For the DZ method, it can completely remove the LCO with a sufficiently large DZ window, sacrificing DC accuracy. The FF method maintains the DC accuracy but also preserves a small LCO.

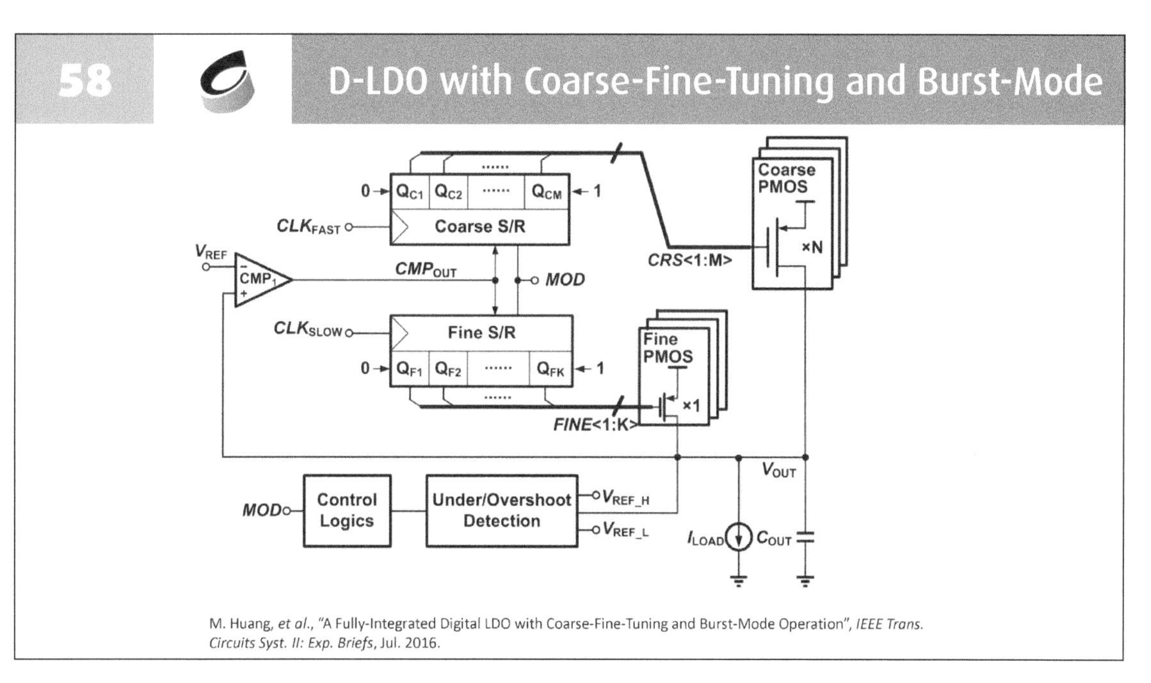

To extend the tradeoff between power and speed, coarse-fine tuning and burst-mode operation was applied to digital LDOs. In this design example, the coarse loop is controlling M switches which have N times strength compared to that of the switches in the fine loop. The voltage undershoot/overshoot detection block has two comparators to compare V_{OUT} with V_{REF_L} and V_{REF_H}, respectively. When V_{OUT} exceeds the boundary, a fast clock will be applied to the coarse-tuning loop for fast transient response.

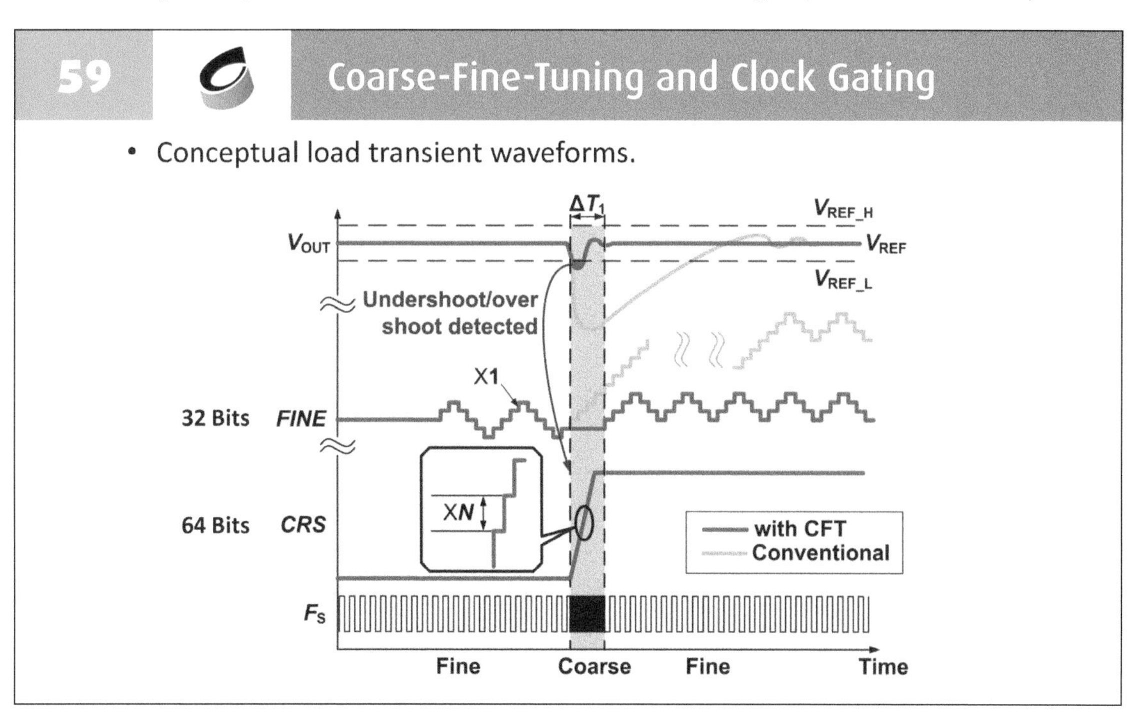

This slide shows the conceptual load transient waveforms. When the load transient is detected, fine-tuning stops and coarse-tuning with a fast clock starts. Therefore, the response time and the recovery time is much faster than the conventional design.

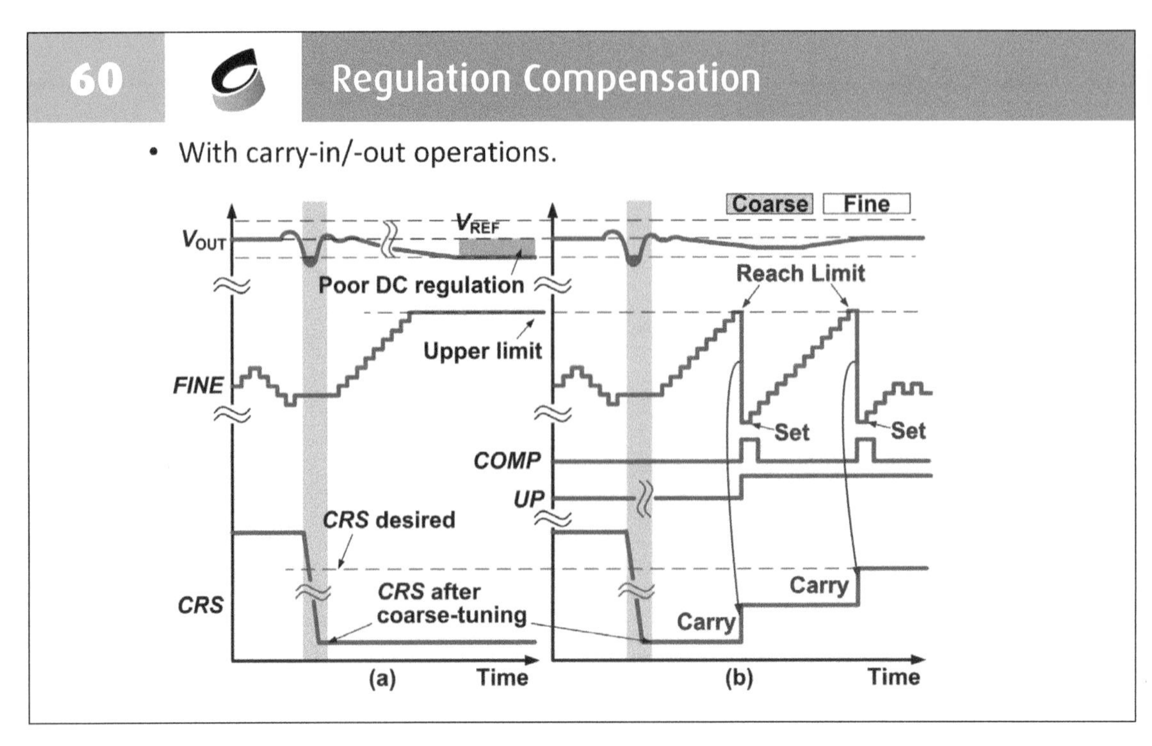

Carry-in/-out operations between the fine word and the coarse word can be performed if the coarse-tuning stops at the an undesired coarse code. This helps V_{OUT} to recover back to the V_{REF} level, improving the DC accuracy.

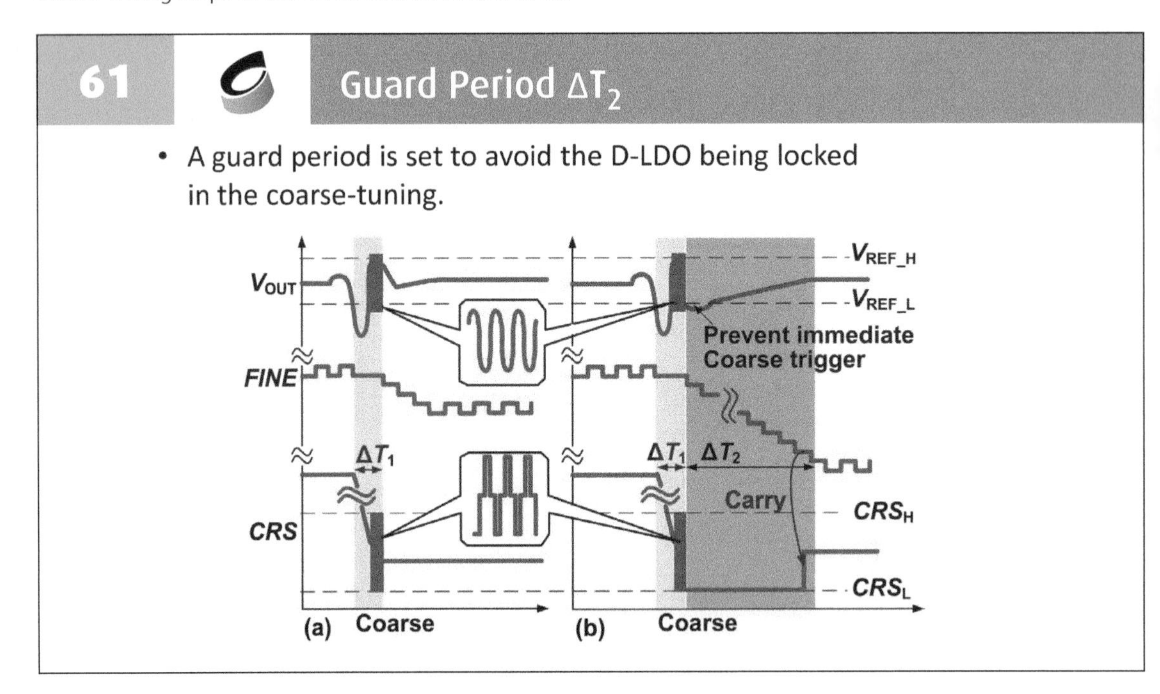

A guard period ΔT_2 is set to avoid the D-LDO being locked in the coarse tuning. Because in coarse tuning, the output ripple could be larger than the boundary window and the D-LDO may be locked in coarse tuning. Therefore, after ΔT_1, the D-LDO has to stay in fine tuning for a while.

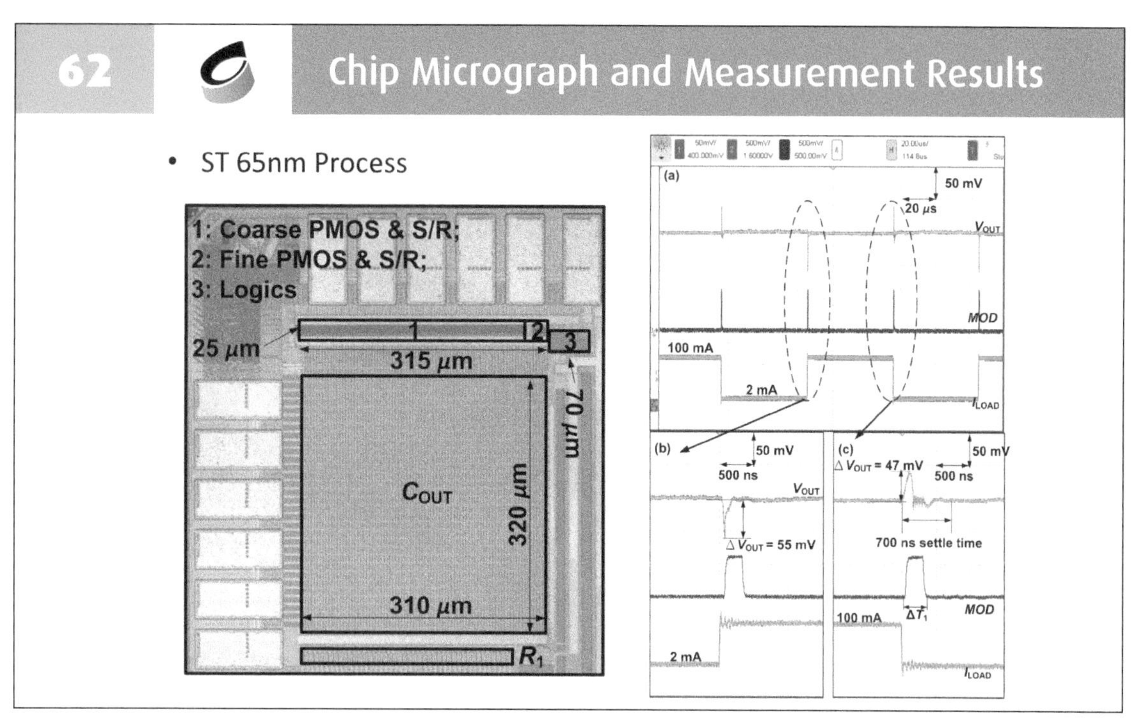

The chip was fabricated in a 65nm CMOS process with 1nF on-chip capacitor. In the measured waveforms, when the load current changes from 2mA to 100mA, the voltage undershoot is only 55mV, thanks to the coarse tuning and burst mode operation.

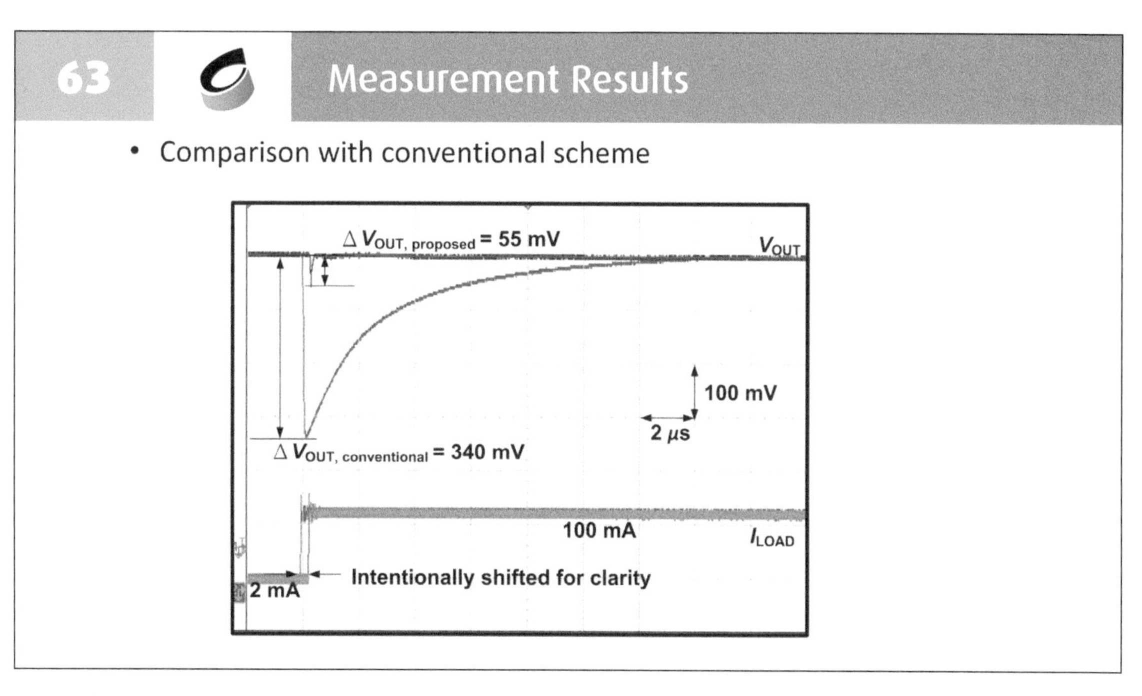

As shown in the measured waveforms, the transient response is much improved comparing to the conventional design.

64  Measurement Results

- DVS function, a.k.a. reference tracking.

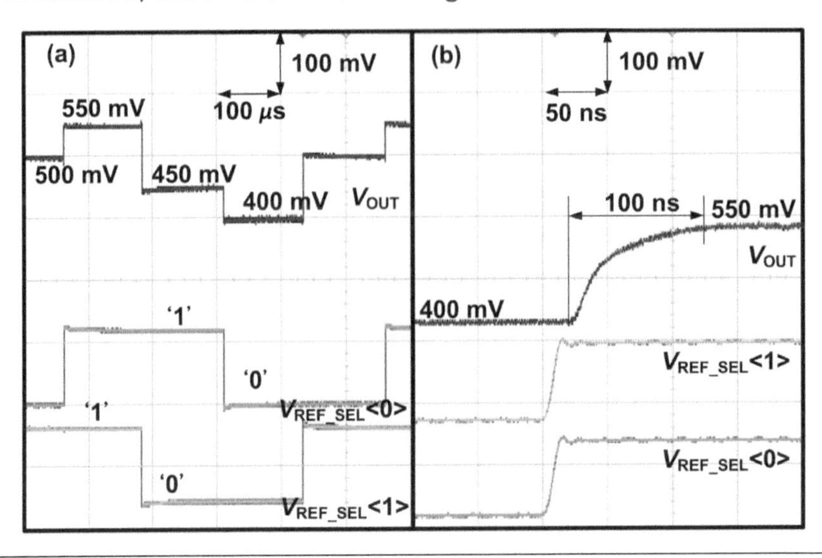

The dynamic voltage scaling (DVS) function, also known as the reference tracking capability, is also measured. It takes about 100ns for the D-LDO output goes from 400mV to 550mV after the V_{REF} has changed.

65  Comparison with Prior D-LDOs

	TPEL 2013	TVLSI 2015	ISSCC 2015	**This work**
Process (nm)	180	110	130	**65**
Active area (mm^2)	0.81	0.04	0.114	**0.01**
V_{IN} range (V)	0.9-1.8	0.6-1.2	0.5-1.2	**0.6-1.1**
V_{OUT} range (V)	0.8-1.5	0.5-0.9	0.45-1.14	**0.4-1**
C_{OUT} (nF)	1000	1	1	**1**
I_Q (μA)	750	32	78	**82**
Peak current eff. (%)	99.6	99.96	98.3	**99.92**
Line reg. (mV/V)	N/A	2	N/A	**3**
Load reg. (mV/mA)	N/A	0.3	<10	**0.06**
Load step (mA)	1-100	80	0.5-2	**2-100**
Trans. edge time (ns)	100	25000	N/A	**20**
ΔV_{OUT} (mV)	70	53	<40	**55**
FOM (ps)	5250	0.26	76.5	**0.43**
Settling time (μs)	4	38	1.1	**0.7***
DVS speed (V/μs)	N/A	N/A	N/A	**1.5**

The comparison table summarized the performances of this design example. It consumes 82μA in the steady-state, and can response to fast load transients with small output variations.

66 

Analog-Assisted (AA) Digital LDO Regulator

- Using a high pass analog path to assist the slow digital loop.

[Huang, ISSCC 2017]

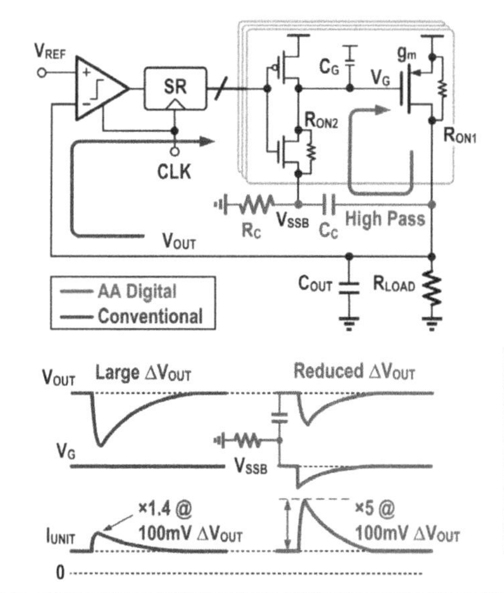

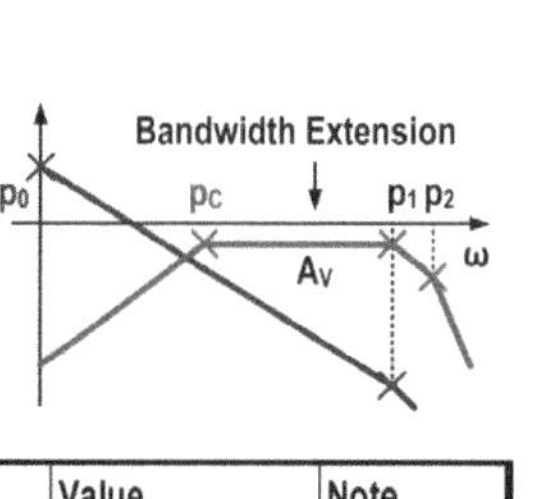

	Value	Note
p_0	DC	SR
p_c	$1/R_C C_C$	AA Path
p_1	$1/(R_{ON1}//R_{LOAD})C_L$	Load
p_2	$1/R_{ON2}C_G$	V_G
A_v	$gm \cdot (R_{ON1}//R_{LOAD})$	Pass Band

The second design example is the analog-assisted (AA) digital LDO work we published on ISSCC 2017. As we mentioned, the digital loop is either power consuming or slow. To break this trade-off, a high-pass analog path is introduced in parallel with slow digital loop. The coupling capacitor C_C will couple the output variations to the V_{SSB} node, while the resistor R_C provides a D_C bias for the V_{SSB} node. Assume V_{OUT} drops due to the load variation, the ΔV_{OUT} will be coupled to the V_{SSB} node. For the turned-on power switches, the NMOS of their driving inverter is on. Therefore, their V_G nodes will also be pulled down by C_C, the V_{GS} of the power switches increases accordingly. Consequently, the LDO output current increases instantly when V_{OUT} drops. In frequency domain, the AA path is a band-pass loop, the lower bandwidth is determined by R_C and C_C, while the high frequency characteristics are affected by the poles on the output and V_G nodes.

67

Power Loss Breakdown of Conventional Design

- High accuracy and wide load range needs more current steps.
- 9 bit resolution needs 512 current steps (shift registers, SRs) in baseline design with thermometer code.

Another problem we want to solve is the quiescent current. In conventional design, a high resolution requires a large number of DFFs which cause large leakage current and dynamic power.

68

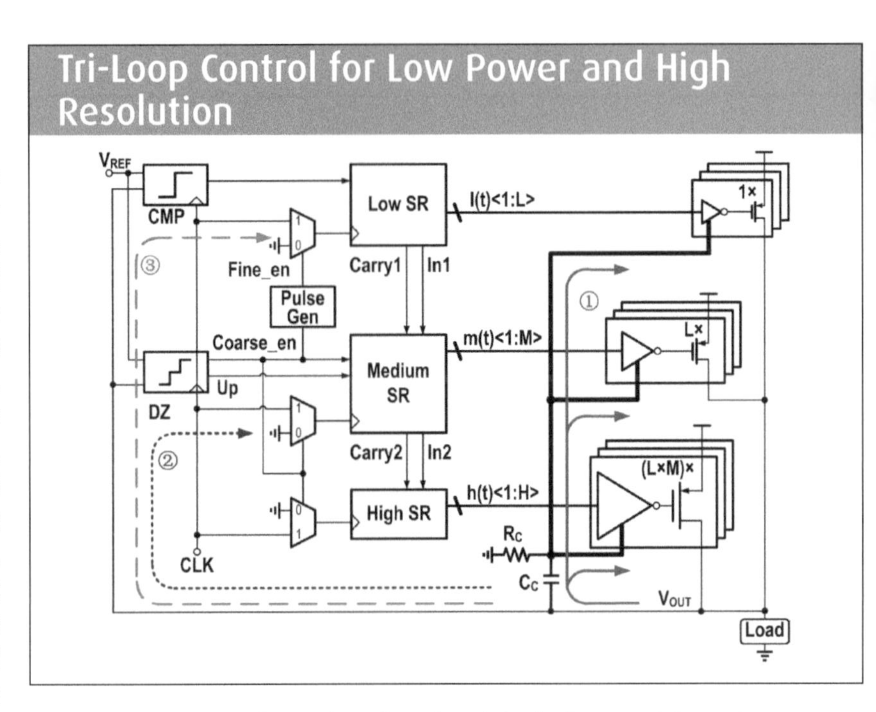

Therefore, we designed three shift-register (SR) sections: Low, Medium, and High SR sections. There are carry-in/-out pins between those sections. The Low section is only used in fine tuning, while the Medium and the High sections are used in coarse tuning. For 512 output current steps, a simple way is to set the SR bit number of each sections to 8, then 8×8×8=512. And only 8+8+8=24 DFFs are required, significantly reduced the leakage and dynamic power.

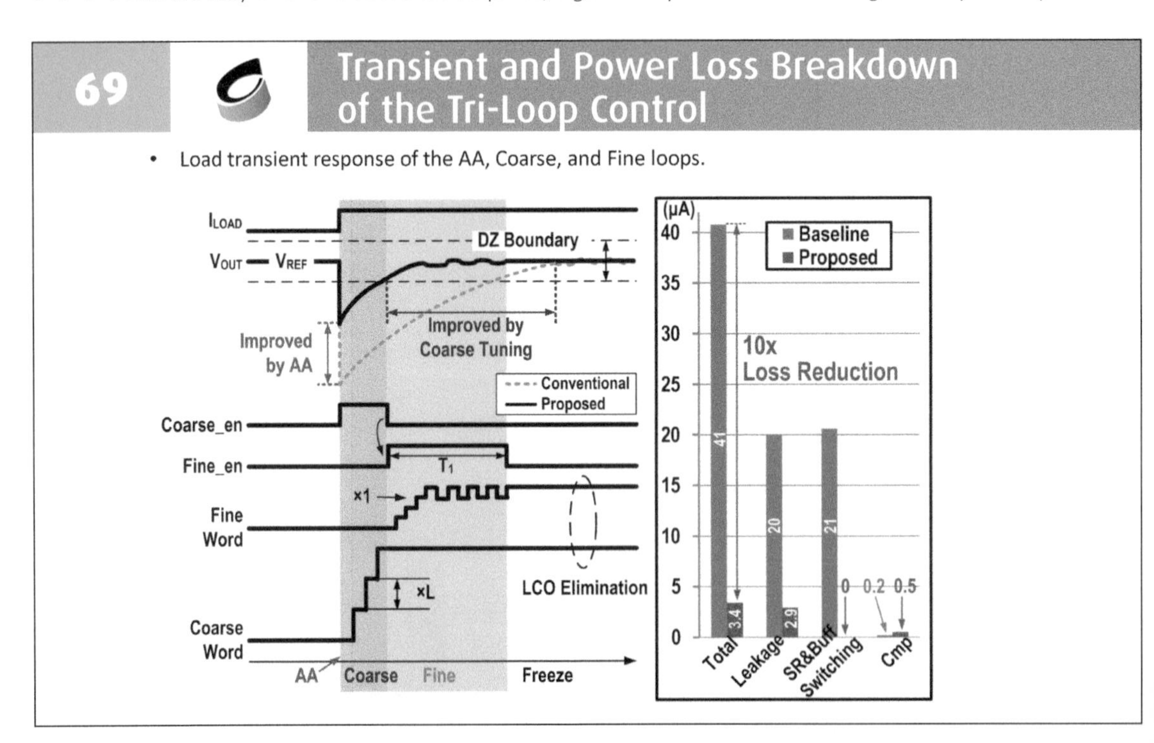

This slide shows the conceptual diagram of the transient response of the proposed D-LDO. When the load current I_{LOAD} changes from low to high, V_{OUT} drops before the D-LDO can response. Instantly, the AA loop reacts first, preventing the output drops too low. Then, because V_{OUT} has exceeded the lower boundary, coarse tuning is enabled. After certain clock cycles, it enters the fine tuning. To remove the LCO, the D-LDO will enter the freeze mode after fine tuning. As shown in the bar diagram on the right, the quiescent of the proposed D-LDO is much reduced to 3.4μA with a 10MHz clock.

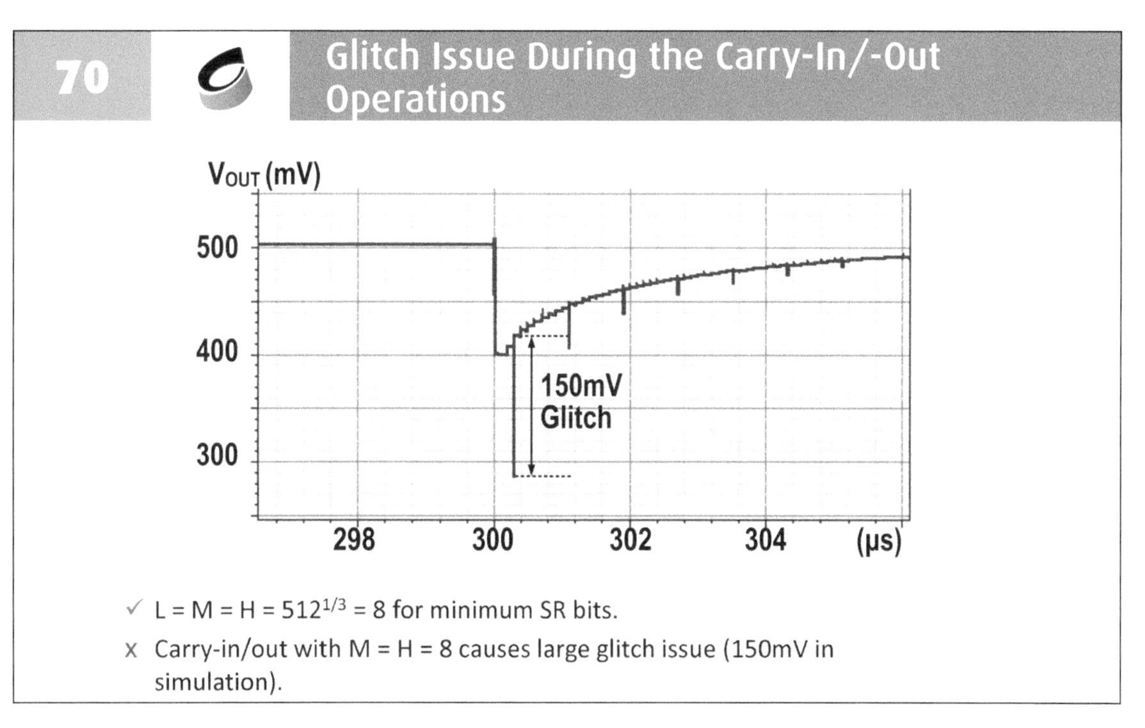

One issue we found during the carry-in/-out operation is the glitch problem caused by delay mismatches between different sections. A large glitch could happen during the carry-in/-out operation between the Medium and the High sections.

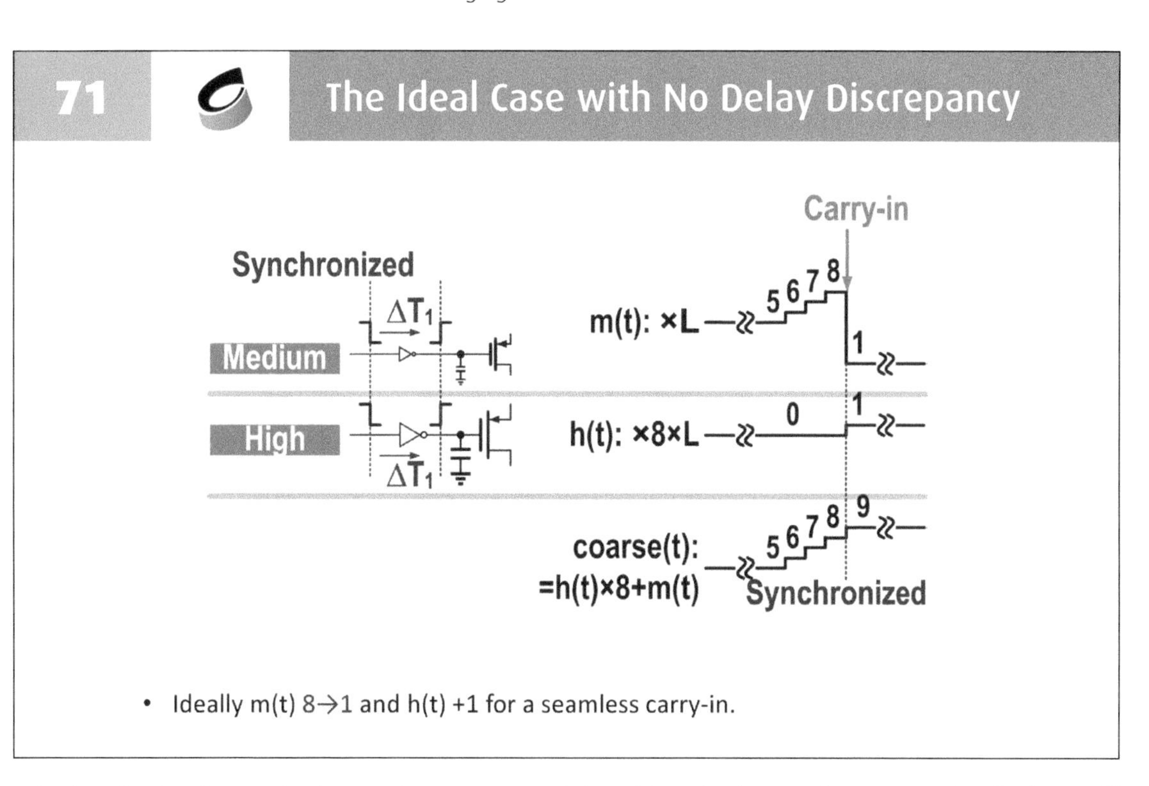

When the Medium code changes from 8 to 1, and the High code changes from 0 to 1, ideally there will be no glitch if the delay matches well.

72 Glitch Issue: Delay Discrepancy (M=8, H=8)

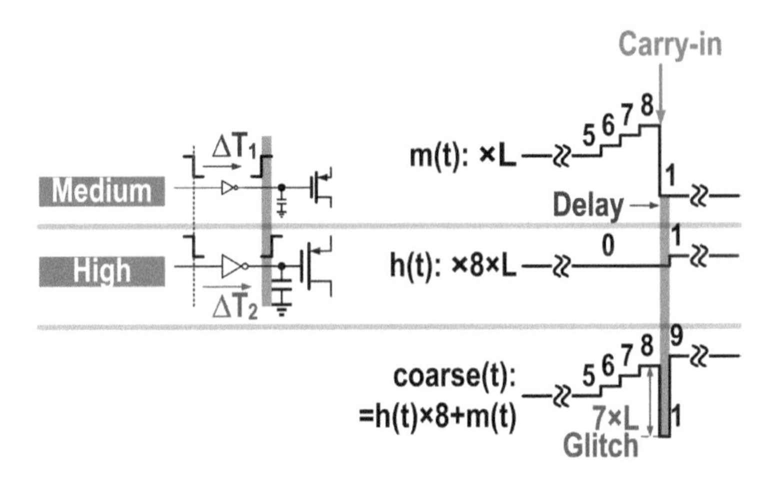

x 8→1→9 instead of 8→9 due to delay discrepancy.
x 7×L Glitch.

However, the Medium and the High codes are controlled by separate digital logics. Usually, the High code lags behind the Medium code. Therefore, a 7×L output current glitch could happen, where L is the strength of the unit current of the Medium section.

73 Glitch Reduction: M=4, H=16

- Halve M and double H.
- m(t) 4→1 when carry-in.

✓ M×H constant (=64).
- M+H (16→20).
x 4→1→5, 3×L Glitch.

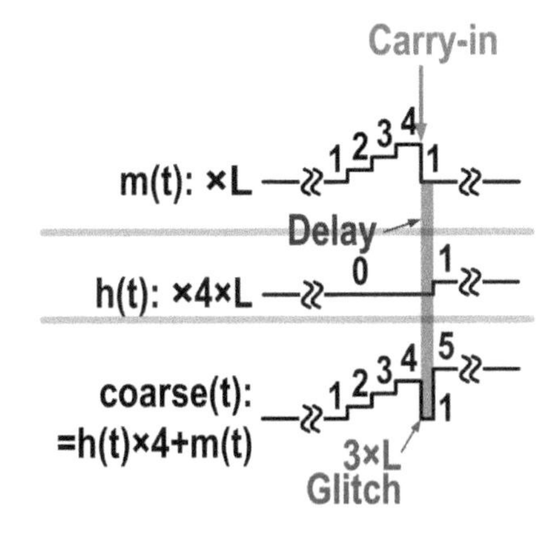

To reduce this glitch, a direct way is to change the number of switches in the Medium section into 4. If the total output steps maintain the same as before, 16 power switches should be used for the High section. Then, the strength of the glitch would be 3×L. By doing so, we have increased the total number of DFFs from 24 into 28, which is acceptable.

74 Glitch Reduction: M=4, H=16 (Proposed)

- Modified carry-in/out.
- m(t) 4→3 instead of 1 when carry in.

✓ 4→3→7, 1×L Glitch.
✓ m(t) 0→1 @ carry-out.
✓ Faster ramping.

Carry-in

m(t): ×L

Delay

h(t): ×4×L

coarse(t): =h(t)×4+m(t)

Faster Ramping

1×L Glitch

To further reduce the glitch, we proposed to use non-linear steps as illustrated in this slide. Instead of reset the Medium code from 4 to 1, we reset it from 4 to 3 during the carry-in operations. There are two benefits of doing so. Firstly, the strength of glitch is reduced to 1×L. Secondly, during load transients, the code goes in one direction for multiple clock cycles, because it needs to recover the large voltage undershoot/overshoot. With this non-linear steps, it recovers favorably faster than using the conventional linear steps.

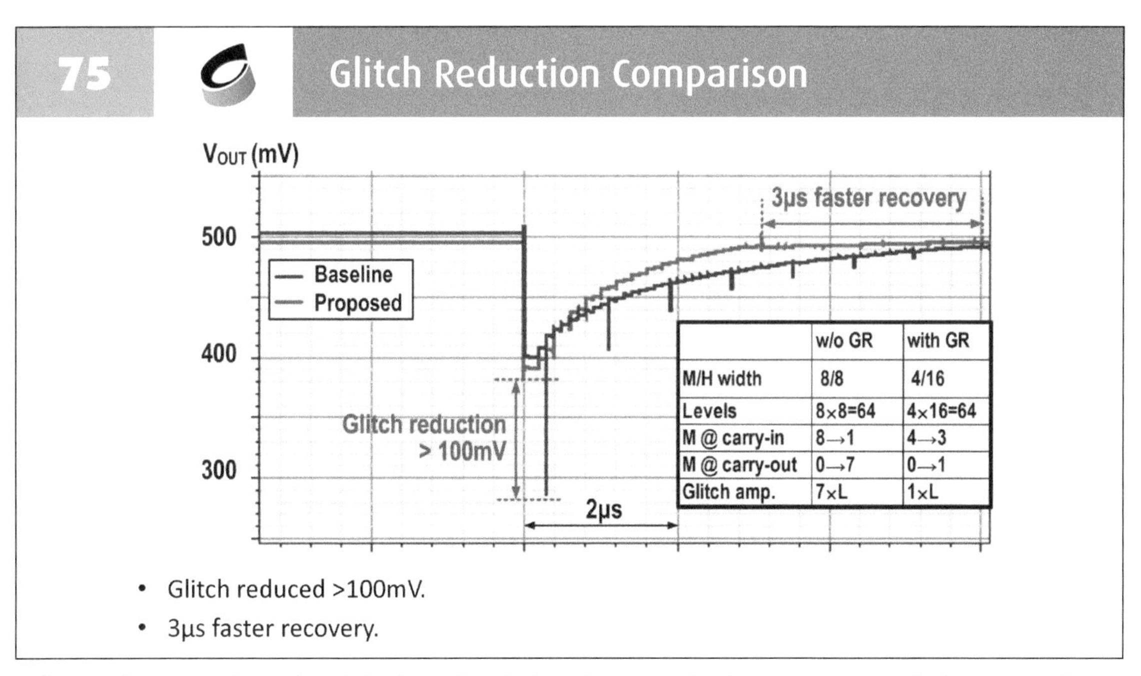

	w/o GR	with GR
M/H width	8/8	4/16
Levels	8×8=64	4×16=64
M @ carry-in	8→1	4→3
M @ carry-out	0→7	0→1
Glitch amp.	7×L	1×L

The simulation results in this slide show the glitch reduction and a faster recovery with the proposed non-linear coarse tuning steps.

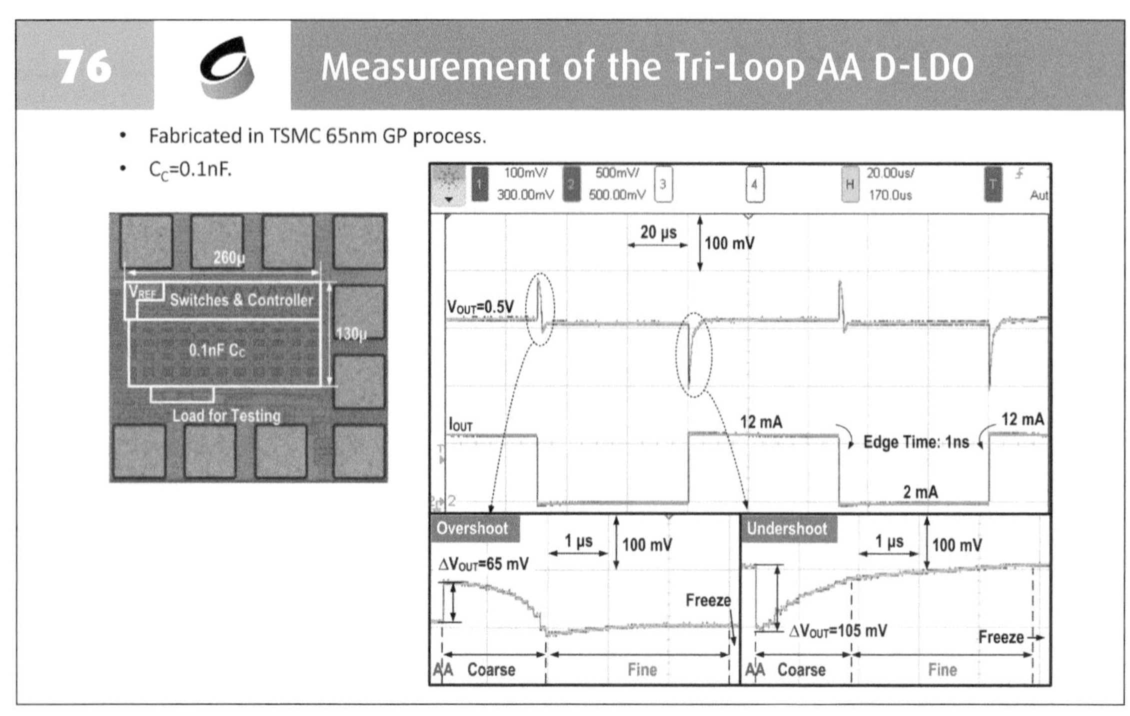

This design example was fabricated in a 65nm process with 100pF on-chip coupling capacitor and zero load capacitor. Operating with a 10MHz clock, when the load changes from 2mA to 12mA with a edge time of 1ns, the voltage undershoot is 105mV determined by the AA loop.

77 Comparison with Recent D-LDO Works

	This Work	Nasir ISSCC15	Lee JSSC16	Kim ISSCC16
Process	**65nm**	130nm	28nm	65nm
Area [mm^2]	**0.03**	0.355	0.021	0.029
Architecture	**SR+AA**	SR Based	SR Based	ADC Based
V_{IN} [V]	**0.5-1**	0.5-1.2	1.1	0.5-1
V_{OUT} [V]	**0.45-0.95**	0.45-1.14	0.9	0.45-0.95
Max. F_{SAMPLE} [MHz]	**10**	400	N.A.	200
Min. I_Q [µA]	**3.2**	24	110	12.5
Res./Total SR	**9bit/28**	7bit/128	6.6bit/25	N.A.
Total Capacitance*	**0.1nF**	1nF	23.5nF	0.4nF
ΔV_{OUT} [mV] @ $\Delta I_{LOAD}/T_{EDGE}$	**105 @10mA/1ns**	90 @1.4mA/N.A.	120 @180mA/4µs	40 @0.4mA/N.A.
FOM** [ps]	**0.23**	76.5	7.75	1.11

*Total Capacitance includes C_{OUT} and C_C. **FOM = $\frac{C\Delta V_{OUT}}{I_{OUT}} \times \frac{I_Q}{I_{OUT}}$. **AA**

This table compares our proposed work with state-of-the-art D-LDOs. Thanks to the AA loop, we obtained a reasonably good undershoot voltage during a sharp load transient event with only small total capacitor and slow clock frequency.

78 Comparison with Recent D-LDO Works

	This Work	Nasir ISSCC15	Lee JSSC16	Kim ISSCC16
Process	**65nm**	130nm	28nm	65nm
Area [mm^2]	**0.03**	0.355	0.021	0.029
Architecture	**SR+AA**	SR Based	SR Based	ADC Based
V_{IN} [V]	**0.5-1**	0.5-1.2	1.1	0.5-1
V_{OUT} [V]	**0.45-0.95**	0.45-1.14	0.9	0.45-0.95
Max. F_{SAMPLE} [MHz]	**10**	400	N.A.	200
Min. I_Q [μA]	**3.2**	24	110	12.5
Res./Total SR	**9bit/28**	7bit/128	6.6bit/25	N.A.
Total Capacitance*	**0.1nF**	1nF	23.5nF	0.4nF
ΔV_{OUT} [mV] @ $\Delta I_{LOAD}/T_{EDGE}$	**105 @10mA/1ns**	90 @1.4mA/N.A.	120 @180mA/4μs	40 @0.4mA/N.A.
FOM** [ps]	**0.23**	76.5	7.75	1.11

*Total Capacitance includes C_{OUT} and C_C. **FOM = $\frac{C\Delta V_{OUT}}{I_{OUT}} \times \frac{I_Q}{I_{OUT}}$.

Tri-loop

By using the tri-loop topology and three-section operation, we achieved 9bit resolution with only 28 DFFs. Therefore, the quiescent current is only 3.2μA.

79 Comparison with Recent D-LDO Works

	This Work	Nasir ISSCC15	Lee JSSC16	Kim ISSCC16
Process	**65nm**	130nm	28nm	65nm
Area [mm^2]	**0.03**	0.355	0.021	0.029
Architecture	**SR+AA**	SR Based	SR Based	ADC Based
V_{IN} [V]	**0.5-1**	0.5-1.2	1.1	0.5-1
V_{OUT} [V]	**0.45-0.95**	0.45-1.14	0.9	0.45-0.95
Max. F_{SAMPLE} [MHz]	**10**	400	N.A.	200
Min. I_Q [μA]	**3.2**	24	110	12.5
Res./Total SR	**9bit/28**	7bit/128	6.6bit/25	N.A.
Total Capacitance*	**0.1nF**	1nF	23.5nF	0.4nF
ΔV_{OUT} [mV] @ $\Delta I_{LOAD}/T_{EDGE}$	**105 @10mA/1ns**	90 @1.4mA/N.A.	120 @180mA/4μs	40 @0.4mA/N.A.
FOM** [ps]	**0.23**	76.5	7.75	1.11

*Total Capacitance includes C_{OUT} and C_C. **FOM = $\frac{C\Delta V_{OUT}}{I_{OUT}} \times \frac{I_Q}{I_{OUT}}$.

AA Tri-loop

With the joint efforts of the AA loop and the tri-loop topology, we achieved the best FOM comparing the other designs.

80  Summary

- Fast-transient circuit techniques for both analog and digital LDOs are introduced.
- The circuit techniques compatible with process scaling are favorable.
- Analog LDOs give excellent performances at nominal supply voltages, while digital LDOs find their applications in low-voltage scenarios.
- Analog-assisted digital LDO is a promising direction.

- Fast-transient circuit techniques for both analog and digital LDOs are introduced.
- The circuit techniques compatible with process scaling are favorable.
- Analog LDOs give excellent performances at nominal supply voltages, while digital LDOs find their applications in low-voltage scenarios.
- Analog-assisted digital LDO is a promising direction.

In summary, this chapter introduced fast-transient circuit techniques for both analog and digital LDOs. In general, we would like to have the circuit techniques that are compatible with process scaling for lower cost and better performances. The analog LDOs give excellent performances at nominal supply voltages, while digital LDOs find their applications in low-voltage scenarios. Last but not least, the analog-assisted digital LDO is a promising direction.

81  References

1. Bang S. Lee, "Understanding the Terms and Definitions of LDO Voltage Regulators," Application Report SLVA079, Texas Instruments, Oct. 1999. Available online: www.ti.com/lit/an/slva079/slva079.pdf
2. W.-H. Ki, L. Der, and S. Lam, "Re-examination of pole splitting of a generic single stage amplifier," *IEEE Trans. on Circuits and Systems I: Fundamental Theory and Applications*, vol. 44, no. 1, pp. 70–74, Jan. 1997.
3. K. N. Leung and P. K. T. Mok, "A capacitor-free CMOS low-dropout regulator with damping-factor-control frequency compensation," *IEEE J. Solid-State Circuits*, vol. 38, no. 10, pp. 1691–1702, Oct. 2003.
4. P. Hazucha, T. Karnik, B. A. Bloechel, C. Parsons, D. Finan, and S. Borkar, "Area-efficient linear regulator with ultra-fast load regulation," *IEEE J. Solid-State Circuits*, vol. 40, no. 4, pp. 933–940, Apr. 2005.
5. M. Al-Shyoukh, H. Lee, and R. Perez, "A transient-enhanced low-quiescent current low-dropout regulator with buffer impedance attenuation," *IEEE J. Solid-State Circuits*, vol. 42, no. 8, pp. 1732–1742, Aug. 2007.

6. R. G. Carvajal, et al., "The flipped voltage follower: a useful cell for low-voltage low-power circuit design," *IEEE Trans. Circuits Syst. I: Regular Papers*, vol. 52, no. 7, pp. 1276–1291, Jul. 2005.

7. T. Y. Man, et al., "Development of single-transistor-control LDO based on flipped voltage follower for SoC," *IEEE Trans. Circuits Syst. I: Regular Papers*, vol. 55, no. 5, pp. 1392–1401, May 2008.

8. J. Guo and K. N. Leung, "A 6-μW chip-area-efficient output-capacitorless LDO in 90-nm CMOS technology," *IEEE J. Solid-State Circuits*, vol. 45, no. 9, pp. 1896–1905, Sep. 2010.

9. M. Ho, K. N. Leung, and K.-L. Mak, "A Low-Power Fast-Transient 90-nm Low-Dropout Regulator With Multiple Small-Gain Stages," *IEEE J. Solid-State Circuits*, vol. 45, no. 11, pp. 2466–2475, Nov. 2010.

10. V. Gupta, G. A. Rincon-Mora, and P. Raha, "Analysis and design of monolithic, high PSR, linear regulators for SoC applications," in *Proc. of IEEE International SOC Conference*, 2004, pp. 311–315.

11. V. Gupta and G. A. Rincon-Mora, "A 5mA 0.6μm CMOS Miller-Compensated LDO Regulator with -27dB Worst-Case Power-Supply Rejection Using 60pF of On-Chip Capacitance," in *IEEE ISSCC*, Feb. 2007, pp. 520–521.

12. M. El-Nozahi, A. Amer, J. Torres, K. Entesari, and E. Sanchez-Sinencio, "High PSR Low Drop-Out Regulator With Feed-Forward Ripple Cancellation Technique," *IEEE J. of Solid-State Circuits*, vol. 45, no. 3, pp. 565–577, Mar. 2010.

13. Y. Lu, W.-H. Ki, and C. P. Yue, "A 0.65ns-Response-Time 3.01ps FOM Fully-Integrated Low-Dropout Regulator with Full-Spectrum Power-Supply-Rejection for Wideband Communication Systems," in *IEEE ISSCC*, Feb. 2014, pp. 306–307.

14. Y. Lu, Y. Wang, Q. Pan, W.-H. Ki, and C. P. Yue, "A Fully-Integrated Low-Dropout Regulator With Full-Spectrum Power Supply Rejection," *IEEE Trans. Circuits Syst. I*, Reg. Papers, vol. 62, no. 3, pp. 707–716, Mar. 2015.

15. Y. Lu, W.-H. Ki, and C. P. Yue, "An NMOS-LDO Regulated Switched-Capacitor DC-DC Converter with Fast Response Adaptive Phase Digital Control," *IEEE Trans. Power Electron.* vol. 31, no. 2, pp. 1294-1303, Feb. 2016.

16. Y. Lu, C. Li, Y. Zhu, M. Huang, S. P. U, and R. P. Martins, "A 312 ps response-time LDO with enhanced super source follower in 28 nm CMOS," *Electronics Letters*, vol. 52, no. 16, pp. 1368–1370, 2016.

17. Y. Okuma et al., "0.5-V input digital LDO with 98.7% current efficiency and 2.7-μA quiescent current in 65nm CMOS," in *IEEE Custom Integrated Circuits Conference (CICC)*, 2010, pp. 1–4.

18. S. Gangopadhyay, Y. Lee, S. bin Nasir, and A. Raychowdhury, "Modeling and analysis of digital linear dropout regulators with adaptive control for high efficiency under wide dynamic range digital loads," in *Design, Automation and Test in Europe Conference and Exhibition (DATE)*, 2014.

19. S. B. Nasir and A. Raychowdhury, "On limit cycle oscillations in discrete-time digital linear regulators," in *IEEE Applied Power Electronics Conference and Exposition (APEC)*, 2015, pp. 371–376.

20. Y.-H. Lee et al., "A Low Quiescent Current Asynchronous Digital-LDO With PLL-Modulated Fast-DVS Power Management in 40 nm SoC for MIPS Performance Improvement," *IEEE Journal of Solid-State Circuits*, vol. 48, no. 4, pp. 1018–1030, Apr. 2013.

21. Y.-C. Chu and L.-R. Chang-Chien, "Digitally Controlled Low-Dropout Regulator with Fast-Transient and Autotuning Algorithms," *IEEE Transactions on Power Electronics*, vol. 28, no. 9, pp. 4308–4317, Sep. 2013.
22. T. J. Oh and I. C. Hwang, "A 110-nm CMOS 0.7-V Input Transient-Enhanced Digital Low-Dropout Regulator With 99.98 #x0025; Current Efficiency at 80-mA Load," *IEEE Trans. VLSI Systems*, vol. 23, no. 7, pp. 1281–1286, Jul. 2015.
23. S. B. Nasir, S. Gangopadhyay, and A. Raychowdhury, "5.6 A 0.13μm fully digital low-dropout regulator with adaptive control and reduced dynamic stability for ultra-wide dynamic range," in *IEEE International Solid-State Circuits Conference (ISSCC)*, 2015.
24. Y. J. Lee et al., "8.3 A 200mA digital low-drop-out regulator with coarse-fine dual loop in mobile application processors," in *IEEE International Solid-State Circuits Conference (ISSCC)*, 2016, pp. 150–151.
25. D. Kim and M. Seok, "8.2 Fully integrated low-drop-out regulator based on event-driven PI control," in *IEEE International Solid-State Circuits Conference (ISSCC)*, 2016, pp. 148–149.
26. M. Huang, Y. Lu, S. W. Sin, S. P. U, and R. P. Martins, "A Fully Integrated Digital LDO With Coarse–Fine-Tuning and Burst-Mode Operation," *IEEE Transactions on Circuits and Systems II: Express Briefs*, vol. 63, no. 7, pp. 683–687, Jul. 2016.
27. M. Huang, Y. Lu, S. W. Sin, S. P. U, R. P. Martins, and W. H. Ki, "Limit Cycle Oscillation Reduction for Digital Low Dropout Regulators," *IEEE Transactions on Circuits and Systems II: Express Briefs*, vol. 63, no. 9, pp. 903–907, Sep. 2016.
28. M. Huang, Y. Lu, U. Seng-Pan, and R. P. Martins, "A digital LDO with transient enhancement and limit cycle oscillation reduction," in *IEEE Asia Pacific Conference on Circuits and Systems (APCCAS)*, 2016, pp. 25–28.
29. M. Huang, Y. Lu, S. P. U, and R. P. Martins, "20.4 An output-capacitor-free analog-assisted digital low-dropout regulator with tri-loop control," in *IEEE International Solid-State Circuits Conference (ISSCC)*, 2017, pp. 342–343.
30. M. Huang, Y. Lu, S. P. U, and R. P. Martins, "An Analog-Assisted Tri-Loop Digital Low-Dropout Regulator," *IEEE Journal of Solid-State Circuits*, vol. 53, no. 1, Jan. 2018.
31. X. Ma, Y. Lu, R. P. Martins, Q. Li, "18.4 A 0.4V 430nA Quiescent Current NMOS Digital LDO with NAND-Based Analog-Assisted Loop in 28nm CMOS" in *IEEE International Solid-State Circuits Conference (ISSCC)*, 2018.

CHAPTER 3

Design and Analysis of Analog and Digital Sub-Sampling Frequency Synthesizers

Wei-Sung Chang and Tai-Cheng Lee

National Taiwan University
Taipei, Taiwan, R.O.C

Sub-sampling-PLLs enable low-noise frequency synthesis because of its high-loop-gain characteristics, leading to low-power analog building blocks. In this chapter, the principle of sub-sampling PLLs would be introduced. Then, the concept is extended to fractional-N synthesis. Detail implementation and analysis would be covered for GHz frequency synthesis.

1  Outline

- **Background (ADI 9884 + X. Gao ISSCC slides)**
- **Fractional-N divider-less PLL**
  - Implementation
  - A model of the nonlinearity noise
  - Measurement Results
- **ADC-based digital PLL**
  - Motivation
  - Implementation
  - Measurement Results
- **Conclusions**

Here is the outline of this talk. I would start from background introduction, based on the datasheet of analog device 9884. Then, I would use some of the reference slides from Dr. X. Gao to give a brief introduction of integer-N sub-sampling PLLs. With the understanding of sub-sampling plls for integer-N operation, I would move to a divider-less fractional-N PLL architecture. It is also based on sub-sampling PLL concept. Its detail implementation will be illustrated. Furthermore, all the non-linearity-induced phase noise model will be built. A prototype in 0.18um CMOS architecture was implemented and measured.

Following that, an ADC-based digital PLL is proposed. I would talk about our motivation. And, its implementation as well as measurement results are shown. Finally, I would draw a conclusion.

2  Outline

- **Background (ADI 9884 + X. Gao ISSCC slides)**
- **Fractional-N divider-less PLL**
  - Implementation
  - A model of the nonlinearity noise
  - Measurement Results
- **ADC-based digital PLL**
  - Motivation
  - Implementation
  - Measurement Results
- **Conclusions**

3

ADI 9884A Data Sheet (I)

ANALOG DEVICES

100 MSPS/140 MSPS Analog Flat Panel Interface

AD9884A

FEATURES
140 MSPS Maximum Conversion Rate
500 MHz Analog Bandwidth
0.5 V to 1.0 V Analog Input Range
400 ps p-p PLL Clock Jitter
Power-Down Mode
3.3 V Power Supply
2.5 V to 3.3 V Three-State CMOS Outputs
Demultiplexed Output Ports
Data Clock Output Provided
Low Power: 570 mW Typical
Internal PLL Generates CLOCK from HSYNC
Serial Port Interface
Fully Programmable
Supports Alternate Pixel Sampling for Higher-Resolution Applications

APPLICATIONS
RGB Graphics Processing
LCD Monitors and Projectors
Plasma Display Panels
Scan Converters

FUNCTIONAL BLOCK DIAGRAM

AD9884A
R_{IN}, G_{IN}, B_{IN}
CLAMP
A/D
R_{OUTA}, R_{OUTB}, G_{OUTA}, G_{OUTB}, B_{OUTA}, B_{OUTB}
HSYNC, COAST, CLAMP, CKINV, CKEXT
CLOCK GENERATOR
DATACK, HSOUT
0.15V
CONTROL
REF
REFIN
FILT SOGIN SOGOUT SDA SCL A_0 A_1 $\overline{PWRDN}$ REFOUT

About 15 years, just right after I got my PhD degree, one of industrial friends gave me a data sheet, Analog device 9884A. This IC is for TV applications with 3-channel 140 MSPS ADCs. Besides, a clock generator is required to clock ADCs. This internal PLL generator from HSYNC also has a peak-to-peak jitter at 400ps. Generally speaking, it translates to around 50ps rms jitter. OK, for a 140 MHz clock, 50 ps rms jitter is around 1% of input clock period. At that time, this number seems not that difficult to me. Then, let's go to the other pages of the datasheet.

4

ADI 9884A Data Sheet (II)

Two bits that establish the operating range of the clock generator.

VCORNGE	Range (MHz)
00	20-60
01	50-90
10	80-120
11	110-140

VCORNGE must be set to correspond with the desired operating frequency (incoming pixel rate).

The power-up default value is VCORNGE = 01.

0C	4-2	CURRENT	Charge Pump Current

Three bits that establish the current driving the loop filter in the clock generator.

CURRENT	Current (μA)
000	50
001	100
010	150
011	250
100	350
101	500
110	750
111	1500

Pixel Rate (MHz)	VCORNGE	K_{VCO} (MHz/V)
20–60	00	100
50–90	01	100
80–120	10	150
110–140	11	180

PV_D
0.039μF C_Z
3.3kΩ R_Z
C_P 0.0039μF
FILT

Figure 10. PLL Loop Filter Detail

Table VI. Pixel Clock Jitter vs Frequency

Pixel Rate (MSPS)	Jitter p-p (ps)	Jitter p-p (% of Pixel Time)
135	350	4.7%
108	400	4.3%
94	400	3.4%
75	450	3.4%
65	600	3.9%
50	500*	2.4%
40	500*	2.0%
36	550*	1.8%
25	1000*	2.5%

*AD9884A in oversampled mode.

This slide shows the filter design. In order for wideband operation, VCO has different setting for different frequency ranges. As far as the charge-pump current, the current range is from 50 uA to 1500 uA, which is pretty large. Interestingly, the capacitor in the loop filter is as large as 39 nF.

At that time, I was thinking why not using smaller current in cooperate with the small cap. Then, the power consumption can be reduced and the loop filter can be integrated into the chip to save the pin counts. I was wondering why ADI analog designers were not so smart. Well, apparently, I am not so correct, right?

So, what's wrong with my design approach? If you look at the multiplication factor of this PLL, it's in the range of thousand. Under such large multiplication factor, the current source in the charge pump would induce huge noise unless you employ large charge © their best to reduce the phase noise by increasing CP current to mA range.

So, I was not aware of the noise of the PLL.

5  ADI 9884A Data Sheet (III)

- **Frequency multiplication table**

Standard	Resolution	Refresh Rate	Horizontal Frequency	Pixel Rate	VCORNGE	CURRENT
VGA	640 × 480	60 Hz	31.5 kHz	25.175 MHz	00	000
		72 Hz	37.7 kHz	31.500 MHz	00	000
		75 Hz	37.5 kHz	31.500 MHz	00	000
		85 Hz	43.3 kHz	36.000 MHz	00	001
SVGA	800 × 600	56 Hz	35.1 kHz	36.000 MHz	00	001
		60 Hz	37.9 kHz	40.000 MHz	00	001
		72 Hz	48.1 kHz	50.000 MHz	00	010
		75 Hz	46.9 kHz	49.500 MHz	00	001
		85 Hz	53.7 kHz	56.250 MHz	01	010
XGA	1024 × 768	60 Hz	48.4 kHz	65.000 MHz	01	010
		70 Hz	56.5 kHz	75.000 MHz	01	011
		75 Hz	60.0 kHz	78.750 MHz	01	011
		80 Hz	64.0 kHz	85.500 MHz	10	011
		85 Hz	68.3 kHz	94.500 MHz	10	011
SXGA	1280 × 1024	60 Hz	64.0 kHz	108.000 MHz	10	011
		75 Hz	80.0 kHz	135.000 MHz	11	100
		85 Hz	91.1 kHz	157.500 MHz*	01	100
UXGA	1600 × 1200	60 Hz	75.0 kHz	162.000 MHz*	01	100
		65 Hz	81.3 kHz	175.500 MHz*	10	100
		70 Hz	87.5 kHz	189.000 MHz*	10	101
		75 Hz	93.8 kHz	202.500 MHz*	10	101
		85 Hz	106.3 kHz	229.500 MHz*	10	110

So, here is the frequency generation table for the entire applications. Depending on the resolution and resolution, horizontal frequencies are in the range of 10's to 100's kHZ and it need to multiply this HSYNC clock frequencies hundreds' to thousands' to generate the pixel clocks.

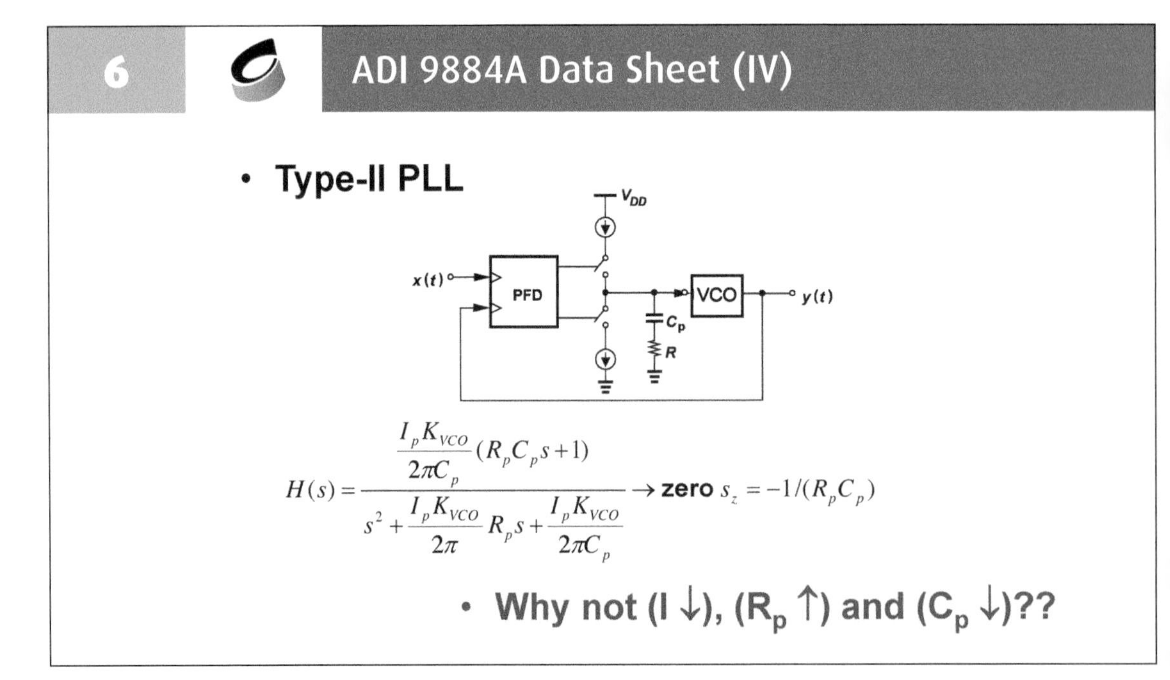

$$H(s)=\frac{\frac{I_pK_{VCO}}{2\pi C_p}(R_pC_ps+1)}{s^2+\frac{I_pK_{VCO}}{2\pi}R_ps+\frac{I_pK_{VCO}}{2\pi C_p}} \rightarrow \textbf{zero } s_z=-1/(R_pC_p)$$

So, this slide just illustrate my description. With a simple 2nd-order PLL, the loop characteristic can be maintained by decreasing charge-pump current, resistor for zero, and decreasing Cp. But, as I mentioned, such scaling would kill the phase noise performance of PLL.

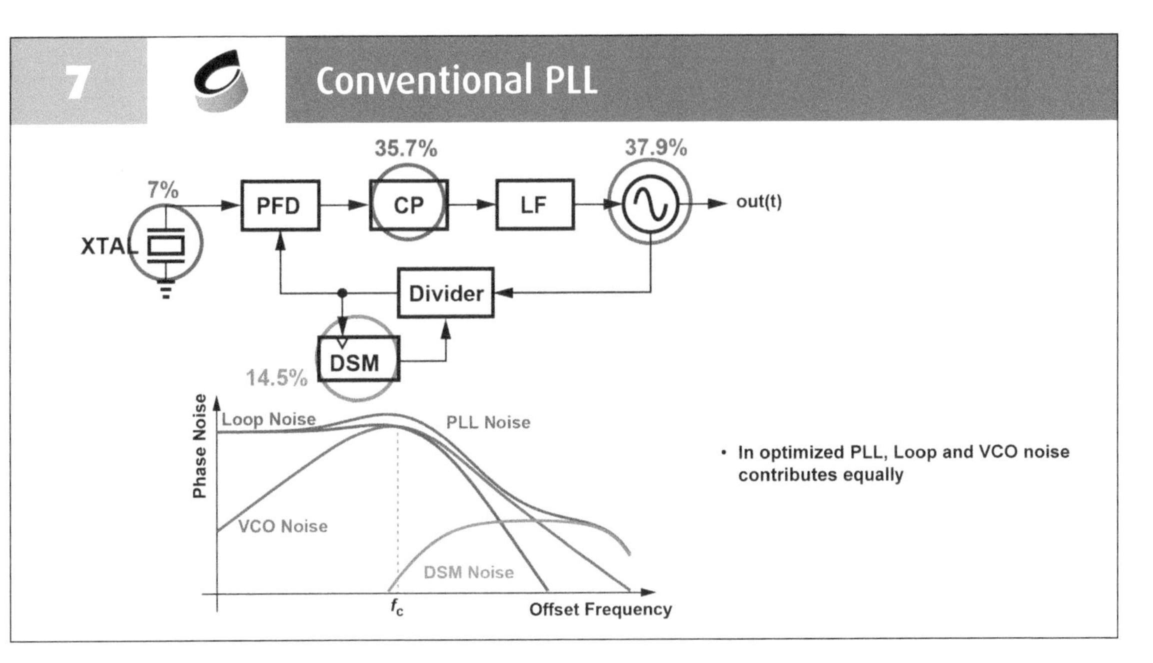

Now, let's spend a few moments in the design of the conventional PLL. A PLL consists of a crystal, a phase frequency detector, a CP, loop filter, VCO and a feedback divider. For a very typical design, both of VCO and charge pump contributes about 1/3 of total phase noise. In this work, we focus on the design of PLL architecture to reduce the noise effect from charge pump.

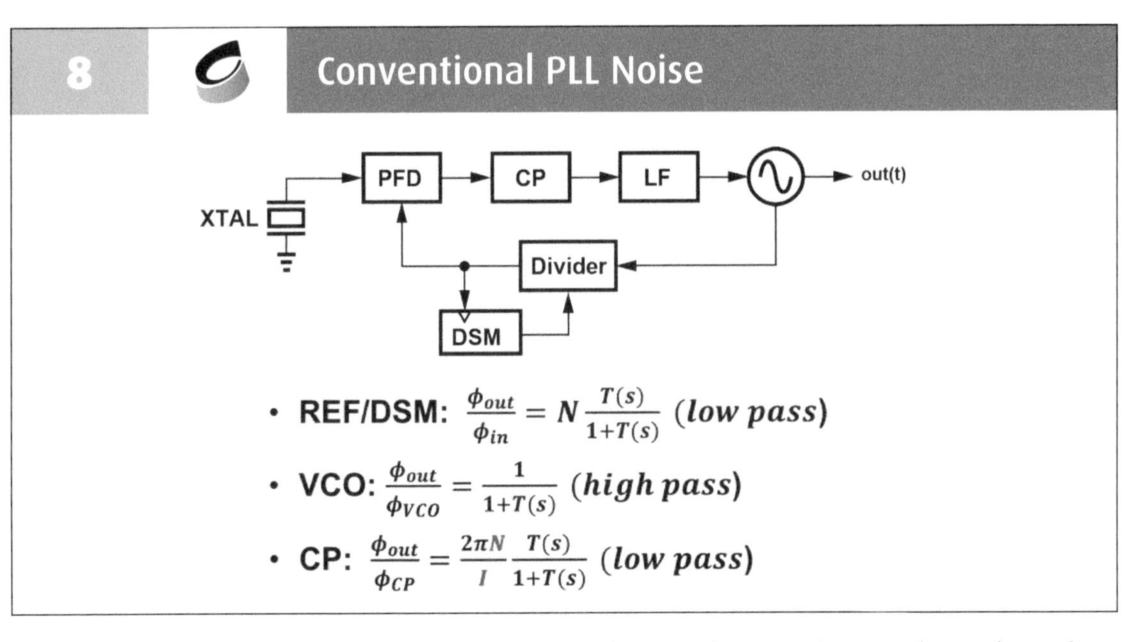

This slide shows the transfer function of four noise sources. Except the noise in VCO, all other noise sources are shaped by the low-pass characteristic of PLL. Note that, TS is loop transmission. Typically, we can tune the bandwidth of PLL to optimize the noise contribution from CP and VCO. The noise of VCO has the stringent design tradeoffs between the Q of the inductor and power of VCO. In this work, we focus on the noise from charge pump. If the pump noise can be greatly reduced, the bandwidth of PLL can be extended to suppress VCO noise. In the following slide, we would do some comparison between different kind of PLL architectures.

9

PLL Comparison Table

	Classical	Injection-locked	Sub-Sampling
Complexity	Good	Fair	Fair
Noise performance	Fair	Good	Good
Multiplication ratio	Good	Poor	Good
Integer implementation	Good	Good	Good
Fractional implementation	Good	Poor	Fair
Robust	Good	Fair	Fair

The noise design constraint has been fully investigated in the literatures. Recently, two different architectures of PLLs have been proposed, which are sub-sampling and injection locked. For these two architectures, the noise of PLL can be greatly reduce with the cost of some design hardware. For instance, for large multiplication factor, injection-locked PLL can not perform well due to the insufficient correction. Furthermore, due to nonlinear behavior of injection, the reported fractional-N injection locked PLL can not perform as well as its integer-N counterpart. In contrast to injection-locked PLL, subsampling PLL only has partial nonlinear behavior, and, we would employ subsampling PLL architecture to implement fractional-N architecture. Finally, for robustness, these two new architectures have very narrow frequency detection range, hence, it cannot be as robust as the conventional one. Luckily, a frequency locked loop can assist to ensure the reliable operation of these two new architectures.

10

SSPLL Background

- **Sub-Sampling based PLL**
- **Very low in-band phase noise**
  - **PD/CP noise not multiplied by N^2**
  - **No divider noise and power**

In ISSCC 2009, Mr. Gao proposed a sub-sampling PLL, as shown in this slide. It removes the feedback divider and replace the conventional PD by a subsampling PD.

There is no divider in the feedback path, so PD and CP's noise transfer function is not multiplied by N square.

Thus, the in-band phase noise due to charge-pump of SSPLL is 20logN lower than conventional one.

Besides, there is no divider noise and power.

This work achieves a very good performance

11

Integer SSPLL Characteristic

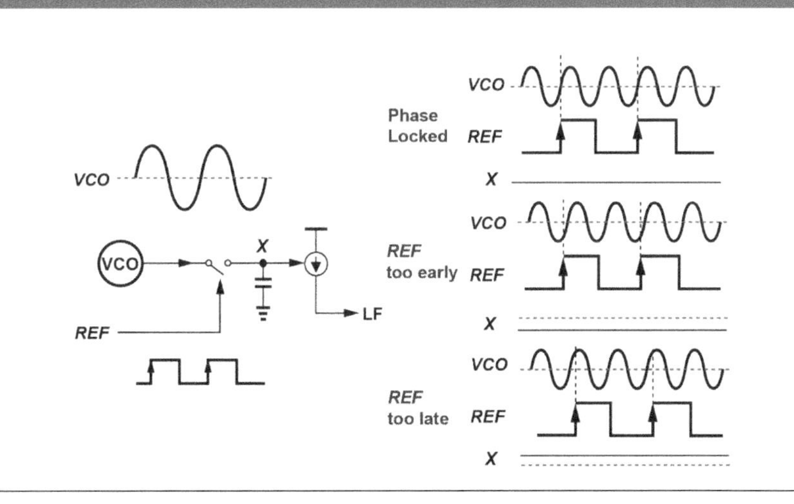

Now, let me briefly introduce the SSPLL operation principle. The VCO output, which is a sine wave, is sampled by a reference clock. When the VCO and Ref are phase aligned and their frequency ratio is an integer, the sampled voltage is a constant value , which is equal to the zero-crossing value.

The zero-crossing value is defined as the DC or common-mode voltage of VCO as shown as the red dotted line.

Whenever phase error occurs between the VCO and Ref, the sampled value will deviate from the zero-crossing value.

The voltage difference between sampled value and zero-crossing value corresponds to the amount of phase error as shown in these two figures.

When the sampled value is less than the zero-crossing value, it means REF is Early. On the other side, it means REF is Late.

SSPLL inherently only operates in an integer-N mode. However, this subsampling techniques cannot be employed in some applications which requires low noise fractional-N synthesizers for better frequency resolution.

12

Sub-sampling PD/CP Transfer Function

$$K_{d,SSPD} = \frac{\Delta I_{out}}{\Delta \phi_{VCO}} = \frac{g_m A_{VCO} \sin(\Delta \phi_{VCO})}{\Delta \phi_{VCO}}$$

$$\approx g_m A_{VCO} = \frac{2 I_{CP}}{V_{ov}} A_{VCO},$$

$$K_{d,PFD} = \frac{I_{CP}}{2\pi N},$$

$$\frac{K_{d,SSPD}}{K_{d,PFD}} = 4\pi N \frac{A_{VCO}}{V_{ov}}$$

- **Compared to conventional tri-state PFD/CP, SS PD/CP has much higher gain.**

This slide show the characteristics of subsampling PD. Shown in this figure, x-axis stands for the phase difference between input and feedback signal. Y-axis is the average current. The transfer characteristics is the output waveform, which is a sinusoidal shape. Note that, its corresponding slope is the PD gain. Shown in the about equation, the center point of the waveform can yield the maximum gain and it is approximately equal to gm times AVCO. In the conventional trip-state PFD, its gain is only equal to ICP over 2PIN. The gain ratio between the convention and the subsampling PDs equal to 4*pi*N*(AVCO/Vov), which is much larger than 1. The large PD gain leads to the large loop gain, which can suppress inband noise.

13

SSPLL Model (ISSCC 09 slide)

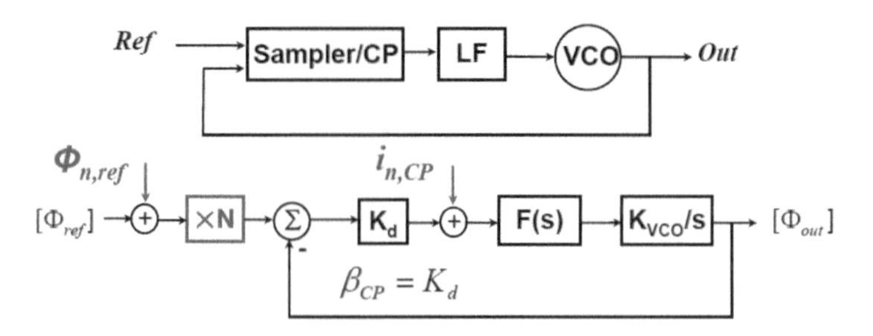

- **No Divider but a virtual Multiplier !**
  - **Fits sub-sampling process**

$$f_{alias} = f_{VCO} - N \cdot f_{ref}$$

  - **Ref noise multiplied by N, same as in classical PLL**

CP noise *NOT* multiplied by N !!

This slide shows the equivalent model for subsampling PLL. It's similar to the conventional PLL except the divider N is moved from the loop to the input path. We can understand this operation in the frequency domain. Basically, the subsampling operation generates the frequency difference between VCO frequency and N times fref. Therefore, subsampling operation cannot reduce the phase noise transfer function from the reference path. But, moving divider to out of loop can prevent CP noise from being multiplied by a factor of N.

14

SSPLL vs. CPLL (ISSCC 09 slide)

- **Divider noise**

SSPLL has no divider noise

- **CP noise**

$$\frac{\beta_{CP,SSPD}}{\beta_{CP,3state}} = 4\pi \cdot N \cdot \frac{A_{VCO}}{2I_{CP}/g_m} = 4\pi \cdot N \cdot \frac{A_{VCO}}{V_{gs,eff}} >> 1$$

e.g. $= 4\pi \times 40 \times \frac{0.4V}{0.2V} \approx 1000$

SSPLL has much larger β_{CP} than classical PLL, suppresses CP noise more

This slide shows the equivalent model for subsampling PLL. It's similar to the conventional PLL except the divider N is moved from the loop to the input path. We can understand this operation in the frequency domain. Basically, the subsampling operation generates the frequency difference between VCO frequency and N times fref. Therefore, subsampling operation cannot reduce the phase noise transfer function from the reference path. But, moving divider to out of loop can prevent CP noise from being multiplied by a factor of N.

15

Strength/ Weakness (ISSCC 09 slide)

- **SSPLL has no divider**
  - **May lock to any integer N**
- **SSPD/CP has very Large β_{CP}**
  - **May need Big Cap for stabilization**

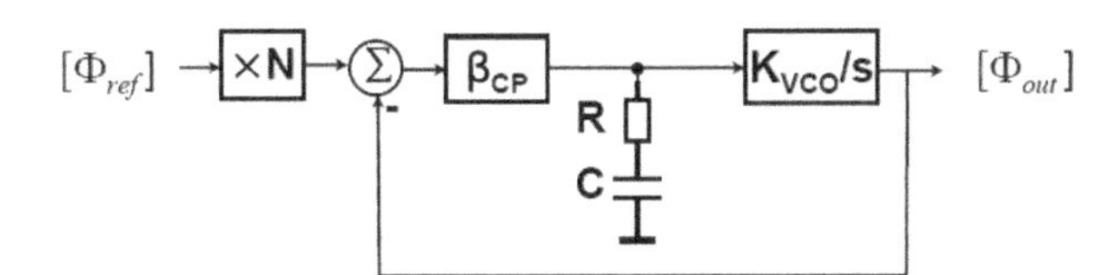

- **For given phase margin and K_{VCO}:** $C \propto \frac{\beta_{CP}}{f_c^2}$

Even though SSPLL has no divider noise, this also indicates that it can lock to any multiple of reference frequency. Typically, we need a frequency-locked loop to get rid of the harmonic locking. Also, the large loop gain needs a greater loop components for loop stabilization. For a typical loop filter, capacitor is proportional to beta divide by the square of bandwidth.

16

Loop Filter Design (ISSCC 09 slide)

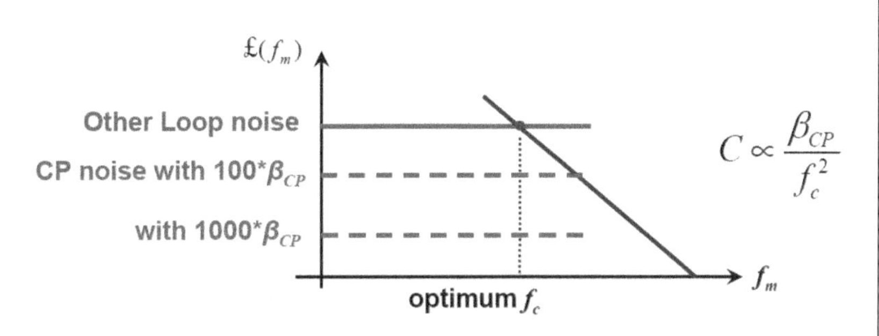

$$C \propto \frac{\beta_{CP}}{f_c^2}$$

- **Once CP noise is negligible, further Larger β_{CP} only wastes chip area (Big Cap)**

β_{CP} control is desired for full integration

The loop filter design in SSPLL needs to take the following into consideration. Even though CP noise can be suppressed greatly, other noise, such as reference noise, remains as the same as that in the conventional PLLs. Therefore, whenever beta is large enough, we shouldn't increase it furthermore. So, how do we tune beta in SSPLL? Well, typically, we add a duty control at the output of CP to inject the required charge to achieve large enough β.

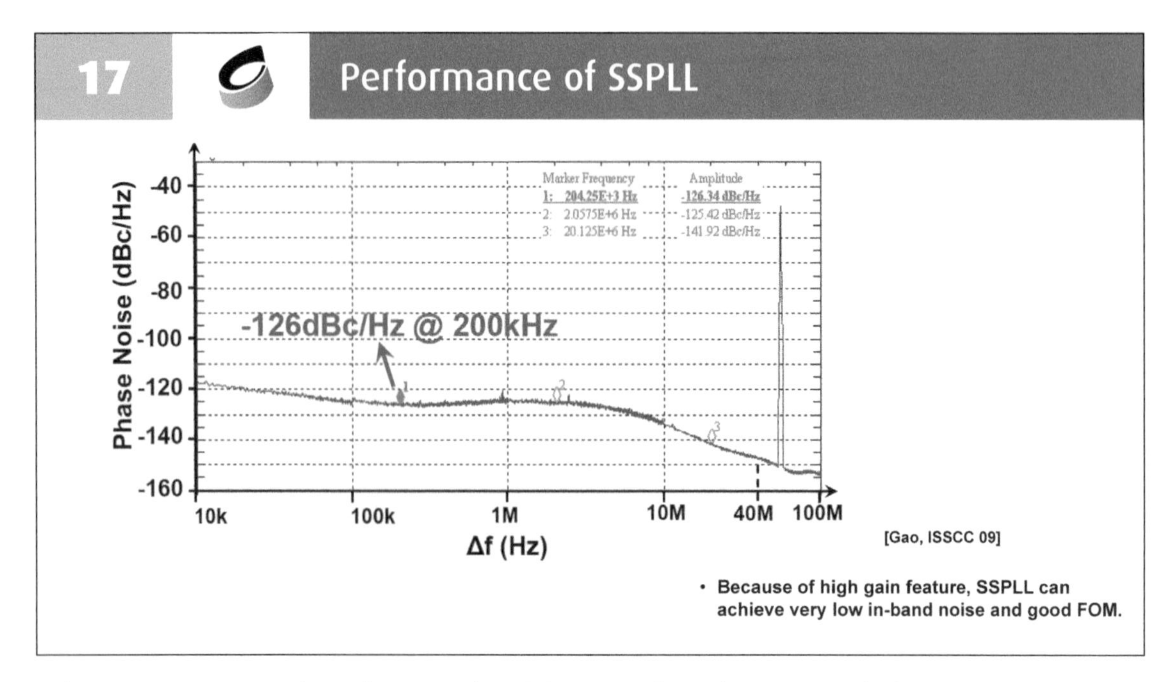

The measurement result is shown in this slide. Due to its high gain in SSPD, SSPLL can achieve excellent phase noise, which is around -126 dBc/Hz @ 200 kHz frequency offset.

18 Outline

- Background (ADI 9884 + X. Gao ISSCC slides)
- **Fractional-N divider-less PLL**
  - Implementation
  - A model of the nonlinearity noise
  - Measurement Results
- ADC-based digital PLL
  - Motivation
  - Implementation
  - Measurement Results
- Conclusions

19

Fractional-N Operation (I)

$F_{VCO} = 1.5 \times F_{REF}$

REF
VCO
$T_{diff.}$

REF
VCO
CP

- Programmable divider required

Before we talk about the proposed fractional-N SSPLL, let's review the basic principle of fractional-N operation.

To achieve fractional-N operation, a programmable divider is a common answer.

Two signals REF and VCO with 1.5 frequency ratio is shown here. Initially, their rising edges are in-phase as shown here. Next, their second rising edges start having phase difference. Then, the phase difference increases with time. Finally, phase difference will become infinite.

So we need a programmable divider to perform a fractional-N operation, it can swallow some pulses of the fast signal. As shown in this figure, these two signals lead or lag each other with time. The charge pump will charge or discharge with the phase difference. In a period of time, the average current is equal to zero. Thus this fractional-N operation can reach a steady state.

So, if we don't have a divider, what can we do?

20

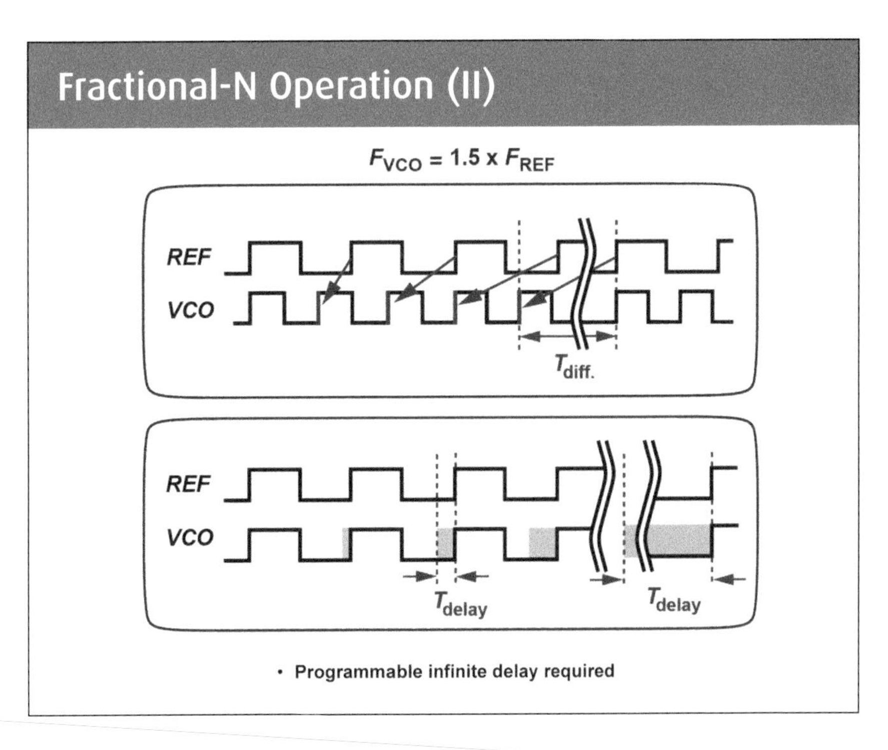

This is the same example as shown in the previous slide. We can only insert some delay to align every edge. After a few edges, phase difference will be infinite, so an infinite delay is required. To reach this steady state, we need a programmable infinite delay line.

21 Fractional-N SSPLL

Can we use the SSPLL in fractional-N mode ?

Therefore, can we design a fractional-N PLL without divider? In practice, we can't realize an infinite delay line.

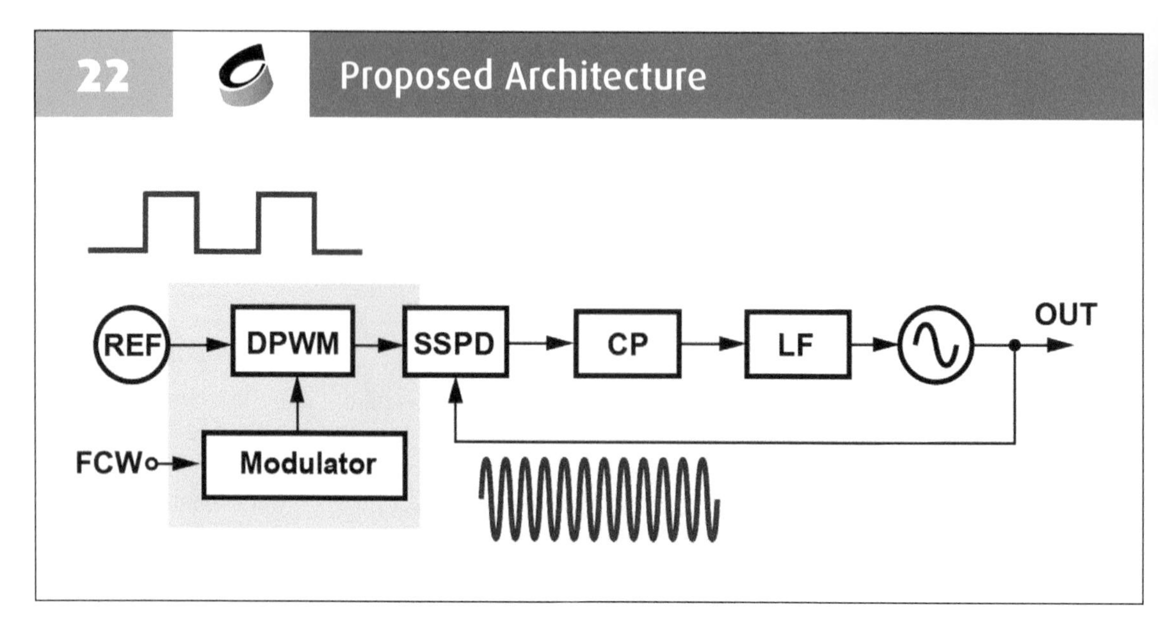

The proposed SSPLL in fractional-N mode is shown in this slide.

Because SSPLL has no divider, it needs a controllable delay in the loop to get fractional operation. There are two choices to insert the delay, right after the VCO or after the reference. If we put the delay line after the VCO, the buffer must exhibit a wide bandwidth to delay the high frequency "sine" wave. The design of a such delay buffer is very challenging and must be power hungry. Thus, we put the delay buffer after the reference. This block is called "DPWM". It can modulate the pulse width of the reference to change the sampling instants.

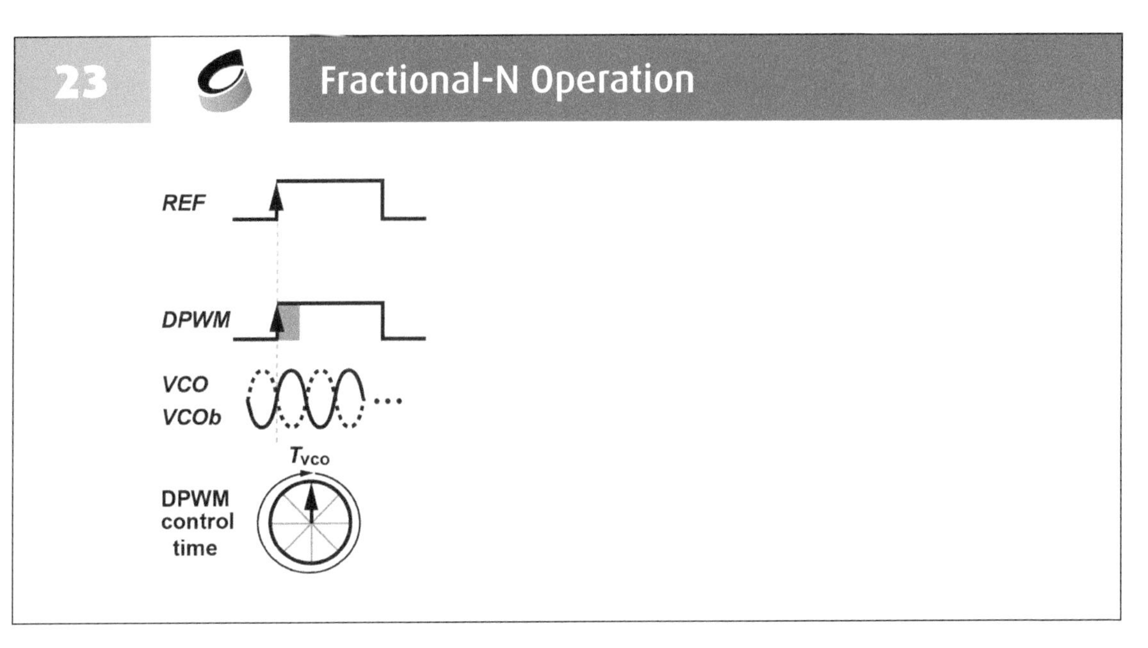

Basically, DPWM takes the input reference clock and shifts its rising edge as shown in the figure. The gray part is defined as the region to which DPWM can shift the rising edge. Let's assume that total gray part is one VCO cycle time. The pointer inside the circle represents every rising edge which is shifted and one circle means a VCO cycle time. Initially, rising edge of the DPWM is not shifted and just aligns the VCO crossing phase. It is the phase locked state.

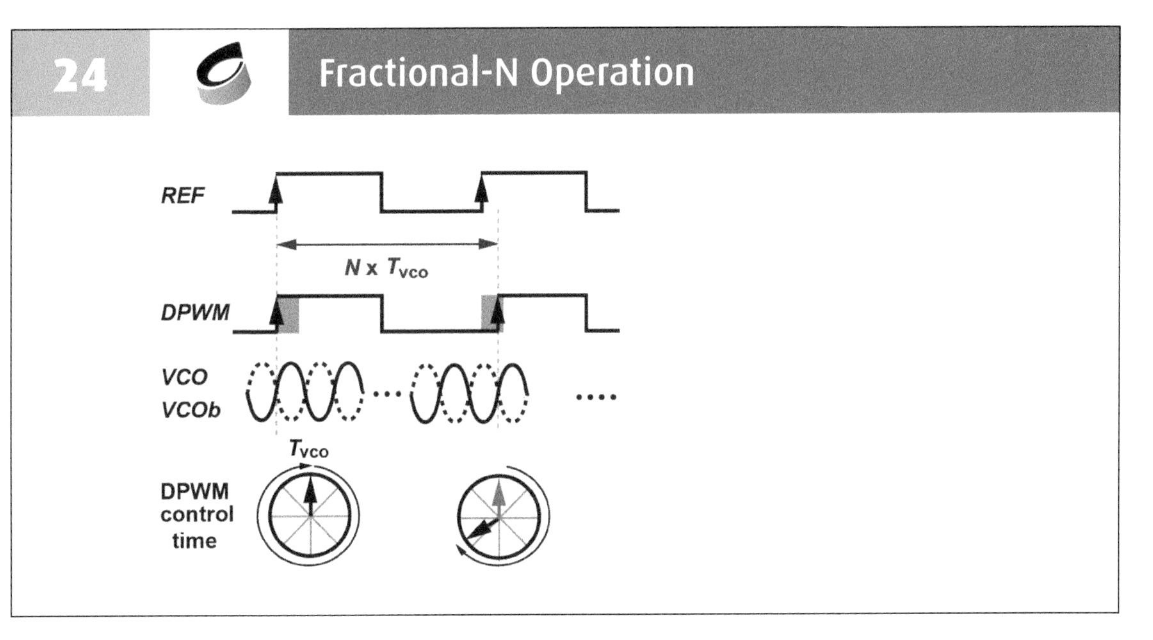

Now, the second edge appears. Because the frequency ratio between ref and VCO is not an integer, DPWM must shift its rising edge to align the VCO crossing phase. We assume that this modulation is larger than half of VCO cycle. The time difference between the first and the second rising edge is N times TVCO.

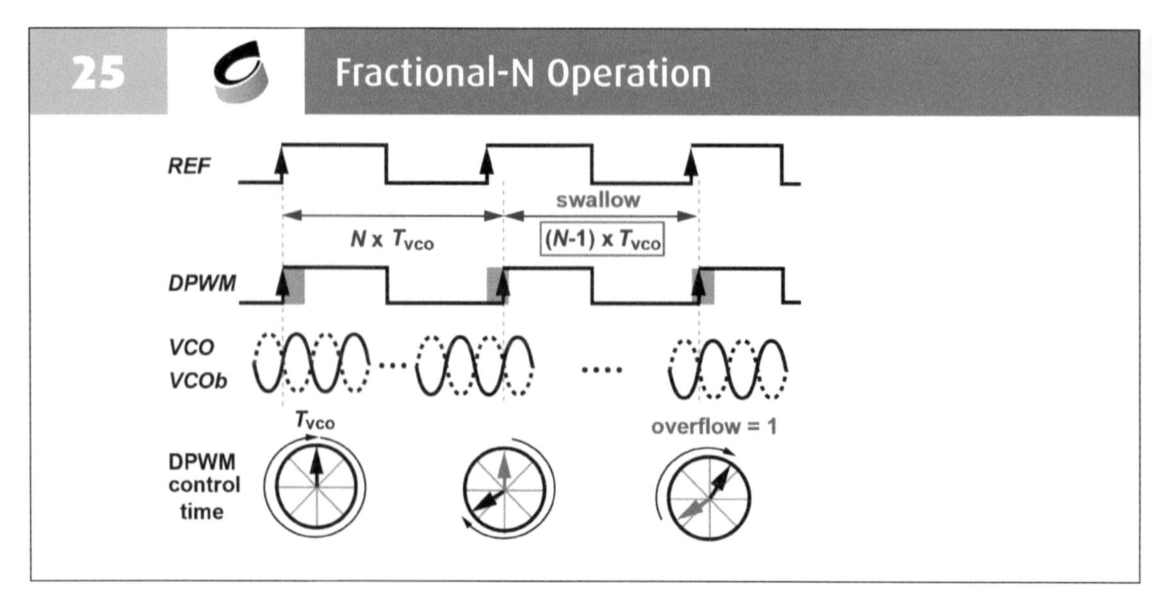

Next, the third edge appears. The time difference between the second and the third rising edge is assumed to be the same with the previous one. (In fact, it would become smaller due to loop tracking) In other words, DPWM must shift the rising edge by more than a TVCO. But we can't continue to increase delay to compensate all time difference, because the total delay shift can be only as large as a TVCO period. The overflow part is calculated from the origin. It means that if the operation of shifting should be more than a TVCO period, it must subtract a TVCO period.

The time difference between the second and the third rising edge is shorter than previous one if a TVCO period is subtracted. Fortunately, VCO waveform is continuous. DPWM can just sample the previous crossing phase of VCO. Thus, this time difference will be (N-1) times VCO period time. When the overflow occurs, this operation is inherently like a swallow divider.

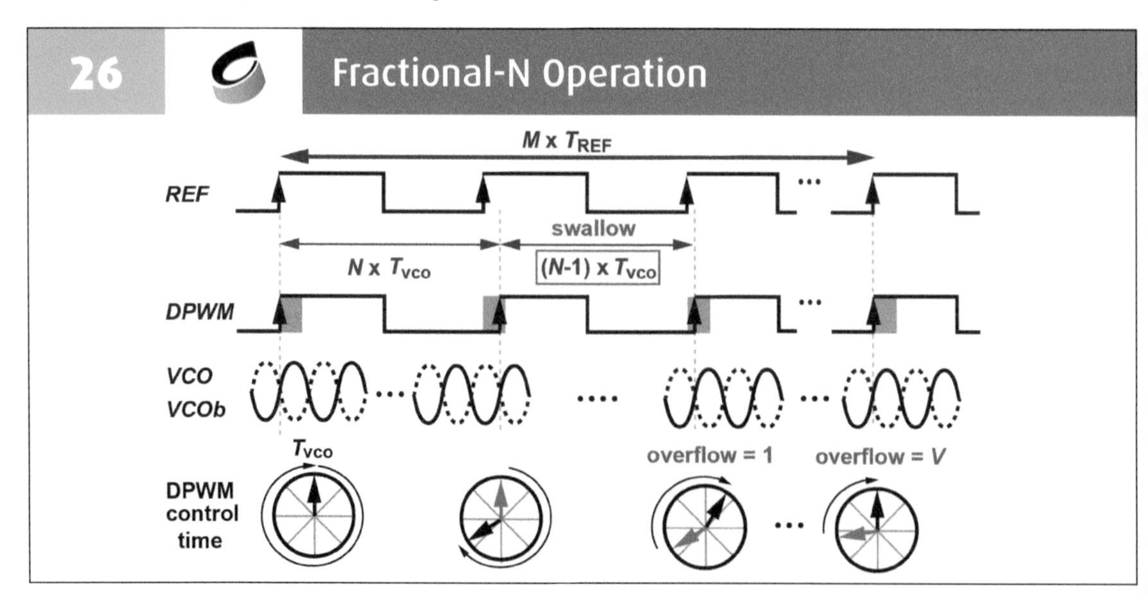

After a time, in M periods of REF, overflow operation occurs V times.

Theoretically, the time difference between two adjacent ref edges is equal to N times TVCO. M TREF is equal to NM TVCO.

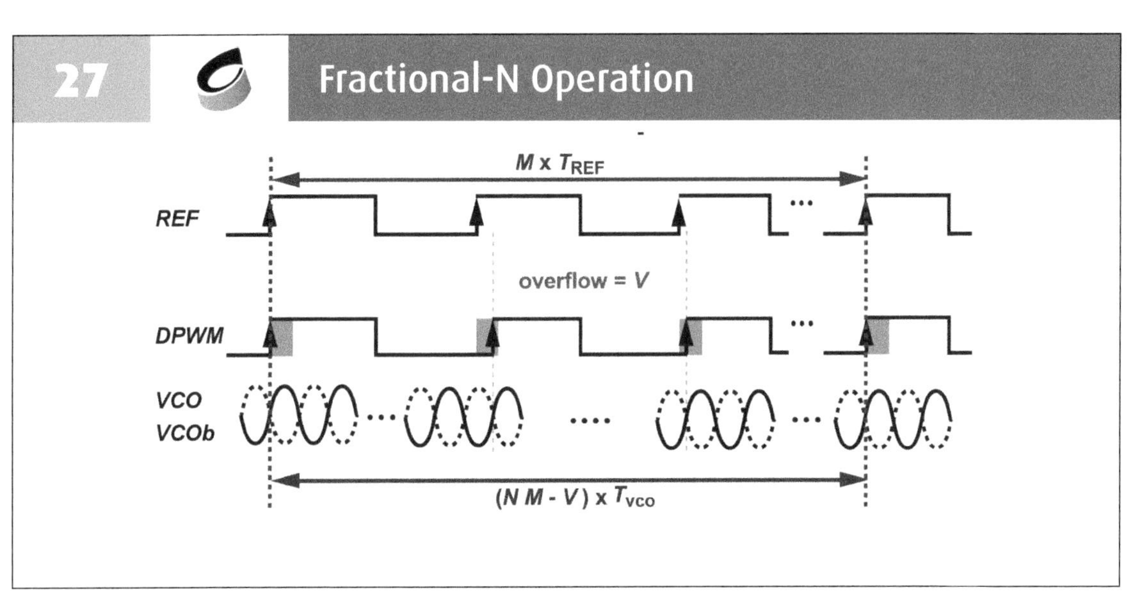

But overflow occurs V times, M TREF must be equal to (NM minus V) times TVCO. We can control the number of overflow by modulator. The frequency ratio between VCO and REF can be an arbitrary fractional number. The swallow-divider-like operation needs no divider and infinite delay line. It only needs a DPWM. The tuning range of the DPWM is one VCO period.

28

Outline

- Background (ADI 9884 + X. Gao ISSCC slides)
- **Fractional-N divider-less PLL**
  - Implementation
  - A model of the nonlinearity noise
  - Measurement Results
- ADC-based digital PLL
  - Motivation
  - Implementation
  - Measurement Results
- Conclusions

29

DPWM Implementation

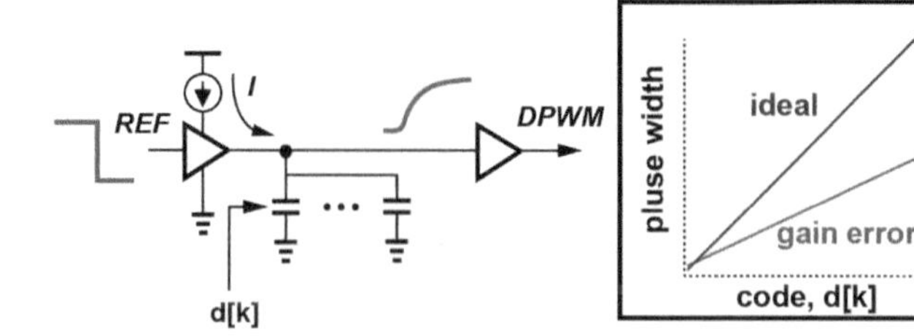

Two important sources of errors

- **Gain error**
- **Random mismatch**

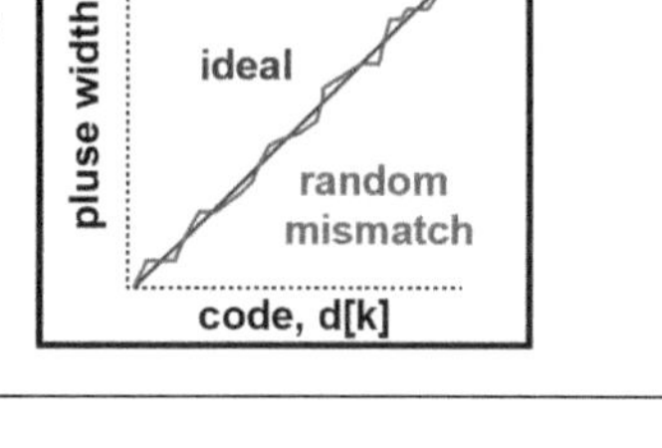

The proposed DPWM is composed of capacitor array and the current DAC. Current DAC is used to control the overall tuning width. Capacitor array is controlled by the digital modulator. The pulse width of each cycle is determined by the number of caps.

The pulse width of the DPWM has two important sources of errors, including a gain error and mismatches.

The previous example is assumed that tuning range is equal to a VCO period. In a real case, it may be larger or smaller than a VCO period. Here, we define it as "gain error".

"Mismatches" means that the cap array has variations. The relationship between the code and the corresponding pulse width is not smooth as shown in this one.

These two errors can't change the number of overflow, so the fractional operation is still correct. However, they can severely degrade phase noise.

30

DPWM Non-Ideal Effect

- **In all cases, CP has 5% current mismatch.**
- **Gain error:**

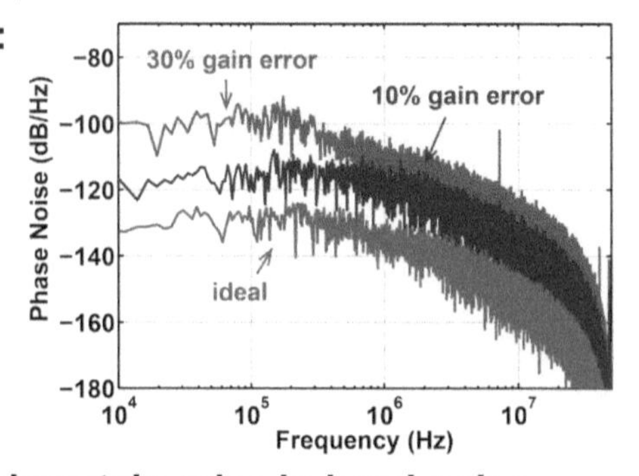

- **Random mismatch: raise in-band noise**

For example, if CP has 5% current mismatch between up and down current in all of the simulation results, the in-band noise is -100 dBc/Hz with 30% gain error. When the gain error becomes10%, the in-band noise improves to -115 dBc/Hz. Similarly, mismatch degrades in-band noise. Therefore, in order to design a fractional-N PLL with -115 dBc/Hz in-band noise, we must resolve with these two errors.

Gain Error in DPWM

How to resolve the gain error of the DPWM?

Next, I'll introduce How to resolve the gain error of the DPWM.

Previously Work of LMS Correlation

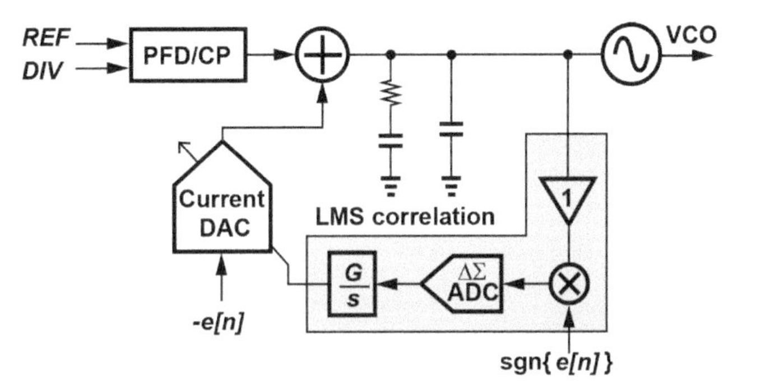

[Gupta, ISSCC 06]

- **A ΔΣ ADC with high dynamic range is required.**

A previously work of LMS correlation is proposed by Mr. Gupta in ISSCC 2006. A DAC is used to subtract the divider ratio error. It is an LMS zero forcing feedback concept. There is mismatch between DAC and CP. Based on multiplication of the sign of phase error from DSM and the voltage variance in the control line, the DAC gain is adaptively trimmed to match the CP gain. A high resolution ADC is used to sense the tiny voltage variance in the control line.

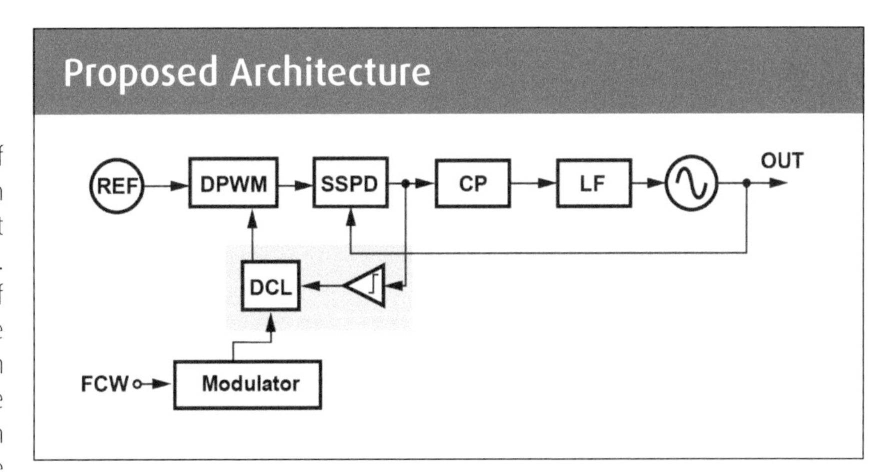

In this work, the DAC of DPWM is in the main path, which is different with the previous work. However, the gain error of DAC will affect the range of voltage variance in control line with the same effect as I mentioned in the previous work. The LMS zero forcing feedback concept can be used.

In this work, based on the sign of the voltage variance in control line, DCL (digital correlation loop) is used to adaptively trim the DAC gain.

34

This slide shows the schematic of SSPD in our work. Track and hold circuits sample the voltage from VCO output. Then, a source-degenerated common-source amplifier is employed to drive a V to I converter. A and B are the sampled waveform of VCO as shown in below.

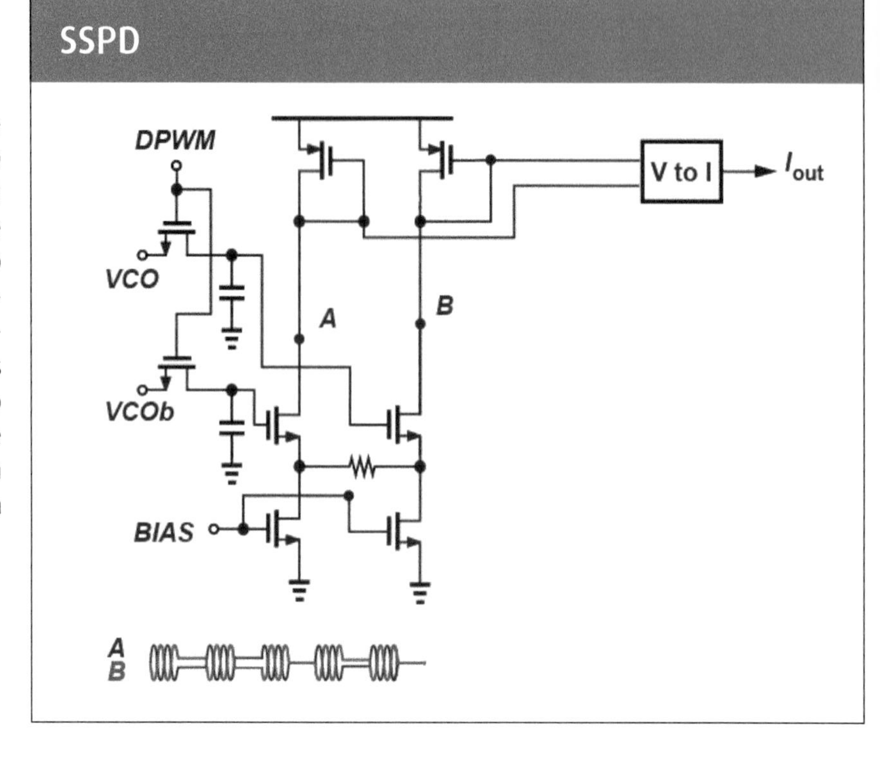

35

To obtain proper sign information, rather than sensing the tiny voltage variance in the control line, the proposed sign extractor can detect the error polarity based on the SSPD output voltage difference, A and B. It uses a re-sampler to filter the VCO signal under the SSPD sampling mode, and then it uses a high-pass filter to remove DC offset, such as CP current mismatch. Thus it can get proper sign information.

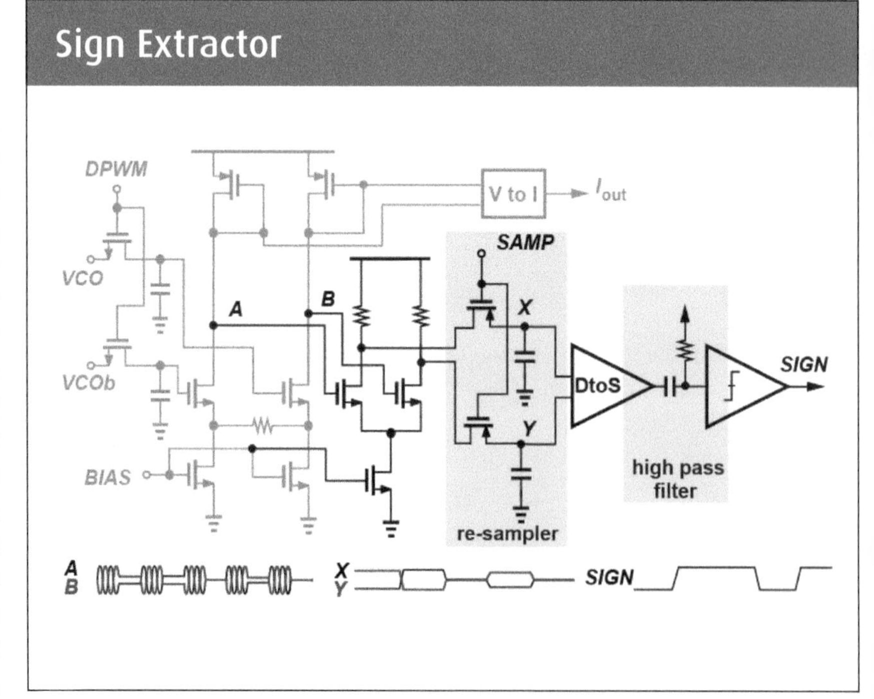

36 

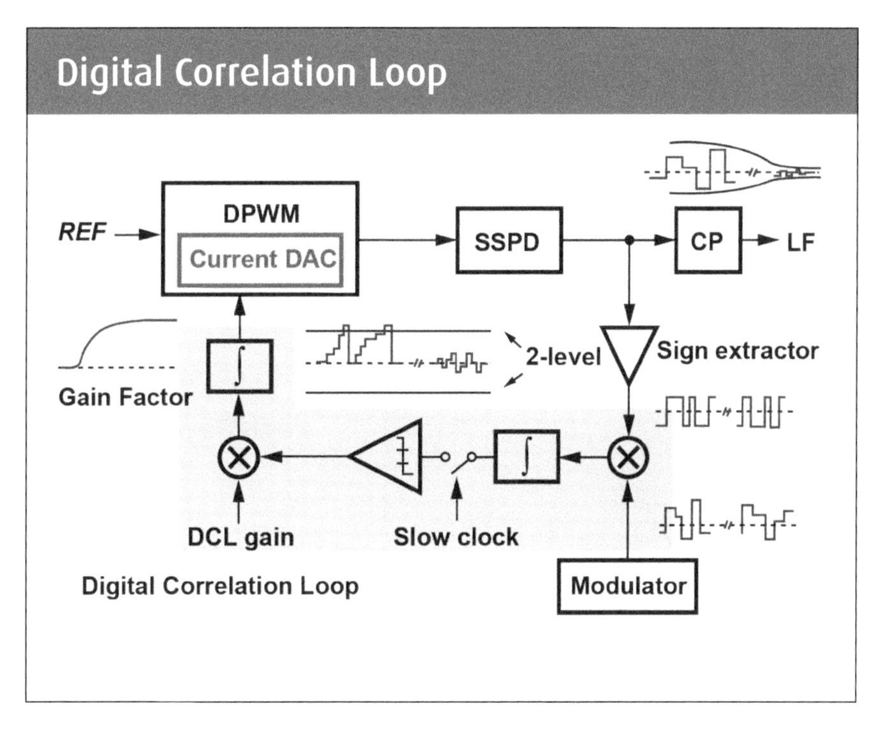

The detail of DCL is shown in this slide. The gray part, which is a digital loop filter, is synthesized by digital standard cells. The output of SSPD is applied to the sign extractor. Then, the sign is multiplied by a modulator to obtain the correction term. The slow clock is used to sample its filtered output. Then, the current DAC in DPWM is tuned. With the convergence of the DCL, the voltage variance in the control line is reduced.

37 

Mismatch in DPWM

How to resolve the mismatch of the DPWM?

After gain error is corrected, let's move to the mismatch of the DPWM effect.

38 

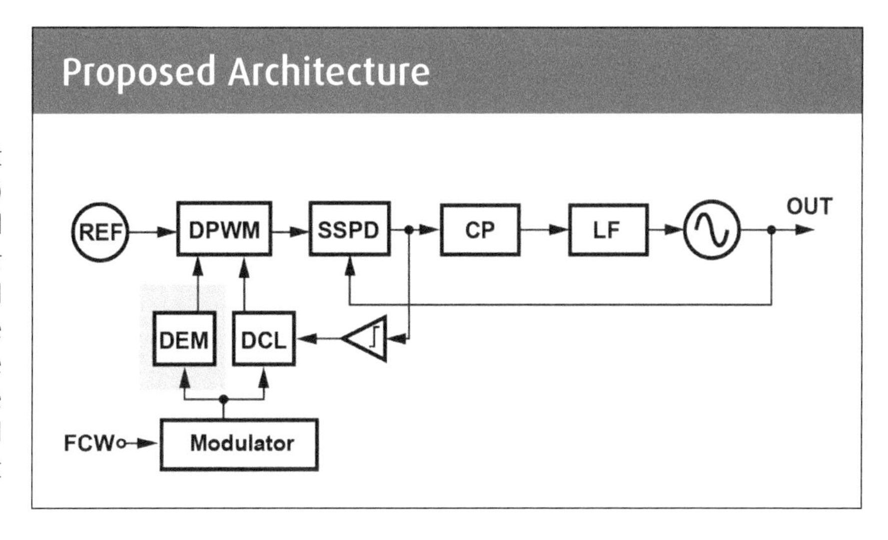

Digital element matching (DEM) is a widely-employed technique in ADC or DAC. A DEM is used to randomize the capacitor to shape the mismatch error. The implementation would be shown in the next slides.

39

DPWM Cap Array

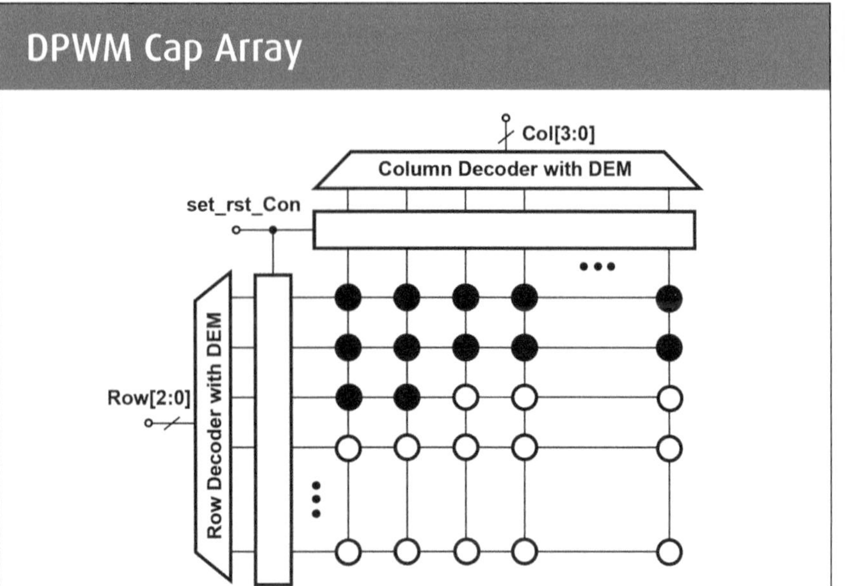

The proposed DPWM is controlled by a 7-b MSB cap array, which is shown in this slide with DEM operation, and 2-b LSB caps. Because a full 7-b DEM needs a very complex digital hardware, the proposed 2-D DEM is composed of row DEM and column DEM. In the matrix, the unit cap cell is controlled by a column word and row word.

40

2-DimentionalDEM

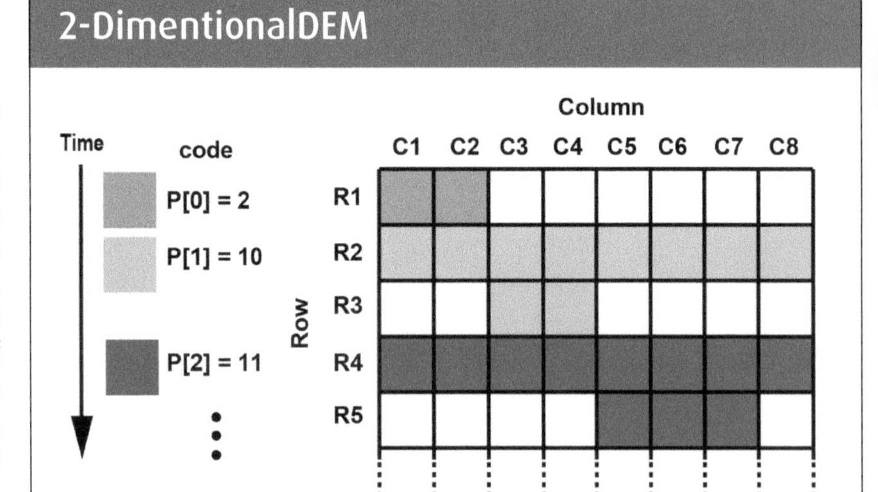

The operation is shown in this slide. The rule of selection is that column starts from the preceding column which is not selected. The row selection is the same. For example, if the DPWM needs two caps in the beginning, these two caps are the intersection of C1, C2 and R1. Next, the DPWM needs ten caps. Because C1, C2 and R1 are used, the column starts from C3. And the row starts from R2. These ten caps are the whole row 2 and the intersection of C3, C4 and R3. Following this rule, a 2-D DEM is achieved. It is hardware efficient.

41 

Finally, because the SSPD has a limited locking range, the remaining blocks are the frequency-locked aiding circuit to ensure the right divider ratio.

Frequency-Locked Loop

Frequency-locked aiding circuit

42 

This slide shows the complete architecture. Frequency-locked loop is added to ensure proper operation. In addition, an automatic frequency control (AFC) loop is used to select the VCO band to extend the output frequency range.

Proposed Architecture

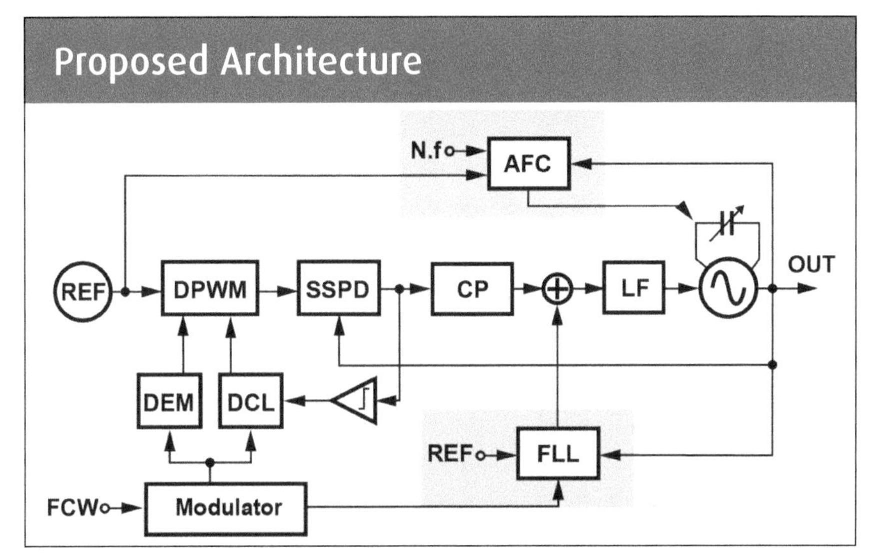

43 

The FLL is a conventional PLL with a dead zone (DZ). The DZ is inserted between the PFD and CP. The divider is modulated by the modulator to perform fractional operation. When the phase error between VCO and Ref is small and falls inside the dead zone. The output current of the CP in the FLL will be zero. On the contrary, the FLL has a large gain. It brings down the frequency difference until the phase difference falls inside the dead zone.

After locking, the FLL can be disabled to save power.

Frequency-Locked Loop

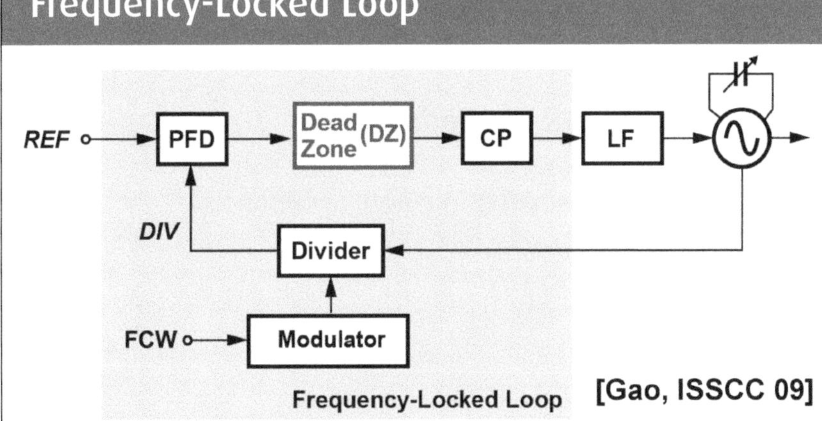

- **Because the SSPD has a limited locking range, a FLL is added.**

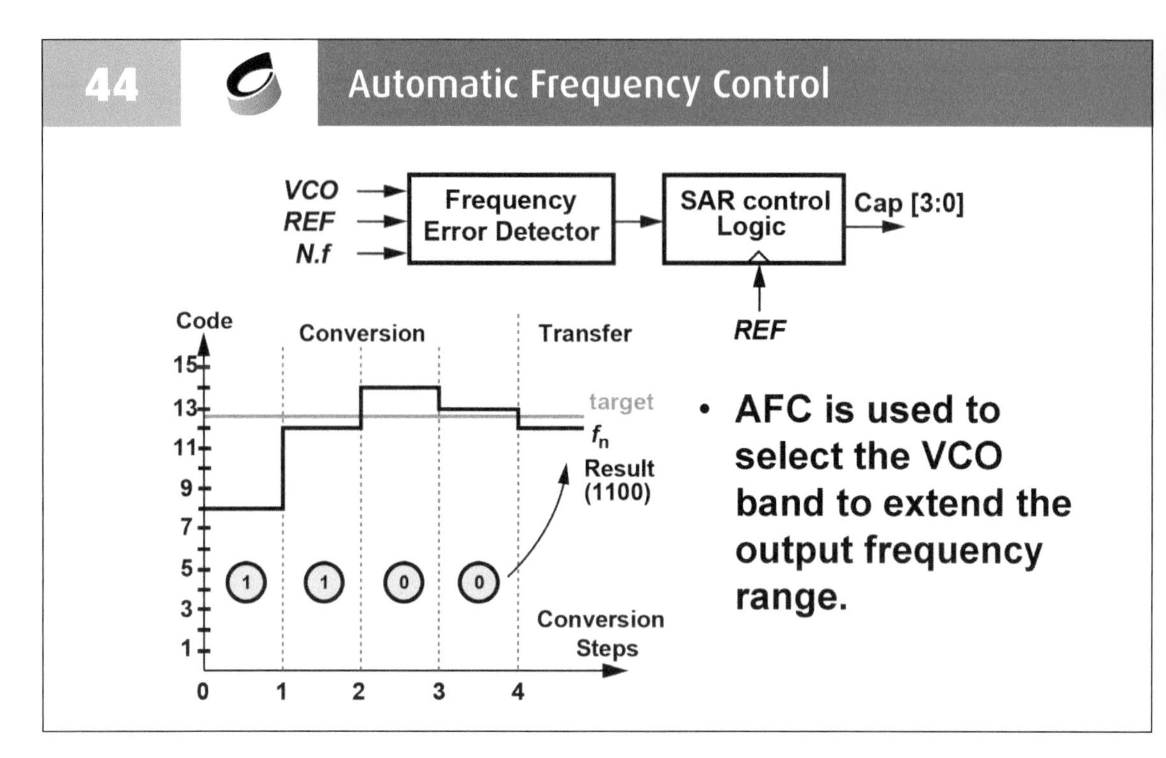

AFC is used to select the VCO band in the initial state. FED is composed of some high-speed flip-flops and DSP circuits. It is used to compare the frequency error between VCO and the target. Then, a SAR control logic is used to choose the frequency band, which is the closest to the target frequency.

45 Outline

- Background (ADI 9884 + X. Gao ISSCC slides)
- **Fractional-N divider-less PLL**
  - Implementation
  - A model of the nonlinearity noise
  - Measurement Results
- ADC-based digital PLL
  - Motivation
  - Implementation
  - Measurement Results
- Conclusions

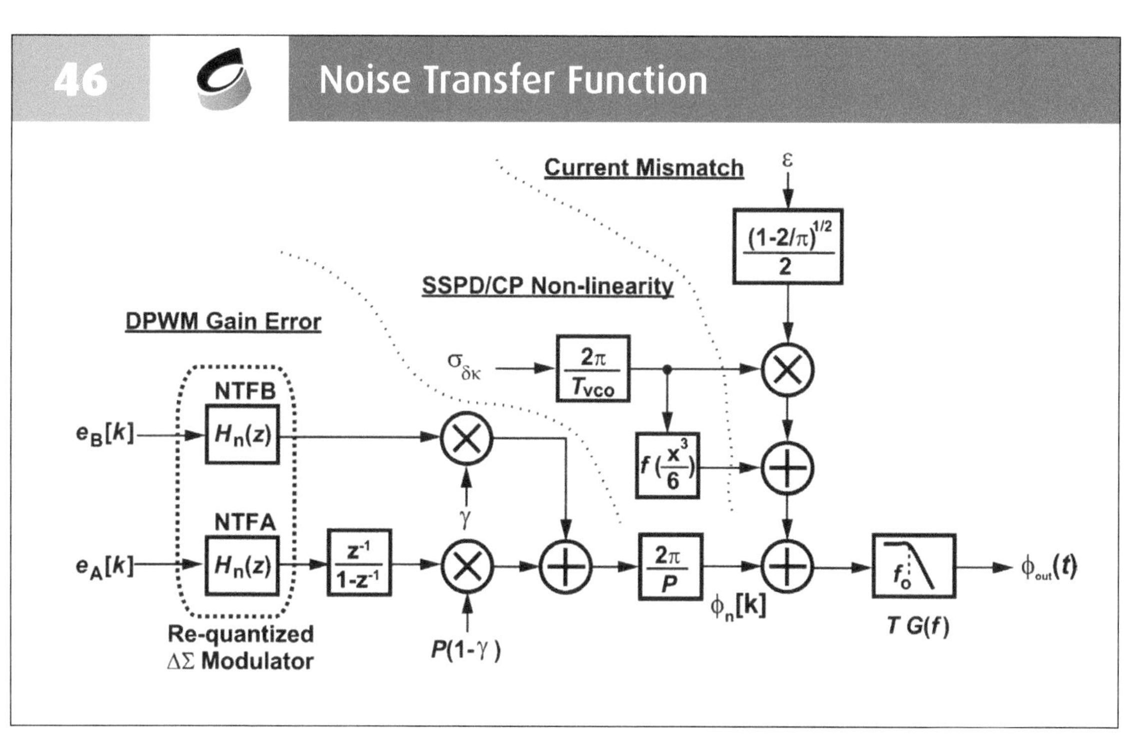

In the following slides, we would study the non-ideal effect on the proposed SSPLL. Three different effects are modeled in this slide, including DPWM gain error, SSPD/CP non-linearity and current mismatch in CP. All of these nonideality folds the noise into low frequency, raising in-band phase noise level. In DPWM control, we use a re-quantized DSM, which will be shown later.

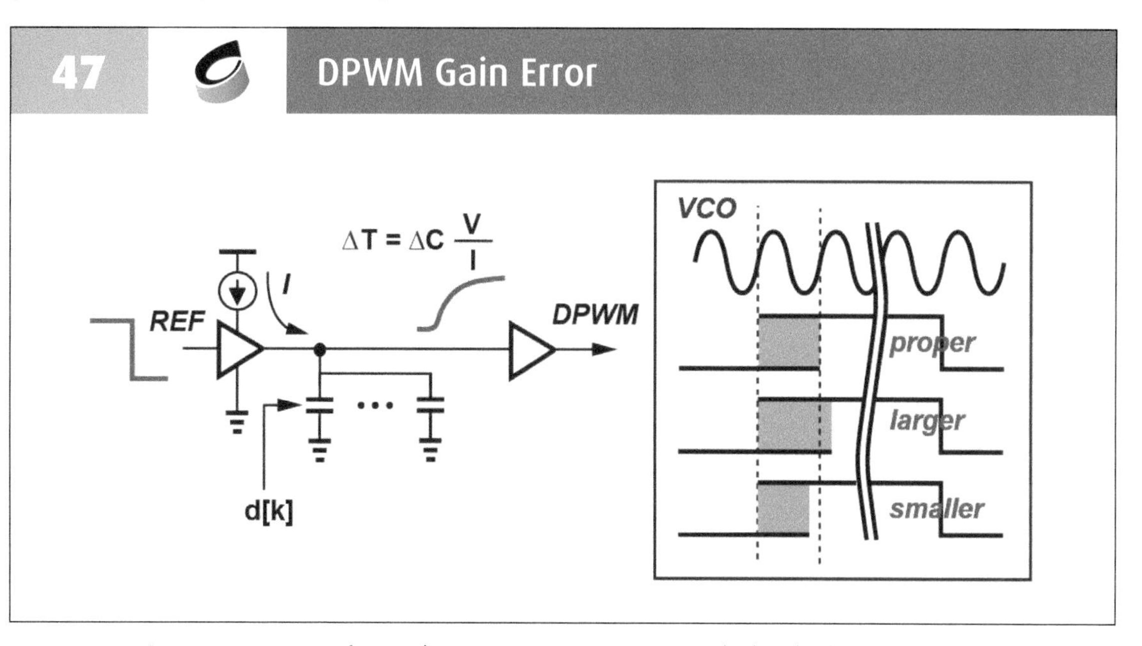

DPWM gain error comes from the process variation, such that delta T cannot be equal to the VCO period. Shown in this figure, with proper gain in DPWM, reference clock can always sample at the common-mode level of VCO signal. However, with gain mismatch, it always induces some phase error at SSPD.

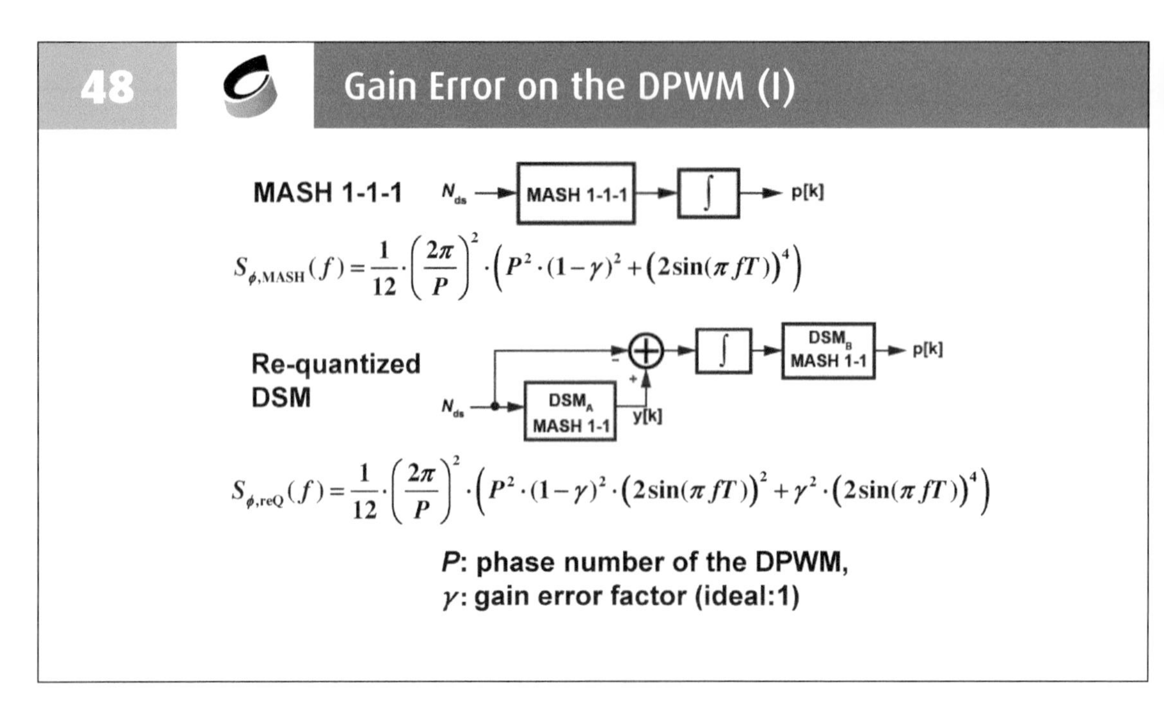

This slide shows the noise shaping functions by using the conventional MASH 1-1-1 and the re-quantized DSM. P is the phase number of the DPWM and gamma is the gain error factor. For a conventional MASH 1-1-1, gain error is not shaped. But, for re-quantized DSM, gain error can be shaped by the sinc filter function.

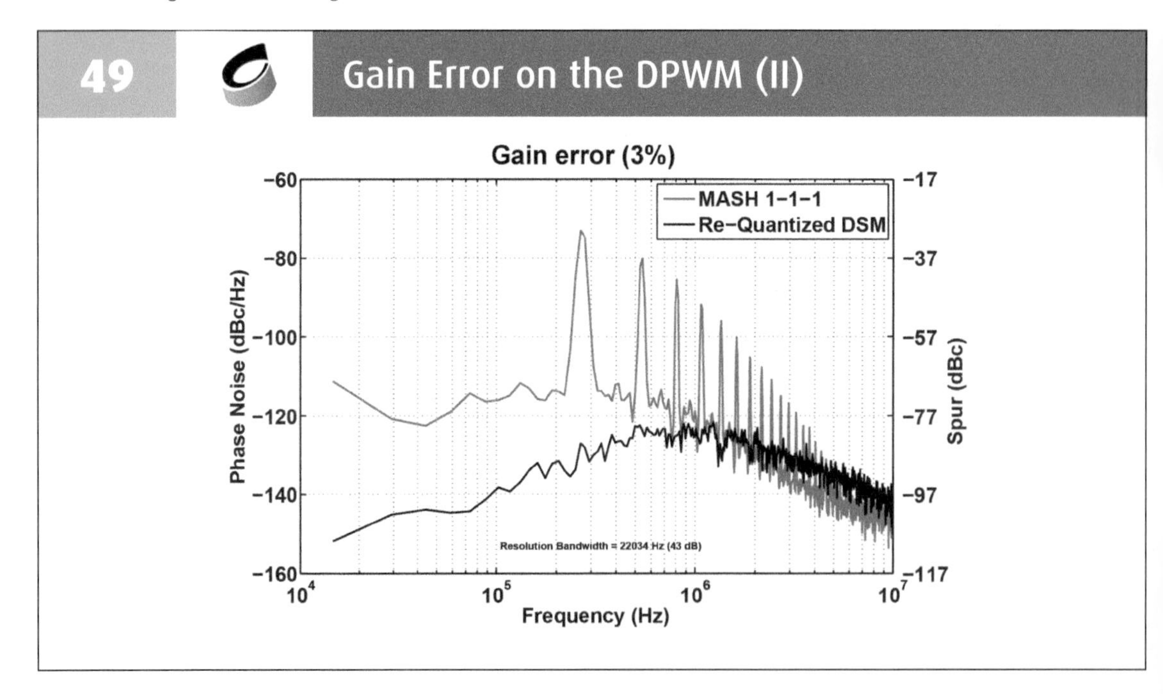

This slide shows the simulated phase noise with 3% gain error in DPWM. Conventional MASH 1-1-1 SDM cannot shaper the induced noise in the low frequency offset. But, with the re-quantized technique, the in-band phase noise can be high-passed filtered.

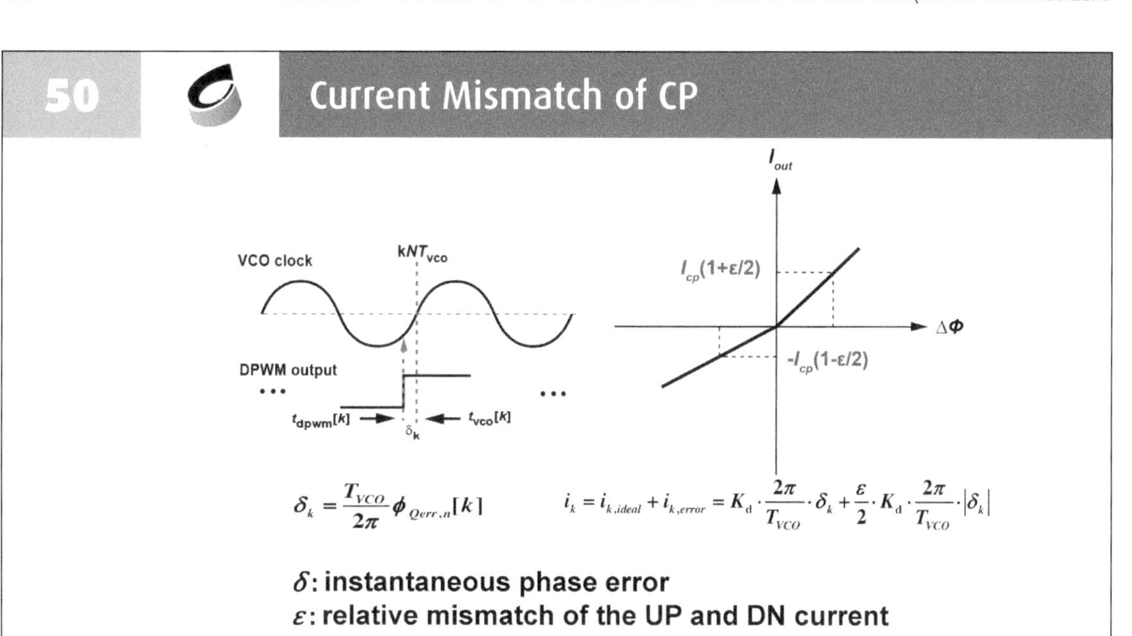

The current mismatch in CP can be illustrated in this slide. And, the equivalent CP current can be decomposed into the ideal one and the mismatch current. Delta is the instantaneous phase error and e is the relative mismatch of UP/DN currents.

51  Current Mismatch of CP

$$S_{i,mismatch}(f) = \left(\frac{\varepsilon}{2}\right)^2 \cdot K_d \cdot \left(\frac{2\pi}{T_{VCO}}\right)^2 \cdot \sigma^2_{|\delta_k|}, \quad \sigma^2_{|\delta_k|} = \sigma^2_{\delta_k} \cdot \left(1 - \frac{2}{\pi}\right)$$

$$\frac{\sigma^2_{\delta_k}}{T^2_{VCO}} = \frac{1}{12} \cdot \left((1-\gamma)^2 \cdot \int_{-1/2}^{1/2} \left(2\sin(\pi x)\right)^2 dx + \left(\frac{\gamma}{P}\right)^2 \cdot \int_{-1/2}^{1/2} \left(2\sin(\pi x)\right)^4 dx \right)$$

DSM orders	1	2	3
$\frac{\sigma^2_{\delta_k}}{T^2_{VCO}}$ (*when* $\gamma = 0$)	$0.17/P^2$	$0.52/P^2$	$1.66/P^2$

- **The folding noise increases with the order of the modulator and gain error of the DPWM delay**

With some derivation, we can obtain the equations shown here. And, the choice of DSM can be shown in this table. With higher DSM order, the quantized noise can be greatly suppressed. But, the folding noise also increase as well. Therefore, we chose 2nd order DSM to make a compromise between noise shaping and noise folding.

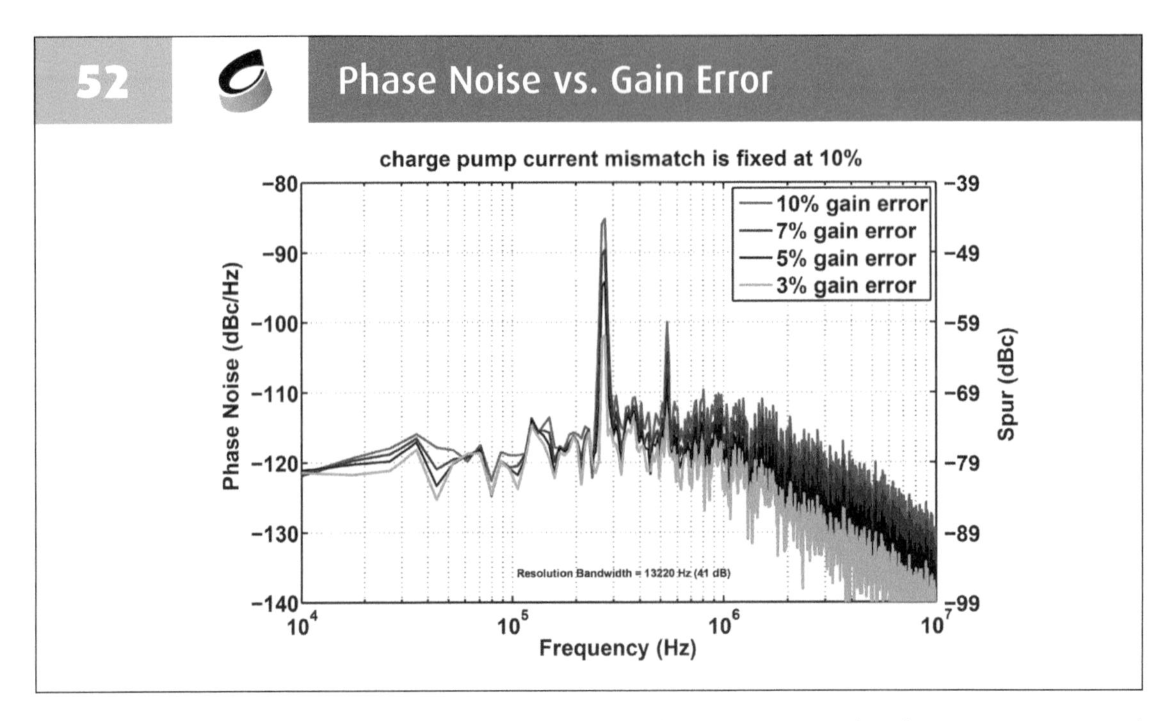

This slide also show the effect of nonideal CP in SSPLL. With 10% gain error, the phase noise is increased by 10 dB.

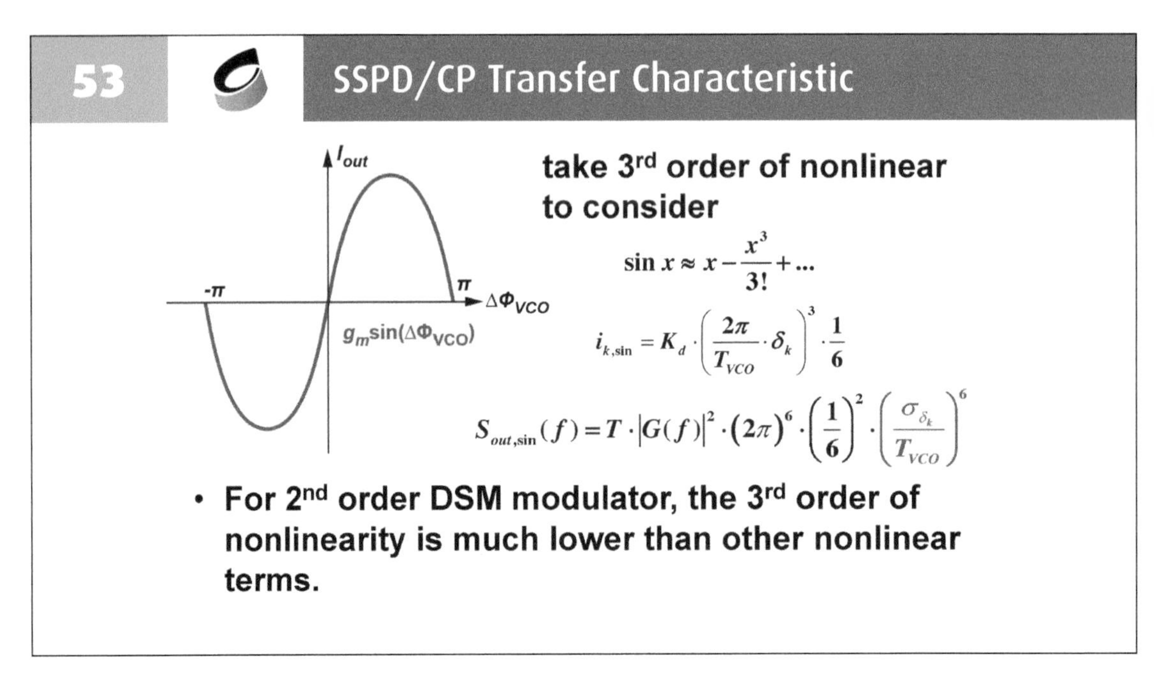

Finally, we come to the non-ideal effect of SSPD/CP transfer characteristics. As we mentioned earlier, the average current vs. phase error is a sinusoidal shape. This indicates that the non-linear behavior around lock. To quantify this effect, Taylor expansion is applied to derive the effect. And, we arrive at the noise shaping equation as shown here. Compared with a 2nd order DSM, 3rd order nonlinearity induce much lower noise than the other two terms.

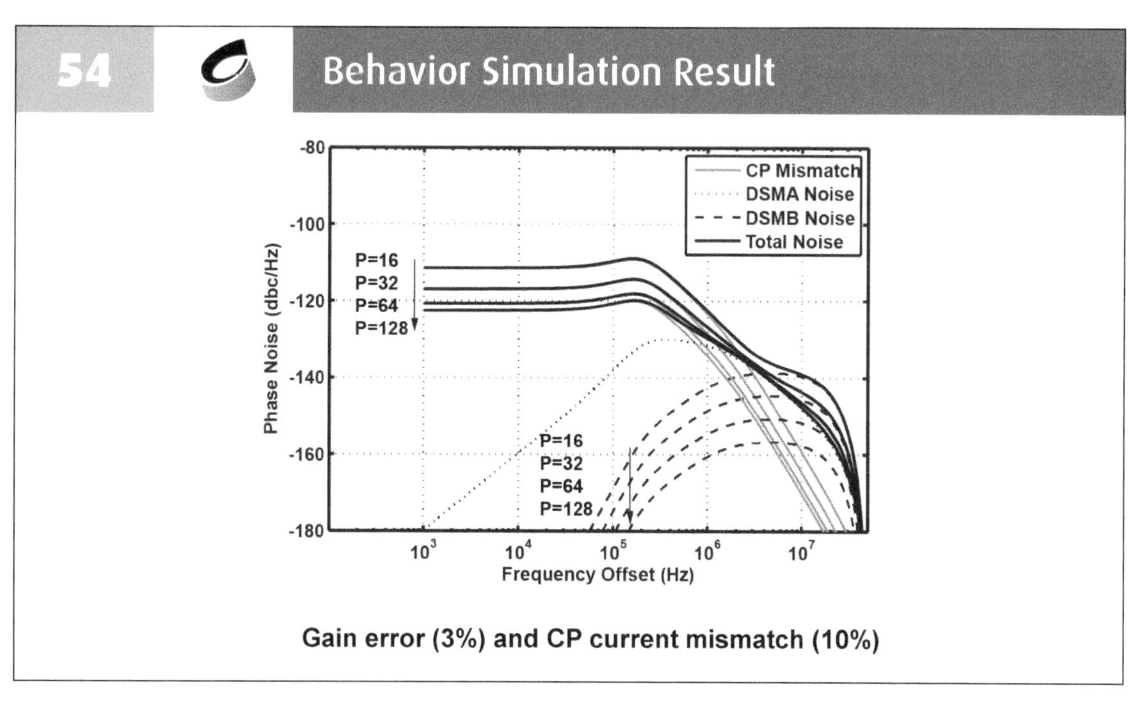

With the consideration of 3% gain error and 10% CP mismatch, the phase noise of SSPLL is simulated with respect to different phase number. To achieve better than -120 dB/Hz in-band phase noise, phase number is chosen as 64.

55 Outline

- Background (ADI 9884 + X. Gao ISSCC slides)
- **Fractional-N divider-less PLL**
  - Implementation
  - A model of the nonlinearity noise
  - Measurement Results
- ADC-based digital PLL
  - Motivation
  - Implementation
  - Measurement Results
- Conclusions

56 Die Photograph

1.5mm

0.9 mm

Loop Filter, FLL, AFC, SSCP, VCO, Digital, DPWM

- **TSMC 0.18 µm CMOS technology**
- **Core area: 1.2 x 0.6 mm²**
- **Supply voltage: 1.8 V**
- **Power: 9.6 mA**

The divider-less fractional-N PLL was fabricated in a TSMC 0.18 µm technology. The active area is 1.2 by 0.6 mm2. We also show layout floor plan in this slide. The power consumption is 9.6mA under 1.8V supply.

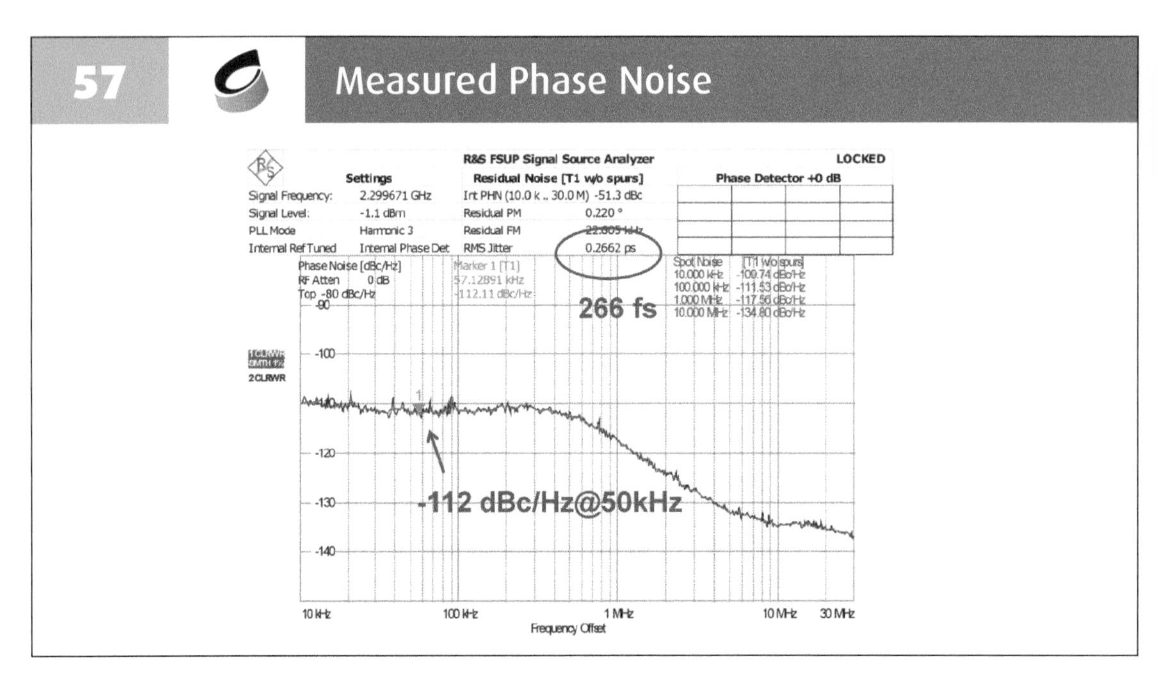

This slide shows the plot of the measured phase noise of the 2.3GHz output. The in-band phase noise is -112dBc/Hz at a 50 kHz offset frequency. The best-case rms jitter is 266fs.

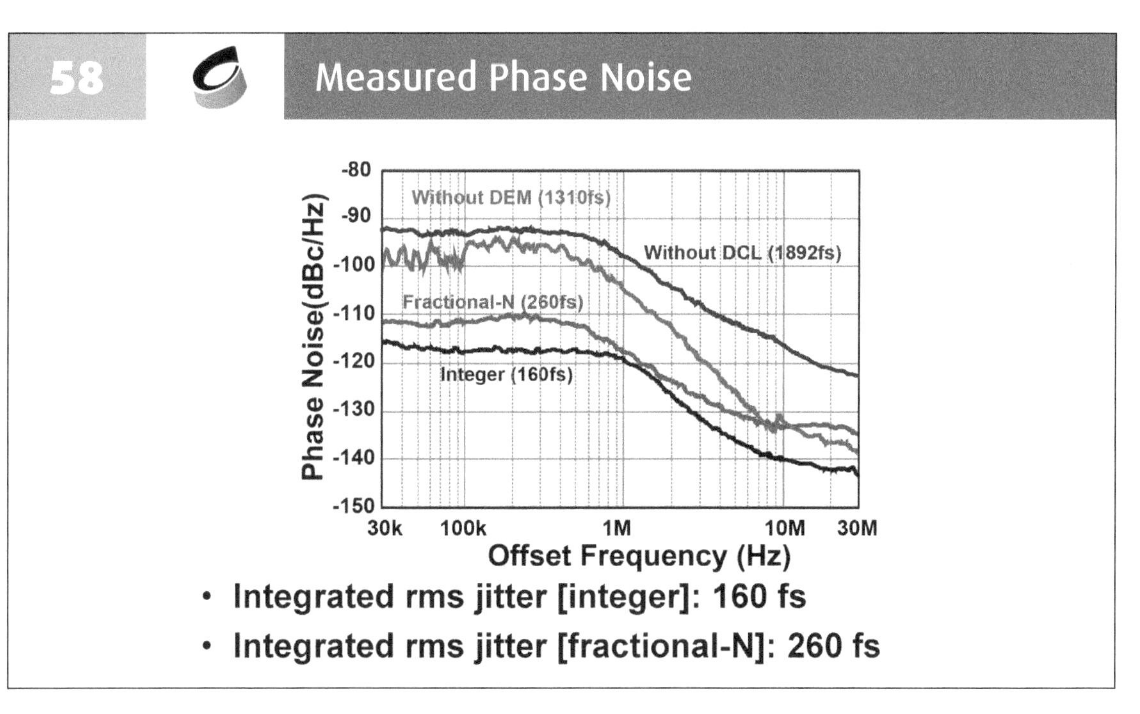

Measurement results show that DCL and DEM both are very important for better phase noise performance. By enabling DEM and DCL, jitter can be as good as 260 fs. Besides, jitter in integer-N mode was also measured at 160 fs.

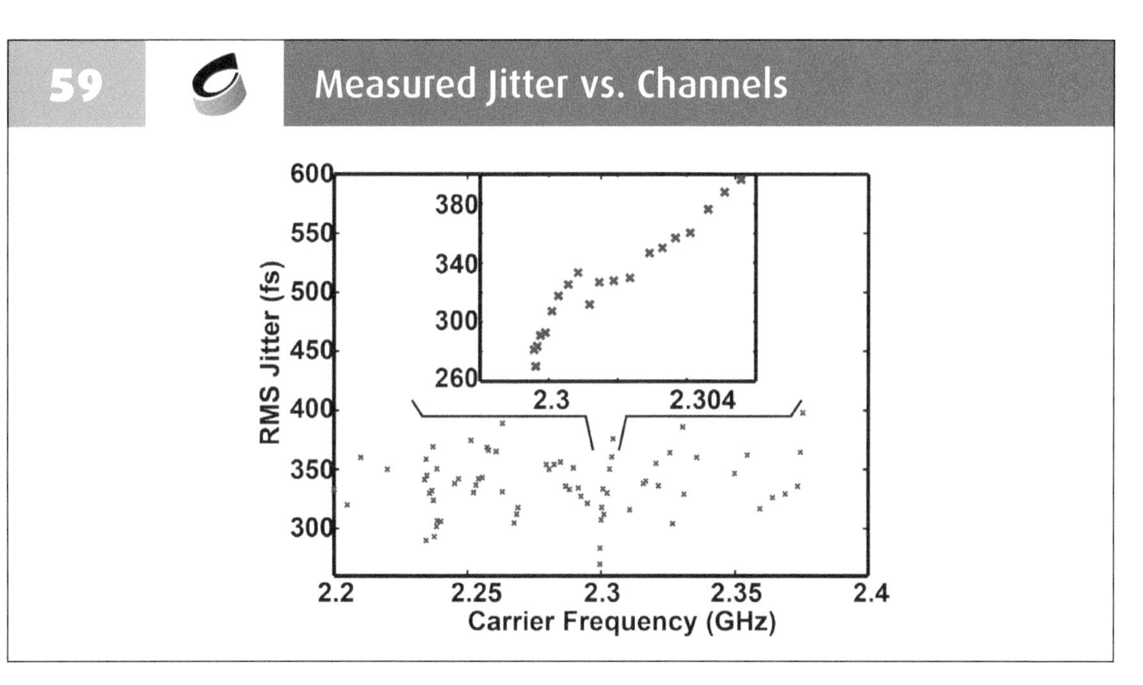

This slide shows the measured jitter over different fractional-N PLL channels. The rms jitter is between 260fs and 400fs for different carrier frequencies.

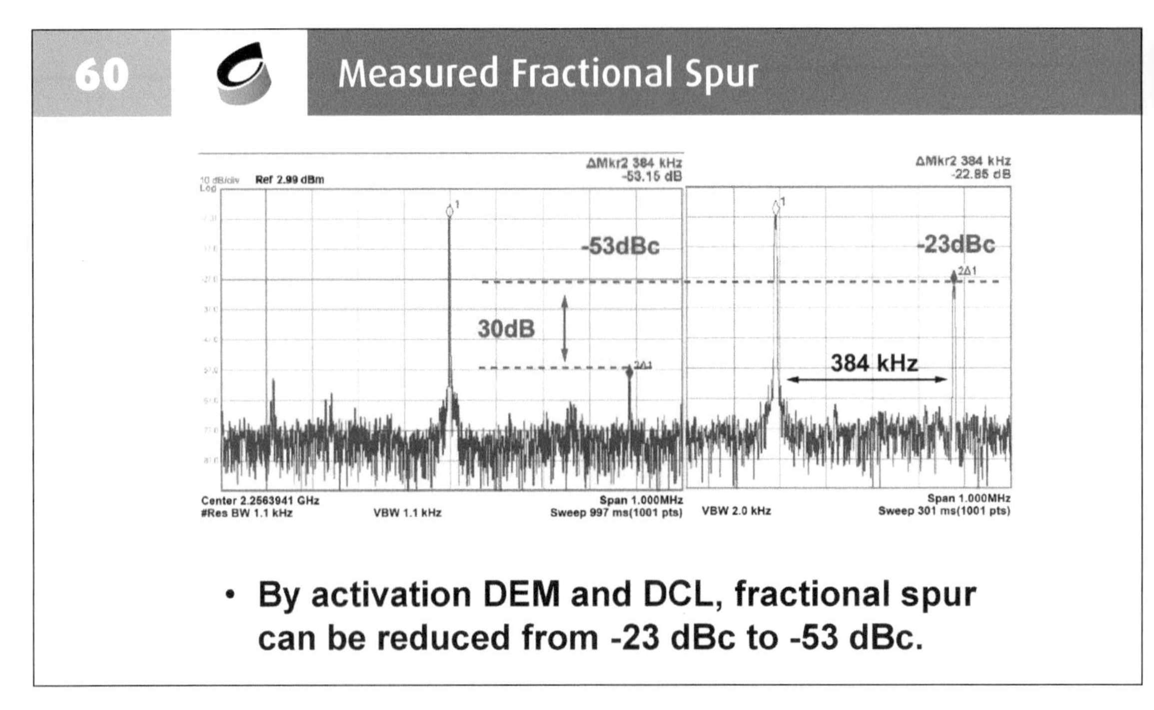

By activation DEM and DCL, fractional spur can be reduced from -23 dBc to -53 dBc at 400 kHz offset.

61 Performance Summary

	ISSCC 08	ISSCC 11	ISSCC 12	This Work
Architecture	Digital	Digital	Analog	Analog
Output (GHz)	3.62~3.67	2.9~4	0.57~0.6	2.12~2.4
In-band Noise (dBc/Hz)	-108 @400kHz	-102 @50kHz	N/A	-112 @50kHz
Integrated Jitter (ps)	0.20~0.30 (1k~40M)	0.40~0.56 (3k~30M)	2.45~4.23 (100~40M)	0.26~0.40 (10k~30M)
Ref. Spur (dBc)	-65	-72	N/A	-55
Frac. Spur (dBc)	-42	-42	N/A	-53
Power (mW)	46.7	4.5	10.5	17.3
Process (nm)	130	65	65	180

To summarize the performance, we compare our circuit to the other's work. Our proposed work is to design a low noise PLL. We achieve the in-band phase noise is -112dBc/Hz at a 50kHz offset frequency. It is one of the best in-band noise performances, in particular for 0.18 μm CMOS technology.

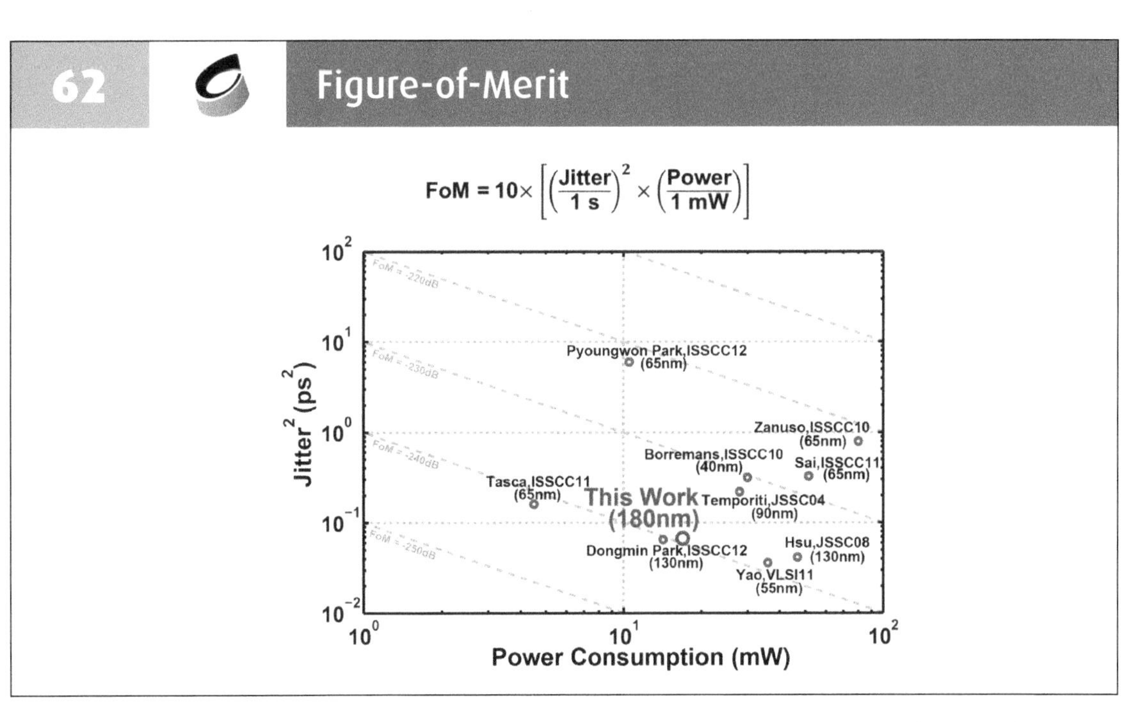

This slide shows FOM. FOM is defined by power and jitter. Better jitter performance is on the bottom of the figure. This work achieves one of the best jitter performances. The FOM of the proposed fractional-N PLL can be as good as -239dB. If finer CMOS technology can be used, we believe that the performance can be greatly improved.

63 Outline

- Background (ADI 9884 + X. Gao ISSCC slides)
- Fractional-N divider-less PLL
  - Implementation
  - A model of the nonlinearity noise
  - Measurement Results
- **ADC-based digital PLL**
  - Motivation
  - Implementation
  - Measurement Results
- Conclusions

64  Analog vs. Digital PLL

	Analog	Digital
Phase conversion	PFD+CP	TDC
	☹ noise ☹ dead zone ☹ up-dn current mismatch and width skew ☹ charge injection ☹ clock feedthrough	☹ resolution ☹ mismatch
Loop filter	Passive components	Logic cells
	☹ leakage current ☹ area ☹ programmable difficulty	☹ power

The major difference between analog and digital PLLs is in Phase comparison and loop filter. Conventional PLLs employ PFD and CP, inducing the problems in noise, dead zone. Also, UP/DN current sources in CP circuit usually have mismatch problem. The digital control toward UP/DN cannot be aligned. Furthermore, charge injection would inject unwanted charge packet into loop filter. Clock feedthrough also creates spurious tones at VCO.

The filter in analog PLL needs passive components, including resistor and capacitor. If capacitors are implemented by the MOS cap, leakage current would affect the sinusoidal jitter significantly. Area is also an issue in analog filter. If PLLs need to operate in wide range, it might be difficult to program the loop filter.

On the other hand, digital PLLs employ a time-to-digital converter to digitize the phase difference. Typically, the resolution of TDC can not be easily to reach pico-second range. Besides, the mismatch in delay cells of TDC would create a potential linearity problem in phase detection. For loop filter design, digital PLLs only need digital cells to fulfil the function.

Finally, power consumption might be high if digital PLLs are not implemented in nanometer technologies.

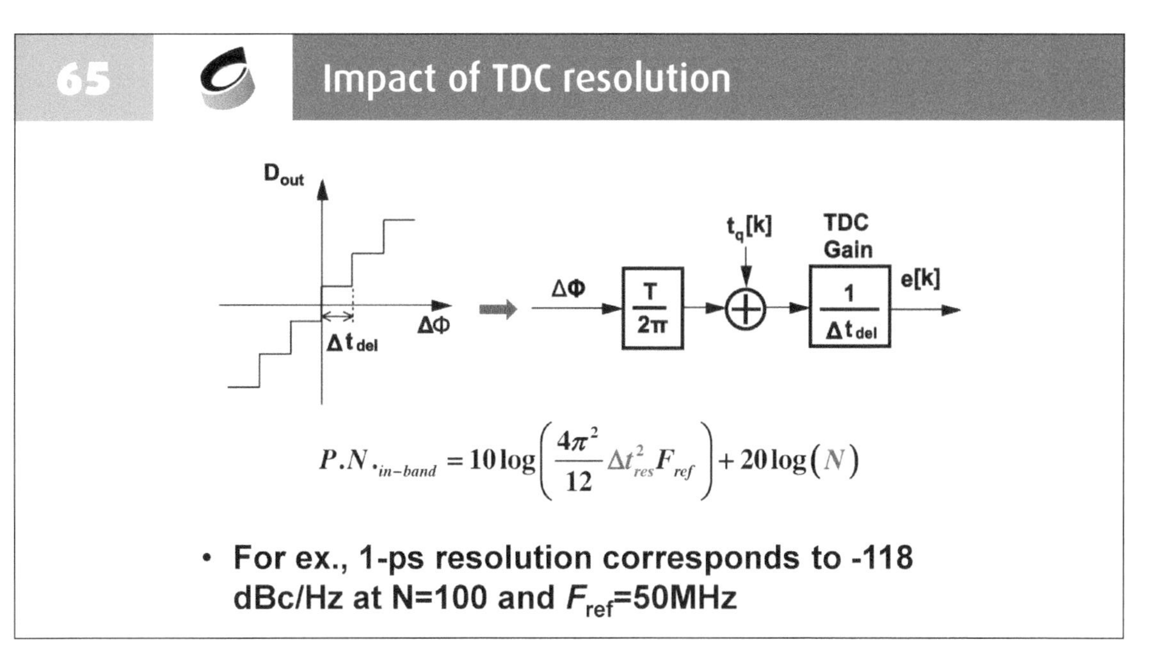

Let's review the impact of TDC resolution. The finite resolution of TDC can be modeled as the quantization noise added into the output of phase difference, just like ADC. Such effect can translate into the phase noise at the output of VCO. Note that for larger multiplication factor, the inband noise would be leveled up by 20 log N. For instance, for 1-ps TDC resolution, it would imply -118 dBc/Hz for N=100 and fref=50 MHz. In practice, 1-ps TDC is very difficult to achieve.

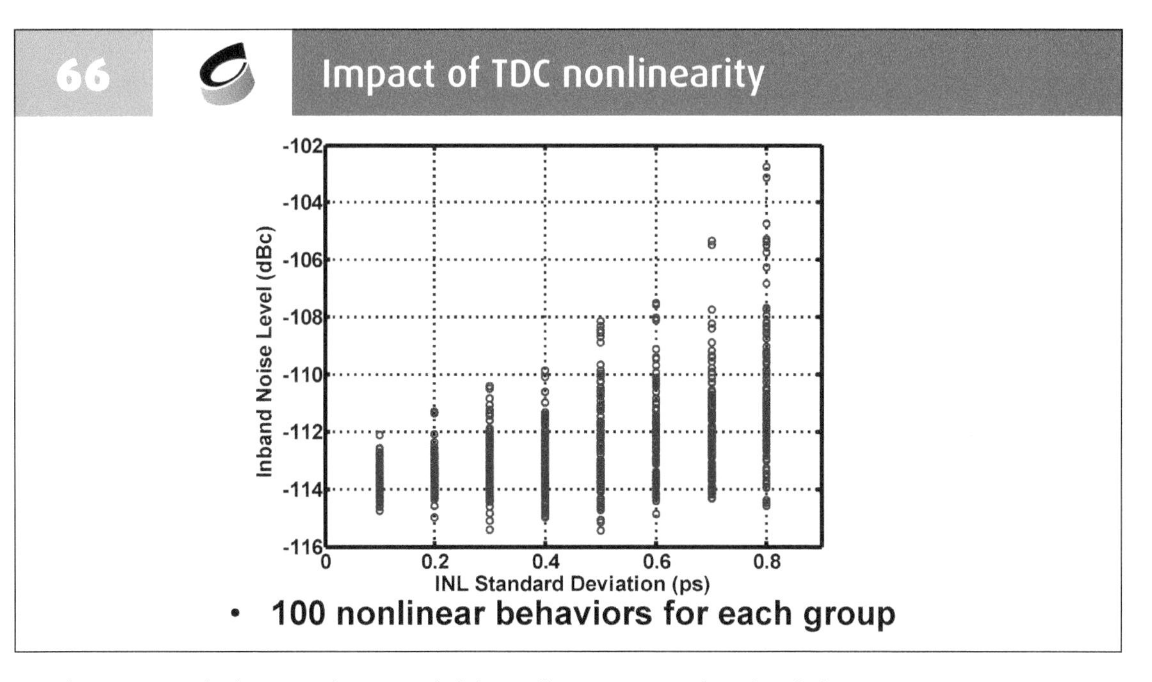

Furthermore, with the consideration of delay cell mismatch in TDC, it would further increase the inband phase noise level by more than 10 dB. For INL 0.8ps, the inband phase noise is as worse as -103 dBc/Hz.

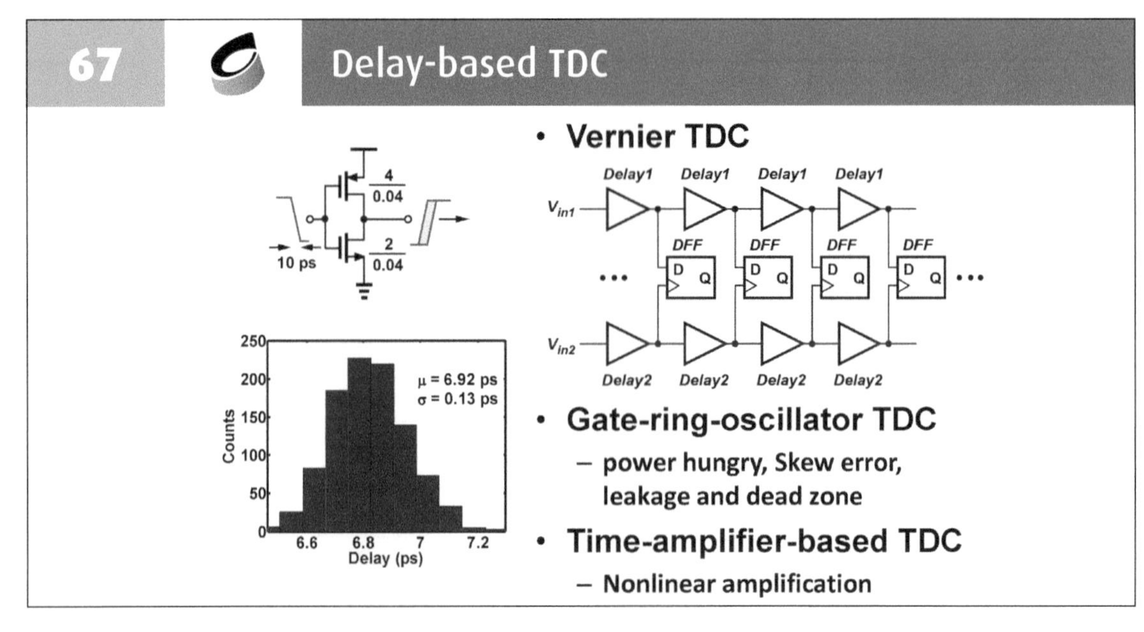

This slide shows you a typical high-resolution TDC, which is so called Vernier TDC. It composes of two delay chains and a chain of D-FF. By designing different delay in Delay 1 and Delay 2, finer resolution TDC can be implemented. Let's use a typical inverter circuit as the delay cells. The mean of the delay is around 7 ps. But, its standard deviation is as large as 0.13 ps.

Other than Vernier TDC, gate-ring-oscillator TDC were proposed but with the problem of power hungry, skew error, leakage and dead zone.

Some paper proposed a time-amplifier-based TDC. But, its nonlinear amplification would cause linearity problem.

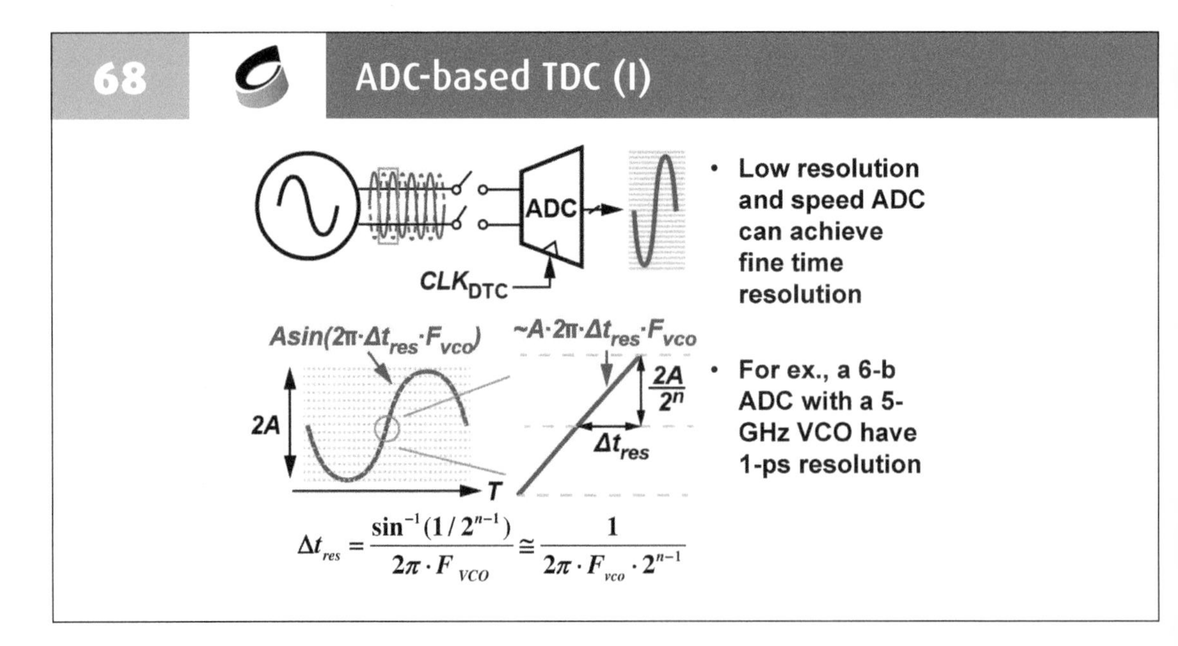

$$\Delta t_{res} = \frac{\sin^{-1}(1/2^{n-1})}{2\pi \cdot F_{VCO}} \cong \frac{1}{2\pi \cdot F_{vco} \cdot 2^{n-1}}$$

Now, we would like to propose ADC-based TDC. From analog SSPLL, the output of the VCO can be sampled as the equivalent indication of phase difference. By looking at this figure, for a sinusoidal signal, delta t resolution is equal to 1/2pi*Fvco*2^n-1). For instance, a 6-b ADC with a 5-GHz VCO, it can easily achieve higher and better resolution at 1ps.

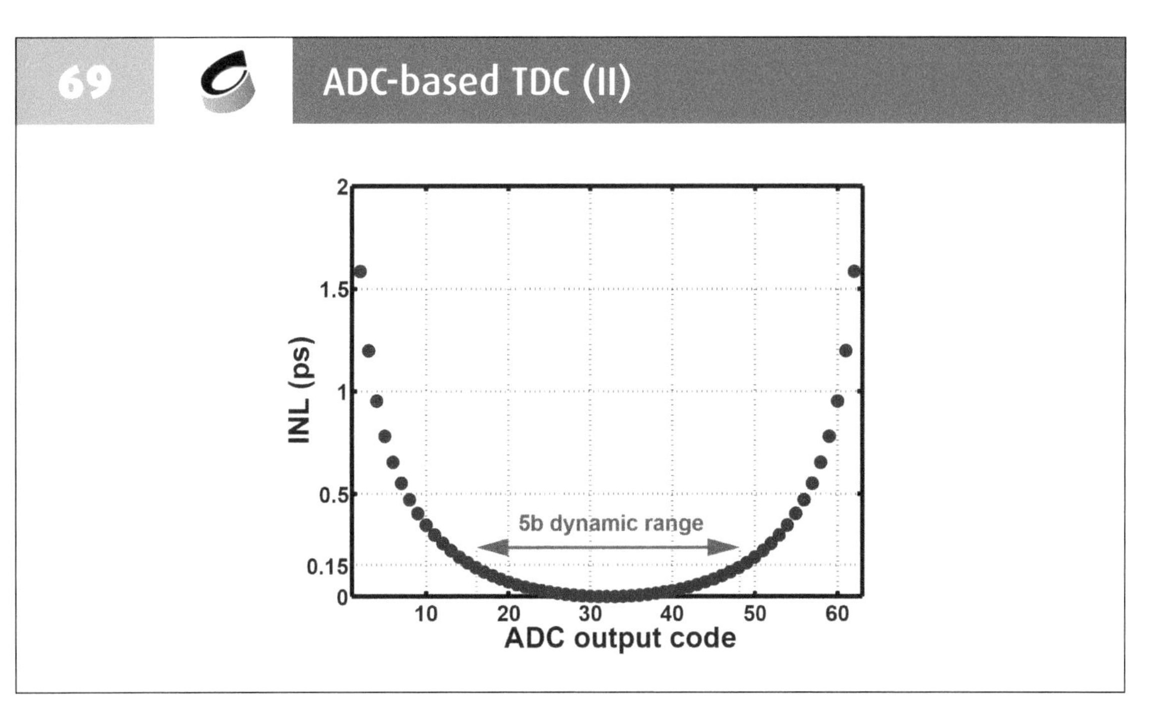

Also, the output waveform of VCO is sinusoidal. For fractional-N operation, we must ensure it can only achieve -120 dBc/Hz. Luckily, around the locked point, with 5b range, the INL is less than 0.2LSB.

70 Outline

- Background (ADI 9884 + X. Gao ISSCC slides)
- Fractional-N divider-less PLL
  - Implementation
  - A model of the nonlinearity noise
  - Measurement Results
- **ADC-based digital PLL**
  - Motivation
  - Implementation
  - Measurement Results
- Conclusions

71 Proposed Architecture

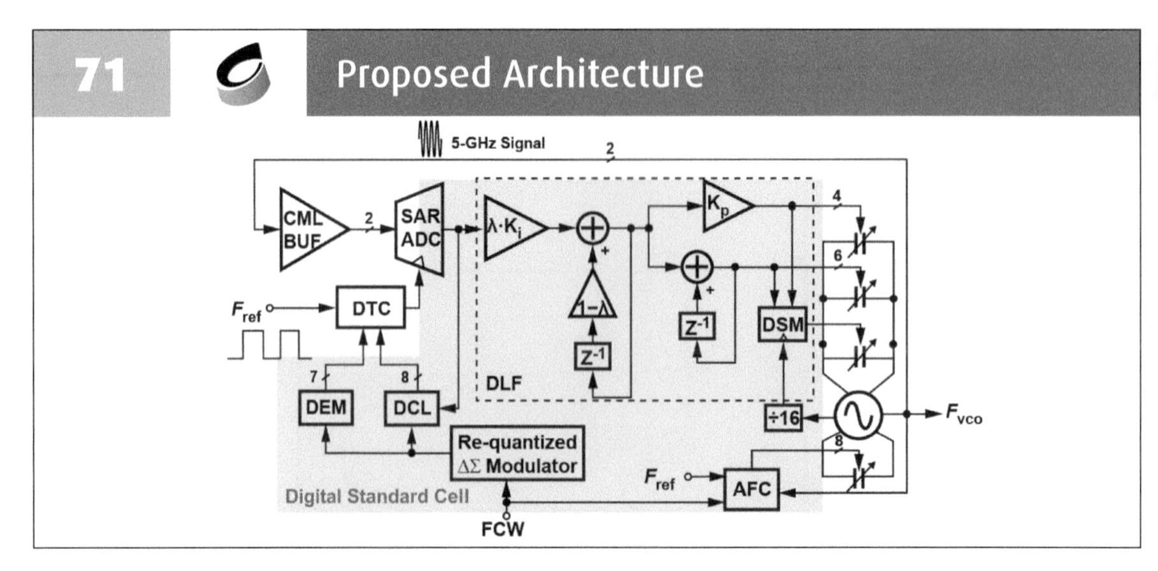

The proposed subsampling PLL contain a SAR ADC to sample the output of VCO by a reference clock. Then, a second order digital loop filter is used to tune the digital DCO. The control of DCO is separated into two paths, including a feedforward-path gain and an integral path. To increase the equivalent resolution of DCO, a first-order DSM clocked by Fvco/16 is employed to enhance its resolution. Furthermore, an AFC circuit is added to extend the operation range of VCO. The digital to time converter controlled by DEM and digital correlation logic is used to perform the digital pulse modulation of Fref.

Note that, to avoid direct coupling from sampling circuit to VCO, a CML buffer is inserted.

72 SAR with Unary Capacitor

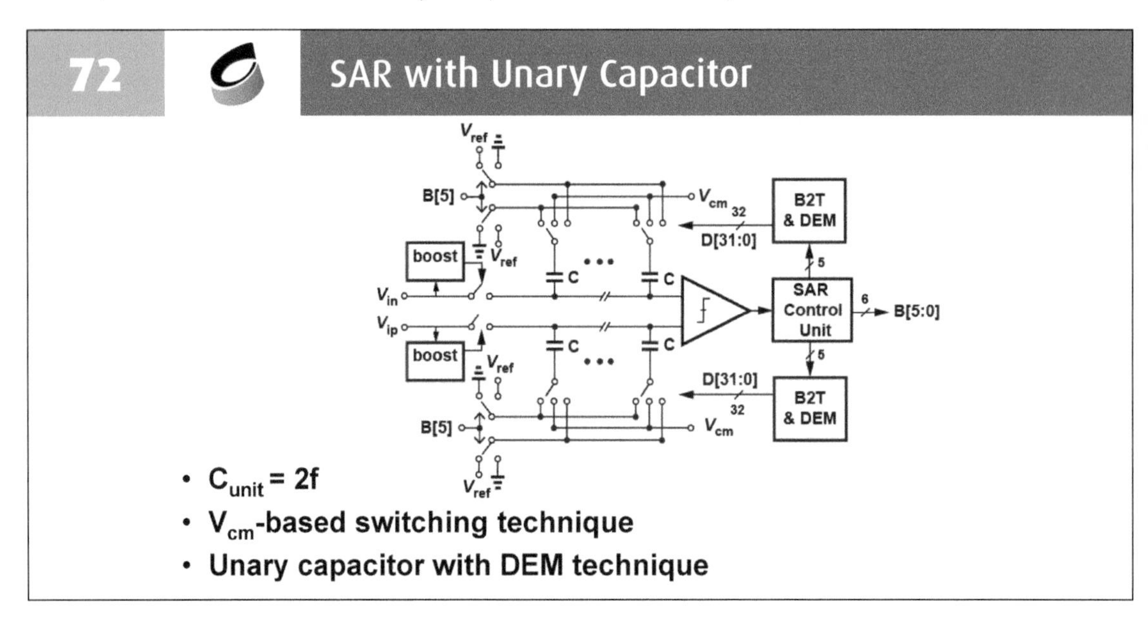

The SAR ADC is 6 bit with sampling rate about 50 MHz. Even though the reference frequency is low, the input frequency is relatively high, 5 GHz. To avoid the nonlinear sampling for such high frequency input, a boost-strapped circuit is used to sample VCO output waveform.

The linearity of SAR is quite important in fractional-N synthesis due to the noise fold-back effect. So, even though the required resolution is low, the linearity is important. Therefore, all the capacitor is implemented by a unary array to guarantee the linear conversion. Such full unary implementation is still manageable for a 6-b resolution. Unit cap in this SAR is 2fF.

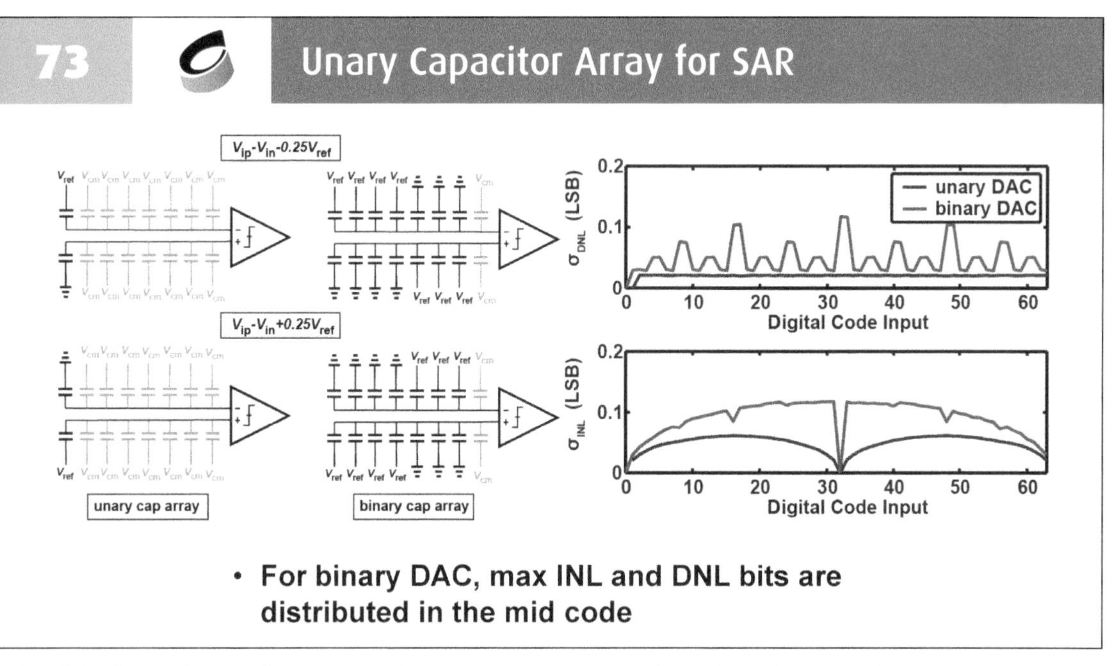

This slide shows the simulation results for unary and binary capacitor array simulation results. For binary cap array, the maximum DNL and INL are centered around mid-code, where the locking condition in the proposed ADC-based PLL. On the contrary, unary cap array exhibit great linearity around mid-code, which ensure the linear phase to control voltage conversion.

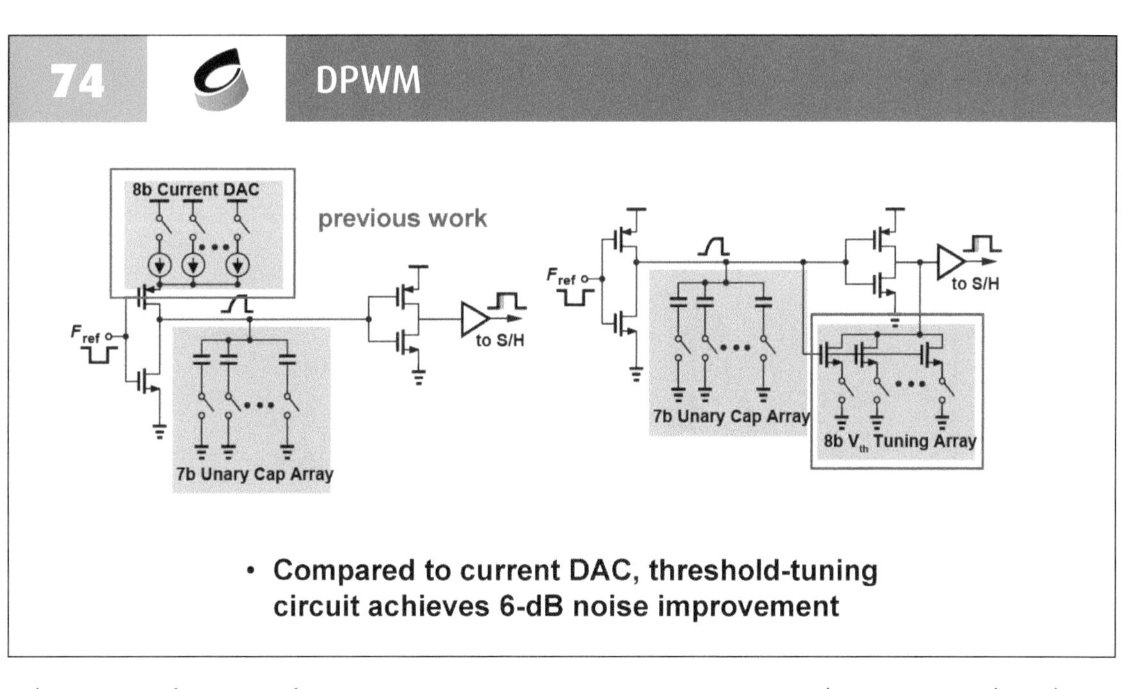

The noise performance of DPWM is important to in-band phase noise because it would be just multiplied by a factor of 20 log N. Our previous work used current source to tune its delay, which induces thermal and flicker noise. In this work, rather using current source, a switch array is tuned to change the gain of DPWM. With such design, 6-dB noise improvement can be obtained with the identical power dissipation.

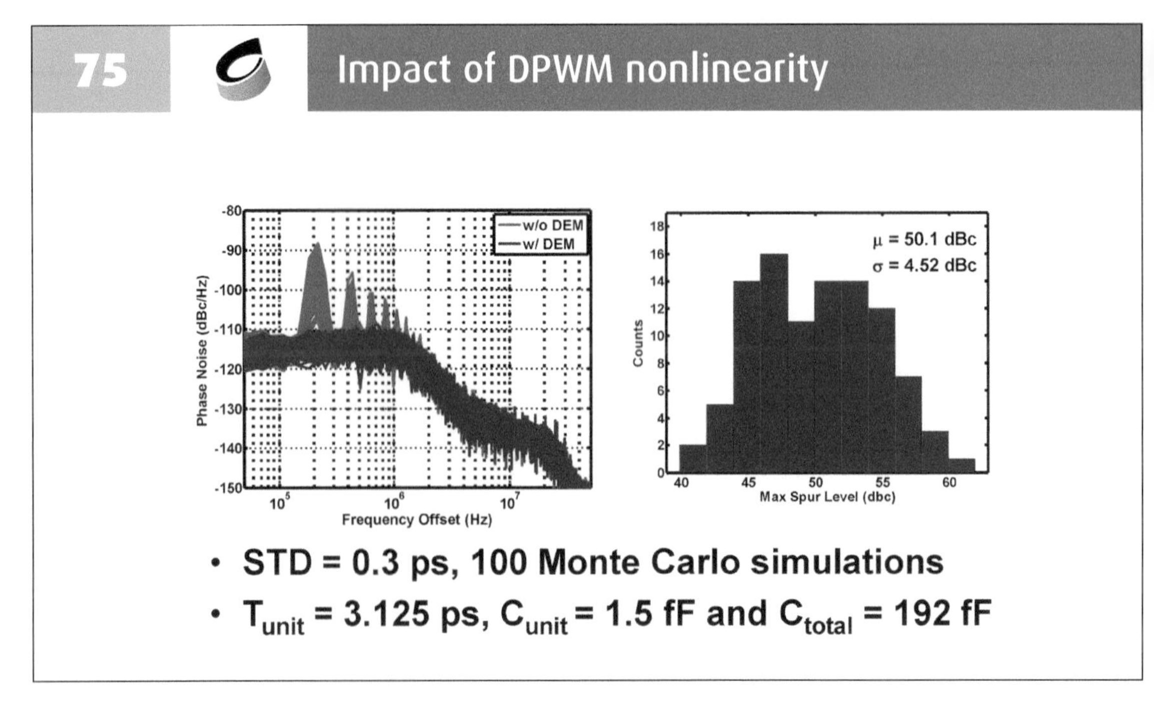

We also ran Monte Carlo simulation for the effect of DPWM. Without DEM, the nonlinearity would create in-band spurious tone. With DEM, it can be effectively suppress the in-band spurs.

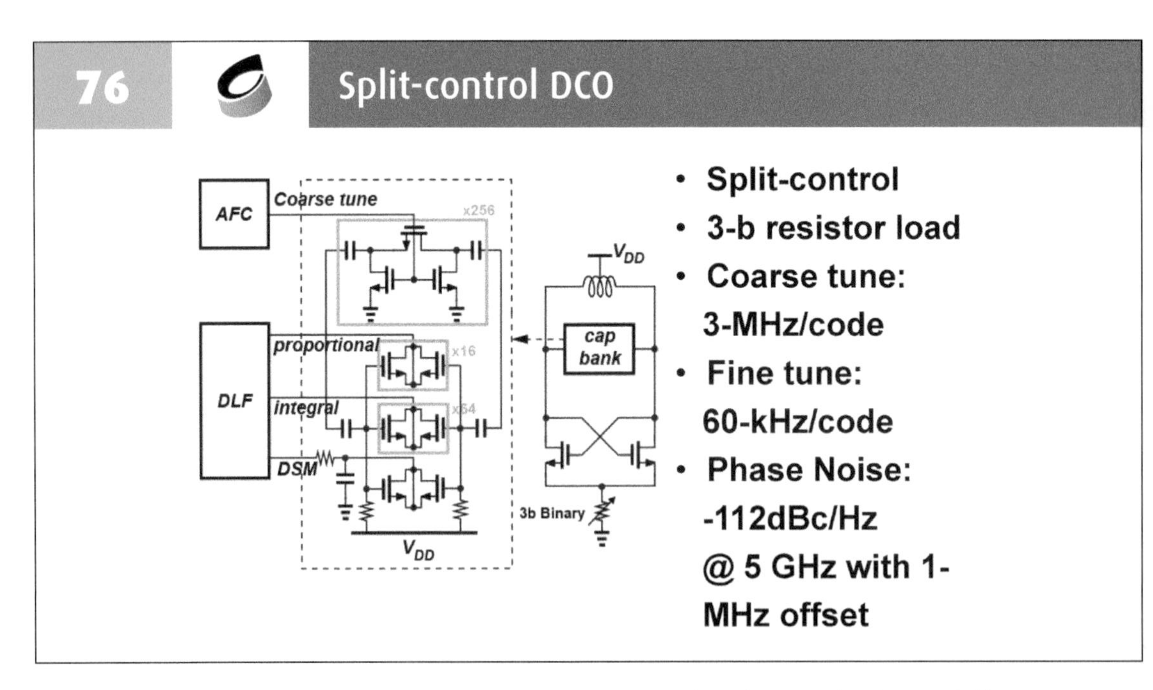

This slide shows you the detail of split-control DCO. As I mentioned earlier, the control paths toward oscillator include a wide-tuning bits by AFC, digital PLL proportional paths and integral paths. Finally, a DSM path with an analog low path filter to suppress the quantization noise of DSM to achieve -112 dBC/Hz phase noise with a 1-MHZ offset. This VCO operates at 5 GHz.

77  Outline

- Background (ADI 9884 + X. Gao ISSCC slides)
- Fractional-N divider-less PLL
  - Implementation
  - A model of the nonlinearity noise
  - Measurement Results
- **ADC-based digital PLL**
  - Motivation
  - Implementation
  - **Measurement Results**
- Conclusions

78  Die Photograph

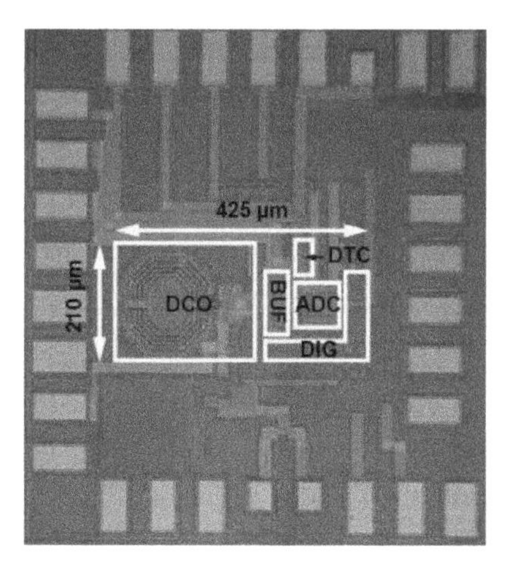

- TSMC 40 nm CMOS technology
- Core area: 0.21 x 0.43 mm^2
- Supply voltage: 0.9 V
- Power: 3.2 mA

The proposed ADC-based PLL has been implemented in a 40nm CMOS technology. The core area is only 0.21 by 0.43 and its supply voltage is 0.9V. The total power consumption is 3.2mA.

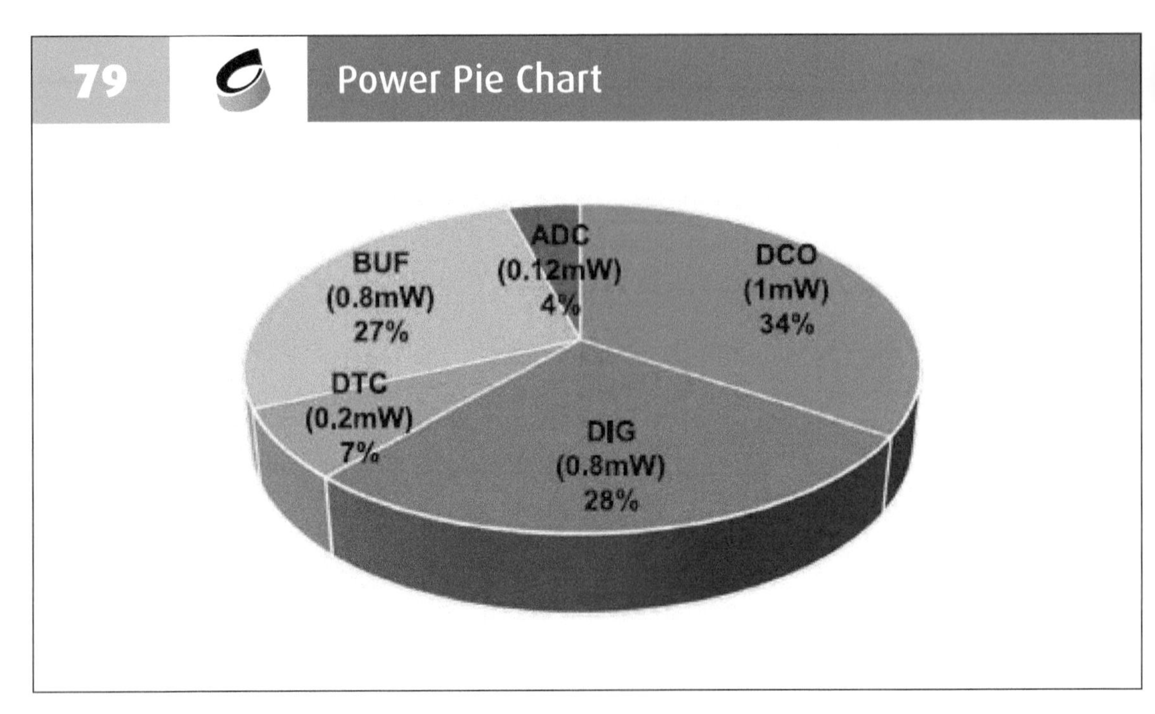

In terms of power breakdown, DCO accounts for 1/3 of power. ADC, which is used for sub-sampling PD, only consumes 0.12 mW, while its BUF need about 0.8mW (very similar to PGA in front of SAR ADCs), which is about the same as DCO. For the rest 1/3 of power, it consumes in digital circuits as well as DTC.

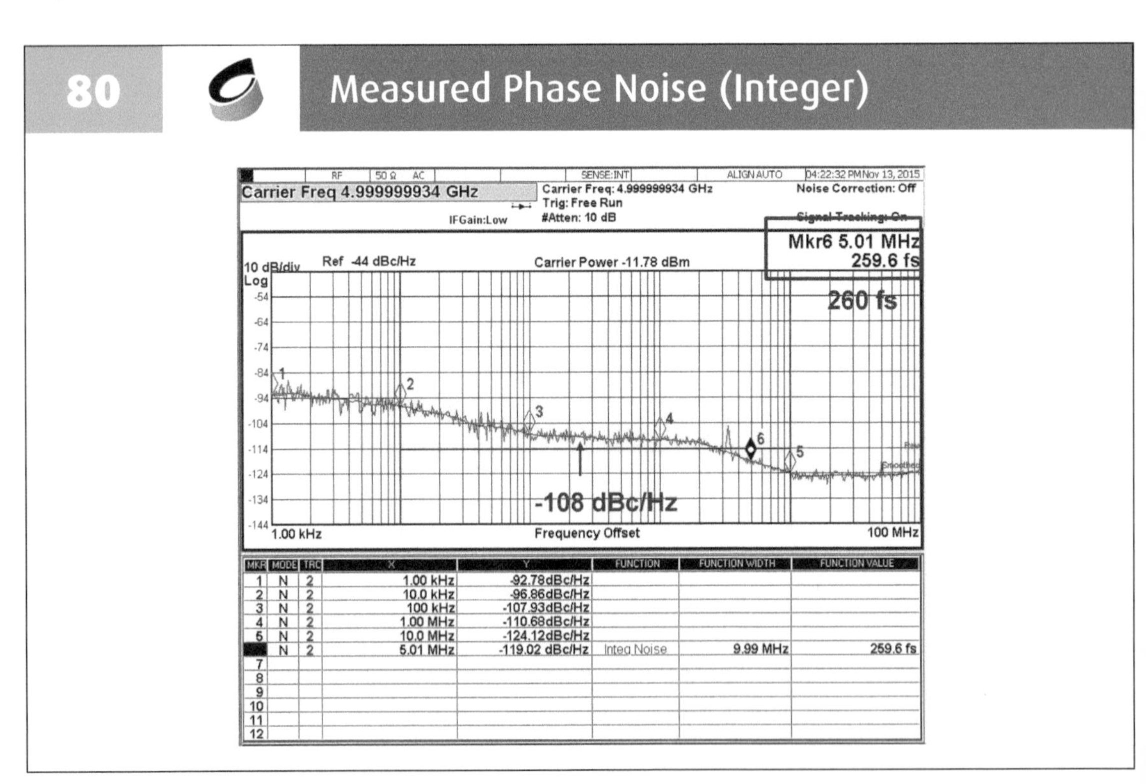

MKR	MODE	TRC	X	Y	FUNCTION	FUNCTION WIDTH	FUNCTION VALUE
1	N	2	1.00 kHz	-92.78dBc/Hz			
2	N	2	10.0 kHz	-96.86dBc/Hz			
3	N	2	100 kHz	-107.93dBc/Hz			
4	N	2	1.00 MHz	-110.68dBc/Hz			
5	N	2	10.0 MHz	-124.12dBc/Hz			
	N	2	5.01 MHz	-119.02 dBc/Hz	Integ Noise	9.99 MHz	259.6 fs
7							
8							
9							
10							
11							
12							

The measured in-band phase noise is -108 dBc/Hz. And, total integrated jitter is around 260 fs in integer-N mode.

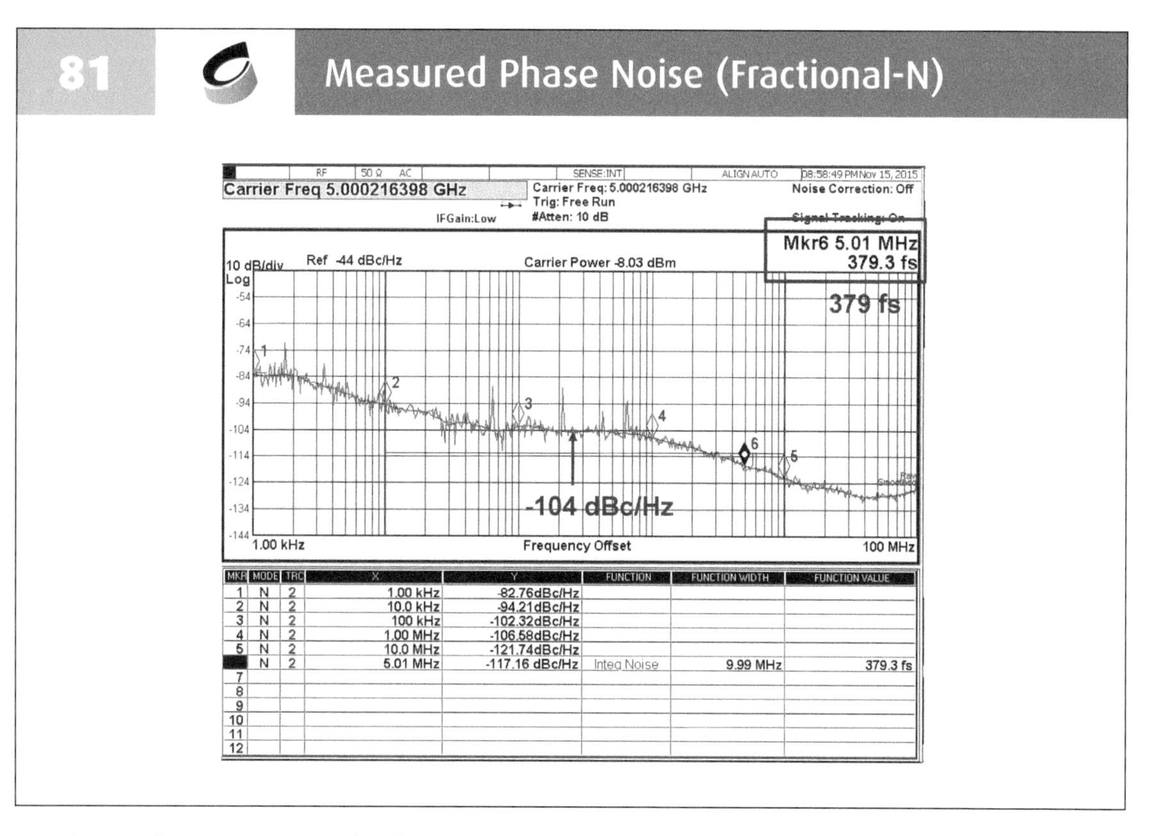

For fractional-N operation, its inband noise is raised upto -104 dBc/Hz. And, total integrated jitter is 379 fs.

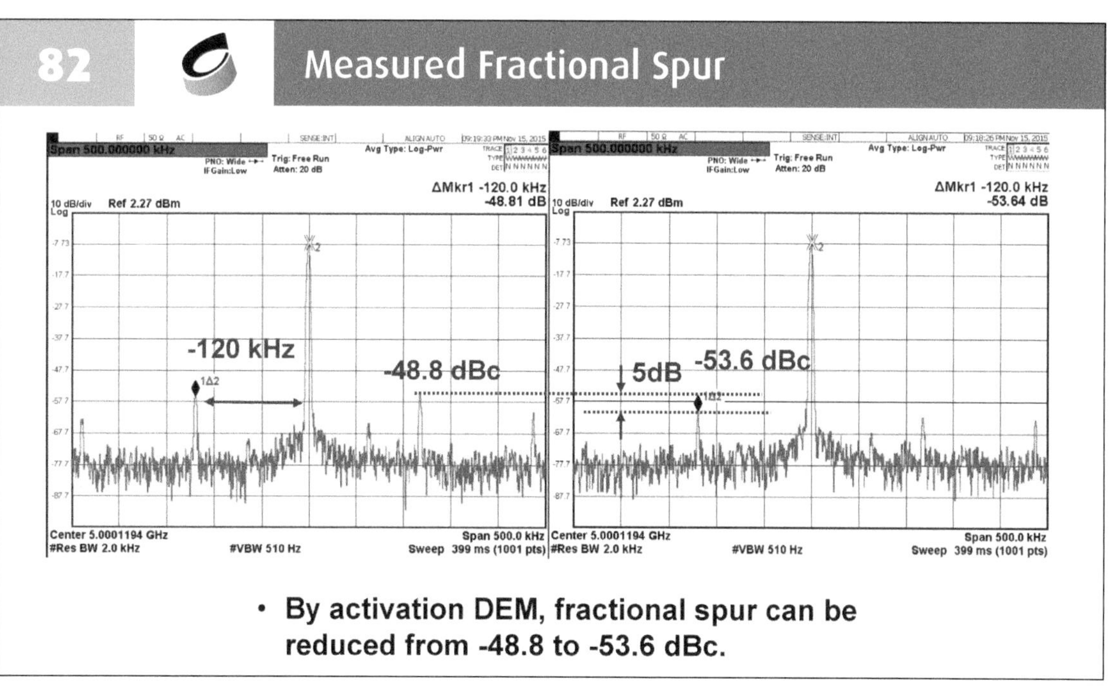

DEM also demonstrated pretty good suppression capability in fractional spurs. 5 dB spurious tone suppression can be observed.

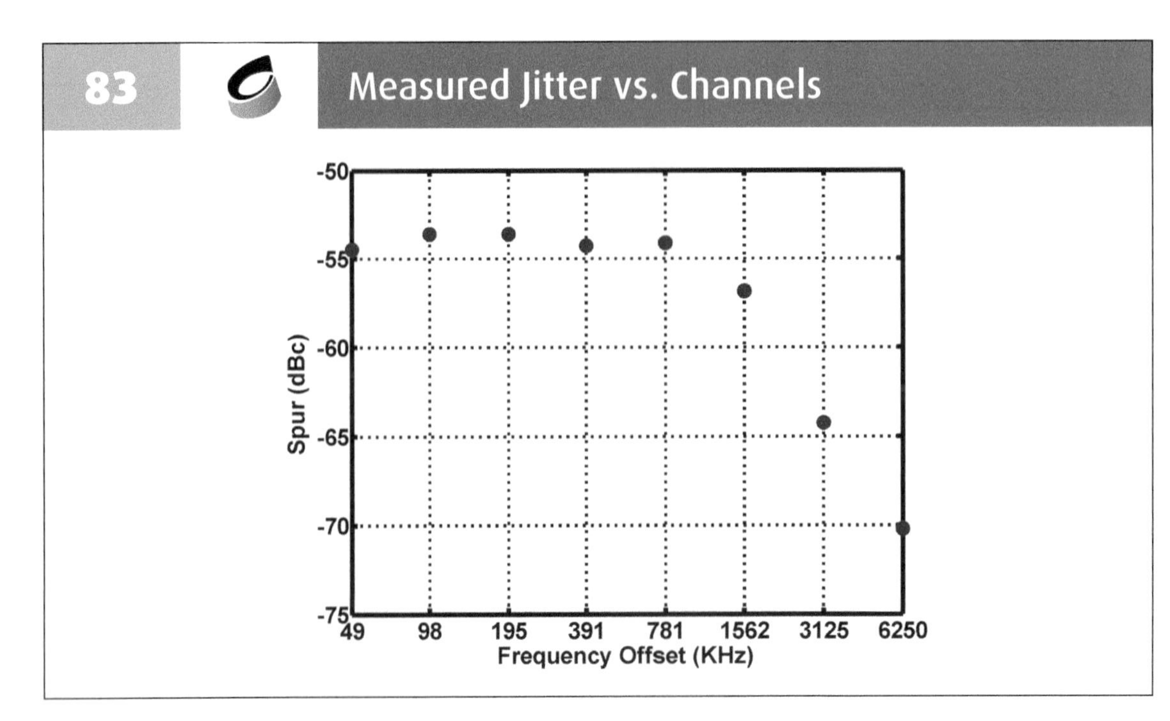

This slide shows the measured jitter vs. different fractional channel. For very small fraction, its measured jitter is large due to in-band spurious tone.

84 Performance Summary

	ISSCC 08	ISSCC 11	ISSCC 15	This Work
Architecture	Digital	Digital	Digital	Digital
Output (GHz)	3.62~3.67	2.9~4	2.6~3.9	4.6~5.4
In-band Noise (dBc/Hz)	-108	-102	-110	-104
Integrated Jitter (fs)	204~300	400~560	226~240	380
Ref. Spur (dBc)	-65	-72	-60	-67
Frac. Spur (dBc)	-42	-42	-62	-54
Power (mW)	46.7	4.5	10.5	2.92
FoM (dB)	-237	-241	-242	-244

This slide shows you the comparison for this work with the other prior arts. For in-band phase noise, its performance is not as good as we simulated due to the coupling between digital and analog domain. However, this work is in a 40 nm CMOS and can achieve very low power. Therefore, in terms of FOM, it is still one of the best.

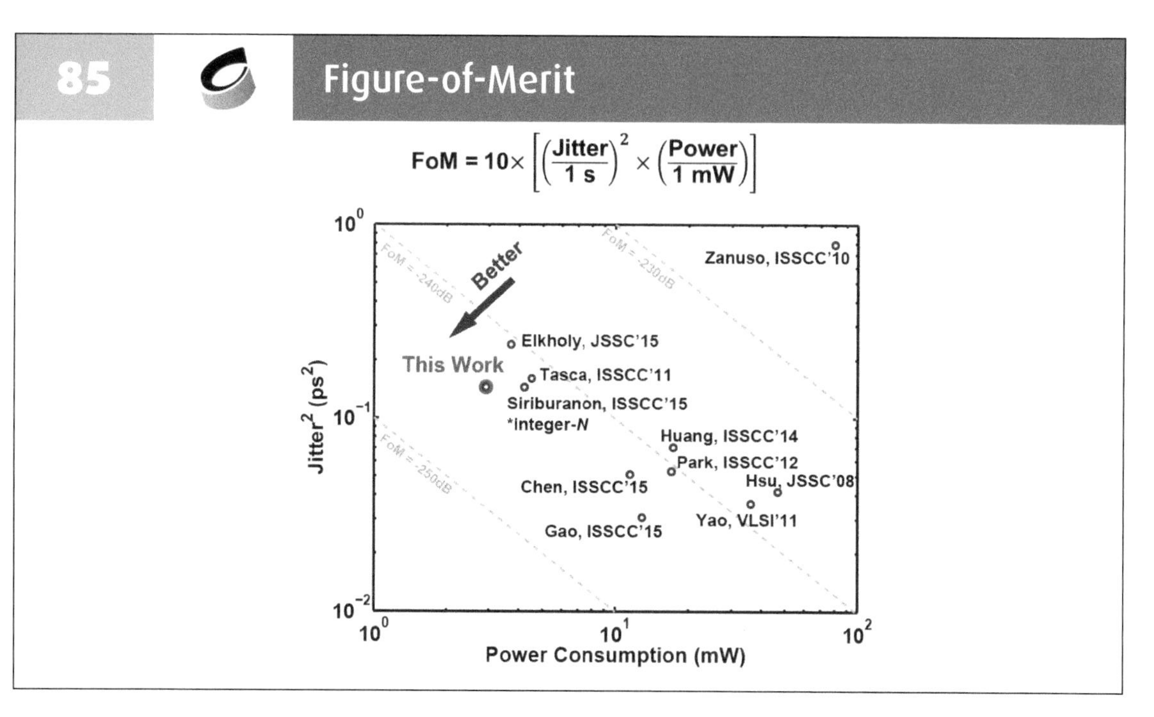

Finally, this FOM plot indicates the proposed ADC-based PLL can achieve state-of-the-arts FOM performance.

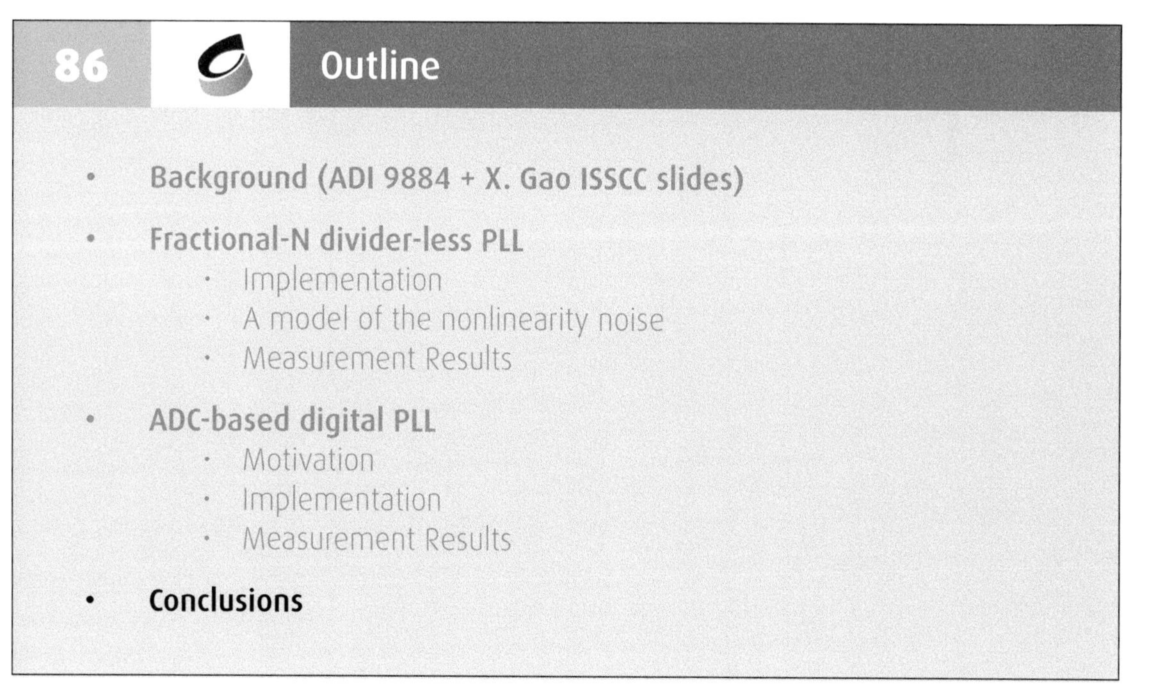

87  Conclusions

- **SSPLL with DPWM**
  - 20logNlower the CP noise
  - Eliminate the divider noise
  - Perform divider-less fractional-N operation
  - A model of the nonlinearity noise
- **ADC-based DPLL**
  - Low speed and low resolution ADC achieves high time resolutionMotivation

In the end, to conclude the presentation, we propose a low-noise divider-less fractional-N frequency synthesizer. The SSPLL architecture lowers the CP noise and eliminates the divider noise. The DPWM modulates the pulse-width of reference clock to realize divider-less fractional-N mode. The DCL reduces the gain error, and the DEM reduces the random mismatch.

It achieves a low in-band noise of -112 dBc/Hz.

In the second work, an ADC-based PLL is proposed by using a 6-b SAR ADC to mimic the digital PLL operation. It can perform TDC-like operations with a picosecond resolution. It achieves decent phase noise with very low power dissipation. Thanks a lot for your attention.

CHAPTER 4

High Performance SAR -Type ADCs & Digital Assisted Techniques

Yan Zhu

University of Macau
Macao, China

SAR ADCs achieve excellent power efficiency due to its simple architecture and dynamic operation, while its conversion speed is limited by its sequential conversion. Hybrid ADC takes the design advantages of multi conventional ADC architecture to optimize the conversion speed, resolution and power dissipation. This chapter introduces a hybrid ADC architecture combining the flash, SAR and pipelined ADC with interleaving and op-amp sharing schemes to achieve near GHz sampling rate and 11-bit resolution. Moreover, some digital assisted solutions are also introduced that fix the conversion errors from offset and DAC mismatches.

1  **Outline**

- **Hybrid ADC Architecture**
  - Conventional ADCs: Flash, SAR, Pipeline, Pipelined-SAR(PSAR)
  - Hybrid ADC: Flash+TimeInterleaved (TI)-PSAR Architecture
  - Mismatches between the subranging ADC
- **Sampling Front-end Design for TI Scheme**
  - Inline Demux Sampling front-end
  - Channel Selection Embedded Bootstrap (CSEB)
- **Gain Error Calibration for Bridge DAC**
  - Matching requirement in binary & bridge DAC
  - Gain Error (GE) in bridge DAC
  - Histogram based GE calibration

This chapter covers several topics for designing the high-speed and -resolution ADCs, including the ADC architecture, sampling front-end design for time-interleaving (TI) scheme and the digital-assisted solutions for accuracy optimization. The design challenges in ADC core as well as its peripheral interface circuities are discussed and some corresponding circuit techniques are proposed to solve the problems.

2  **Outline**

- **Hybrid ADC Architecture**
  - Conventional ADCs: Flash, SAR, Pipeline, Pipelined-SAR(PSAR)
  - Hybrid ADC: Flash+TimeInterleaved (TI)-PSAR Architecture
  - Mismatches between the subranging ADC
- **Sampling Front-end Design for TI Scheme**
  - Inline Demux Sampling front-end
  - Channel Selection Embedded Bootstrap (CSEB)
- **Gain Error Calibration for Bridge DAC**
  - Matching requirement in binary & bridge DAC
  - Gain Error (GE) in bridge DAC
  - Histogram based GE calibration

3 

Applications

➢ Broadband communications and Cable receiver demand ADC spec. >300MS/s & ~12bit

Resolution > 9b, fs>300MS/s

TI-SAR
TI-Pipeline
Pipeline
Pipelined SAR

$FOM_{W,hf}$ [fJ/conv-step]

1.E+04
1.E+03
1.E+02
1.E+01

1.E+08
1.E+09
1.E+10

f_{snyq} [Hz]

Boris Murmann: ADC Survey"https://www.stanford.edu/~murmann/adcsurvey.html".

For wireless and wireline broadband communications and cable receiver applications demand the specification of the ADC with sampling rate near 300MS/s and 12bit resolution. If we do the survey of the state-of-the-art ADCs with resolution not less than 9 bit and sampling rate larger than 300MS/s, the potential ADC architectures can achieve this specification are TI-SAR, TI-pipeline, pipeline and Pipelined-SAR (PSAR) ADCs, among this the PSAR ADC obtained best power efficiency.

4 

Conventional Flash ADC

V_{IN} V_{REF} f_s

Encoder

B-bits

Thermometer code

➢ RDAC generates 2^n-1 level compare with V_{in}

➢ ½ cycle compare

½ decode

❑ **Advantages**

❖ **High Speed (~GHz)**

❑ **Disadvantages**

❖ **Large design complexity**

❖ **Large power-RDAC,2^n-1Comp.**

❖ **Low resolution (5-8b)**

The flash ADC can achieve highest speed with single channel mostly targeting for GHz range and low to medium resolutions. It consists of a resistive DAC, 2^n-1 comparators and a decode logic. The input signal is compared with the 2^n-1 reference levels and the outputs of the comparators are passed through the decoder to convert the thermometer code to binary one. It need half conversion cycle for comparing and another half for decoding, thus it can achieve very high conversion speed. However, it also suffers some design limitations for higher accuracy. Firstly, the RC mismatch of the input network limits the sampling accuracy. Typically the S/H circuit need to be implemented. Secondly, the non-idealities from the comparators due to the offsets and kick-back noise limit the conversion accuracy to not more than 10 bit. To suppressed them the pre-amplifiers are required consuming large power dissipation.

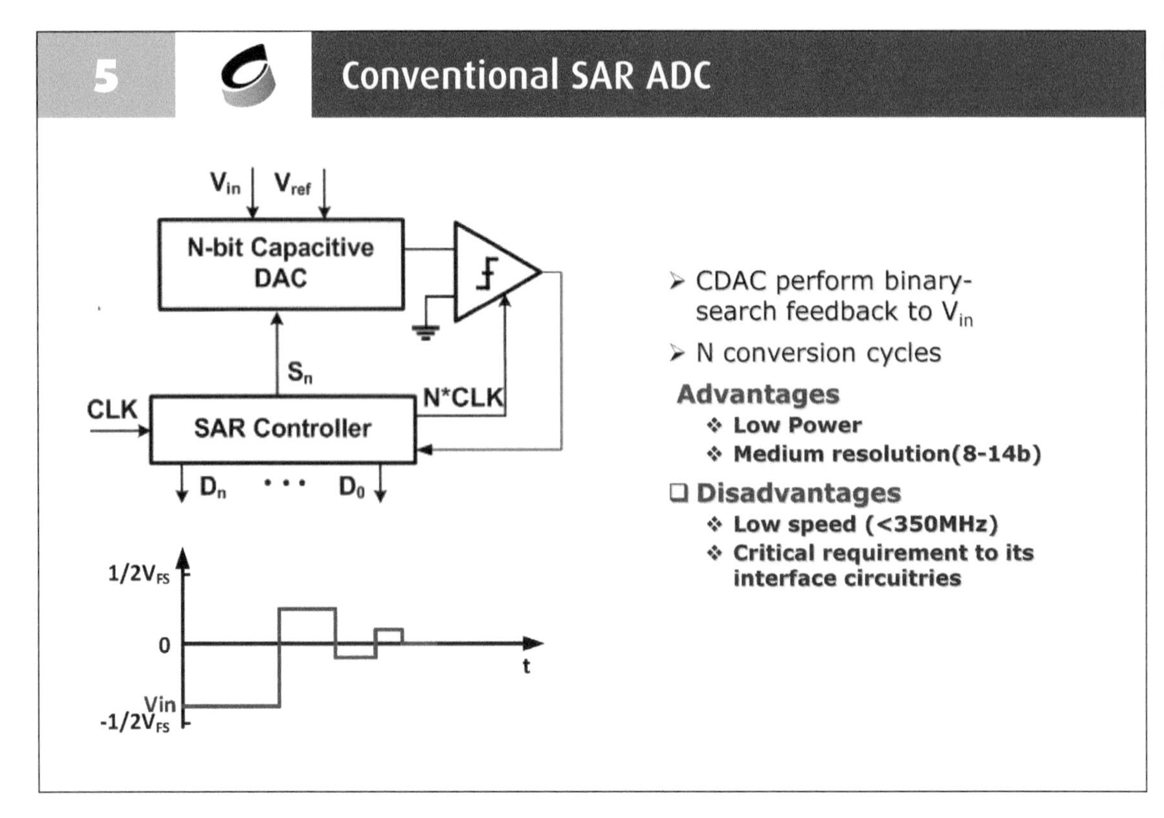

The SAR ADC benefit from it simple configuration and mostly dynamic operation exhibiting excellent power and area efficiency under technology down scaling. The overall ADC architecture consists of a capacitive DAC array, a comparator and a SAR controller. The DAC array is used to sample the input signal and subtract the reference voltages with a binary-searched approximation. The SAR controller includes the bit register and the switch drivers, performs the corresponding switching method. Its conversion accuracy is limited by the comparator thermal noise, the reference error from fast switching transient and the DAC mismatch. Also, the sequential conversion feature limits the speed of the single channel SAR with 10b resolution at 320 MS/s [3] in the advanced technology node of 20 nm CMOS. Since the switching noise in a high speed SAR ADC is usually in the GHz range, in practice, a reference buffer is required to attenuate it. However, such buffer implies a trade-off between the output impedance and the power consumption, to provide a fast switching transient.

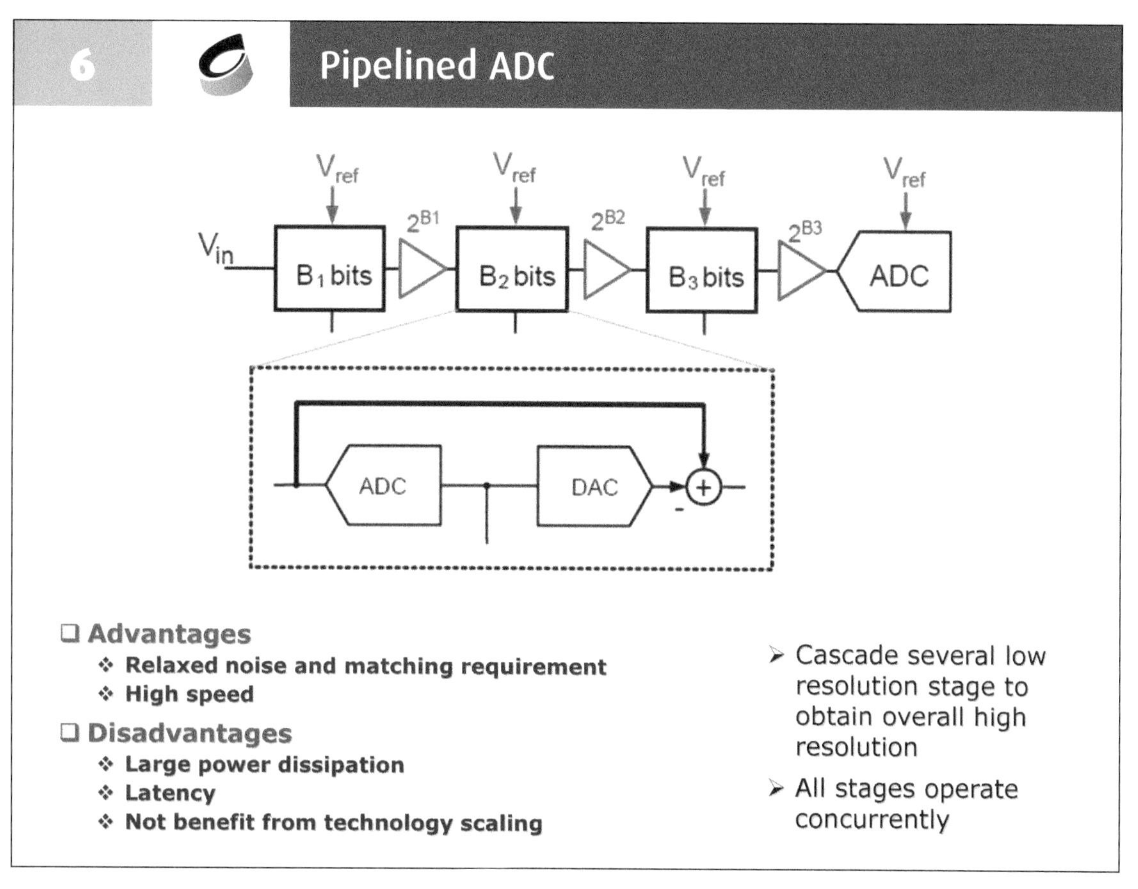

The pipelined architecture cascades several low resolution stage to obtain overall high resolution and each stage operate concurrently, thus its conversion speed is also higher than the SAR ADC. The stage gain relaxed the quantizer accuracy from the noise and matching requirement, thereby, it is usually built with flash architecture. Obviously, this architecture would be power hungry as the large number of opamps are required consuming static power dissipation. Also, as the technology down scaling, the shrink of the supply and reduced intrinsic gain of the device further limit the power efficiency of this architecture.

7

Pipelined-SAR (PSAR) ADC

CLK | SAR LOG | m | SAR Control & Digital Error Correction | M+K-1

V_{in} | M-bit Binary CDAC | x8 | K-bit Binary CDAC

±Vref | ±1/4Vref

- Two-stages quantize 5-6bits
- Low stage-gain reduce opamp requirement for low power

❑ **Advantages**

- ❖ **High resolution(10-14bit)**
- ❖ **Excellent power efficiency**

❑ **Disadvantages**

- ❖ **Medium speed**

The pipelined-SAR ADC maintains its superiority in power efficiency for high resolution even in deep sub-micron technology. The implementation of low stage-gain relaxes the desired noise requirement in the 2nd stage comparator and balances the trade-off between the power dissipation and gain-bandwidth product (GBW) requirement of the Op-Amp. Also, since a less number of opamps is required, hence its power dissipation is no longer dominating, typically occupying around 20% ~30% of the total ADC power consumption . As the 1st-stage SAR ADC quantizes a larger number of bits, the residue for amplification is comparatively small. This results in a smaller output swing, thus offering better amplification linearity. Therefore, the pipelined-SAR architecture has the potential to achieve higher resolution with excellent power efficiency.

8

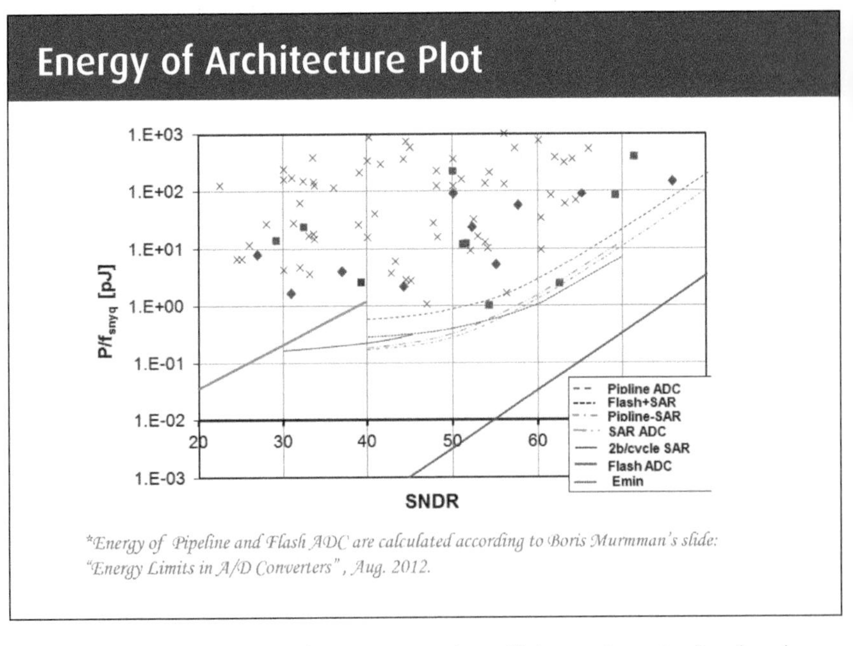

**Energy of Pipeline and Flash ADC are calculated according to Boris Murmman's slide: "Energy Limits in A/D Converters", Aug. 2012.*

According to the survey of SNDR versus the FoM under different ADC architectures, it can be found the flash ADC exhibits good power efficiency at 2 to 4 bit resolution, while the SAR ADC has the better energy efficiency at more than 4 bit resolution due to its digital overhead limiting the performance at low resolution. As the SNDR increasing the SAR-type ADC like flash-SAR or pipelined SAR ADC demonstrate better conversion efficiency than pipelined and pure SAR ADCs.

9

Proposed Hybrid SAR ADC

- High speed:
  - Flash+TI SAR@1st stage
- High resolution:
  - Inter-stage gain of 4
- Low power:
  - Less No. of opamps with low gain
  - Multi-shared elements to relax calibration effort & save area

Yan Zhu, Chi Hang Chan, Seng-Pan U, R. P. Martins, "An 11b 450 MS/s 3-way Time-Interleaved Sub-ranging Pipelined-SAR ADC in 65nm CMOS", IEEE Journal of Solid-State Circuits, Feb-2016.

The overall ADC architecture is depicted in this slide, consisting of a 3-way TI sub-ranging pipelined-SAR ADC operating at 150 MS/s for an aggregate of 450 MS/s. The flash ADC exhibits good power efficiency at 2b to 3b resolutions, while the SAR ADC consumes larger dynamic power at the leading bits transitions. To reduce the number of transitions in leading bits that cause reference ripple, a 2b flash ADC with 3-way TI pipelined-SAR architecture is proposed. The 1st and the 2nd-stages determine the coarse 6 bits and the fine 6 bits, respectively. One bit overlapping between two stages is designed for relaxing the sampling and comparison accuracies between the flash ADC and the TI sub-ADCs. The flash ADC resolves 2 leading bits, and the control logic passes the code to the corresponding sub-SAR ADC that resolves the remaining 4 bits. The residue is then amplified by 4 to the 6b SAR ADC in the 2nd-stage. The Op-Amp is shared by the sub-ADCs, and its bandwidth determines the optimum number of TI channels.

10

1st Stage Flash & SAR Implementation

- Signal coupling & Reference interference
- Mismatches between Flash & SAR ADCs

The detailed implementation of the 1st stage is shown in this slide, which consists of a 6b binary-weighted DAC array associated with 2×32C units. Two portions contain the same total unit of 64C for different functions. The 6b DAC performs binary-searched conversion, while the 2×32C units are kept connected to V_{cm} that serves as the capacitive-dividing by 2 of the reference voltages. The feedback capacitor C_{FB} defines the stage-gain of 4, as its value is 1/4 of the total array capacitance. However, there are two main non-idealities impairing the SNDR of this ADC. Firstly, the input signal feedthrough to C_F and sub-ADC and the reference interference among the channels. Secondly, the mismatch between the flash and SAR conversion in the 1st stage.

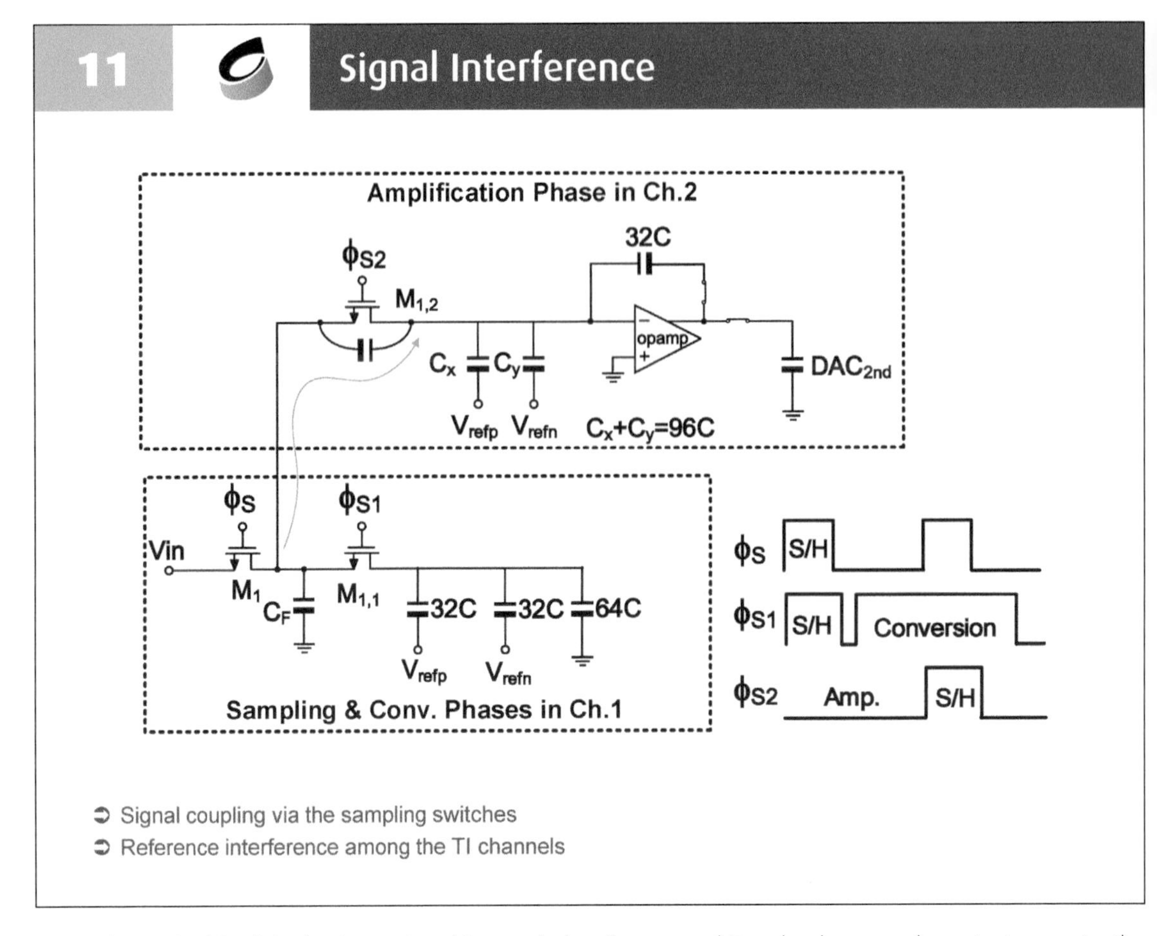

As shown in this slide the input signal is coupled to the top-plate of C_F through C_{DS} of the master switches (M1). The error causes only sub-ranging mismatches, by using the cross-coupled dummy switches the error can be suppressed. However, the signal coupling via the capacitance of slave switches (M1,1&M1,2) to the sub-SAR ADC would be more critical. When the channel 1 is sampling, channel 2 is performing the residue amplification. The signal coupled to the MDAC via the parasitic capacitance C_{DS} of slave switch in channel 1 degrades amplification accuracy, which needs to fulfill the overall ADC precision. Therefore, the signal feedthrough-compensation to the slave switches in each sub-SAR ADC is also used to provide differential isolations.

Second issue is the reference noise due to switching transients in the sub-SAR ADCs. Benefiting from one bit redundancy and a ×4 stage-gain the required accuracy of the reference voltages for the 1st and the 2nd-stage SAR conversion should be limited within $\pm V_{ref}/2^7$ and $\pm V_{ref}/2^{10}$, respectively. Since the accuracy of the residue amplification is required to be the overall ADC precision. The reference interference among the TI channels will degrade the SNDR, as ch.2 is amplifying the residue when the ch. 1 is performing the conversion that cause reference ripples. One of the advantages in using this hybrid structure is the reduction in the number of switching transitions in the leading bits. From the timing diagram in slide 9 it can be observed that before the sub-ADC1 completes its amplification the 1st-stage of the other sub-ADCs is either at the end of the SAR cycling or waiting for outputs from the flash side. The reference noise due to the switching transient is comparatively smaller.

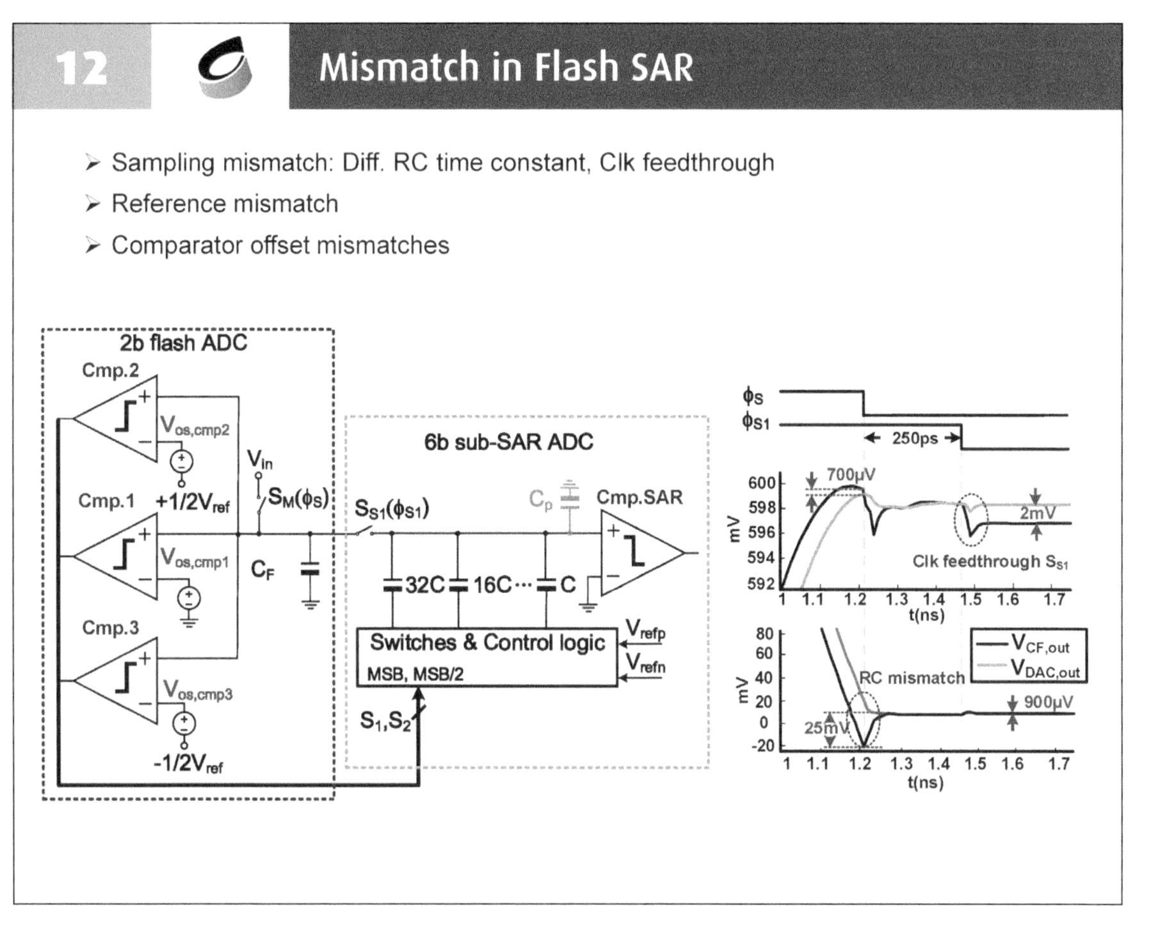

First, the mismatch comes form the different RC time constants of the sampling networks, as the T/H capacitor C_F is much smaller than the DAC in the sub-SAR channel. The mismatch is maximized at the middle of the input, where the slope of the signal is the highest. According to post-layout simulation the maximum error is near ±25 mV. However, as the slave switch S_{S1} is turned off 250 ps after the master switch S_M, it provides sufficient time for C_F and DAC to settle to the same value. Second, the clock feedthrough of the slave switch. For the second term the mismatch is minimized at the middle and maximized at the two sides of the input. The sub-ranging architecture is sensitive to the sampling mismatch near the MSB and MSB/2 transitions, where the mismatch in this design is ≈ ±90 μV (±0.01 LSBs) and ±0.5 mV (±0.05 LSBs) respectively. As the mismatch is quite small and well within the error tolerance range, the clock feedthrough imposed on the sub-ranging mismatch is not problematic. Moreover, the reference and the offset mismatches in flash and SAR cause conversion errors. However, all these mismatches eventually lead to a code dependent offset error that can be corrected by digital error correction.

13

Mismatch Analysis

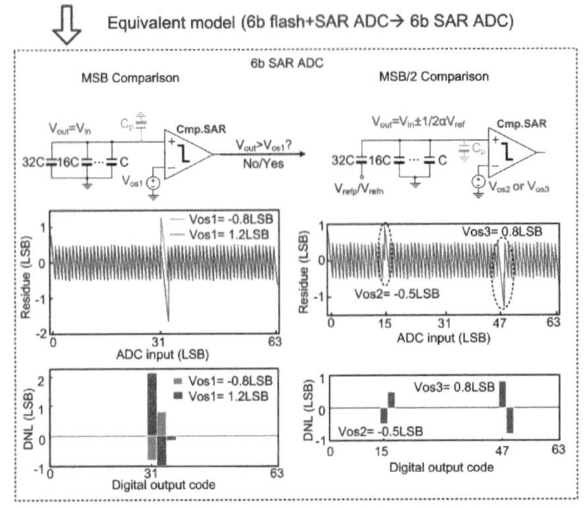

- Mismatch result in code dependent offset errors
- Errors need to be suppressed with in redundancy range

The overall mismatches from sampling, reference and offset can be seen as an input referred offset to the MSB and MSB/2 comparisons in a 6b SAR ADC. As shown in this slide the MSB transition directly compares two differential inputs that is reference uncorrelated, the mismatches, mainly contributed from sampling and comparator offset, can be equivalent to an offset voltage (V_{os1}) at the negative side of the comparator leading to DNL errors at the code MSB-1 (D31) and MSB (D32). Similarly, the MSB/2 transition is reference-correlated, where the decision thresholds of the flash ADC are set by the reference voltages in Cmp.2 and Cmp.3. The conversion is also modeled as input referred offset V_{os2} or V_{os3} during the comparisons. The mismatches are transferred to a bit-dependent offset causing DNL errors near MSB/4 and 3 MSB/4 digital outputs.

As a conclusion the mismatches in flash+SAR subranging architecture cause only bit-dependent offset errors, which can be tolerated later by redundancy.

14

Digital Error Correction (DEC)

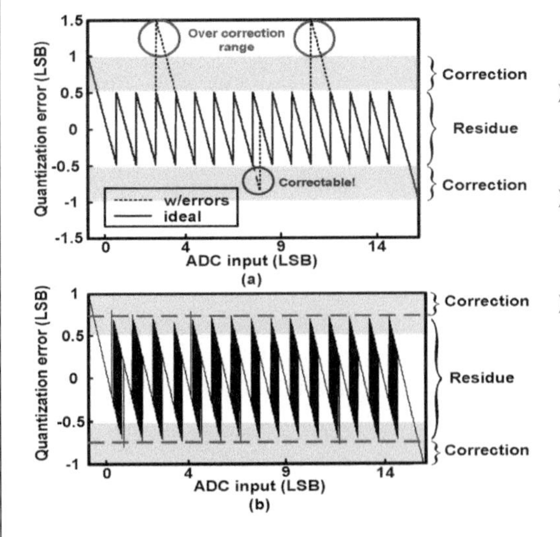

- DEC relax the 1st stage conversion accuracy
- Error within the correct range can be fixed
- Comparator noise occupy DEC range

Since the 1st conversion is relaxed by the DEC, the error within the tolerance range can be recovered in the later conversions. However, considering the noise from the comparator, the total input referred offset voltages need to be suppressed below ±0.25% of error tolerance region.

15 

Die Chip Photographs

320μm
220μm
Cal. Logic
1st-stage
CLK
Opamp
2nd-stage

- 1st design @ 450MS/s 11bit
- Total active area is 0.07mm^2 including offset cal.
- Reference voltages are external
- 300pF on-chip decoupling capacitor consuming 0.028mm^2

An 11-bit 450MS/s sub-ranging pipelined-SAR ADC was fabricated in a 1P7M 65 nm CMOS process and Metal-oxide-Metal (MOM) capacitors. This slide shows the die photograph of the design; the active area is 0.07 mm^2 (320 μm×220 μm) including on-chip offset calibrations. The input and reference buffers are not implemented in this design. The reference voltages are generated externally and 300pF on-chip decoupling capacitor is placed consuming an area of 0.028mm^2.

16 

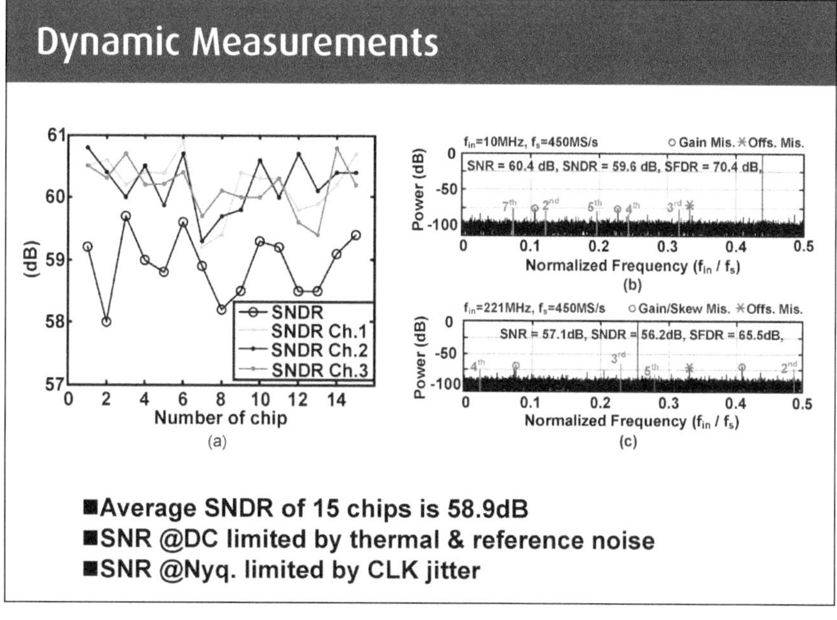

Fig. (a) illustrates the measured performance of total 15 chips @10 MHz input and 450 MS/s sampling rate. The achieved average SNDR of the ADC and each sub-channel is 58.9 dB and 60.2 dB respectively, where the mismatch elements among the sub-channels degrade the SNDR by 1.3 dB. After offset calibrations

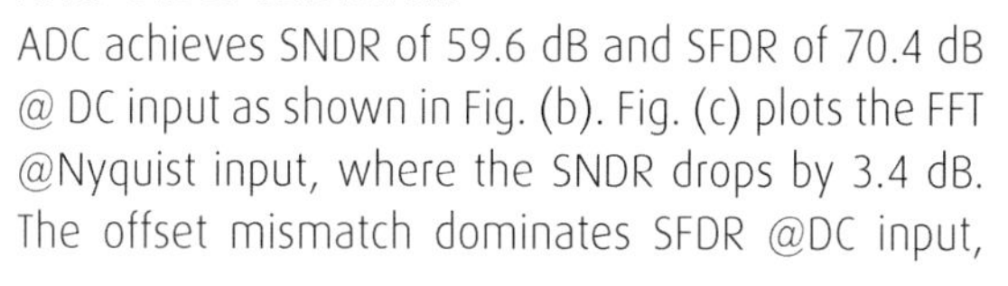

ADC achieves SNDR of 59.6 dB and SFDR of 70.4 dB @ DC input as shown in Fig. (b). Fig. (c) plots the FFT @Nyquist input, where the SNDR drops by 3.4 dB. The offset mismatch dominates SFDR @DC input, as the offset calibrations run at foreground by only once. The noise could potentially affect its accuracy. At Nyquist input the 3rd harmonic dominates the SFDR.

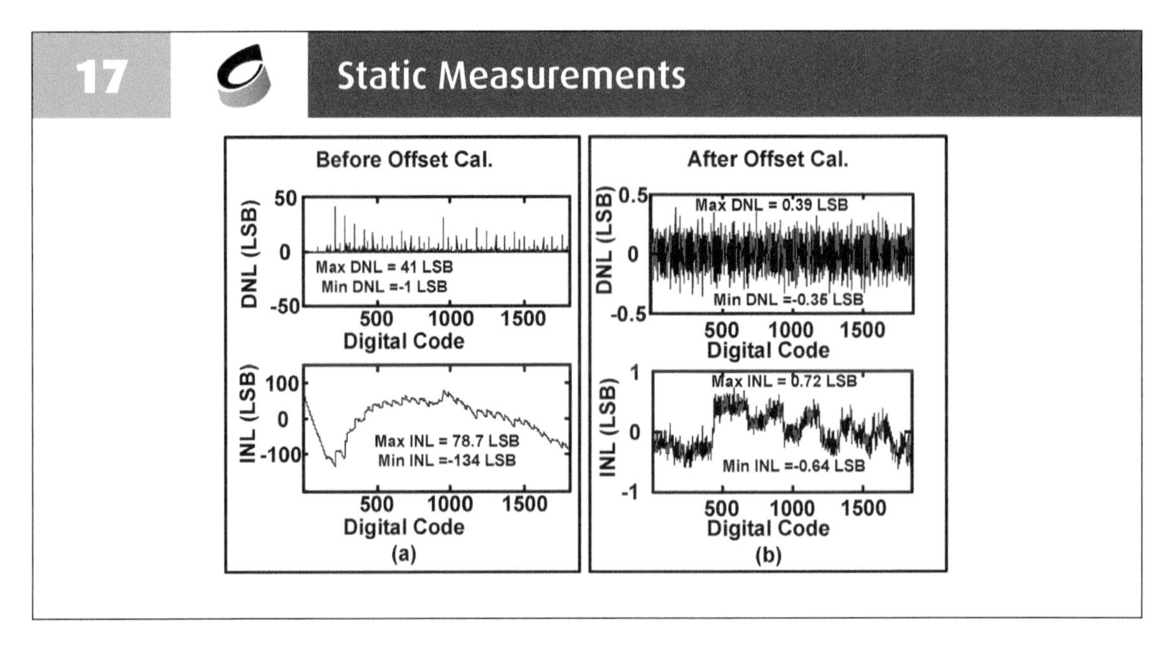

The static performances before and after offset calibrations are shown in this slide. Before offset calibration, the digital output contains large DNL and integral-nonlinearity (INL), as the 2nd-stage SAR operation is fully saturated by the offsets from the 1st-stage comparators and the Op-Amp. Once the offset calibration is active, the DNL and INL is compensated within 0.39 LSBs and 0.72 LSBs, respectively.

18 Performance Summary & Comparisons

	CICC' 14	VLSI' 14	VLSI' 14	This Work
Architecture	Pipelined-SAR	Pipelined-SAR	Pipelined-SAR	Pipelined-SAR
Technology (nm)	65	65	40	65
Resolution (bit)	13	12	12	11
No. of Channels	1	2	1	3
Sampling Rate (MS/s)	160	210	160	450
Supply Voltage (V)	1/1.2	1	1.1	1.2
Input Swing (V_{p-p})	1.2	1.6	2	1.2
SNDR @DC (dB)	68.3	63.5	65.3	60.8 (peak)
SNDR @Nyq. (dB)	66.2	60.1	66.5	56.2
DNL/INL (LSB)	N/A	0.7/1.5	N/A	0.4/0.7
Area (mm^2)	0.09	0.48	0.042	0.07
Power (mW)	11.1	5.3	5	7.4
FoM @DC (fJ/conv.step)	32.6	20.5	17.7	21
FoM @Nyq. (fJ/conv.step)	41.6	30.5	20.7	32
Calibration (on-chip)	No	No	No	Yes

The performance summary and comparison with State-of-the-art ADCs are shown in Table I. The design implements all the calibrations on-chip and achieves > 2× conversion speed than those prior arts. Also, considering that the full-scale of this design is well within the supply rail, the achieved conversion accuracy is still comparable.

19  Outline

- Hybrid ADC Architecture
  - Conventional ADCs: Flash, SAR, Pipeline, Pipelined-SAR(PSAR)
  - Hybrid ADC: Flash+TimeInterleaved (TI)-PSAR Architecture
  - Mismatches between the subranging ADC
- **Sampling Front-end Design for TI Scheme**
  - Inline Demux Sampling front-end
  - Channel Selection Embedded Bootstrap (CSEB)
- Gain Error Calibration for Bridge DAC
  - Matching requirement in binary & bridge DAC
  - Gain Error (GE) in bridge DAC
  - Histogram based GE calibration

20  Motivation & Challenges

- Application: Diversity multiband, multimode digital receivers 3G/4G, TD-SCDMA, W-CDMA, GSM, LTE General-purpose software radios requires ADC >800MS/s,~12 bit
- 2-way TI Hybrid ADC for 900MS/s

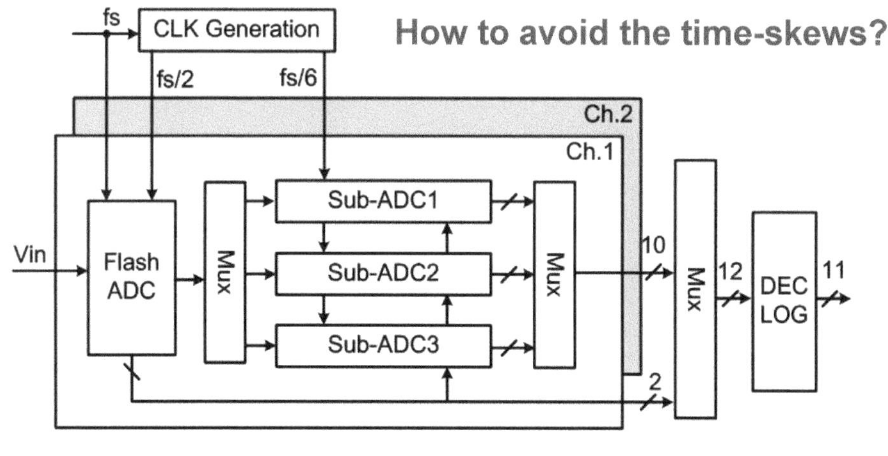

Y. Zhu, et al., "An 11b 900 MS/s Time-Interleaved Sub-ranging Pipelined-SAR ADC," in Proc. of IEEE ESSCIRC, pp. 211-214, Sep. 2014.(Best paper award)

The conversion rate can be boosted either by using multi-bit-pre-cycle or TI schemes. By interleaving the previous 450MS/s ADC by 2, we target the sampling rate up to 900MS/s. However, the timing mismatch between the TI channels becomes critical.

21 Design Specifications

Design Target: 900MS/s, 11b(6b+6b,1b DEC), 1.2V_{P-P}

➲ Total Offset at 1st-stage < 4.5mV
✓Offset of Comparators
✓Residue amplifier Op-amp's Offset

➲ Timing Mismatches
✓Timing-Skew < 140fs (Clock Gen, Clock Bus)

Inter-stage Gain	Opamp			2nd-stage SAR		
	Finite DC Gain (dB)	GBW (GHz)	Output Swing(mV)	Conversion Accuracy	Offs. Mis. σ_{os} (mV)	Gain Mis. σ_{gmis} (%)
8x	48	2.2	150	9b	2.4	0.2
4x	42	1.1	75	10b	1.2	0.1

Some design specification of the ADC are listed in this slide. The offset mismatch is not problematic, as its requirement is relaxed by the DEC that can be calibrated simply. The comparator offset and op-amp will be amplified to the 2nd stage occupying the DEC range and potentially saturates the 2nd stage conversion. The timing mismatch needs to be suppressed below 140fs being the design limitation.

22 Existing Solutions

The conversion rate is boosted by using highly time-interleaved (TI) schemes, while the most critical design challenge is the time spurs that degrade both SNDR and SFDR. The timing skews can be reduced and tolerated by design constraints by using active T/Hs with precise clock distributions or timing-calibrations.

➲Active T/H circuit with precise clock distributions
- Large power dissipation [ISSCC'06 S. Gupta]

➲Timing Calibrations
- Correlation-based
  a. w/ reference channel- [JSSC'11 M.El-Chammas]
  ☹ Ref. Channel increase power burden to input buffer
  b. w/o reference channel [JSSC'14 B. Razavi]
  ☹ Require a wide-sense stationary input
- FIR filter- based [ISSCC'14 N.L. Dortz]
  ☹ Cal. cost large digital overhead and power

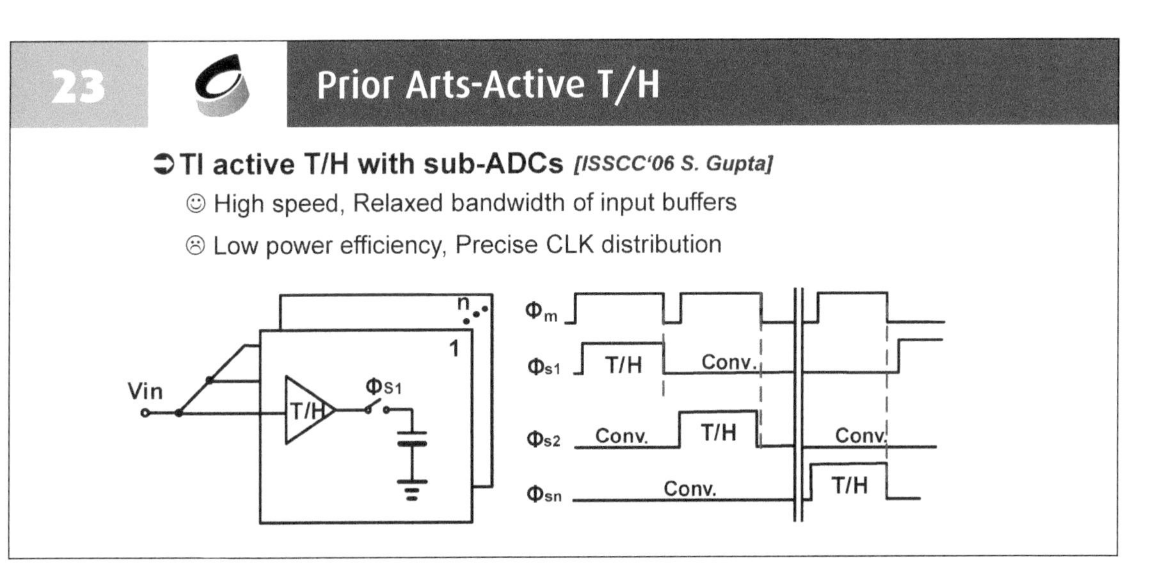

Interleaving the T/H with the sub-ADC channels relaxes the bandwidth of the T/Hs. Comparing to the conventional solution it combines better power efficiency with high speed. However, its power efficiency is not competitive with the state-of-the art TI ADC with timing calibration, as the T/H still consumes static power and the clock buffers to provide precise clock distribution to sampling switches are the most power hungry parts.

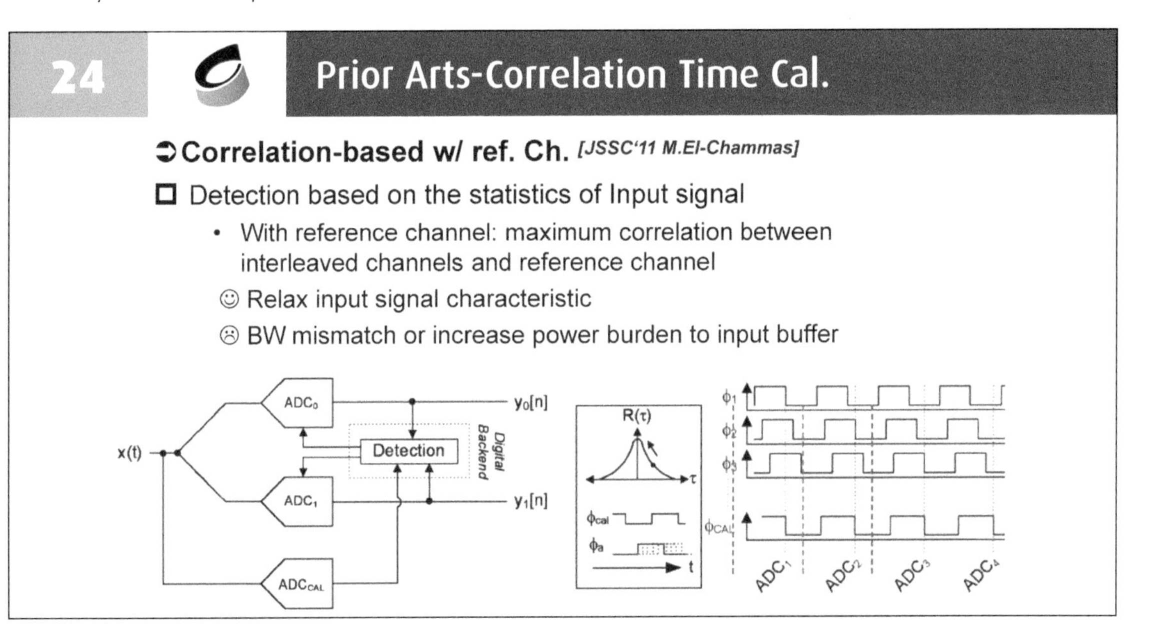

The correlation based timing calibration maximize the correlation coefficient between each interleaved channel with one reference channel to cancel the timing mismatches. The timing mismatch is compensated by adjusting the delay of the TI sampling clock via a feedback loop. The solution relaxes input signal characteristic, which can be extended to wide-sense cyclostationary (WSCS) signals. However, the main limitation of the calibration is that the reference channel need to operate at full speed to match with the sub-ADCs. If it works at low speed, there exists bandwidth limitation due to different loading to the input buffer once the reference channel is overlapped with the sub-channel sampling. Additional dummy channel need to be added to suppress the effect that increases the loading of the input buffer.

25  Prior Arts Correlation Time Cal.

➲ **Correlation-based w/o ref. Ch.** ***[JSSC'13 B. Razavi]***

☐ Based on a wide-sense stationary has constant variance

☺ Avoid BW limit, better power efficiency

☹ Depend on input characteristic

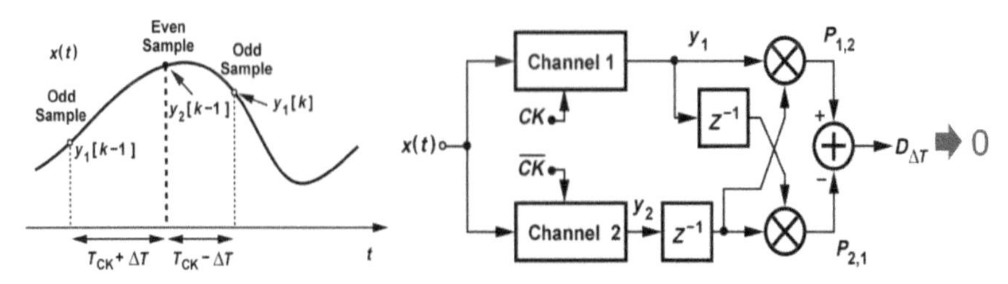

Another correlation based timing calibration does not require reference channel, which estimates the timing mismatch by calculating the correlation among the channels. With a wide-sense stationary input whose variance should be a constant value, therefore the difference of the correlation of each channels should approach 0 if there occurs no timing mismatches. Since it no needs dummy channel, avoiding the bandwidth limitation as mention in previous solution. The detection is based on statistics consuming long converging time and hardware overhead.

26  Prior Arts FIR Filter Based Time Cal.

➲ **Using Differentiating FIR filter**

☐ Based on derivative of Input signal ***[ISSCC'14 N.L. Dortz]***

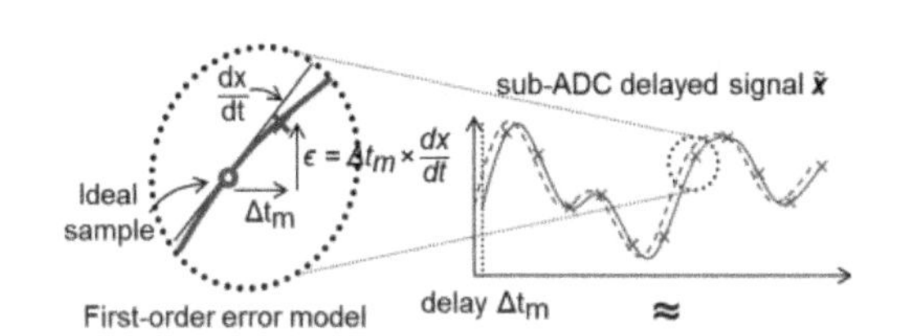

$$\Delta x = \Delta t \times \frac{dx}{dt}$$

$$\tilde{\Delta t}_m = \overline{\tilde{x} \times \frac{dx}{dt}} / \overline{(\frac{dx}{dt})^2}$$

☺ No feedback loop- better jitter performance

No input constrain

☹ Implementation of FIR filter-large power, BW limitation

Offset & gain mismatches affect the Cal. accuracy

Another time calibration uses the differentiating filter to calculate the derivative dx/dt of the outputs, thus, estimating the timing error. The calibration is performed in digital domain requiring no time adjustable feedback leading to better jitter performance. The bandwidth limitation due to use the FIR filter can be reduce by increasing the number of filter tap, while it will consumes more power dissipation. Also, the offset and gain mismatches need to be suppressed to sufficiently low first, otherwise, they may affect the calibration accuracy. The calibrations occupies large total ADC power.

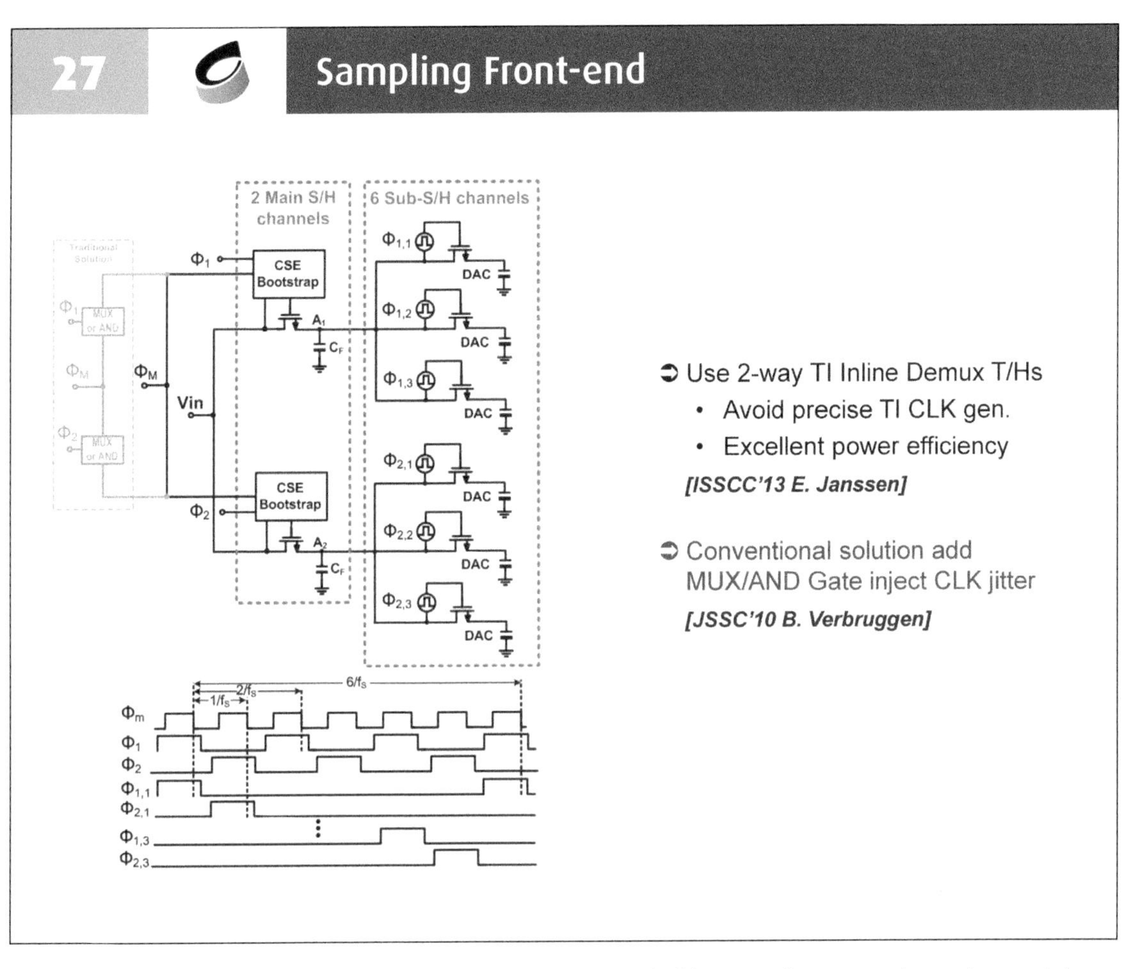

The overall TI sampling front-end is depicted in this slide, which consists of 2-way TI main S/H operating at 450MS/s for an aggregate 900MS/s. Each channel is built with 1 to 3 inline demux sub-S/H. The input signal is sampled passively onto the capacitor C_F and the DAC according to the time-sequence. The strategy used in this design is similar to [11] that synchronizes the sampling instances of n TI-channels with a full speed master clock, avoiding the active T/H and precise TI clock for sub-channel sampling. The series connection between the master and slave switches avoids the time skews among the sub-channel's sampling. The penalty is the reduced bandwidth, which can be traded by using larger sized sampling switches.

To suppress the timing mismatch of the main S/H the concept in [12] is demonstrated in parallel with the proposed solution, of which the clock path is highlighted in gray. The master clock Φ_M is selectively applied to two time-interleaved master bootstrap circuits via a MUX or AND gate controlled by the clock signals Φ_1 and Φ_2. The extra circuits in the main clock path add additional clock jitter.

We propose a channel-selection-embedded (CSE) bootstrap, which minimizes the master clock path to the bootstrap terminal by simply performing the channel selection in the bootstrap itself.

28 Channel Selection Embedded Bootstrap

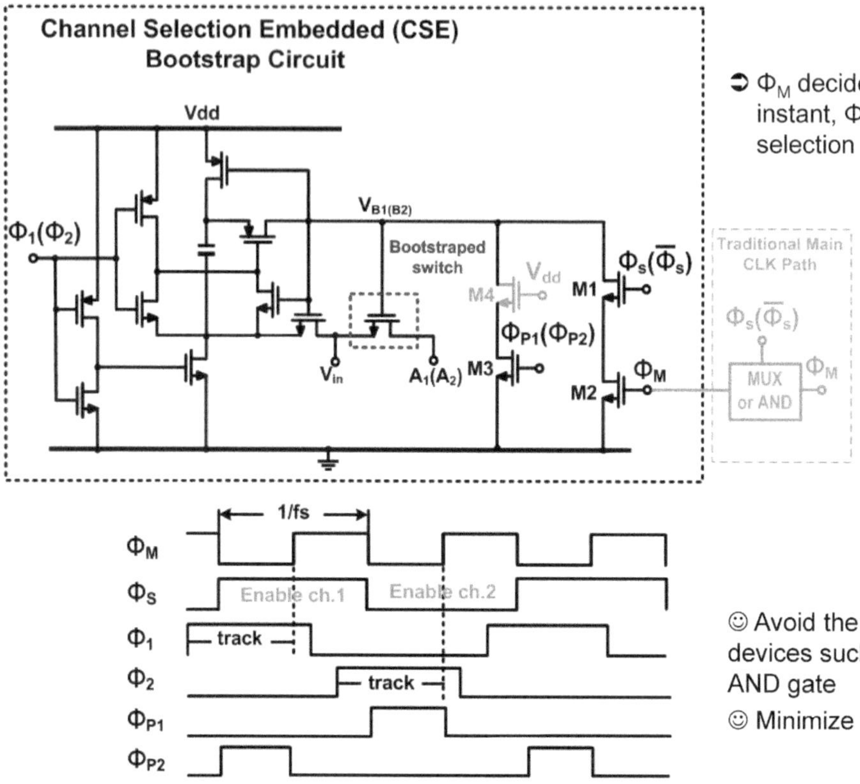

Channel Selection Embedded (CSE) Bootstrap Circuit

➲ Φ_M decide sampling instant, Φ_S channel selection

☺ Avoid the additional devices such as MUX or AND gate

☺ Minimize the clock jitter

The sampling instant in the main channel is defined by a common-master clock Φ_M, which is applied directly to the transistor M2. The transistor M1, of which the gate is usually connected to V_{dd}, now is used to enable the channel according to the clock signal Φ_S. The solution avoids the additional devices such as MUX or AND gate implemented in series with the master clock signal, which minimizes the clock jitter injected in the main clock path to the bootstrap terminal. The 1st(2nd) channel starts to track the input signal, when $\Phi_1(\Phi_2)$ is high. Either of the channels stops tracking, while M1 and M2 are both turned on. The sampling instant is determined by the rising edge of Φ_M that pulls down the gate voltage of the sampling switch from $V_{in}+V_{dd}$ to Gnd. The simulated spread on the falling edge of $V_{B1(B2)}$ is reduced to 320 fs by using a 6μm device for both M1 and M2. The transistor M3 is assisted to avoid the floating of $V_{B1(B2)}$, which is not necessary, if the floating duration is short. Also, the transistor M4 could be added to avoid the overstress of M3.

29 Die Chip Photographs

430μm
Ch.1
Ch.2
2nd-stage
1st-stage
340μm
CLK Gen.
Output Buffers

- 2nd design @ 900MS/s, 11bit
- Active area is 0.146mm² including offset cal.

Y. Zhu, et al., "An 11b 900 MS/s Time-Interleaved Sub-ranging Pipelined-SAR ADC," in Proc. of IEEE ESSCIRC, pp. 211-214, Sep. 2014.(Best paper award)

An 11-bit 900MS/s 2-way TI sub-ranging pipelined-SAR ADC was fabricated in a 1P7M 65 nm CMOS process and Metal-oxide-Metal (MOM) capacitors. The ADC occupies 0.146mm² (430μm×340μm) of core area.

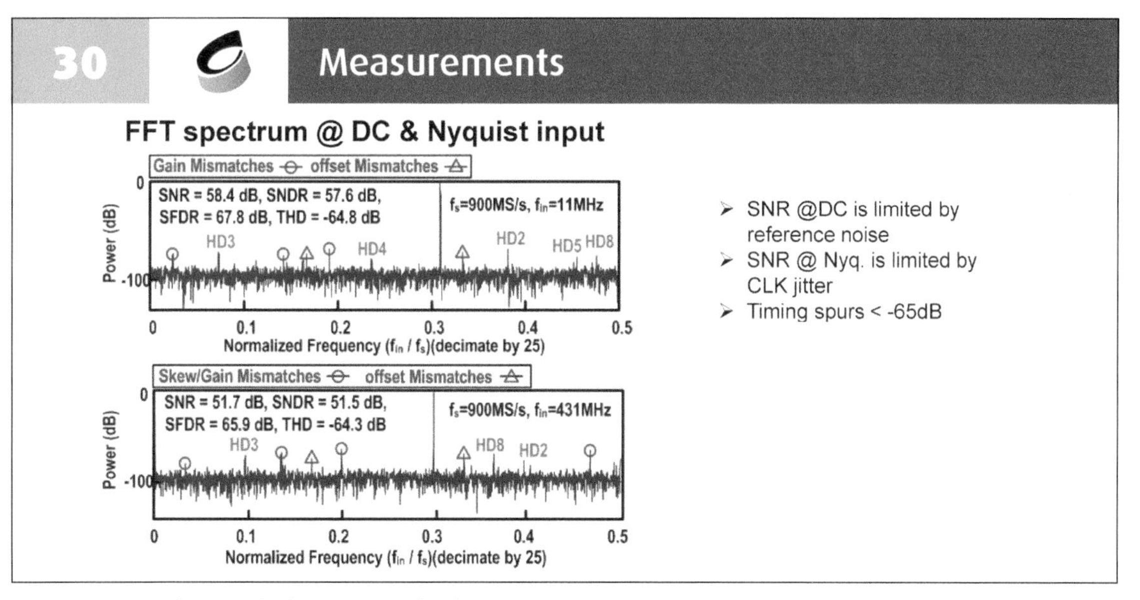

The measured FFT of the ADC with the input signal frequencies at DC and near Nyquist are shown, where the SNDR is only limited by the noise instead of the mismatches in the sub-channels. The spurs due to the gain mismatch between the sub-channels are below -67 dBFS, thus the gain calibration is not necessary. When the input signal goes up to Nyquist, the noise from clock generator limits SNR to 51.7dB, with only a slight drop in SFDR and THD. This SNR degradation is mainly limited by the jitter from the clock generator corresponding rms value of 650 fs. The jitter is larger than the design expectation, which can be significantly improved by using a differential clock generator instead of a single-ended.

31  Performance Summary & Comparisons

	ISSCC' 14	ISSCC' 14	JSSCC' 13	VLSI' 13	This Work	
Architecture	TI-SAR	TI-SAR	TI-SAR	Pipeline	TI-Pipelined-SAR	
Technology (nm)	40	65	65	65	65	
Resolution (bit)	10	10	10	10	11	11
Sampling Rate (MS/s)	1.62	1	2.8	0.8	0.9	1.1
Supply Voltage (V)	1.1	1	1	1.2	1.2/1.2	1.2/1.3
Input Swing (V_{p-p})	1	N/A	N/A	1.8	1.2	1.2
SNDR @DC (dB)	51	53.5	53.5	51	57.6	56.2
SNDR @Nyq. (dB)	48	51.2	51.2	48	51.5	50.7
SFDR @Nyq. (dB)	62	60	55	N/A	65.9	64
DNL/INL (LSB)	N/A	0.1/0.1	N/A	0.7/1.8	0.66/1.5	0.69/1.6
Area (mm^2)	0.83	0.78	1.7	0.18	0.15	
Power (mW)	71	19.8	44.6	19	15.5	18
FoM @DC (fJ/conv.step)	150	51	56	53	28	32
FoM @Nyq. (fJ/conv.step)	210	62	78	71	56	58
Require Timing Correction	Yes	Yes	Yes	No	No	
Calibration (on-chip)	Offset, gain, time	Offset	Offset, time	No	Offset	

The performance summary and comparison with the most advanced and recent high-speed and high-resolution ADCs are listed in Table, which confirms that this design achieves a better SFDR without the timing correction. Also, as the design requires less calibration efforts, it leads to a much smaller area and the best FoM.

32  Outline

- **Hybrid ADC Architecture**
  - Conventional ADCs: Flash, SAR, Pipeline, Pipelined-SAR(PSAR)
  - Hybrid ADC: Flash+TimeInterleaved (TI)-PSAR Architecture
  - Mismatches between the subranging ADC
- **Sampling Front-end Design for TI Scheme**
  - Inline Demux Sampling front-end
  - Channel Selection Embedded Bootstrap (CSEB)
- **Gain Error Calibration for Bridge DAC**
  - Matching requirement in binary & bridge DAC
  - Gain Error (GE) in bridge DAC
  - Histogram based GE calibration

33

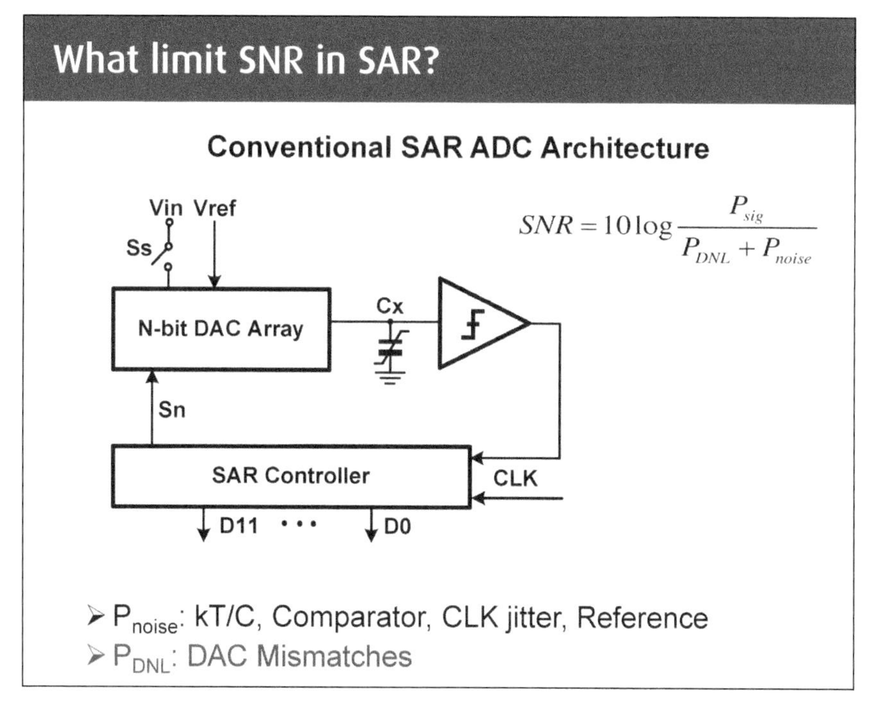

The conventional SAR ADC consists of a comparator, a CDAC and SAR logic. The non-idealities from the DNL error and circuit noise dominate the SNR of the ADC. The noise are mainly contributed from kT/C, comparator thermal noise, clock jitter and the reference noise due to switching transient. The DNL error comes from the error in DAC array, therefore, in this section we seek for the digital assisted solution to fix the problem.

34

Trade-offs Using Binary & Bridge DACs

11b Binary Weighted DAC

Vout, 2^{10} C, 2^9 C, C, C, S_{11}, S_{10}, S_1, S_0, Vin, +Vref, -Vref

(a)

11b Bridge Structure DAC

$C_{out} \approx C$, 6b MSB Array, C_a, 5b LSB Array, Vout, 2^5 C, 2^4C, C, 2^4C, C, C, S_{11}, S_{10}, S_6, S_5, S_1, S_0, Vin, +Vref, -Vref

(b)

12b kT/C &1.2V_{pp} → $C_T \approx 1.2$pF

For BW, C_0=0.6fF, $\frac{\sigma_0}{C_0} < 0.55\%$

For BS, C_0=19fF, $\frac{\sigma_0}{C_0} < 0.1\%$

➢ $\sqrt{32}$ times better unit element matching in BS cancels out

➢ Bridge-DAC: less interconnection and parasitices

The conversion linearity of an SAR ADC relies basically on the capacitive DAC, where the binary-weighted and bridge structures are commonly used. The bridge-DAC has less total unit capacitors compared with the binary-weighted DAC under the same total capacitance according to the kT/C noise requirement. For example, an 11-bit DAC with binary-weighted structure requires a total of 2048 units, while the bridge-DAC obtains a minimum of 97 units, implementing 6b and 5b in the MSB array and the LSB array, respectively. The better unit element matching cancels out, since the matching requirement of unit in split DAC is $32^{1/2}$ times critical than the binary one. Therefore, the bridge-DAC exhibits less interconnection and overhead area loss that has more potential structure for high speed.

35 

Nonlinearity in Bridge DAC

Nonlinearity

- **Inner node parasitic C_{PB}**
- **Mismatch of none unit Ca**

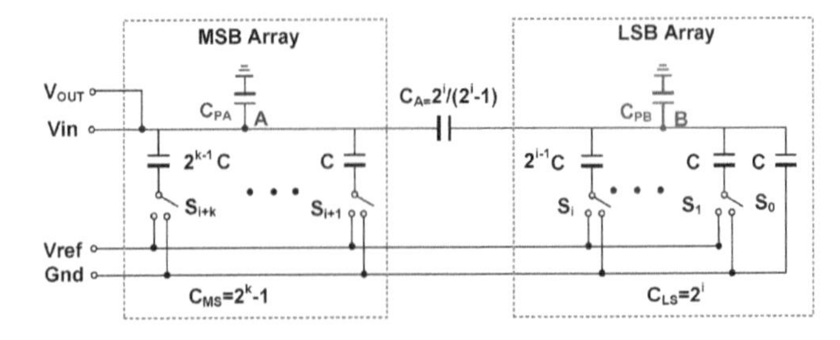

$$V_{out} \approx \frac{1}{\delta}(\underbrace{\frac{2^iC+C_{PB}+C_a}{C_a}\sum_{j=1}^{k}2^{j-1}CS_{i+j}}_{\text{MSB Array}}+\underbrace{\sum_{j=1}^{i}2^{j-1}CS_j}_{\text{LSB Array}})\cdot V_{ref}$$

$$\delta=(2^i+2^k-1)C+2^{i+k}C/C_a$$

However, the bridge-DAC suffers conversion error due to the mismatch of the non-unit attenuation capacitor Ca as well as the top-plate parasitics at the inner node of the bridge-DAC. The analog output Vout of a bridge-DAC, including the parasitics C_{PA} and C_{PB}, can be calculated as shown in this slide, where S_j equal to 1 or 0 represents the DAC connecting V_{ref} or Gnd for bit n. The C_{PA} and C_{PB} is correlated with the bit decision of the MSB array causes a gain error between the MSB array and the LSB array.

36 

Nonlinearity in Bridge DAC

Design example of a 4b(2b+2b) Bridge-DAC

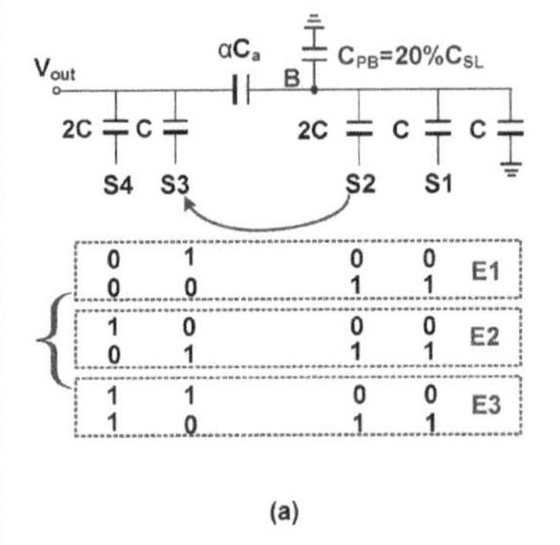

(a)

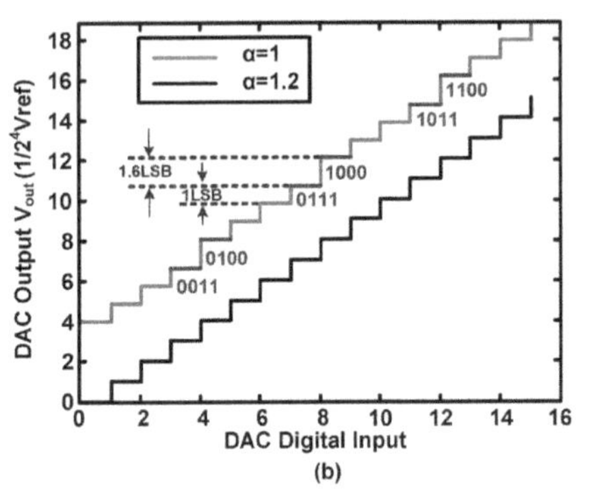

(b)

Correct the nonlinearity by increasing α, according to

$$\frac{\alpha C_a}{\alpha C_a+C_{SL}+C_{PB}}=\frac{C_a}{C_a+C_{SL}}\Rightarrow\alpha=1+\frac{C_{PB}}{C_{SL}}.$$

As shown in this slide a 4-bit (2b MSB array + 2b LSB array) example of a bridge-DAC, with a 20% top-plate parasitic capacitance C_{PB} of the LSB array is considered, and DAC is otherwise ideal. The least significant bits of the MSB array and LSB arrays are S3 and S1, respectively. Ideally, the ratio between two outputs $V_{out}(0100)$ and $V_{out}(0001)$ is 4:1. As there is C_{PB}, the ratio increases to 4.6:1. This ratio mismatch causes periodical large DNLs at 3 pairs of two adjacent inputs E1, E2 and E3. Each contains the carry from the LSB array to the MSB array (S2 to S3). The corresponding DAC outputs versus its digital inputs are plotted in Fig. (b). At 3 transitions of E1,E2 and E3 the DNL is 1.6LSB. To compensate the nonlinearity the step size of the LSB should be increased simultaneously, which can be achieved by enlarging the ratio **α** of Ca, where C_{SL} is the total capacitance in the LSB array. Since C_{PB} is assumed as 20% of C_{SL}, the ratio **α** should be 1.2.

37

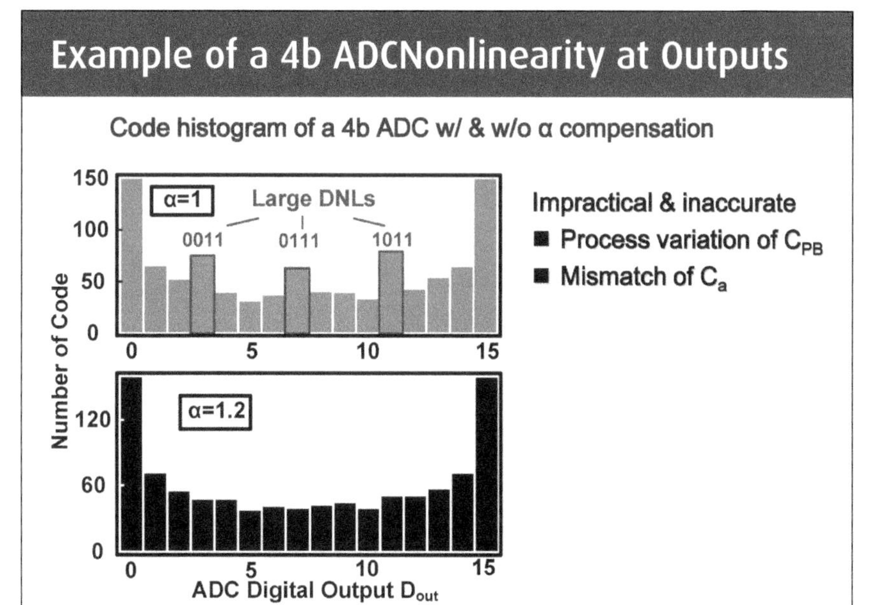

The code histogram of the 4b example is illustrates in this slide. Without compensation (α=1) there exists large peaks distributed periodically. The error codes corresponding to the input of the DAC where the nonlinearity occurs in previous slide. The peaks can be compensated by enlarge the ratio α to 1.2. However, in practice it is difficult to guarantee the compensation accuracy of Ca, because of the process variation value of the C_{PB} and mismatch of non-unit capacitance of Ca. For 12b accuracy it is required that the variation of Ca should be less than ±0.9% leading to the SNDR drop of <1 dB.

38

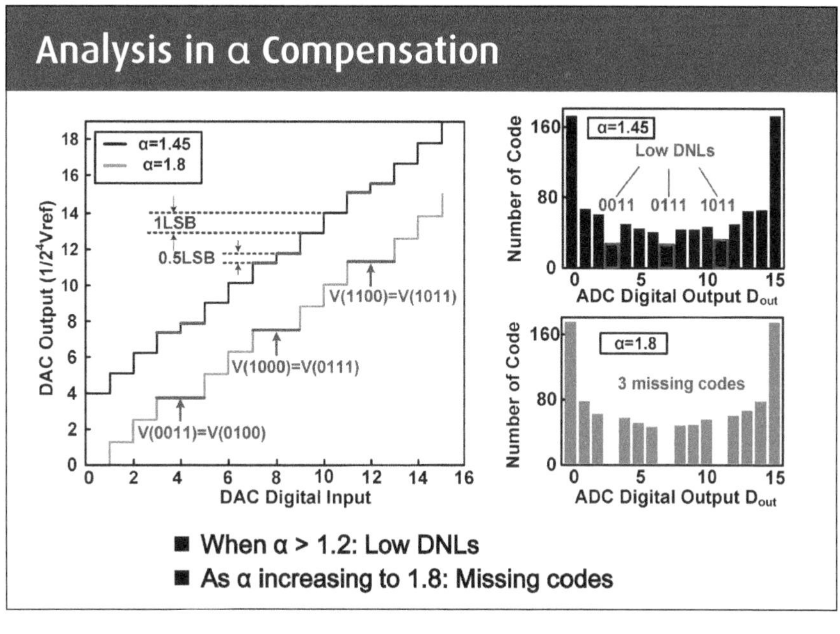

If the ratio α of Ca increases beyond the compensation value 1.2, the nonlinearities in the DAC appear in the opposite direction. The error transition points are decreased by enlarging the ratio α to 1.45. Accordingly, that results in correspondingly negative DNL of −0.5LSB at the output of the ADC. Since the DNL is within ±0.5LSB, it is acceptable. However, if the ratio α is further increasing, it will cause DNL less than−0.5LSB or missing codes. E.g. if α is increased to 1.8, the output voltage of the DAC at codes 0011, 0111, and 1011 is equivalent to their corresponding right adjacent codes 0100, 1000 and 1100, which causes missing codes at the output of the ADC. The code histogram of the ADC with an α of 1.8 is shown. As expected, the three digital codes are missing due to the non-idealities in the DAC.

39 Analysis in α Compensation

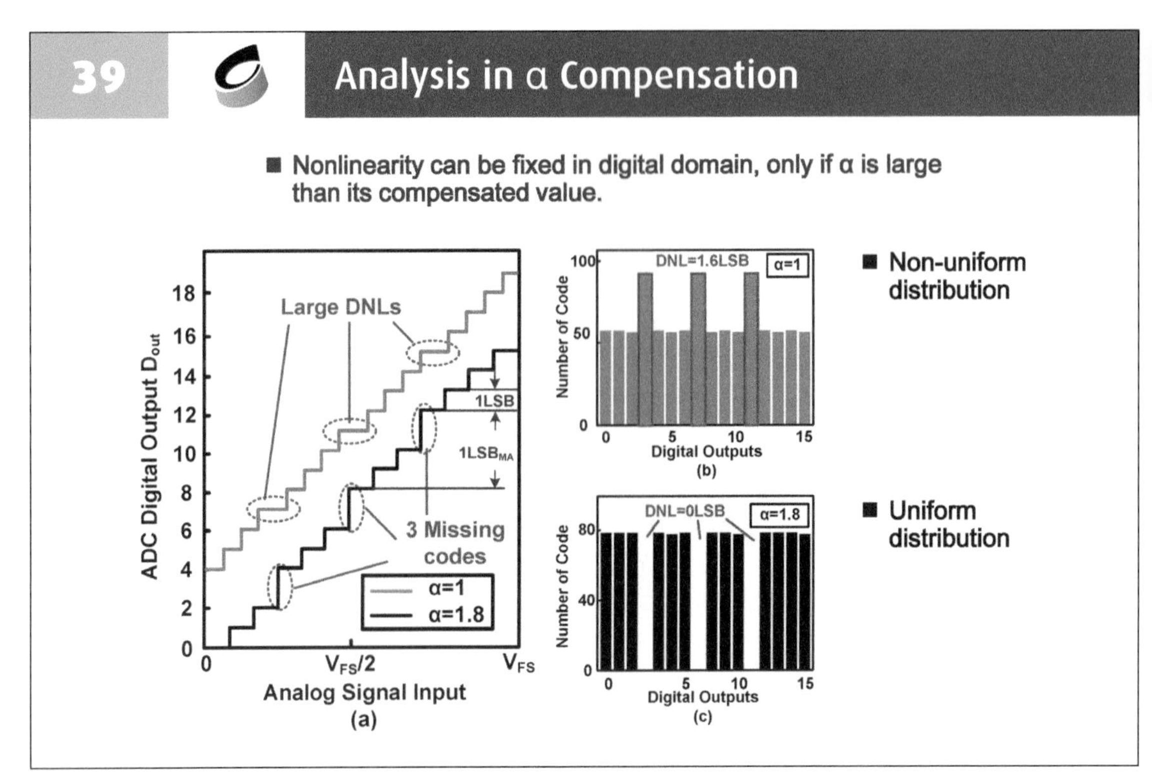

The proposed HBRM calibration corrects the conversion nonlinearities in the digital domain, which does not rely on the compensation accuracy between Ca and C_{PB}. The attenuation capacitor Ca is initially designed to be larger than its compensation value. Therefore, even when there is variation of C_{PB}, the nonlinearities at the ADC output are systematically negative DNL < 0 LSB or missing codes. This is the key requirement to perform the calibration, because the positive DNL cannot be fixed in the digital domain. The output characteristics of the 4-bit ADC with respect to **α** of 1 and 1.8 are plotted. When **α** =1, there exists positive DNL. These codes accumulate a larger number of the hits than the others. A larger number of analog inputs are represented by the same digital code, then, impossible to be differentiated in the digital domain. However, for the case that contains the missing codes, the number of hits for each code is approximating (more uniformly distributed). Thus, its nonlinearities can be corrected accordingly by the proposed calibration.

40  Histogram Based Ratio Mismatch(HBRM) Cal.

Its nonlinearities can be corrected accordingly through the multiplication of all the bits in the LSB array by a gain factor **β**, where B_n equal to 1 or 0 represents the digital output. Once **β** is applied to the digital output, the D_{out} at the left adjacent of the missing code will overflow. Therefore, the digital outputs reacquire continuity, as the missed steps are compensated by enlarging the step size to **β** LSB. When normalized the output to 1LSB, the missing codes are shifted to the upper boundary of the output.

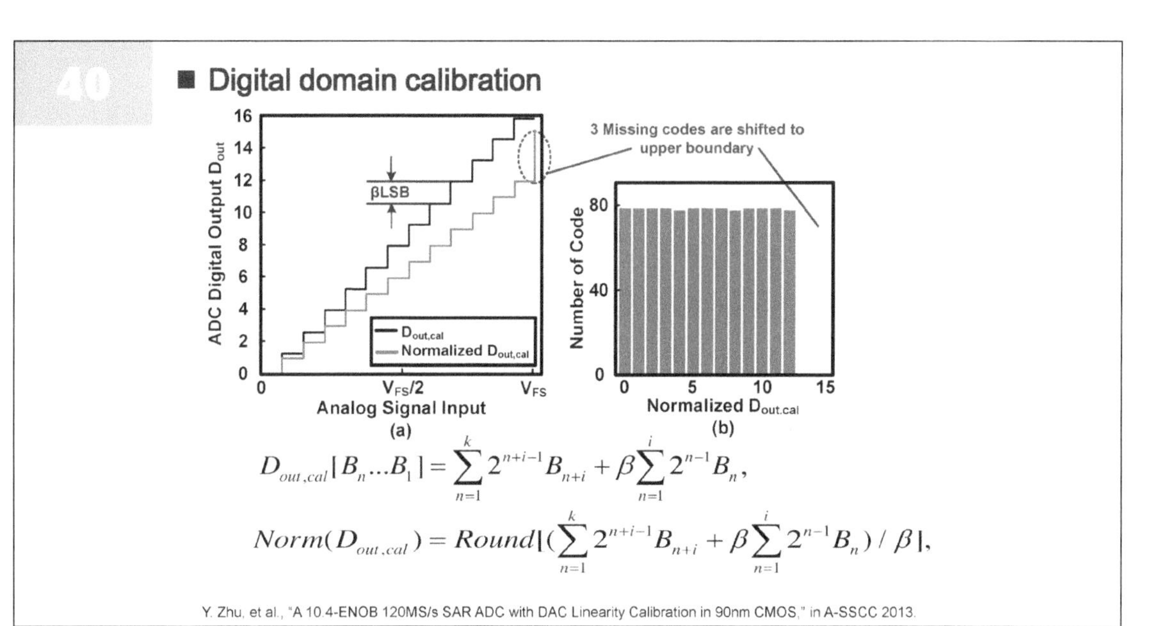

41 

Estimation of β

- **The gain factor β is estimated via codes histogram statistic**

$$\beta = \frac{2^i}{N_m}$$

- The 2^i is ideal No. code in i-bit LSB array
- N_m sum of measured code density

Steps:

1) Consecutive 2^i code in middle
2) Collect total No. of k codes ($N_{avg}=k/2^i$)

If $N(D_{out}) \geq N_{avg}/2$, then $C_D(D_{out})=1$

If $0 \leq N(D_{out}) < N_{avg}/2$, then $C_D(D_{out})=N(D_{out})/N_{avg}$

$N_m=C_D(1)+...+C_D(2^i)$

Ideally, with i-bit in LSB array the bits full-scale of the LSB array is 2^iLSB, which matches with the calibrated full-scale. Accordingly, β can be derived as $\beta = 2^i/S_m$, where S_m is the measured step sum, which is estimated by the codes histogram statistics. Since the nonlinearity has a periodical interval of 2^i, to simplify the statistical analysis, the code range evolves only in 2^i consecutive outputs. As with a sine-wave input, the digital outputs are more uniformly distributed in the mid-code region than those at the two sides, the statistical range is set from $D_{out}(2^{n-1}+10)$ to $D_{out}(2^{n-1}+2^i+10)$. Assuming the average code count is N_{avg}, the threshold for erroneous output is defined as half of the average value ($N_{avg}/2$). As mentioned before, the digital outputs contain only the negative DNLs. If $N(D_{out}) \geq N_{avg}/2$ (DNL≥−0.5LSB), which indicates less conversion nonlinearity, the code will be counted as 1LSB to S_m. If $0 \leq N(D_{out}) < N_{avg}/2$ (DNL<−0.5LSB), the code contains the error that is added as a fraction of LSB, where the fraction is defined as $N(D_{out})/N_{avg}$. The comparison repeats 2^i times. Once the value of S_m is decided according, the gain factor β can be estimated and applied to the i LSBs in the subsequent digital outputs to perform the correction. As β is only applied to the LSBs instead of to the full n bits, the implementation of the digital multiplier can be much more relaxed.

42 

Circuit Implementation

- 11b (6b+5b) bridge DAC is implemented
- C_{PB}/C_{SL}≈3% according to layout extraction
- C=20fF, C_a=27fF (α=1.35)

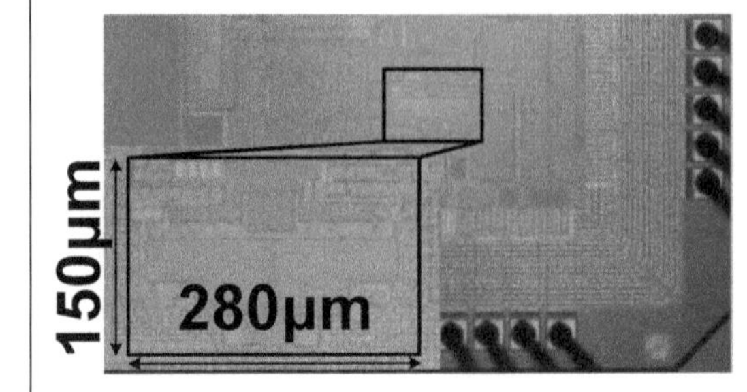

- 12bit 120MS/s SAR in 90nm CMOS
- Active Area:0.042mm²
- Input Cap. 1.28pF
- Power:3.2mW,1.2V
- HBRM is off-chip

The 12-bit SAR ADC was fabricated in a 1P9M 90-nm CMOS process with MOM capacitors. The die photograph of the design with Ca of 27 fF; the active area is 0.042 mm². The analog power consumption is 0.8 mW and the digital power including the SAR logic and clock generator is 2.4 mW. The total power consumption is 3.2 mW at 120 MS/s from 1.2 V supply. The digital outputs are post processed according to the HBRM calibration algorithm. The calibration is controlled by MATLAB running on a PC, but the controller could be integrated in VHDL.

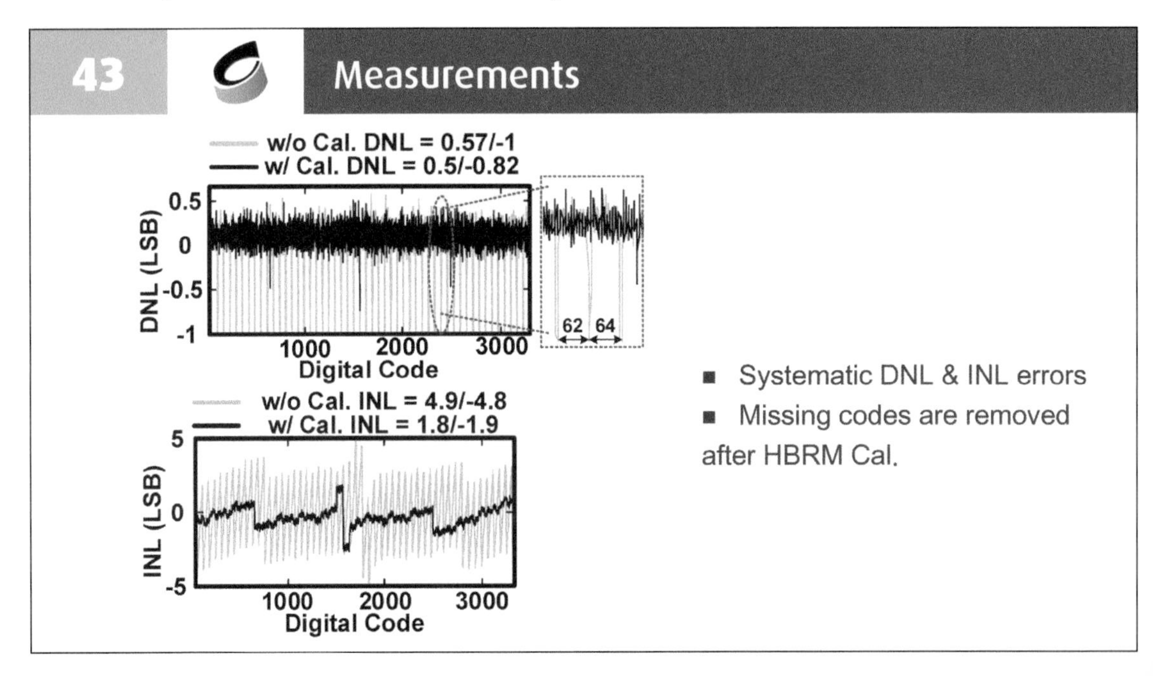

The measured static performances with and without the HBRM calibration are shown in this slide. Based on the measurement, the DNL before calibration has the systematic comb-like pattern due to a large number of missing codes. Since 6 bits are determined in the LSB array, the error is periodic and have an interval of near 64 codes. In addition, the INL contains the systematic sawtooth pattern that reflects the intrinsic conversion nonlinearity relevant to the capacitor mismatches due to the use of the split DAC. When the calibration is active, the missing codes in DNL are all removed and the INL is improved to within 2LSB, correspondingly.

44 

Measurements

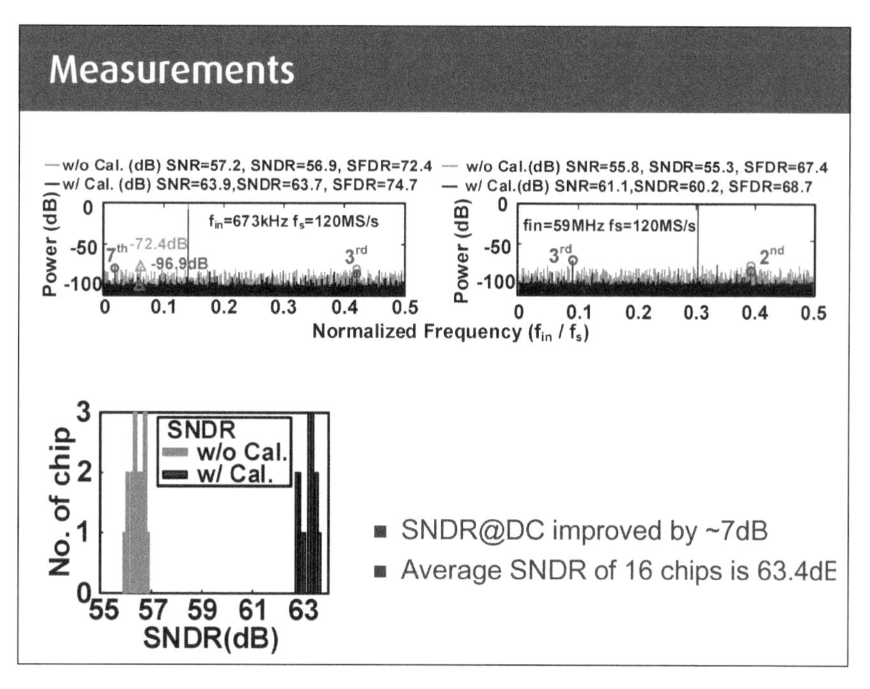

This slide shows the measured FFT at 673 KHz and at Nyquist frequency (59 MHz) with a conversion rate of 120 MS/s. Before calibration, the ratio mismatch causes a bunch of spurs spreading among the whole spectrum and the rise to the third harmonic. This limits the SNR and the SFDR to 57.2 and 72.4 dB, respectively. Once the calibration is active, the spurs are removed. The spur limiting the SFDR before calibration is suppressed to lower than−96.9 dB. In addition, the third harmonic is below −81 dB. Consequently, both SNR and SNDR are improved by near 6.7 dB. The seventh harmonic on the left-hand side dominating the SFDR is potentially caused by the capacitor mismatches rather than the ratio mismatch of the bridge-DAC. As the input frequency increases to Nyquist, the sampling distortions originate the third harmonic that dominates the SFDR before and after calibration. Even though the SFDR is slightly improved by 1.3 dB, the SNDR can be increased by 4.9 dB. The measured dynamic performance of the total available 16 chips at dc input and 120 MS/s sampling rate demonstrate that the design achieves average SNDR of 56.4 and 63.4 dB, respectively, before and after calibration.

45 

Conclusion

- **Architecture Optimizations**
  - Hybrid structures-High speed, better noise performance
  - Multi-shared elements-Relax Cal., small area
- **Accuracy Optimized Techniques**
  - CSE Bootstrap -Improve timing skews
  - HBRM Cal.-Improve SNDR
- **Implementation of multiple Designs (65nm)**
  - Achieved the Excellent FoM for High-Speed

46  Reference

1. Boris Murmann: ADC Survey"https://www.stanford.edu/~murmann/adcsurvey.html".
2. Yan Zhu, Chi Hang Chan, Seng-Pan U, R. P. Martins, "An 11b 450 MS/s 3-way Time-Interleaved Sub-ranging Pipelined-SAR ADC in 65nm CMOS", *IEEE Journal of Solid-State Circuits*, Feb-2016
3. V. Tripathiand B. Murmann, "A 160 MS/s, 11.1 mW, Single-Channel Pipelined SAR ADC with 68.3 dB SNDR," in *Proc. of IEEE Custom Integrated Circuits Conference (CICC)*, pp. 1-4, Sep. 2014
4. C. Y. Lin and T. C. Lee,"A12-bit 210-MS/s 5.3-mW Pipelined-SAR ADC with a Passive Residue Transfer Technique," in *Symp. VLSI Circuits Dig. Tech. Papers*, pp.1-2, Jun. 2014.
5. Y. Zhou, B. W. Xu and Y. Chiu, "A 12b 160MS/s Synchronous Two-Step SAR ADC Achieving 20.7fJ/step FoM with Opportunistic Digital Background Calibration," *in Symp. VLSI Circuits Dig. Tech. Papers*, pp.1-2, Jun. 2014.
6. Y. Zhu, et al., "An 11b 900 MS/s Time-Interleaved Sub-ranging Pipelined-SAR ADC," in *Proc. of IEEE ESSCIRC*, pp. 211-214, Sep. 2014.
7. S. Gupta; M. Choi; M. Inerfield; JingboWang ,"A 1GS/s 11b Time-Interleaved ADC in 0.13/splmu/m CMOS," *IEEE ISSCC*, Feb-2006.
8. M. El-Chammas,B. Murmann, "A 12-GS/s 81-mW 5-bit Time-Interleaved Flash ADC With Background Timing Skew Calibration," *IEEE Journal of Solid-State Circuits*, 2011
9. H.G. Wei, P. Zhang, B. D. Sahoo, B. Razavi, "An 8 Bit 4 GS/s 120 mWCMOS ADC," *IEEE Journal of Solid-State Circuits*, 2014
10. N. L. Dortz, et al., "A 1.62GS/s Time-Interleaved SAR ADC with Digital Background Mismatch Calibration Achieving Interleaving Spurs Below 70dBFS," in *ISSCC*, Feb. 2014.
11. E. Janssen, et al., "An 11b 3.6GS/s Time-Interleaved SAR ADC in 65nm CMOS," in *ISSCC*, Feb. 2013.
12. Bob Verbruggen, et al., "A 2.6 mW6 bit 2.2 GS/s Fully Dynamic Pipeline ADC in 40 nm Digital CMOS," in *IEEE JSSC*, Oct. 2010.
13. SunghyukLee, et al., "A 1GS/s 10b 18.9mW Time-Interleaved SAR ADC with Background Timing-Skew Calibration," in *ISSCC*, Feb. 2014.
14. Simon M. Louwsma, et al., "A 1.35 GS/s, 10 b, 175 mWTime-Interleaved AD Converter in 0.13 µm CMOS," in *IEEE JSSC*, vol. 43, no. 4, pp. 778 –787, Apr. 2008.
15. S. H. W. Chiang, et al., "A 10-Bit 800-MHz 19-mW CMOS ADC," in *Symp. VLSI Circuits Dig. Tech. Papers*, pp.100-101, Jun. 2013.
16. C. C. Liu et al., "A 10b 100MS/s 1.13mW SAR ADC with binary-scaled error compensation," in *ISSCC Dig. Tech. Papers*, pp. 386-387, Feb. 2010.
17. C. Y. Lin and T. C. Lee,"A12-bit 210-MS/s 5.3-mW Pipelined-SAR ADC with a Passive Residue Transfer Technique," in *Symp. VLSI Circuits Dig. Tech. Papers*, pp.1-2, Jun. 2014.

CHAPTER 5

CMOS Image Sensors and Biomedical Applications

Jun Ohta

Nara Institute of Science & Technology
Japan

This chapter introduces recent advancement of CMOS image sensors. The fundamental operation principle is mentioned in detail as well as some recent topics including shrinking pixel pitch, high sensitivity, high speed, etc. Finally, biomedical applications with CMOS image sensor technology are demonstrated such as retinal prosthesis, fluorescence detection for ELISA, and implantable micro imaging devices.

1  Outline

- **Image sensor technologies**
  - Trend of CMOS image sensors
  - Fundamental characteristics of CMOS image sensors
  - Classification of image sensors for biomedical applications
- **Fluorescence detection**
  - Introduction
  - Voltage sensitive dye: Measurement of cell activities through fluorescence with high speed CMOS image sensor
  - FLIM (fluorescence lifetime imaging microscopy): Measurement of fluorescence lifetime with ultrafast CMOS image sensor
- **Retinal prosthesis**
  - Introduction
  - Implantable stimulators in subretinal space
- **Summary**

This is the outline of my talk. First, I will introduce the image sensor technology, especially for the CMOS image sensors, including the trend and fundamental characteristics of CMOS image sensors. I will also discuss on some classification criteria of image sensors for biomedical applications. Then I will show you two examples of CMOS image sensors for biomedical applications. The first example is fluorescent detection, which is very important in life science, and I will show 2 to 3 examples of image sensor applications in this aspect. The second example is retinal prosthesis. Retinal prosthesis is a remedy/regain of vision for blind patients. For instance, micro chips can be implanted into the eye directly to stimulate the patient's retina to vision restoration. I will introduce some of my research as examples in retinal prosthesis using the CMOS image sensor technology. And finally, I will summarize my talk.

2 

Outline

- **Image sensor technologies**
  - Trend of CMOS image sensors
  - Fundamental characteristics of CMOS image sensors
  - Classification of image sensors for biomedical applications
- **Fluorescence detection**
  - Introduction
  - Voltage sensitive dye: Measurement of cell activities through fluorescence with high speed CMOS image sensor
  - FLIM (fluorescence lifetime imaging microscopy): Measurement of fluorescence lifetime with ultrafast CMOS image sensor
- **Retinal prosthesis**
  - Introduction
  - Implantable stimulators in subretinal space
- **Summary**

3 

What is an image sensor?

What is an image sensor? In fact, you should all be very familiar with the image sensor in your smart phone. This is a low cost webcam which you can buy easily. You just need to connect it to a PC through the USB port to directly get an image. If you take away the cover and then disassemble the lens, you can see the image sensor inside. By placing the image sensor under an optical microscope, you can see the imaging plane with integrated circuit. And when the magnification is increased, you can see an array of pixels as shown. These pixels can detect the light, and each pixel has its own color filter, in this case red, green and blue. By using these color filters, the image sensor can sense the color, the intensity profile and the distribution of the incoming light. You can then obtain a clear image similar to the case for your smart phone.

4 

Trend in Image Sensors

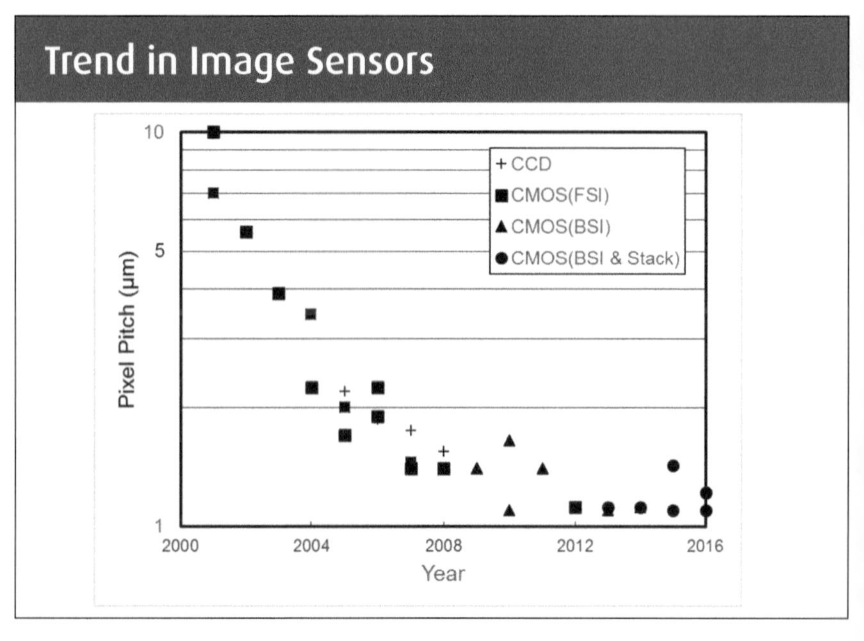

Such image sensors are installed for many different applications, such as your smart phones, high-end digital cameras, and also professional broadcast cameras. They are also popular for medical use, where an image sensor can be mounted at the end of a probe to view the inside of your stomach, i.e. endoscopy. This slide shows the recent trend of image sensors. You can see the pixel pitch, which is equivalent to the pixel size of the image sensor, decreases dramatically over the years, from about 10 μm in 2000 to roughly 1 μm where it settles. When the pixel/photodiode size reaches 1 μm, it is almost equal to the wavelength of light. As a result, the image sensor performance is mainly limited by the diffraction limit, making it very difficult for accurate light detection.

5 

More Moore and More Than Moore

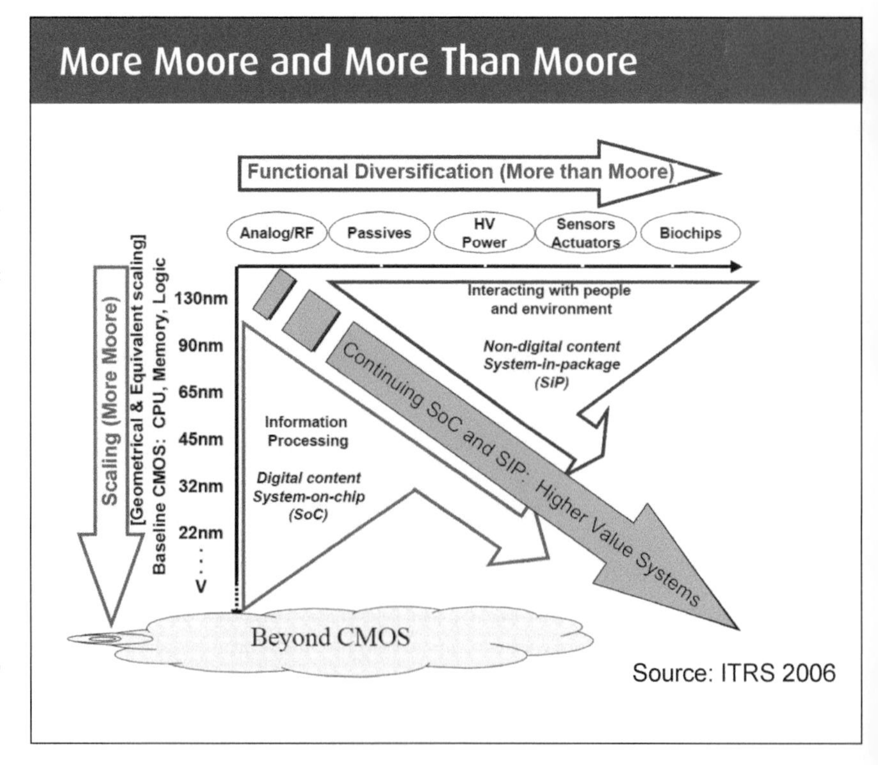

This slide shows the Moore's law. As you know, the Moore's law must continue, but there should be two aspects of this law. These include technology scaling, which is referred to as more Moore, and also the more than Moore, where the fine process technology is utilized to achieve more functional diversification, for examples analog, RF, sensors, and also biomedical applications.

6 

Although the pixel pitch for the fine process has reached the flat region in CMOS image sensors, there are possibilities for application extensions in other applications. We are not limited to just use CMOS image sensors to take a pictures, but also apply them to view the inside of your body, or detect the fluorescence, or monitor the pH and many more. So you can use this technology for other fields.

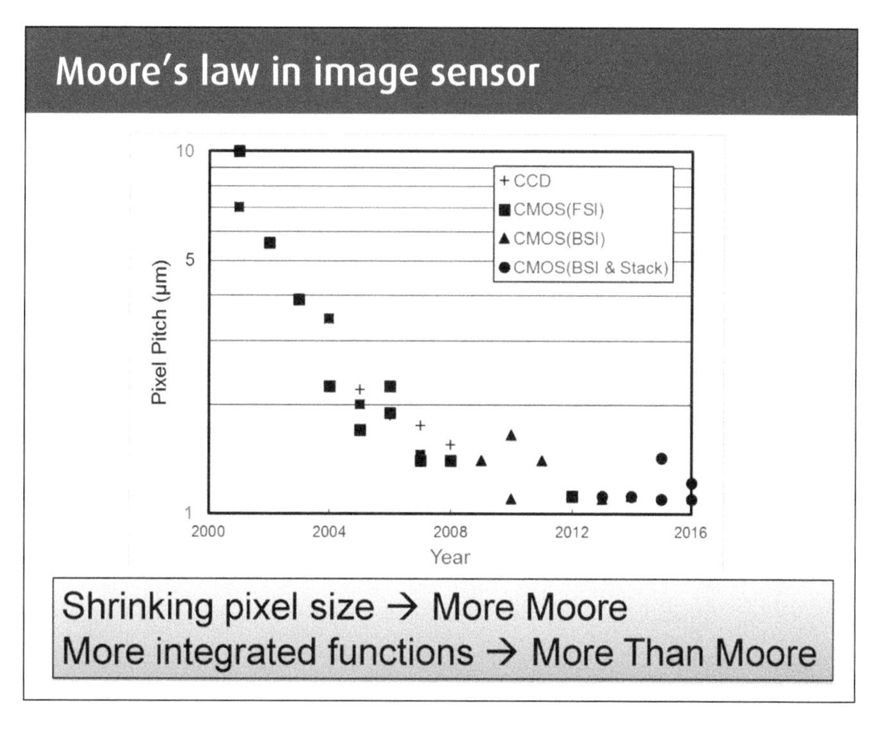

7 

This slide shows a very basic design of CMOS image sensor. As discussed, CMOS image sensor consists of an array of pixels. Each pixel has a photodiode, a buffer and some switches, and it is interfaced with the column and row lines. The buffer is simply a source follower transistor. When the horizontal control line is exerted, the buffered photodiode signal flows into the vertical readout line. This signal is then readout using the current load and stored using the sample and hold circuit. As a result, you can output the photodiode voltage to the vertical output line with almost unity gain. So in this stage, you may have some questions, as a photodiode basically generates photocurrent, but not a voltage. It is true that when you shine light onto a photodiode, it generates a photocurrent.

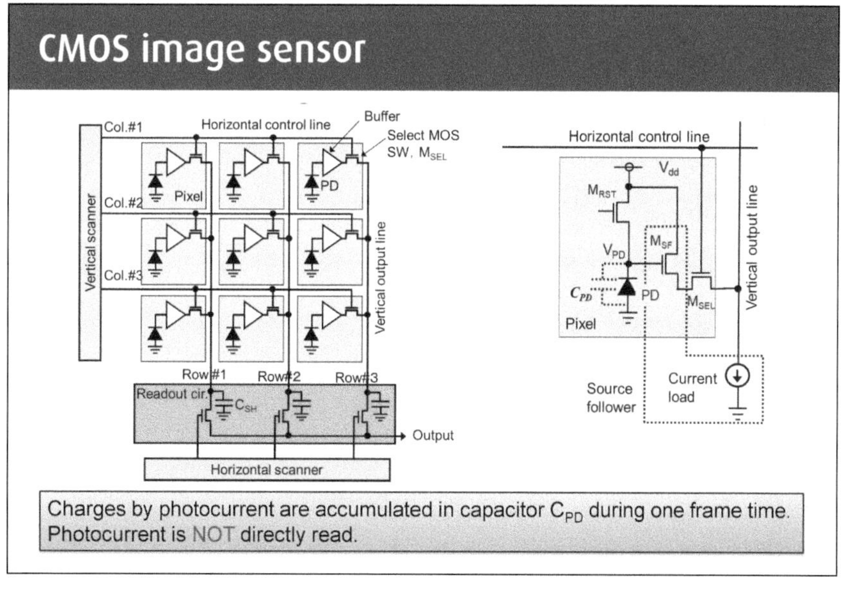

For image sensor, you usually do not use the photocurrent, instead you use the charge produced by the photodiode. Within the pixel, the charge induced by the photocurrent is accumulated by C_{PD}, which is the capacitance of the photodiode. The accumulation/integration time limits the frame rate of the image sensor, which is roughly 1/30 seconds, or about 33ms per frame. So we readout the stored charge produced by the photocurrent for an integration time of 33 ms.

8 

Why is charge accumulation required in CMOS image sensor?

- PD size in CMOS image sensor: small
  → Photocurrent: small
  - 10 μm square pixel → PD size < 100 μm²
  - Lecture room ~500 lux → ~50 μW/cm²
  - PD sensitivity ~0.6 A/W (Si: ideal ~1 A/W)→~ 30 μA/cm²
  - Photocurrent from one pixel →30 pA
    → Difficult to detect all of the pixel output current precisely
- PD capacitor in CMOS image sensor: small (~PD area)
  → Voltage: large (V=Q/C)
  - C_{PD}~100 fF, I_{ph}~30 pA
  - Accumulation time: t=1/30 s
  - Voltage in C_{PD}
    → $V_{PD} = I_{ph}t/C_{PD}$~10 V!

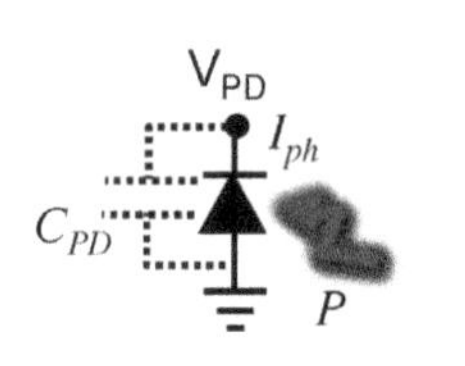

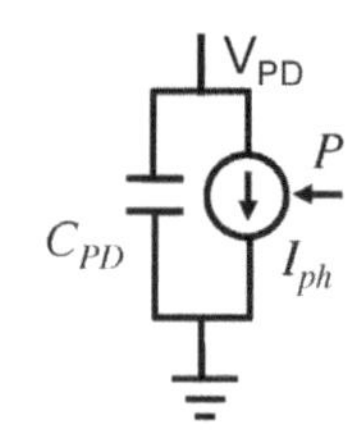

So why is charge accumulation required in CMOS image sensor? The photodiode size in CMOS image sensor is of course very small, in the order of few μm². Consequently, for a square pixel with 10 μm on each side, the photodiode size is less than 100 μm², which is a very small size. As an example, the illumination level in this lecture room is about 500 lux, which can be converted to roughly 50 μW/cm². With a typical silicon photodiode sensitivity of 0.6 A/W, you can get a photocurrent of about 30 μA/cm². You can then calculate that for a photodiode size of 100 μm², you can only get 30 pA, which is very small. Of course, we can measure this small current of 30pA using ammeter with pA precision fairly easy. But why are we not using this in CMOS image sensor? First, image sensor has an array of pixels, so that each photodiode produces this very small current. Obviously, they do not produce the same current. One can produce 30pA, another can produce 25pA, and the next 10pA. So you need to measure this small current very precisely in every pixel. For example, the image sensor in your smart phone can have over 10M pixels or so. It's almost impossible to measure this large amount of photocurrent precisely. To solve this problem, we can use the stored charge by using the photodiode parasitic capacitance, which is very small and is almost the same area as the photodiode. The size of C_{PD} is in the order of 100 fF. We can obtain the voltage of the photodiode by using this simple equation, i.e. calculate the stored charge by multiplying the current and the integration time, then divide it by the photodiode capacitance. As a result, for a photocurrent of about 30pA and an accumulation time of 33 ms for one frame, you can get a voltage of almost 10V. This value can be easily measured using the source follower configuration. The key here is the small photocurrent and small capacitance in each pixel, and we have some time to accumulate enough charge before we readout the next frame. This is the equivalent circuit of the photodiode, which is composed of a current source controlled by the light intensity and a photodiode capacitance connected in parallel. When the photodiode is off/open circuit, the photocurrent gradually charges up this capacitor, and finally lead to some measurable value, which represents the intensity of the incoming light.

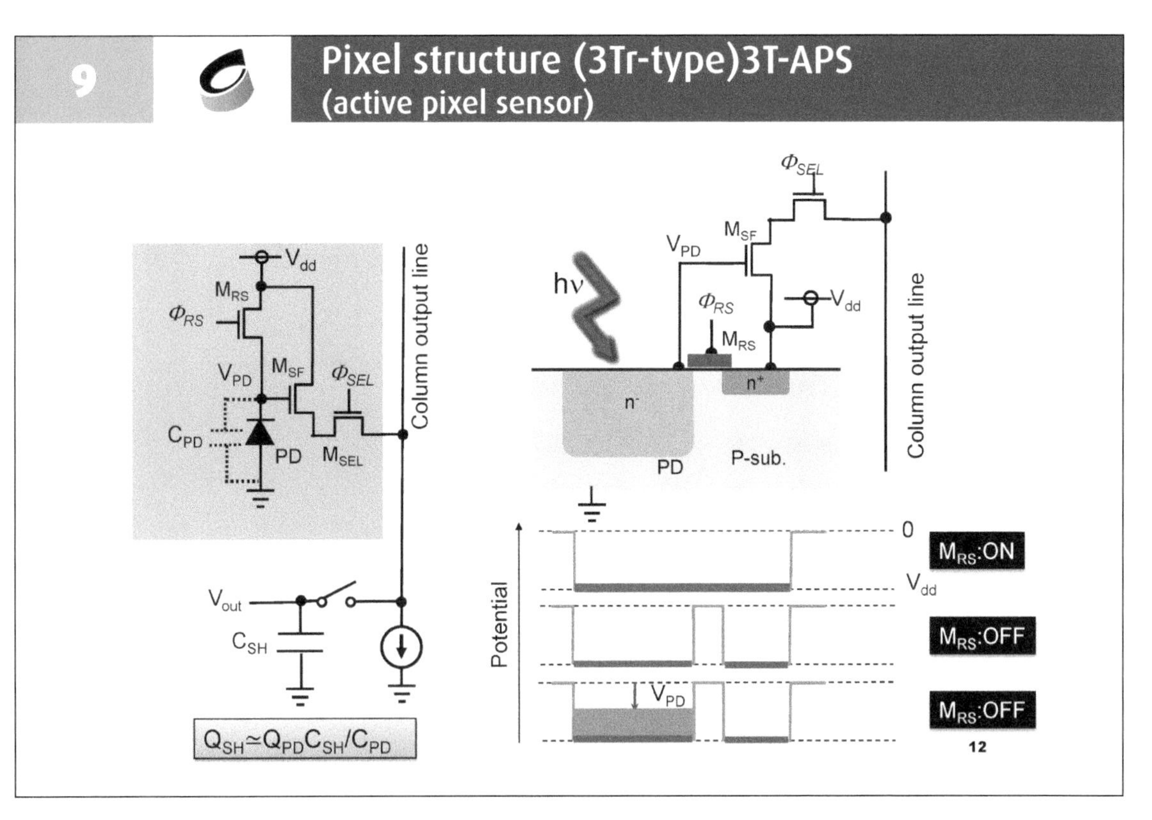

This slide shows the circuit implementation for one pixel. So you can see the photodiode here, the SF and some switches, including the reset and select transistors. If you disconnect the reset transistor, the photodiode is floating, which means that photodiode is charging up C_{PD}. Here shows the timing sequence for the pixel operation, and the schematic with partial cross section of the pixel circuit inside the substrate. The cross section shows the photodiode region which is the p-n junction here, and the reset transistor with the drain and source regions here. In this case, the large source region of the reset transistor is also the photodiode. The cathode of the photodiode is connected to the source follower transistor, which is further connected to the column output line through the select transistor here. The drain of the reset transistor is connected to the supply voltage V_{dd}. This shows the potential diagram for the cross section, bounded by the substrate ground (0 V) and the supply voltage (V_{dd}). The colored block shows the change in potential due to electron charges. Notice that the potential of electrons is lower than 0 V, and 0 V is a high potential to electron when compared to V_{dd} because electrons has negative charges. For the pixel operation, when the reset transistor is on, V_{dd} is connected to the photodiode, and the potential of this entire region is V_{dd}. When light shines onto the photodiode, charge is produced. But as the photodiode is still connected to V_{dd}, all of the photocurrent produced by light is drained to the supply. As a result, the potential is maintained at V_{dd}. When we turn off the reset transistor, a very high potential is produced is underneath the gate, leading to a change in the potential diagram like this. Note that charges cannot be transferred as the photodiode is completely floating. The produced charge by light, which are basically electrons, gradually piles up within the photodiode region, leading to a decrease of potential. We can then measure this voltage by this source follower transistor. So in the initial state, the photodiode potential is V_{dd}. When the light shines onto the photodiode, its potential gradually decreases. We can measure the difference in photodiode potential from V_{dd} and the final value with the column output line using the sample and hold circuit easily.

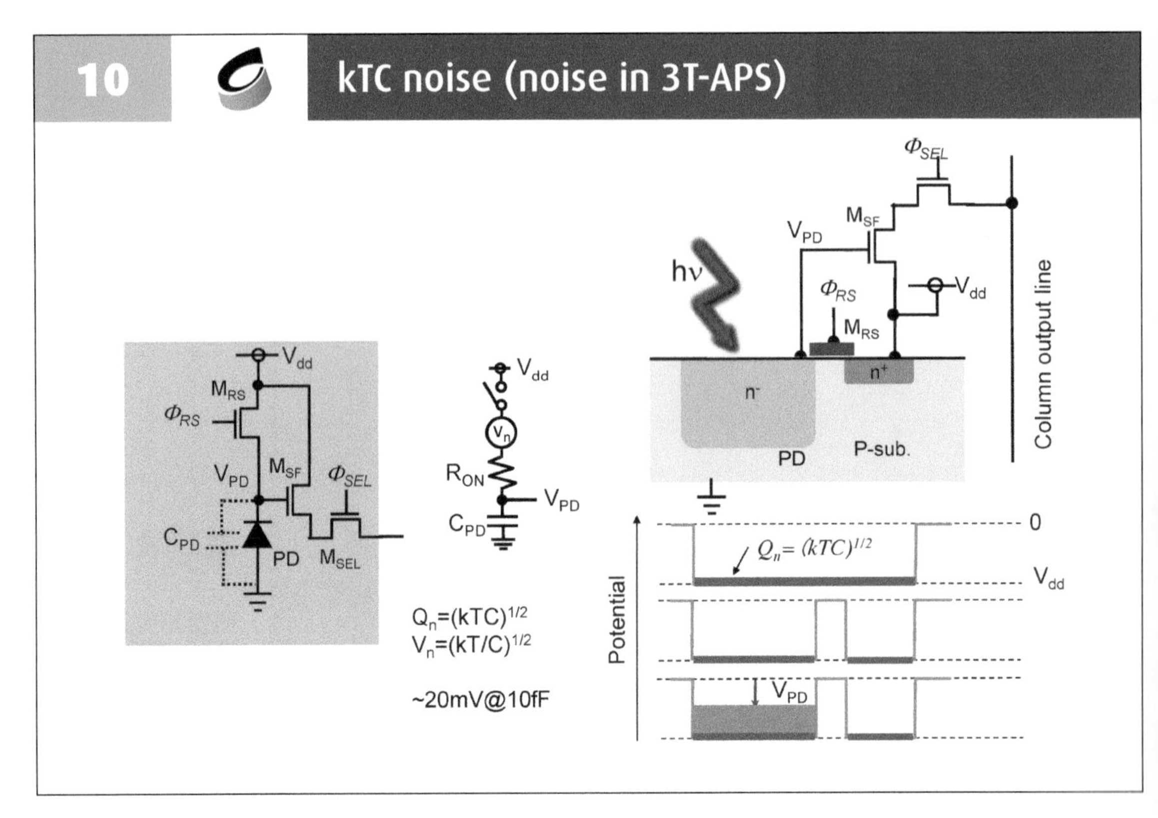

As discussed, we can measure the charge in the photodiodes in all the pixels using this method. But there is a very big noise in this case, that is the kT/C noise appearing due to the reset action. When the reset transistor is off, its finite resistance produces thermal noise, which in turn produce some fluctuation in the photodiode reset voltage. This kT/C noise is induced when the reset transistor is turned off, and can corrupt the signal induced by the photocurrent. It also fluctuates from pixel to pixel. For example, the reset voltage in one pixel is 3.3V, and the other 3.4V, and the next 3.2V. You can find out the kT/C noise like this, and the voltage noise is equal to the square root of kT/C. As an example, with a photodiode capacitance of 10fF, the kT/C noise can be up to 20mV, which is not small. This problem becomes worse as a result of pixel size shrinkage. As the photodiode capacitance is also reduced, the kT/C noise becomes more prominent.

11 Pixel structure (4Tr-type)4T-APS (active pixel sensor)

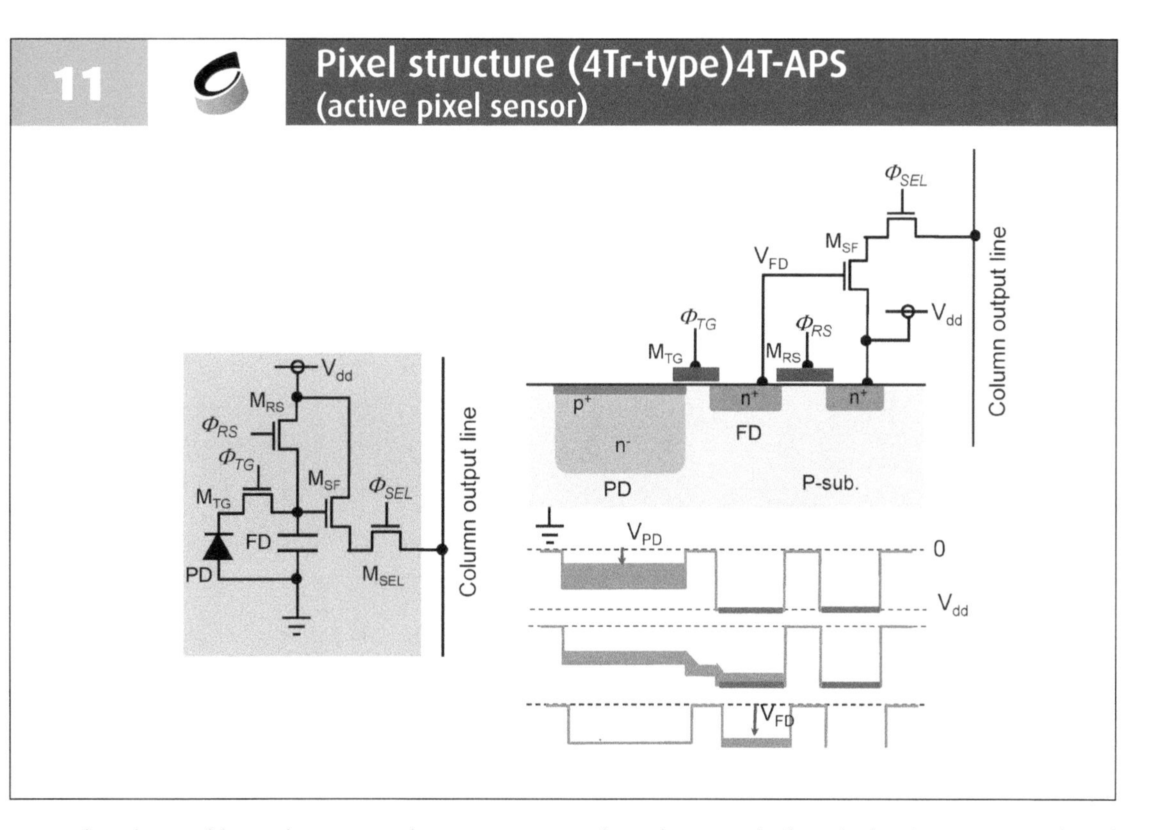

To solve this problem, the 4-T pixel structure is developed. The photodiode charge/voltage is still measured by the source follower configuration. The major difference is that another capacitance, called the floating diffusion, is added. This is achieved by including one more pixel transistor known as the transfer gate. You can see the photodiode here and the floating diffusion here. When the transfer gate is turned on, the stored charge is completely transferred from the photodiode capacitance to the floating diffusion capacitance. By using this configuration, we can eliminate kT/C noise. Here is the explanation. The floating diffusion is connected to the reset transistor. So when we reset the pixel, kT/C noise is added onto the floating diffusion like this. When the stored photodiode charge is transferred to the floating diffusion, the voltage corresponds to the summation of both the stored charge and kT/C noise. But just before the transfer of the photodiode stored charge, we can first measure the kT/C noise by using the source follower configuration. We can then transfer the charge to the floating diffusion, and readout the voltage at the floating diffusion node. As a result, if we can measure the difference between the kT/C noise and the stored charge plus kT/C noise, and obtain the signal charge only and completely eliminate the kT/C noise by this simple method. This is a very big advantage for achieving high quality imaging in CMOS image sensor.

12 CDS circuits to eliminate kTC noise

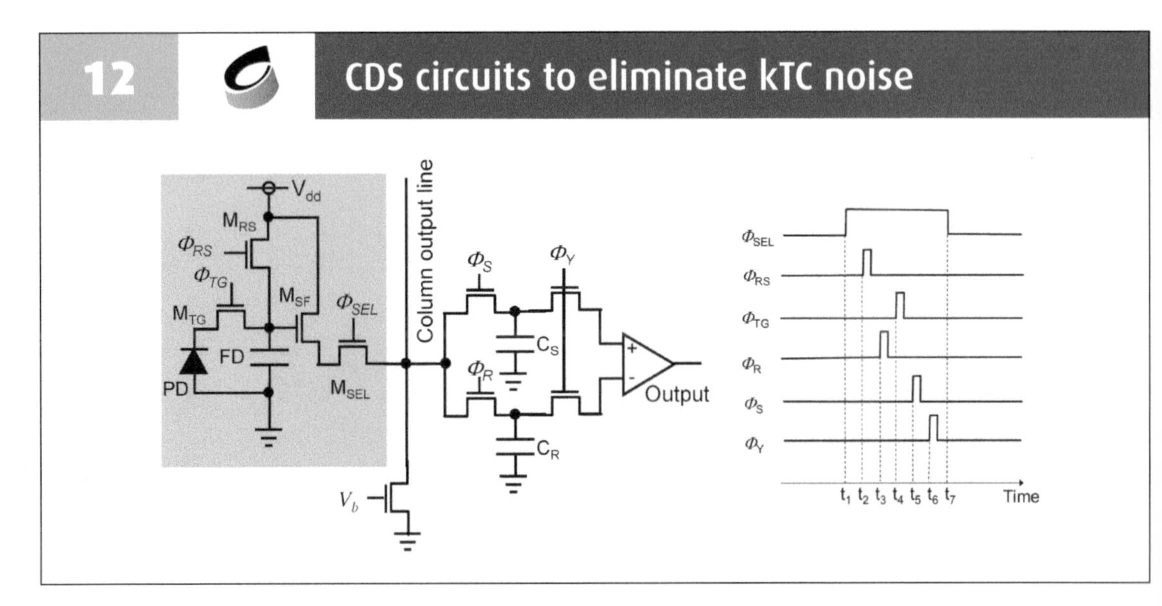

When you use the 3T configuration, you cannot eliminate the kT/C noise. This is because light is still shining onto the photodiode when we measure kT/C noise in this case. When the light is strong, the measured value will then include both the kT/C noise and the signal charge. Consequently, you cannot precisely measure the kT/C noise. But in the 4T case, you can, as the floating diffusion region cannot produce light induced charge. This is the major difference between 3T and 4T.

13 Pixel Sharing

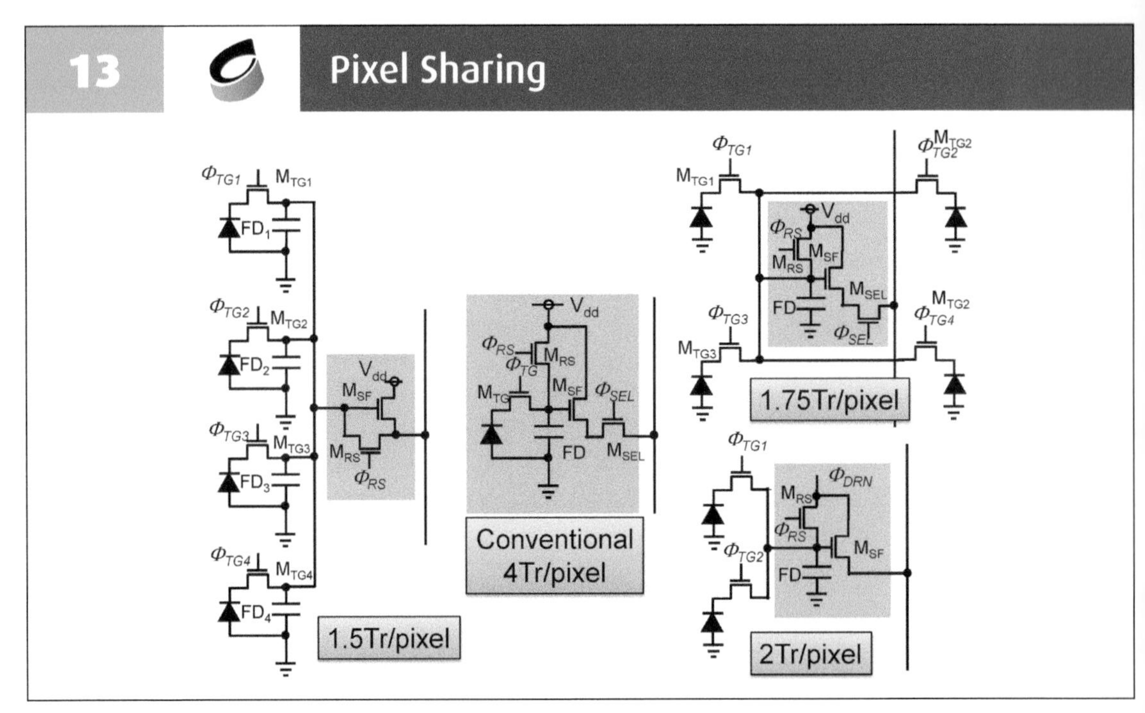

To reduce the pixel size, we can use the pixel sharing technique. This shows the conventional 4T pixel. By sharing transistors among different pixels, you can reduce the transistor count per pixel. For example, you can share the source follower and the select transistor among 2 pixels to achieve 2 transistors per pixel here. You can further reduce the transistor count to 1.75 or even 1.5 transistors per pixel by using pixel sharing. As a result, you can easily shrink the pixel size by using this technology.

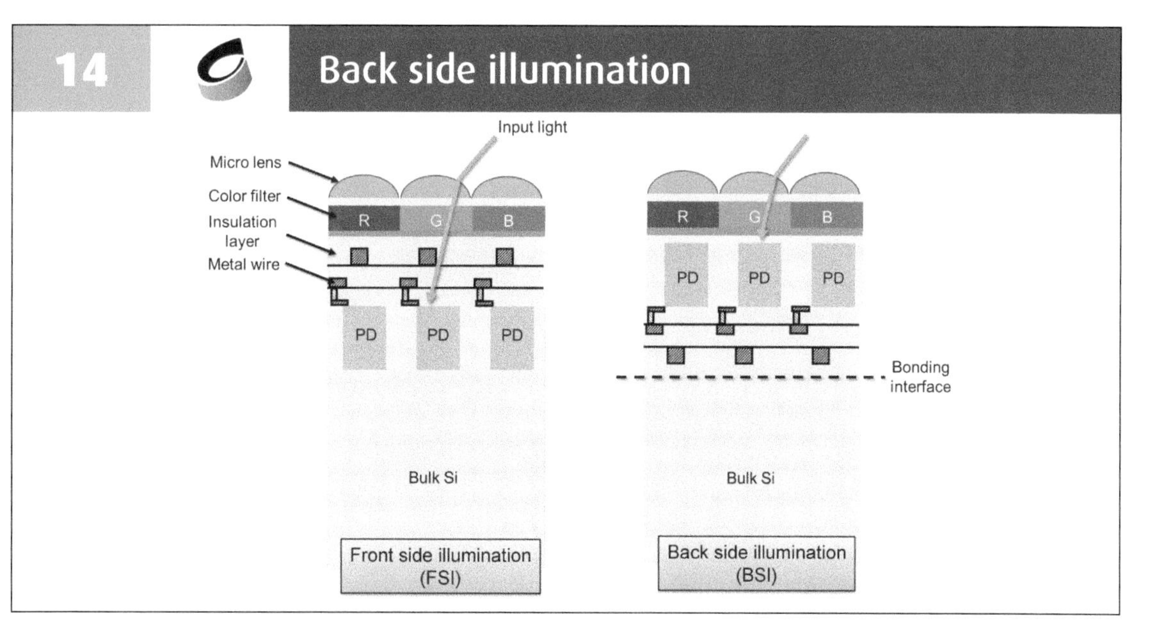

Another issue for reducing the pixel size is the cross talk from the nearby pixels. So this is the cross section of the image sensor using front side illumination (FSI). In each pixel, there are routings for connections to the photodiode. When the pixel pitch shrinks, this light path sometimes hit the nearby photodiode, leading to optical cross talk. To avoid this problem, back side illumination (BSI) is developed. In this case, this light is hit on the back side of the image sensor. As the distance between the photodiode and the micro-lens is much closer in this case, optical cross talk is greatly reduced. In the present state, most image sensors in cameras and smart phones use the BSI technology.

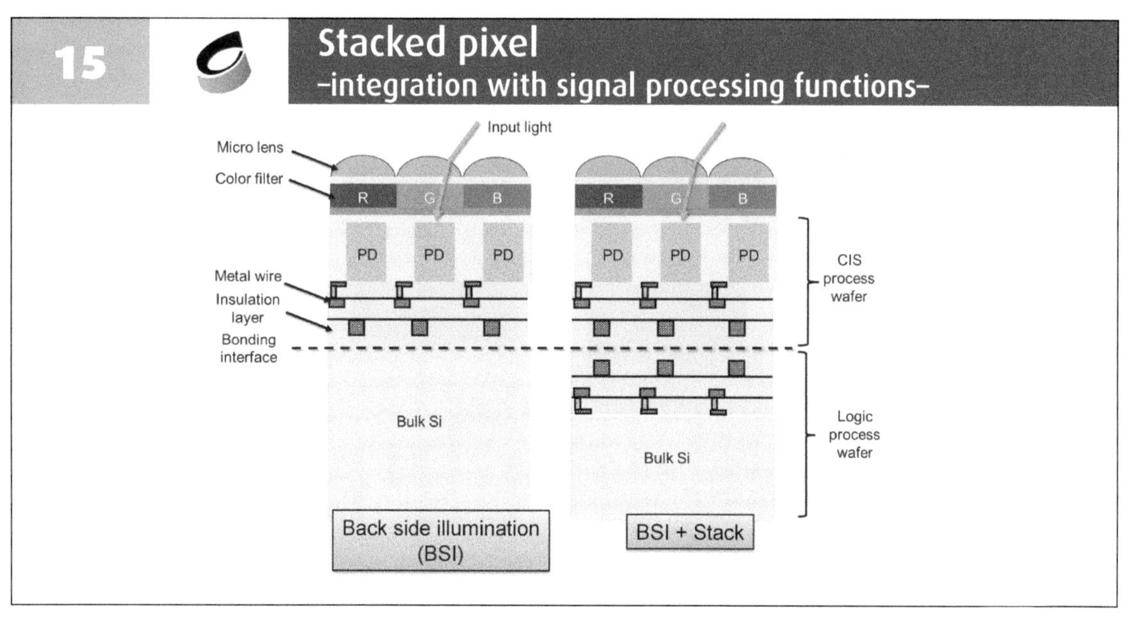

Also, recently another technology, called the stacked pixel, is developed. In this case, the CMOS image sensor and some logic cells are stacked for 3D integration. As a result, you can expand the signal processing ability of CMOS image sensors. This 3D integration technology is used by Sony and other companies.

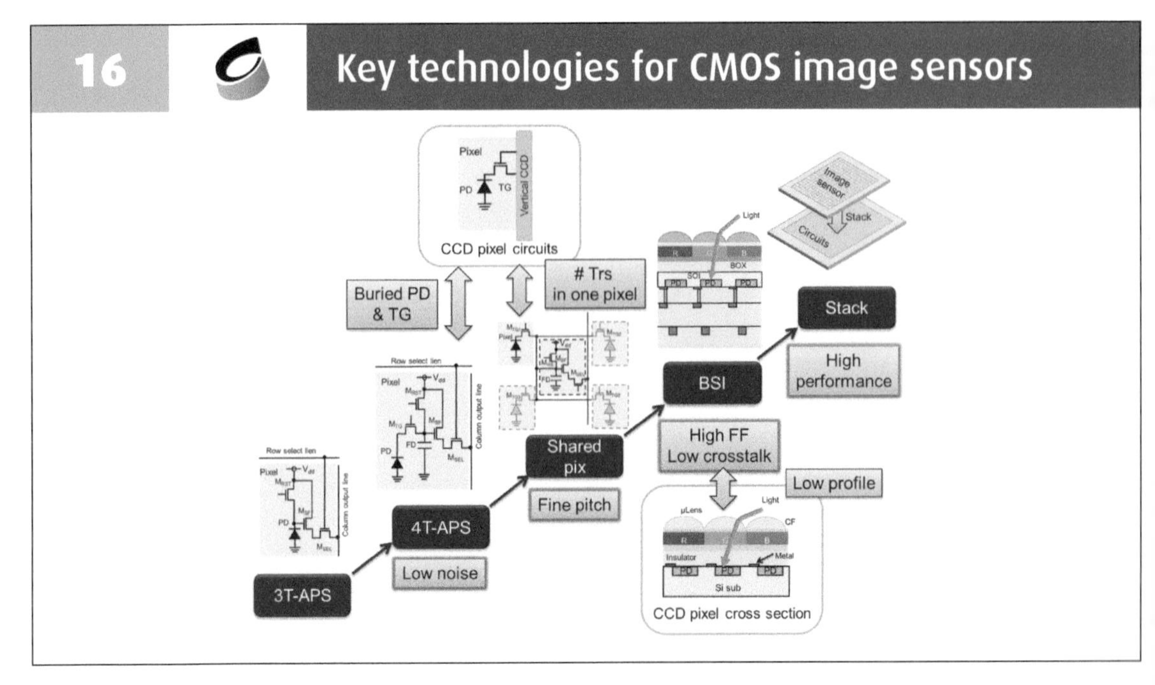

By using the above key technologies, that is 4T, shared pixel, BSI and stacked pixel, the performance of CMOS image sensor has drastically improved in the present state. In this case, we can use such an advanced CMOS image sensor technology for other reasons, instead of just taking pictures using our smart phones.

CLASSIFICATION OF IMAGE SENSORS FOR BIOMEDICAL APPLICATIONS

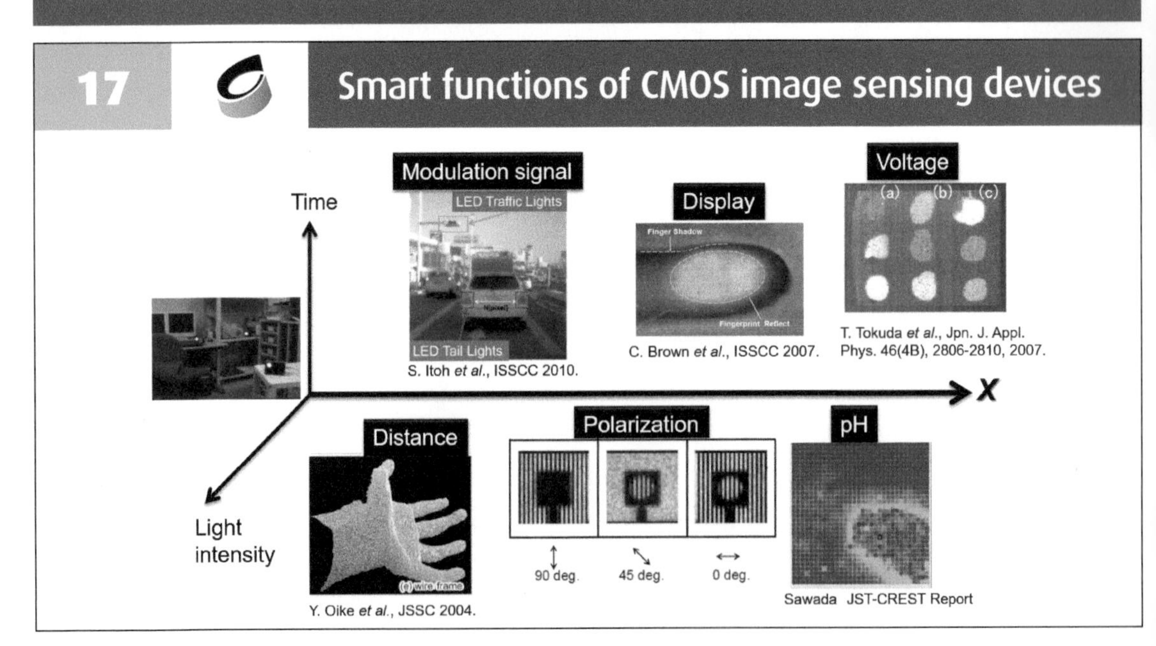

This slide shows some smart functions of CMOS image sensing devices. At present, the image sensor is normally used to measure the light intensity. However, we can expand their usage in another dimension, like distance measurement, pH measurement and polarization.

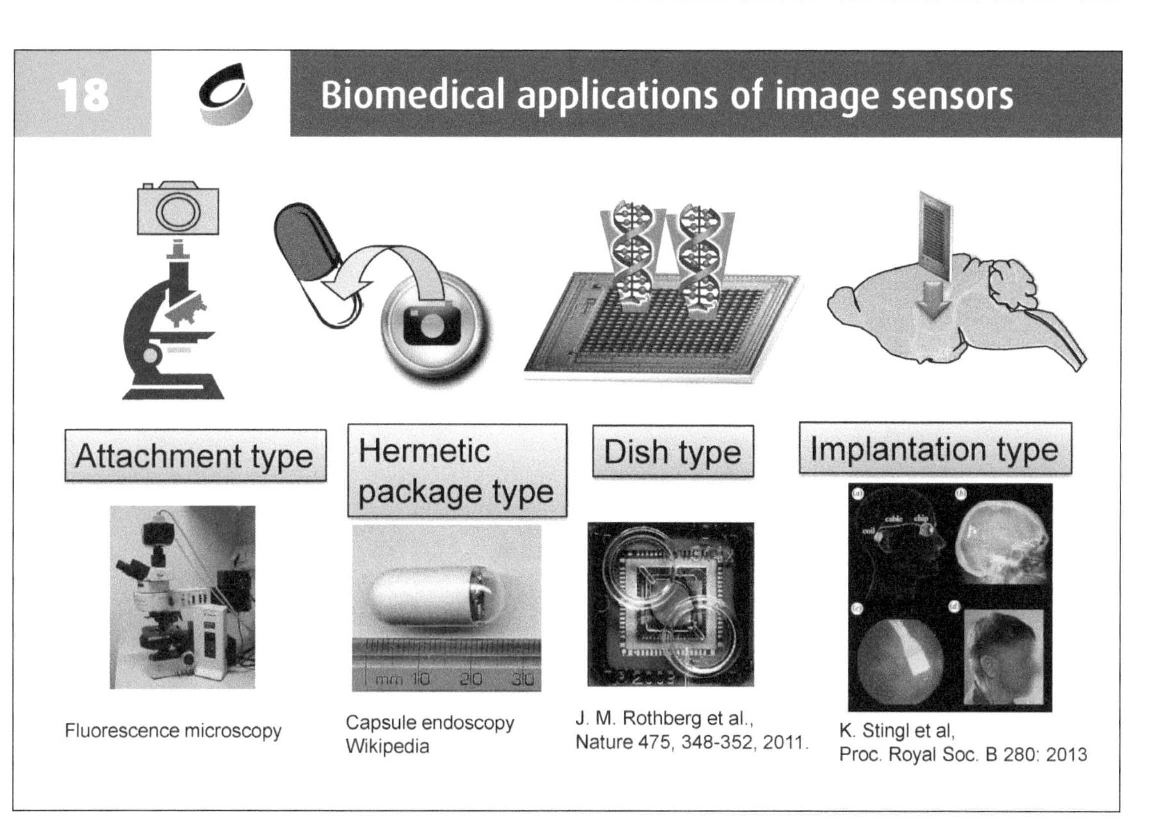

Here, I'll introduce the biomedical applications of image sensors. This slide shows some potential image sensor usage for biomedical applications. A very simple case is just to attached a CMOS image sensor to the instrument to form an optical fluorescence microscope like this. The second case is the hermetic package type, where we install the image sensor inside a hermetic package. This package can be in the form of a capsule which we can swallow. In this capsule, all the camera system including illumination, analog and RF functions are installed. As a result, images inside our body can be transmitted for diagnosis purpose. Notice that a hermetic package is required to protect the system against water and/or humidity. The third case is the dish type. In this case, some biomedical substance is attached on top of the image sensor. The fourth case is direct implantation. In this case, the image sensor chip is directly implanted into the brain or the eye. We will discuss these cases using some examples as follows.

19 Outline

- **Image sensor technologies**
  - Trend of CMOS image sensors
  - Fundamental characteristics of CMOS image sensors
  - Classification of image sensors for biomedical applications
- **Fluorescence detection**
  - Introduction
  - Voltage sensitive dye: Measurement of cell activities through fluorescence with high speed CMOS image sensor
  - FLIM (fluorescence lifetime imaging microscopy): Measurement of fluorescence lifetime with ultrafast CMOS image sensor
- **Retinal prosthesis**
  - Introduction
  - Implantable stimulators in subretinal space
- **Summary**

20 Fluorescence

1. Wavelength
   $\lambda_{Fluorescence} > \lambda_{Excitation}$
   → Stokes shift
2. Intensity
   $I_{Fluorescence} > I_{Excitation}$
   $10^{-3} \sim 10^{-5}$
3. Photobleaching
   - Fluorescence material in excitation state is more unstable than that in ground state due to chemically activated
4. Fluorescence lifetime
   - $< 10^{-4}$ sec
   - $> 10^{-4}$ sec → phosphorescence
   - FRET : Fluorescence Resonance Energy Transfer
   - FLIM : Fluorescence Lifetime Imaging Microscopy

VSD (Voltage sensitive dye)
GFP (Green fluorescence protein)
GCaMP

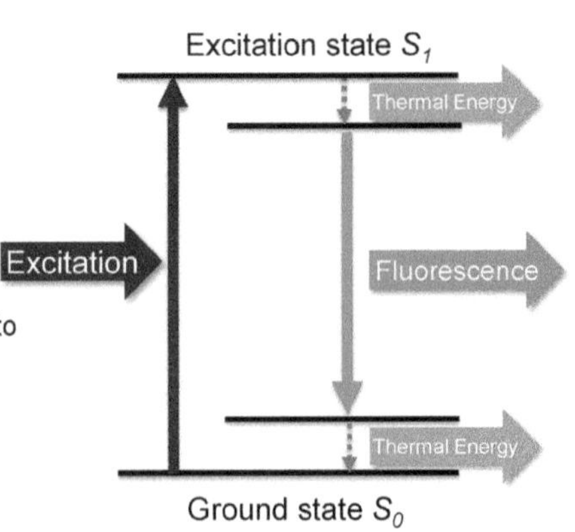

This is a brief introduction in fluorescence. Here, you need an excitation light. When some objects are excited using the excitation light, fluorescence is emitted. Notice that fluorescence is at a lower energy level than the excitation light, and is frequently used in life science.

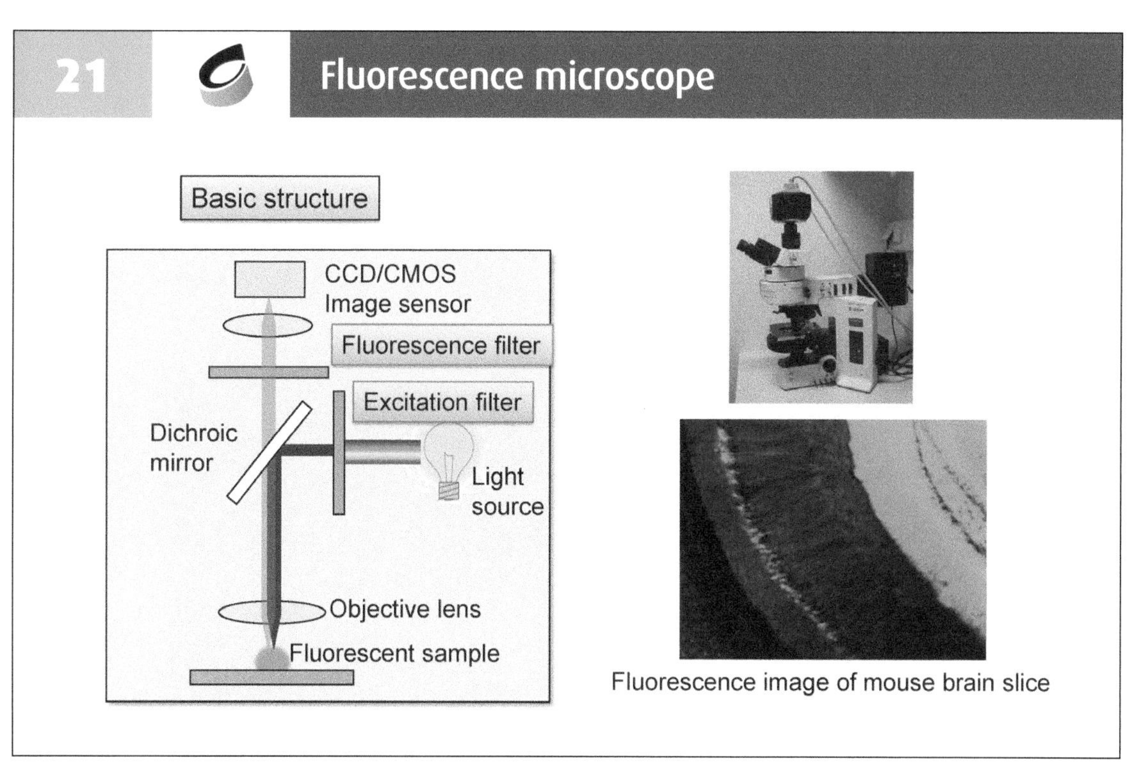

This is an example of fluorescence microscope. It is a very simple configuration. The excitation filter is used for filtering other wavelengths except from the excitation one. The fluorescence light/green light emitted by the sample is basically gathered by the image sensor, with the fluorescence filter cutting off the excitation/environmental light. You can see here the neurons through fluorescence.

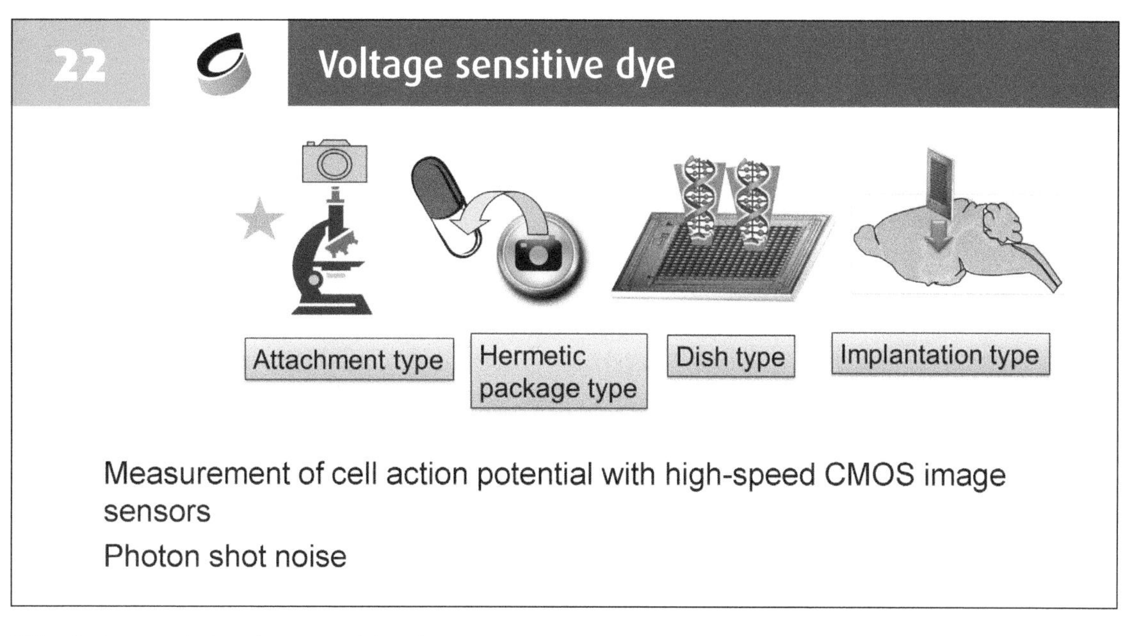

The first example is voltage sensitive dye.

23 Activity of cell: action potential

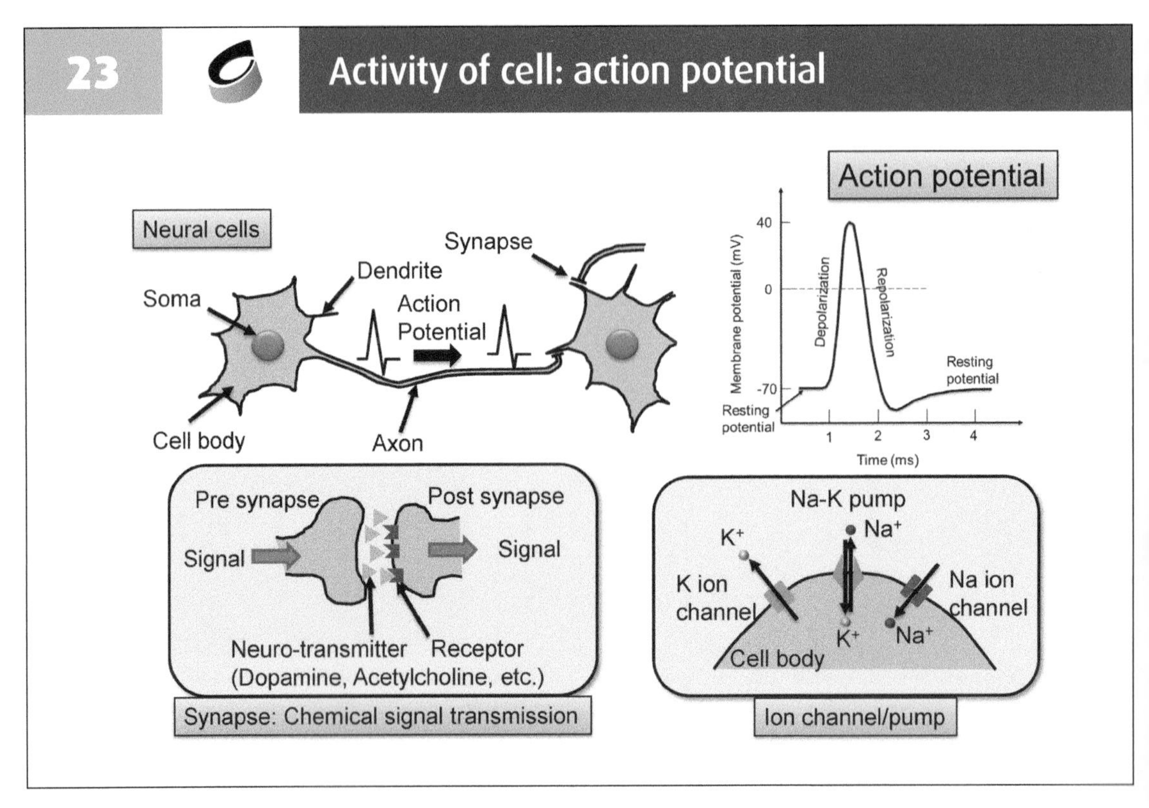

Before, I start talking about voltage sensitive dye. I will briefly show you the activity of cells. The neural cells communicate with each other through electrical pulse signals called the action potentials. Each neuron is connected through wirings called axons. An Axon collects the action potential, which is a very short electrical pulse, with a pulse width of about 1ms and an amplitude of 40 µV. Such a short pulse is transmitted through the Synapse. Anyway, to make an action potential is very important in life science. Voltage sensitive dye can be introduced in the neural cells. When an action potential (which is a voltage) is produced, it is converted into fluorescence. In this way, we can measure the fluorescence which corresponds to the action potential of the neural cells. This is especially useful because when you try to measure the action potential by electrical methods, it is a pin point measurement. But if you use an optical microscope like this, you can measure a very wide area of neural activities simultaneously. Nowadays, many people use voltage sensitive dye to measure the neural activity in wide areas.

24  Measurement of neural activities with VSD (voltage sensitive dye)

One drawback is that the background fluorescence is very strong, but the change of fluorescence is very small, typically 1%. In this case, the very large photon shot noise due to strong fluorescence is a very big problem, as the near saturation photon shot noise is ultimately limiting the fluorescence detection, and in this case voltage measurement. If the full well capacity which indicates the maximum stored charge in the photodiode is 10 ke-, then the SNR we can obtain is 60dB. But if the photodiode is in the near saturation region, the photon shot noise is becoming larger. As the photon shot noise is related to the square root of the charge, the SNR will drop to 40dB. In this case, you need to increase the full well capacity to 100 ke- so as to restore the SNR to 60 dB again.

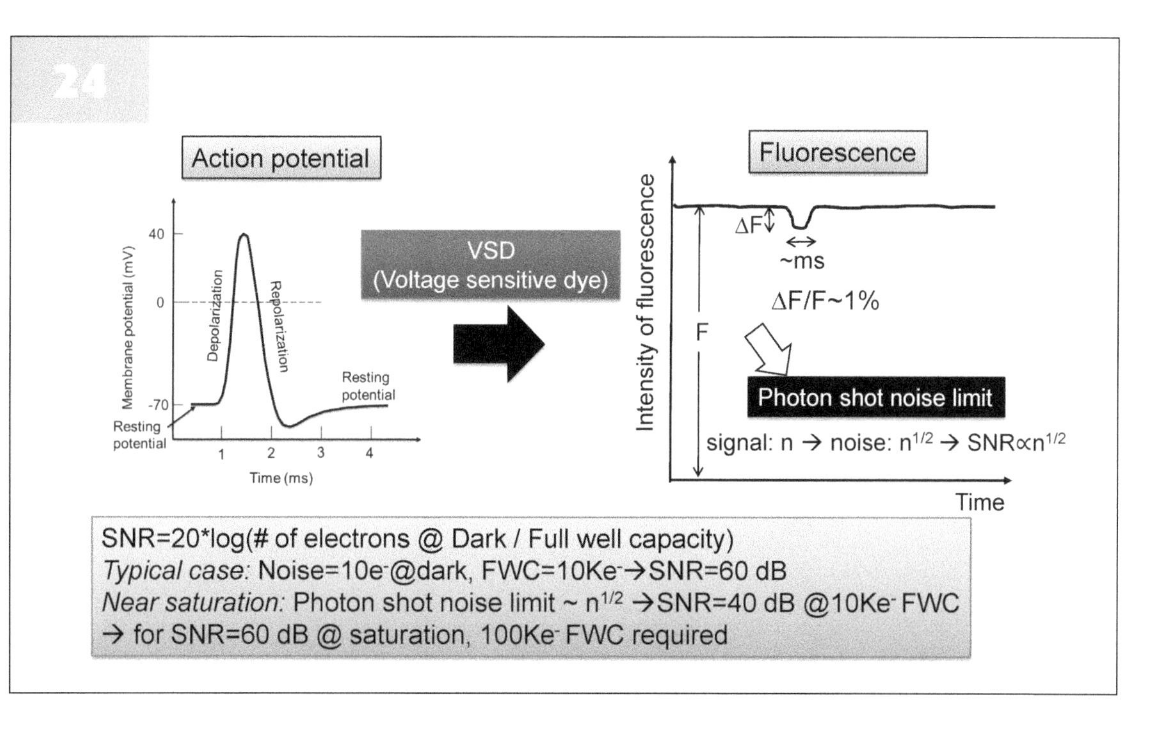

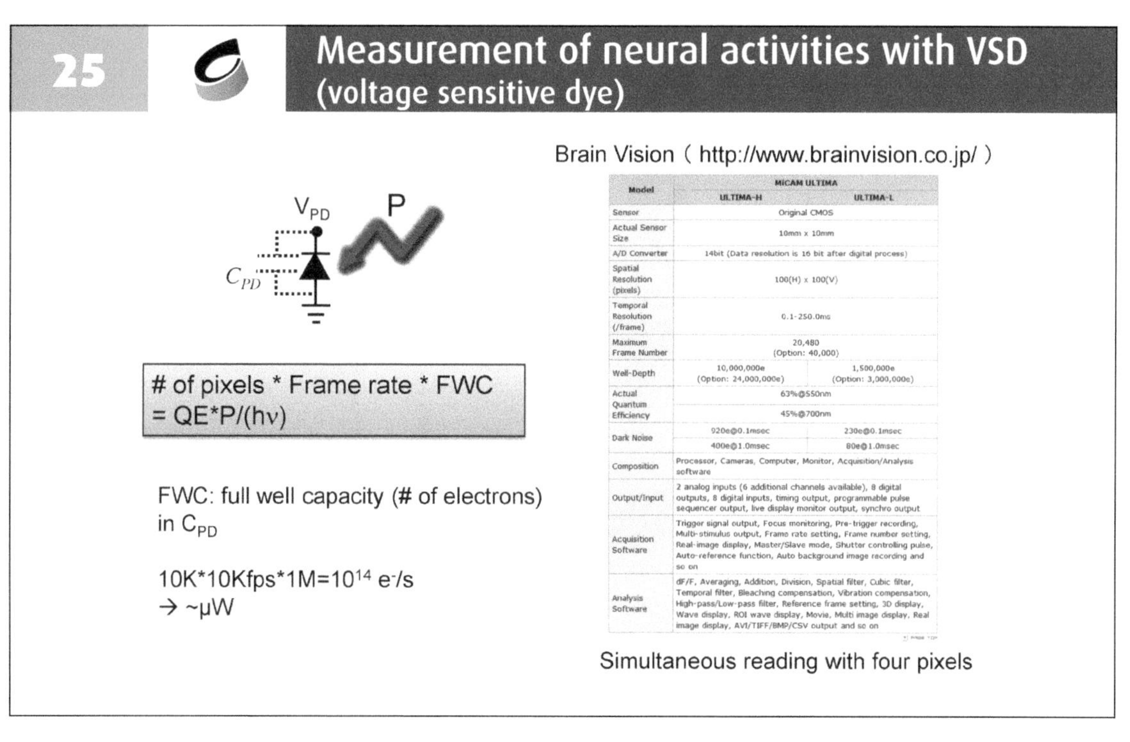

Model	MiCAM ULTIMA	
	ULTIMA-H	ULTIMA-L
Sensor	Original CMOS	
Actual Sensor Size	10mm x 10mm	
A/D Converter	14bit (Data resolution is 16 bit after digital process)	
Spatial Resolution (pixels)	100(H) x 100(V)	
Temporal Resolution (/frame)	0.1-250.0ms	
Maximum Frame Number	20,480 (Option: 40,000)	
Well-Depth	10,000,000e (Option: 24,000,000e)	1,500,000e (Option: 3,000,000e)
Actual Quantum Efficiency	63%@550nm	
	45%@700nm	
Dark Noise	920e@0.1msec	230e@0.1msec
	400e@1.0msec	80e@1.0msec
Composition	Processor, Cameras, Computer, Monitor, Acquisition/Analysis software	
Output/Input	2 analog inputs (6 additional channels available), 8 digital outputs, 8 digital inputs, timing output, programmable pulse sequencer output, live display monitor output, synchro output	
Acquisition Software	Trigger signal output, Focus monitoring, Pre-trigger recording, Multi-stimulus output, Frame rate setting, Frame number setting, Real-image display, Master/Slave mode, Shutter controlling pulse, Auto-reference function, Auto background image recording and so on	
Analysis Software	dF/F, Averaging, Addition, Division, Spatial filter, Cubic filter, Temporal filter, Bleaching compensation, Vibration compensation, High-pass/Low-pass filter, Reference frame setting, 3D display, Wave display, ROI wave display, Movie, Multi image display, Real image display, AVI/TIFF/BMP/CSV output and so on	

Simultaneous reading with four pixels

As the full well capacity is almost proportional to the size of the photodiode, you need to increase the pixel size to obtain good SNR. In the measure of voltage sensitive dye, the required pixel size can be very large, in the order of several tens of μm². This is very big when compared to existing CMOS image sensors, which is in the order of 1 μm². Even though the pixel size is larger, we need this to increase the capacitance so as to improve the SNR.

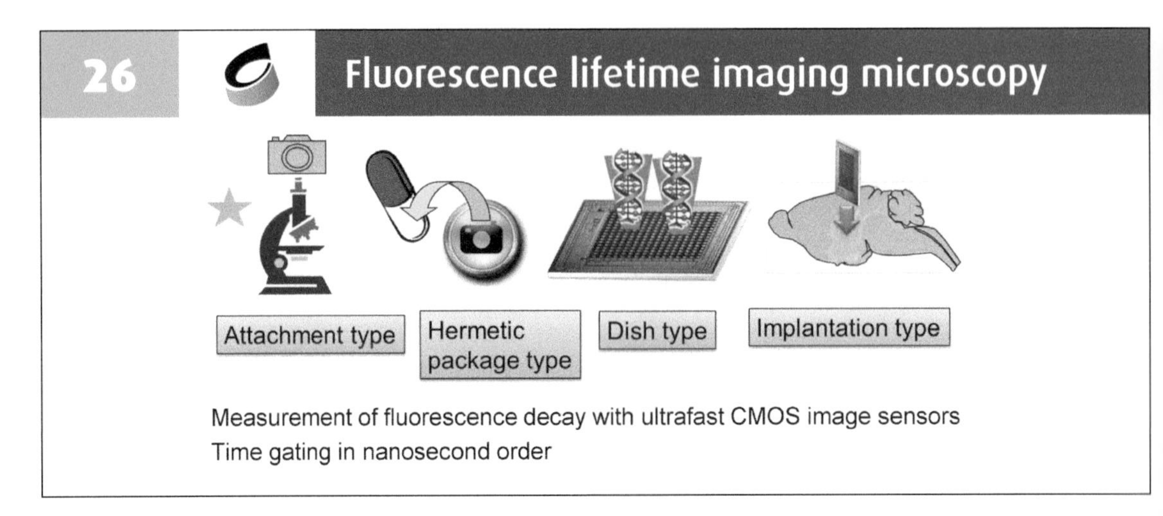

Another example for fluorescence measurement is fluorescence lifetime imaging microscopy.

27 

FLIM: Fluorescence Lifetime Imaging Microscopy

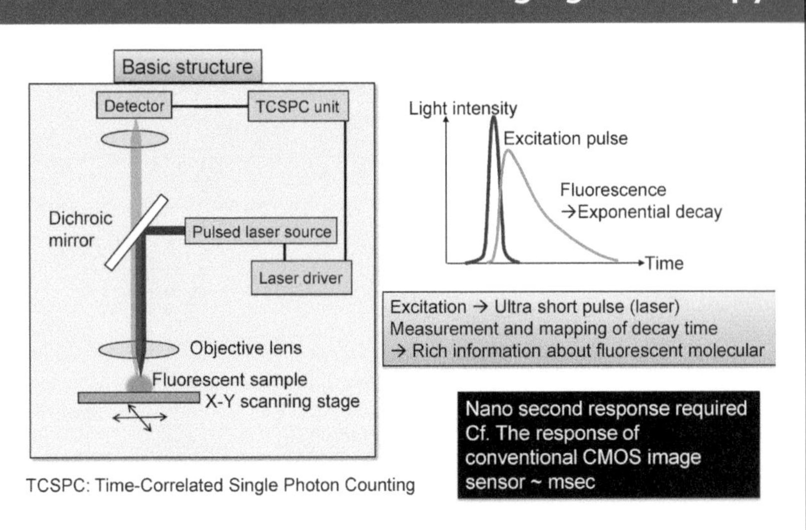

In applications like to voltage sensitive dye, we need to measure the change of fluorescence. However, sometimes we need to measure the decay of the fluorescence light as in FLIM. This figure shows the basic structure for FLIM measurement. This is the excitation light pulse, which emits only very short light pulses. After the fluorescence is triggered by the excitation light pulse, it starts to decay and finally diminishes. This fluorescence decay time is a very important measure for the molecular information of neural cells, which helps us to understand their corresponding molecular configurations. As the decay time is normally very short, in the order of ns, high speed imaging with ns response time is required. Yet, we know that the response time in conventional image sensor is ms, meaning that it is very difficult to measure such a short decay time using conventional CMOS image sensors. A conventional FLIM microscope usually uses one detector, which is generally an avalanche photodiode (APD) with a very fast response time. The time-correlated single photon counting technique is employed to measure the correlation between the measured pulse and the fluorescence decay as well as the distribution of the fluorescence lifetime. Obviously, it is a very time consuming action. If we use a CMOS image sensor instead of just one detector, we can obtain the distribution of fluorescence lifetime in real time. The major problem is that the response time of conventional CMOS image sensor is very slow.

28 

How to enhance the response speed of CMOS image sensor

- The main limitation
  - 3T-APS → Response speed of PD
  - 4T-APS → Transit time to FD
- Solutions
  - Fast detector → APD
  - Introducing drift mechanism in transfer action

So how can we enhance the response speed of CMOS image sensor? The main speed limitation for a 3T-APS is the response time of the photodiode, which ultimately limits the overall readout speed. For a 4T-APS, the transition time for the charge to transfer from the photodiode capacitance to the floating diffusion can result in extra delay for the signal readout, and determines the speed of the CMOS image sensor. To enhance the speed of CMOS image sensor, one solution is to use a fast detector, which is the APD. Another solution is to introduce drift mechanism during charge transfer.

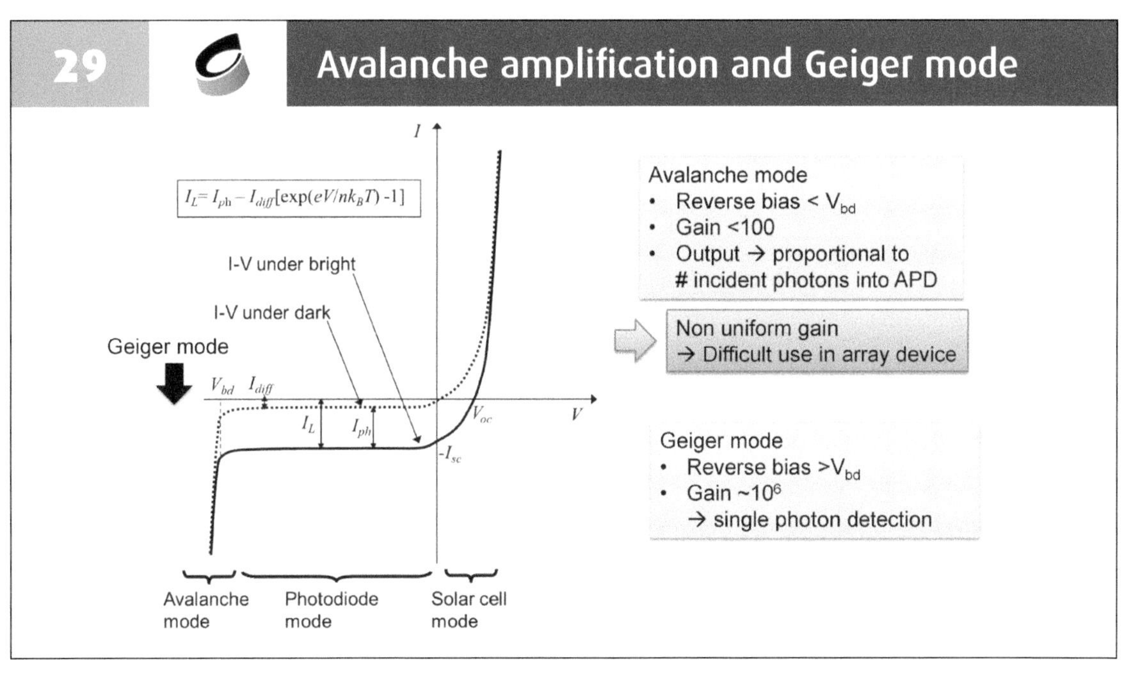

This slide shows the I-V characteristics of a photodiode. If you negative bias the photodiode, avalanche occurs which can enable very fast operations. We can achieve ns response time very easily if the photodiode is operating in the avalanche mode. One major issue is that the gain within the avalanche mode for different photodiode can be different, and we cannot have a precise gain in each pixel. For example, some pixels may have a gain of 100, while other pixels may have a gain of 110, and others maybe 90. This makes devices operating in avalanche mode difficult to use for image sensing. The solution is to operate the photodiode in the Geiger mode by reverse biasing the photodiode far less than the breakdown voltage in avalanche mode. In this case, the gain is almost infinity. Whenever a photon is introduced in this mode, a massive amount of photocurrent is produced very quickly. As a result, the response time can be very fast, and short pulses can be produced across the sensor array.

30

SPAD (Single Photon Avalanche Diode)

V_o
Rq
SPAD
Rs
p+
Multiplication region
p+ n+ p-well p-well n+ p+
deep n-well
p-sub
Cross section
Current
(2)
(3)
(1)
Applied bias V_b Vo
(1) Apply Vo > Vb
(2) Avalanche occurs by single photon→ photocurrent flows
(3) Voltage drop by Rq → Applied voltage becomes less than Vb (Quenching)
~ nsec response
V_{dd}
SPAD
V_p
Pixel circuits
Column output line
34

We call this device a single photon avalanche diode, SPAD. Its configuration is shown here. Each pixel is composed of a SPAD and a quenching resistor, which can be implemented by the channel resistance of a transistor. We can apply a breakdown voltage across the SPAD. When avalanche occurs, a large photocurrent flows through the device. This large current also passes through the quenching resistor. As a result, the applied voltage across the SPAD becomes less than the breakdown voltage, and the avalanche action is then quenched quickly, generating a very short pulse. This pulse signal can be readout using the column output line, and a response time in the order of ns can be easily obtained.

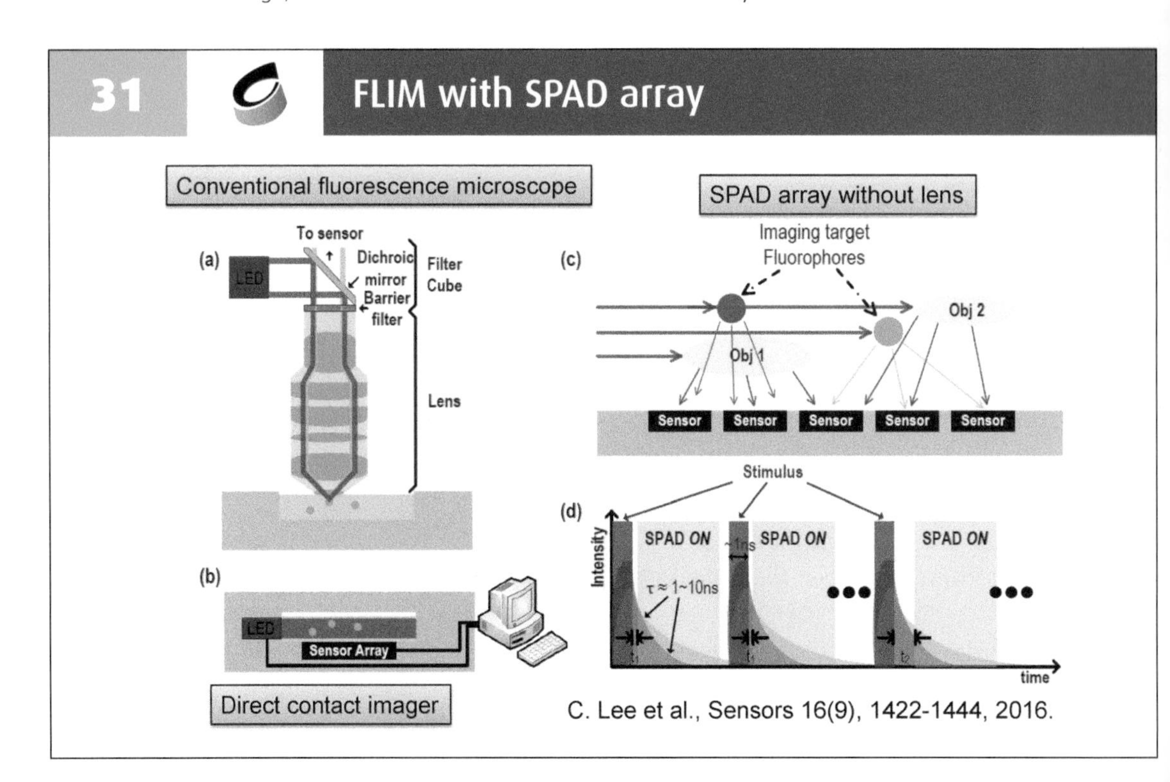

And by using an array of SPAD sensors, you can obtain very fast response time, and the FLIM signal can be measured through the delay of these short pulses.

32 

How to enhance the response speed of CMOS image sensor

- The main limitation
  - 3T-APS → response speed of PD
  - 4T-APS → Transt time to FD
- Solutions
  - Fast detector → APD
  - Introducing drift mechanism in transfer action

Except from using SPAD, another method is by introducing drift mechanism during the transfer action.

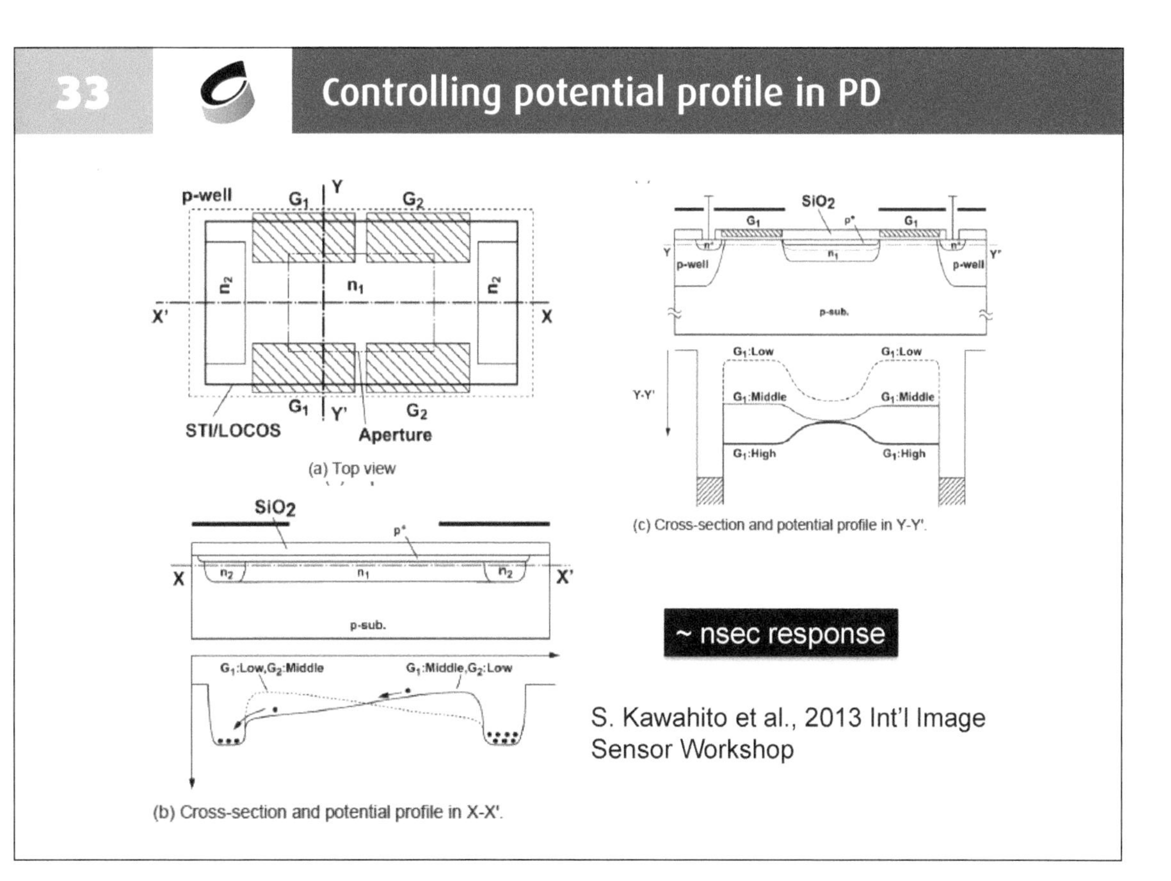

Here we show the top view of the photodiode, and these are the cross sections. In each pixel, we can use 4 electrodes to apply different voltages to control the potential profile. By changing the potential inside the photodiode, we can either prohibit charge flow, or enhance charge flow using the drift mechanism so as to drastically increase the charge transfer time to the ns regime. By measuring the charge contents at different time instances, we can easily calculate the delay of the fluorescence.

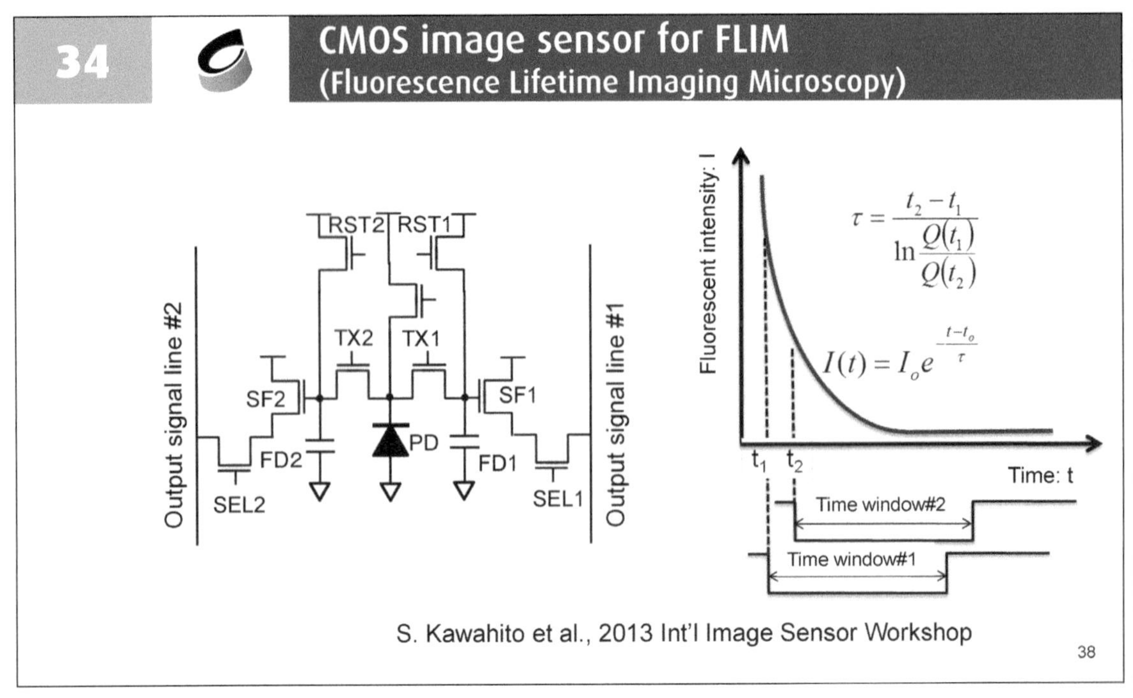

By using differential circuits, we can measure the ultra-fast fluorescence decay so as to realize FLIM using a CMOS image sensor.

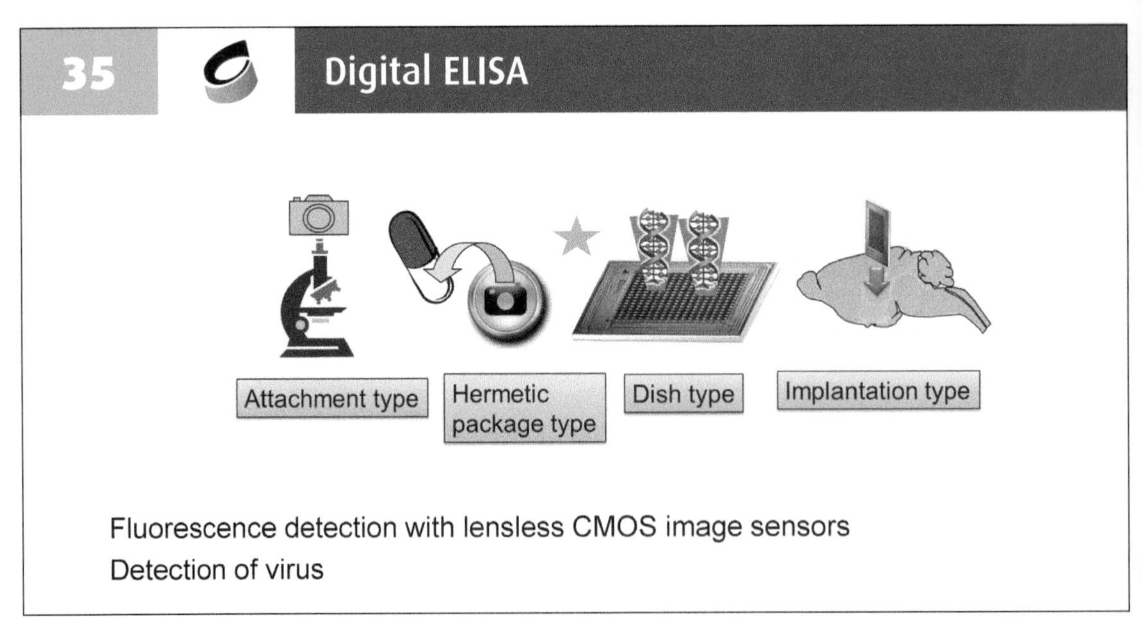

The third example for using fluorescence detection is lensless CMOS image sensor. In this case, the fluorescence sample is located on top of the CMOS image sensor.

36

Background

ELISA : Enzyme-linked Immunosorbent Assay → Virus inspection

Antigen-antibody reactions

3 μm
20 nm
Microbead
Antibody
Target Antigen (Virus)
2nd Antibody
Enzyme

Enzyme-catalyzed fluorescent reaction

Nonfluorescent
FDG
SβG (Enzyme)
Fluorescent
Fluorescein
+ 2
Ex. 470 nm
Em. 525 nm

In life science, enzyme-linked immunosorbent assay (ELIZA) is a frequently used method to detect a particular virus. It's a very simple method. If you would like to measure a virus or a target antigen, you can utilize the antigen-antibody reactions, with the antibody specifically corresponds to this specific virus/antigen. If the antibody is fixed by some kind of micro-bead, the target antigen can be connected to the antibody. After that, we can inject a second antibody, which can connect to this enzyme. This enzyme is responsible for fluorescence reaction characterizations. If the particular virus/antigen is not present, then there will no fluorescence reactions. In other words, we know that a particular antigen is present when fluorescence occurs. We can then measure the intensity of the fluorescence, and that is basically the operation of ELIZA.

37

ELISA

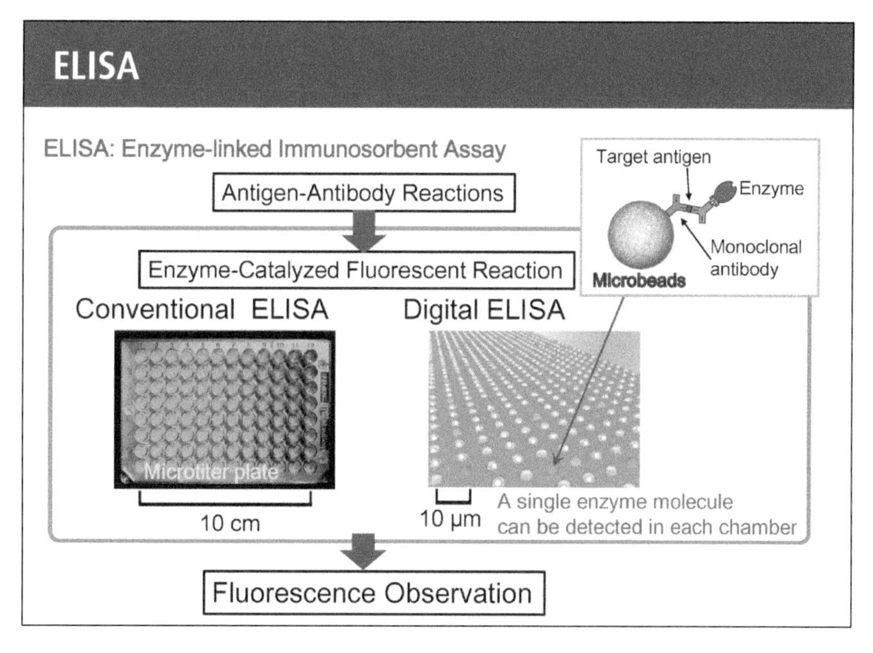

Conventionally, ELISA are measured using a micro-liter plate. On each plate, there are some wells, and we can measure the fluorescence concentration in different wells simultaneously. In this case, the size of each well is in the order of microliter, and there are many viruses/antigens within one well. But what if the size of the sample volume is drastically reduced from micro-liter to femto-liter? In such a scenario, only one virus/antigen may exist in one well. We can then just count the number of fluorescence cells as the representation of the concentration of the virus, which can be measured very precisely. This method is called digital ELISA, and the detection limit is much improved when compared with the conventional case.

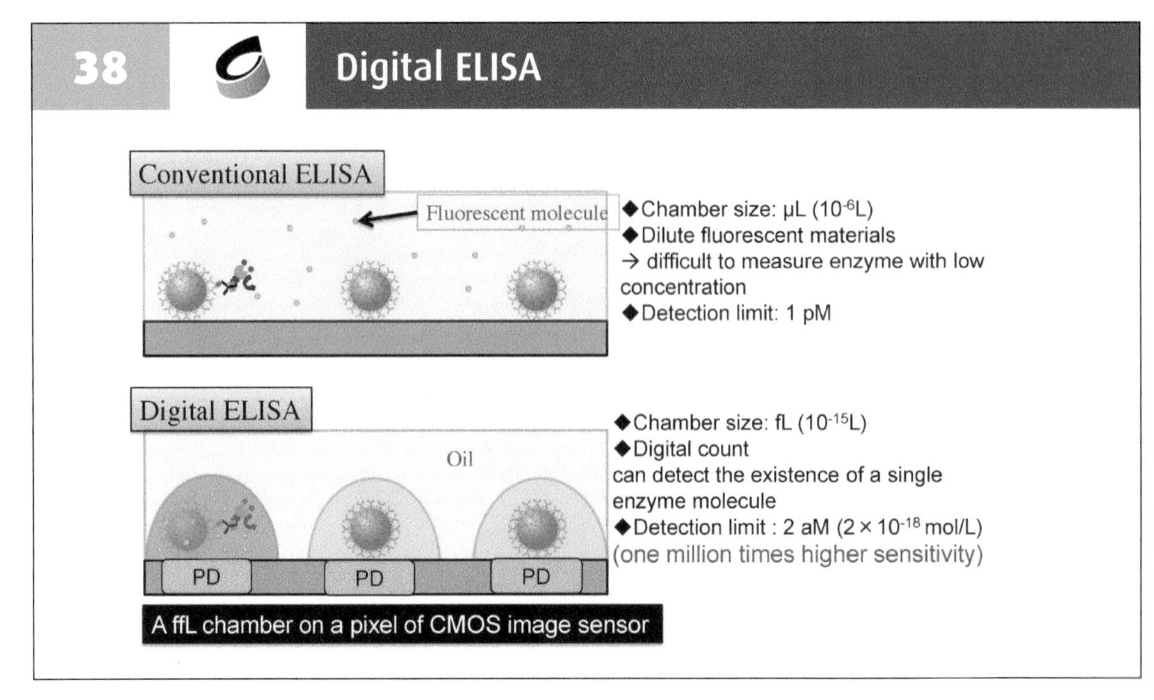

For the conventional ELISA with a very deep well, the virus is very small. When the sample volume reduces to femto-liter, the well becomes so big and the virus is so small. By putting a photodiode under each small well, you can directly measure the number of fluorescence wells. Consequently, digital ELISA is very compatible with CMOS image sensors, with one pixel per well, and there can be many pixels to support massive measurements. Also, we do not need a microscope, and we can just put the samples on top of the image sensor. However, we need to take care of the fluorescence cross talk. As fluorescence can emit in every direction, it is very difficult to measure fluorescence only in one photodiode. Also, the excitation light can penetrate the well, and some portion of the excitation light can be detected by the photodiode. This can be a problem as the excitation light is very strong when compared to fluorescence light.

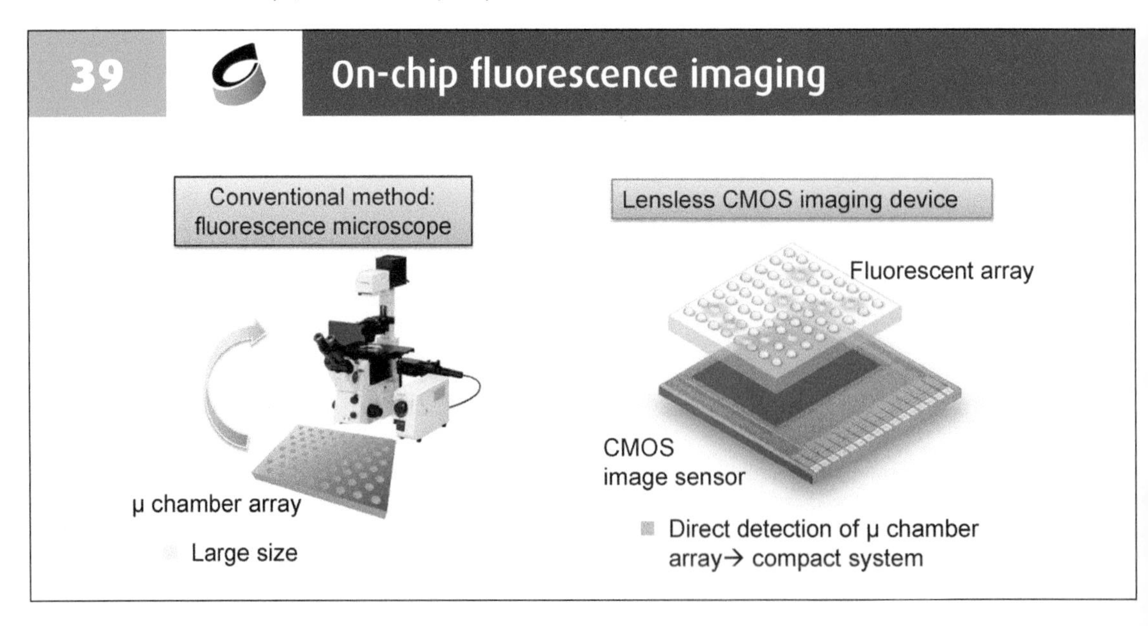

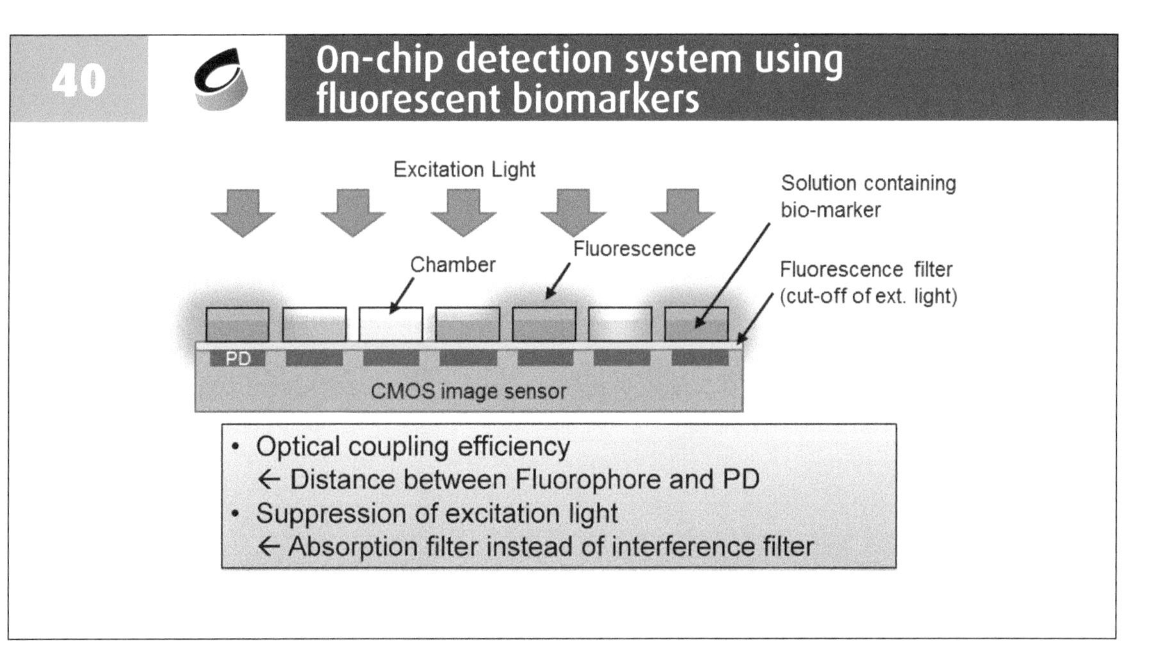

To eliminate the excitation light from the fluorescence light, we can use the dependence of absorption efficiency (i.e. penetration depth) and different wavelength.

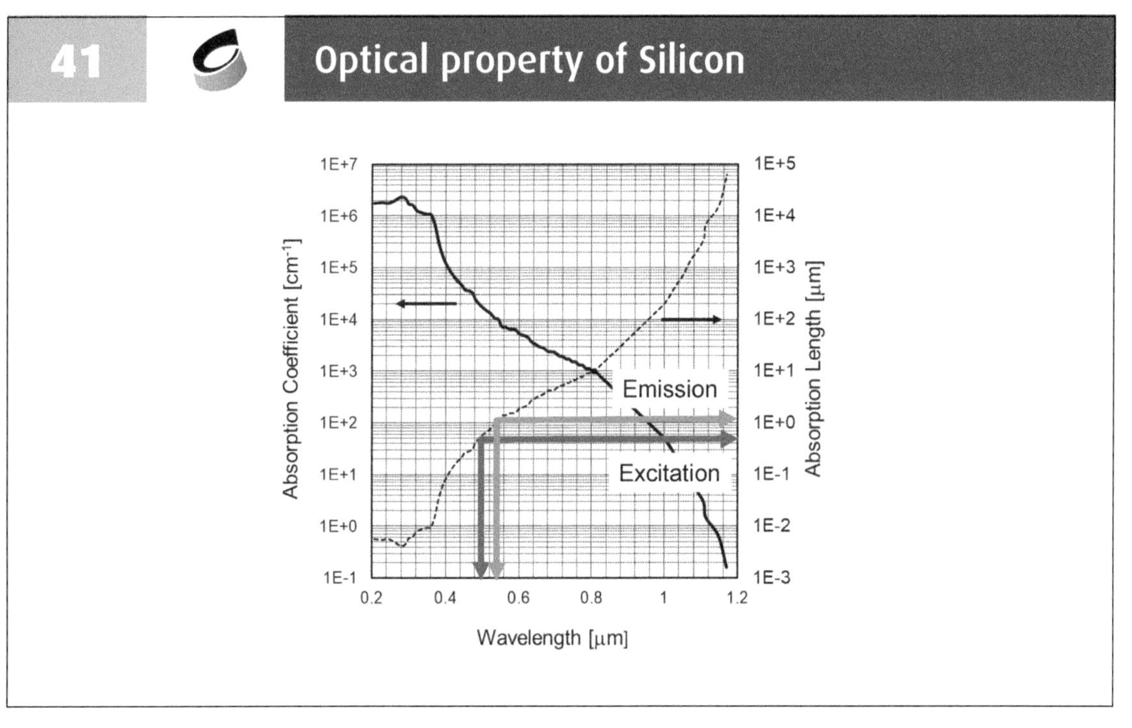

This plot shows the optical property of silicon. This vertical axis shows the absorption coefficient, and this shows the penetration depth. For example, if the excitation light is at 470 nm, the penetration depth is about 1 μm. Due to the lower energy, the fluorescence is having a longer wavelength than the excitation light. As a result, the penetration depth is longer/deeper than that of the excitation light, that is excitation light is absorbed at a shallower region, and fluorescence light is absorbed at a deeper region.

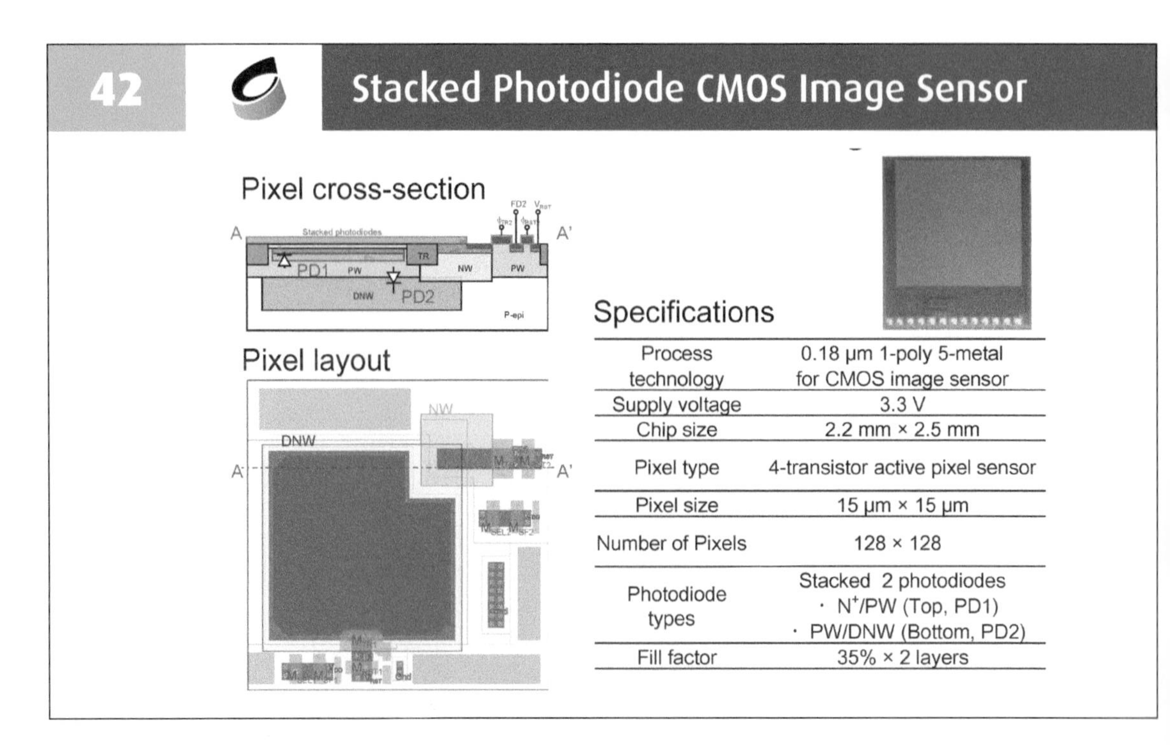

Process technology	0.18 µm 1-poly 5-metal for CMOS image sensor
Supply voltage	3.3 V
Chip size	2.2 mm × 2.5 mm
Pixel type	4-transistor active pixel sensor
Pixel size	15 µm × 15 µm
Number of Pixels	128 × 128
Photodiode types	Stacked 2 photodiodes · N^+/PW (Top, PD1) · PW/DNW (Bottom, PD2)
Fill factor	35% × 2 layers

By exploiting this penetration depth dependency, we produced two stacked photodiode implementations which have absorption regions located at different depths. This photodiode is at a shallower region, which is used to measure the excitation light. And this deeper photodiode can be used to measure fluorescence light. By doing so, you can subtract the effect of the excitation light intensity. We fabricated a fluorescence detection system like this, and the number of fluorescence in digital ELISA can be correctly measured using this method.

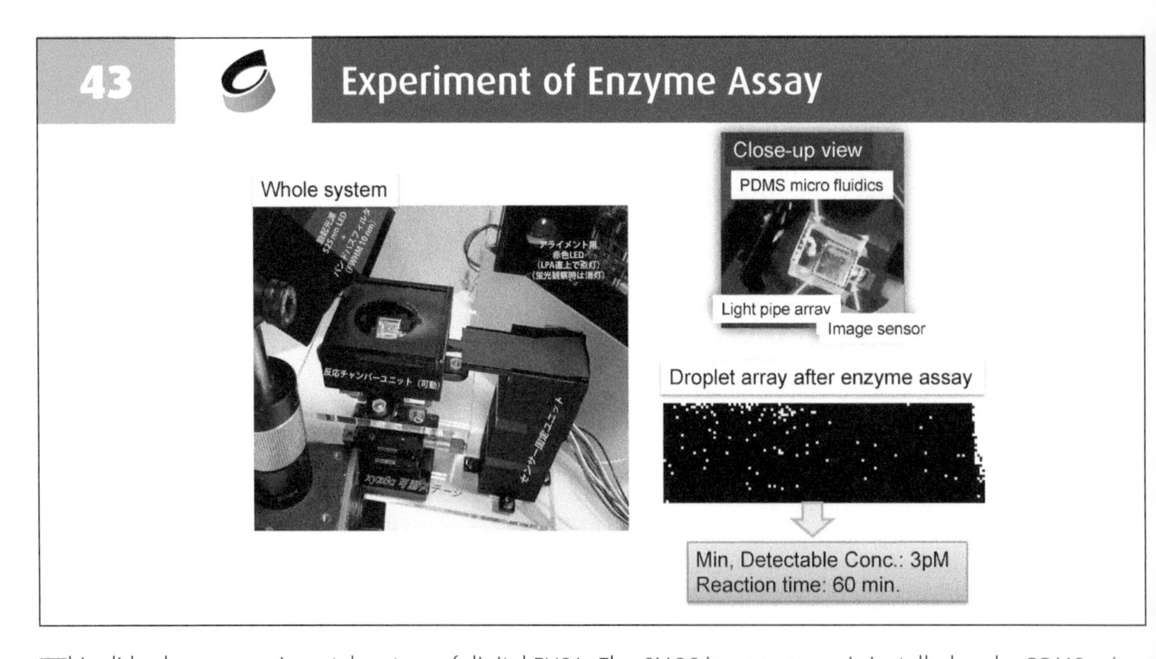

This slide shows experimental system of digital ELISA. The CMOS image sensor is installed under PDMS micro fluidics.

44 Potential control to extract fluoresce from excitation light

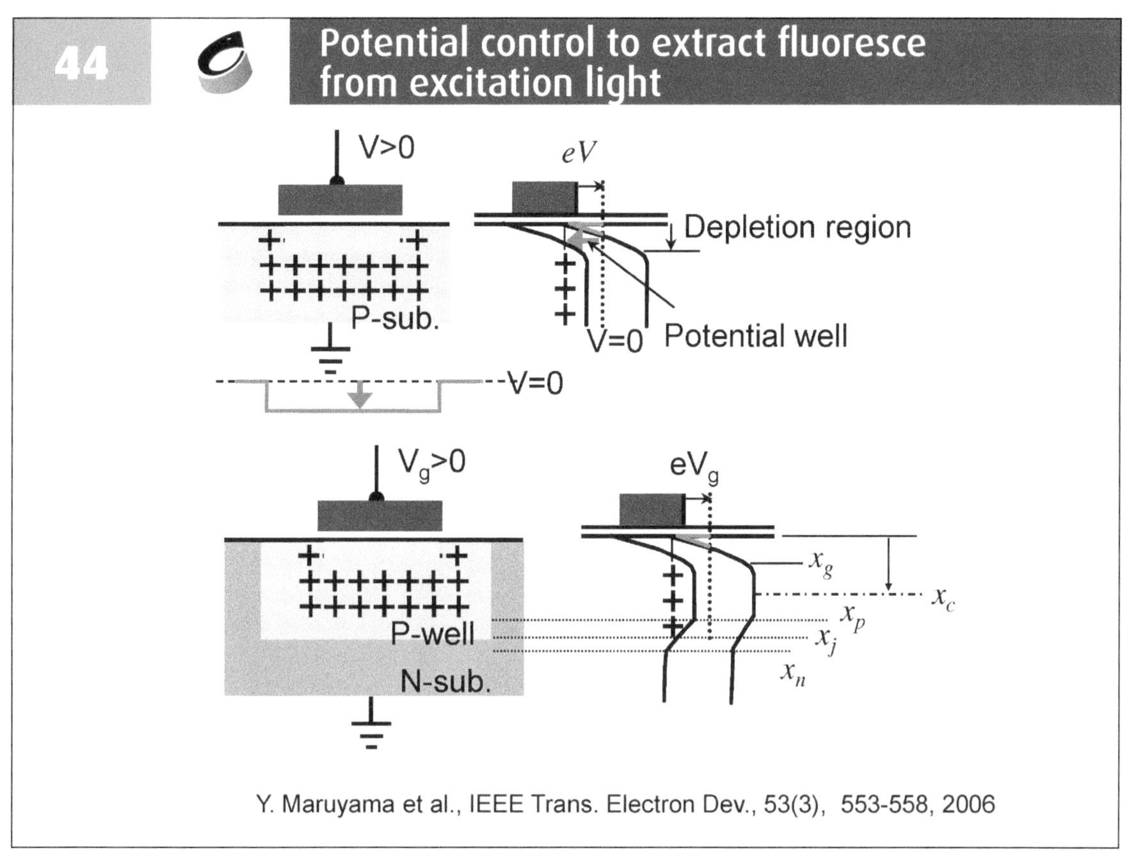

In our method, there are only two stacked photodiodes with their location of absorption fixed. By introducing gate controls as shown here, we can then apply different gate voltages to control the potential profile within the photodiode. As shown here, if we apply a positive voltage, the absorption place becomes deeper. We can then better distinguish the excitation and fluorescence light output, and measure the fluorescence light intensity more precisely.

45 Implantable micro CMOS image sensors

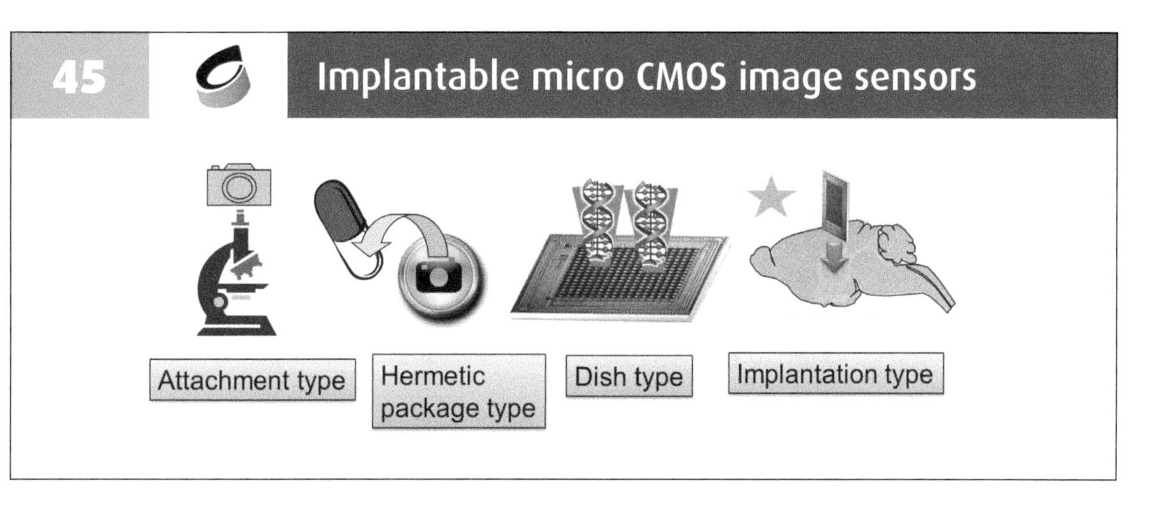

The final example is implantable micro CMOS image sensor, where the device is directly implanted into the brain.

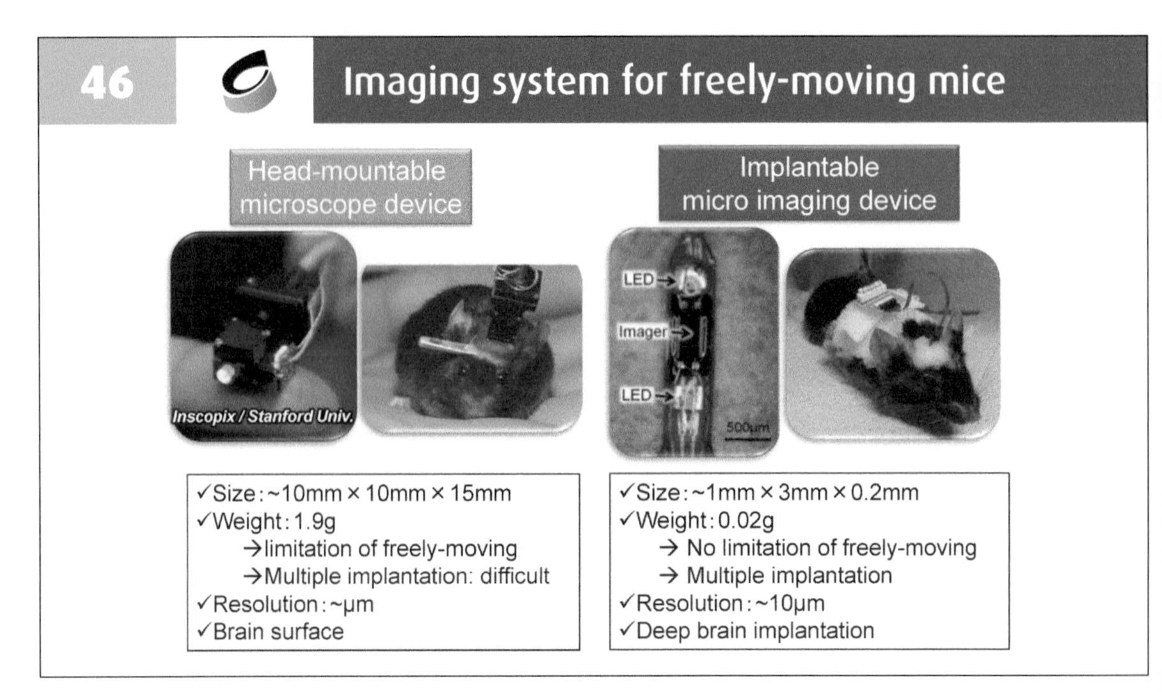

Here we show imaging systems for freely moving mice, which are very rare in prior works. This one on the left is developed by Inscopix and Stanford. In their implementation, a very tiny microscope is installed on the head of the mouse. In our case as shown on the right, a very small size image sensor module is directly implanted into the brain of the mouse. So here shows the implantable device. Here is the sensor and photodiode array. The CMOS image sensor can then record neural activities inside the brain. Typically, the weight of a mouse is roughly 20g, so the system is about 10% of the total mouse weight for the Inscopix and Stanford case. With such a big camera module, it is difficult for the mouse to move freely. This issue is much alleviated in our case, where the system weight is so light.

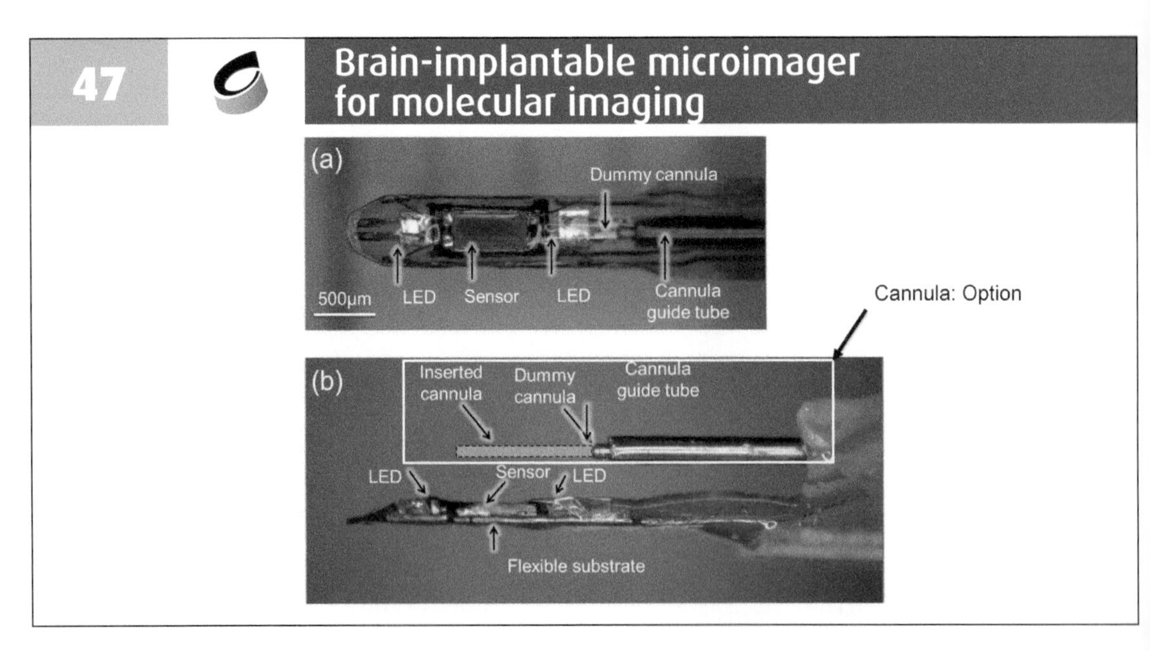

This slide shows close-up views of the brain implantable micro imager for molecular imaging.

48 Sensor Specifications

Type	Needle	Planar
Die size	320 μm x 1025 μm	1000 μm x 3500 μm
Pixel number	30 x 90	120 x 268
Array size	225 μm x 675 μm	900 μm x 2010 μm
Pixel size	7.5 μm x 7.5 μm	
Fill factor	35%	
Pixel structure	3T-APS	
Technology	AMS 0.35 μm CMOS 2P3M	

We have developed two types of sensors. One is a needle shape for measuring neural activities in deep brain region, and the other is a planar shape for measuring neural activities in brain surface.

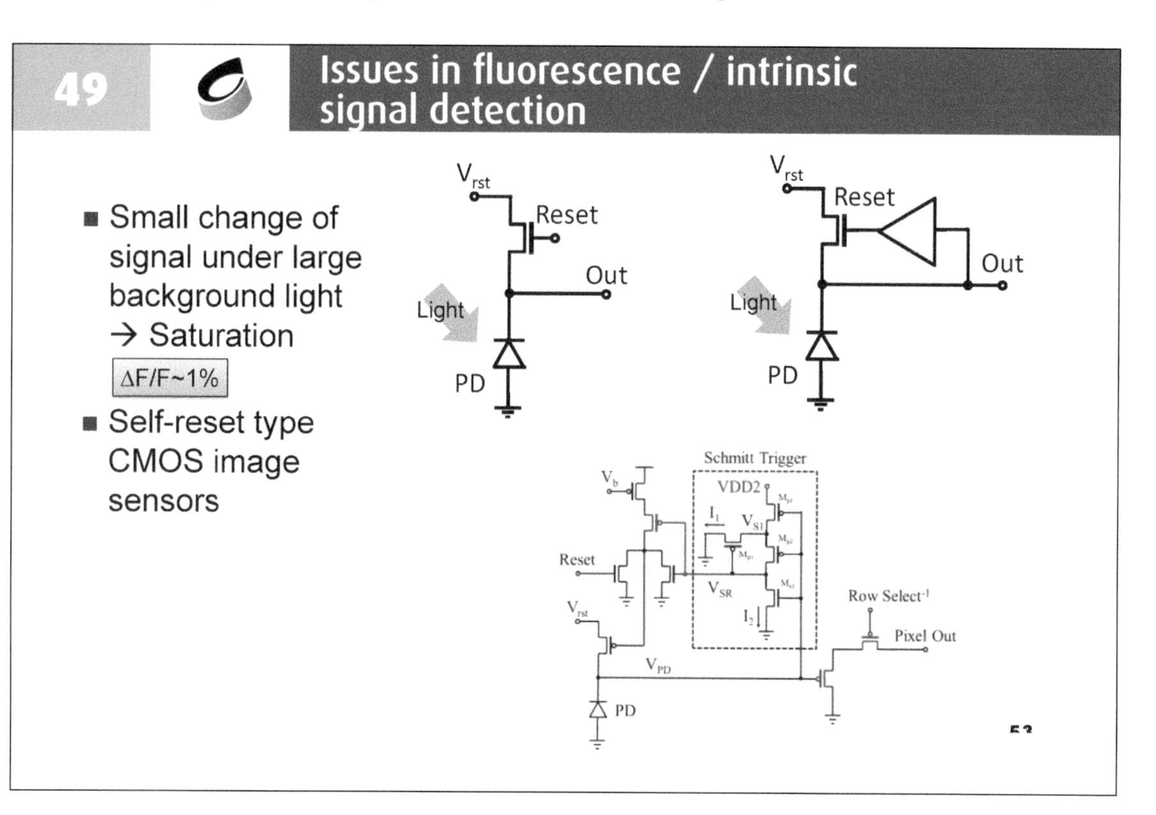

Here, we discuss the issue of fluorescence signal detection in this implantable application. As we mentioned before, the fluorescence is so strong, and its change is very small. As a result, sometimes the photodiode is saturated due to the fluorescence light. To solve this problem, we have developed a self-reset type CMOS image sensor.

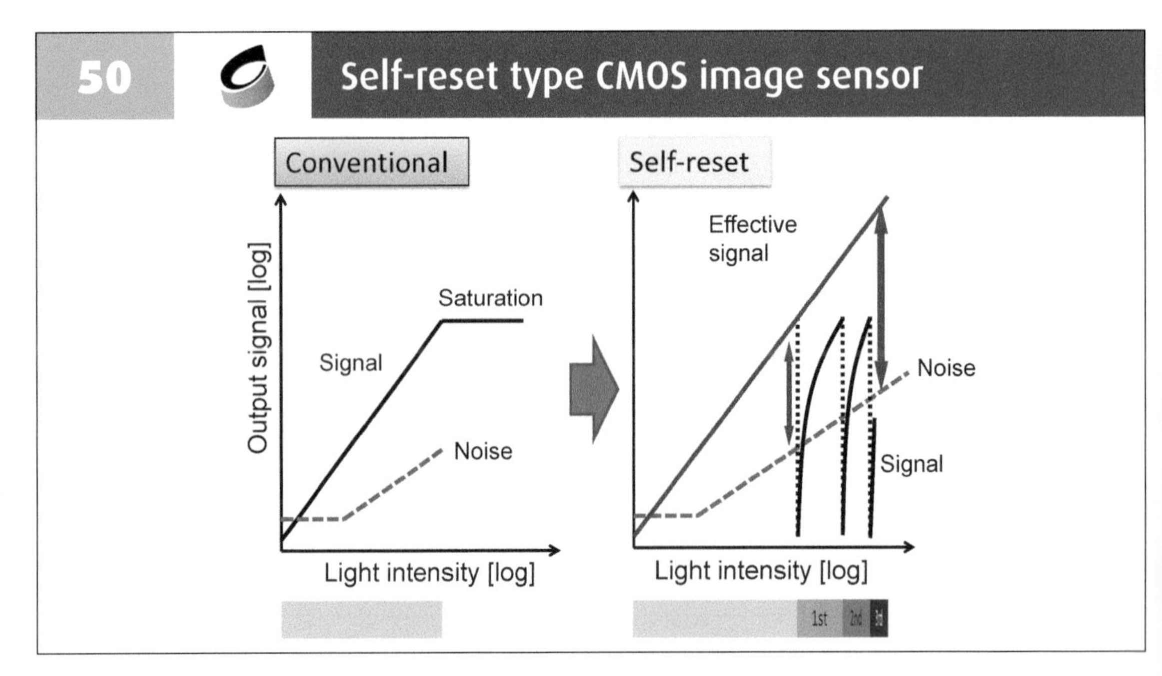

The principle of the self-reset type CMOS image sensor is very simple. When the light is very strong, the signal increases and finally saturates due to the limited well capacity/stored charge. Consequently, we cannot measure the signal over this intensity level. Of course, the shot noise also increases. For a self-reset type image sensor, if the light causes well-saturation, photodiode resetting automatically occurs by using a feedback path. Particularly, when the photodiode voltage is very low, the feedback path triggers the reset transistor and restores the photodiode to its reset state to effectively increase the well capacity.

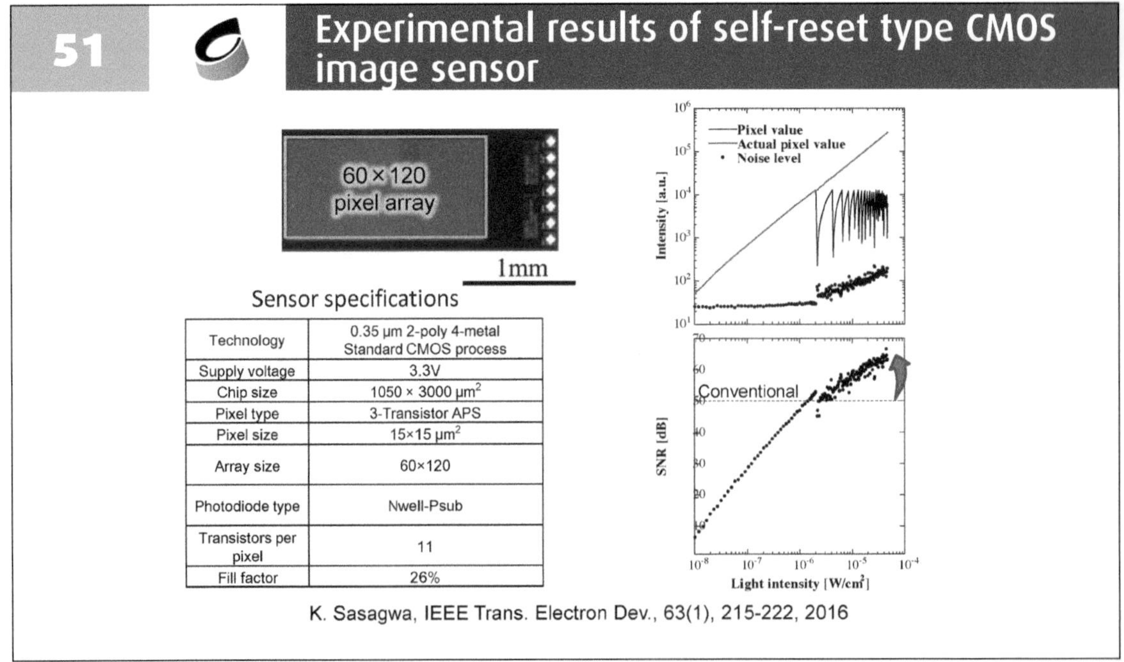

Sensor specifications

Technology	0.35 μm 2-poly 4-metal Standard CMOS process
Supply voltage	3.3V
Chip size	1050 × 3000 μm^2
Pixel type	3-Transistor APS
Pixel size	15×15 μm^2
Array size	60×120
Photodiode type	Nwell-Psub
Transistors per pixel	11
Fill factor	26%

We have implemented the self-reset type image sensor as shown here. It can be observed that the measured SNR can be extend by over 10dB when compared to the conventional image sensor.

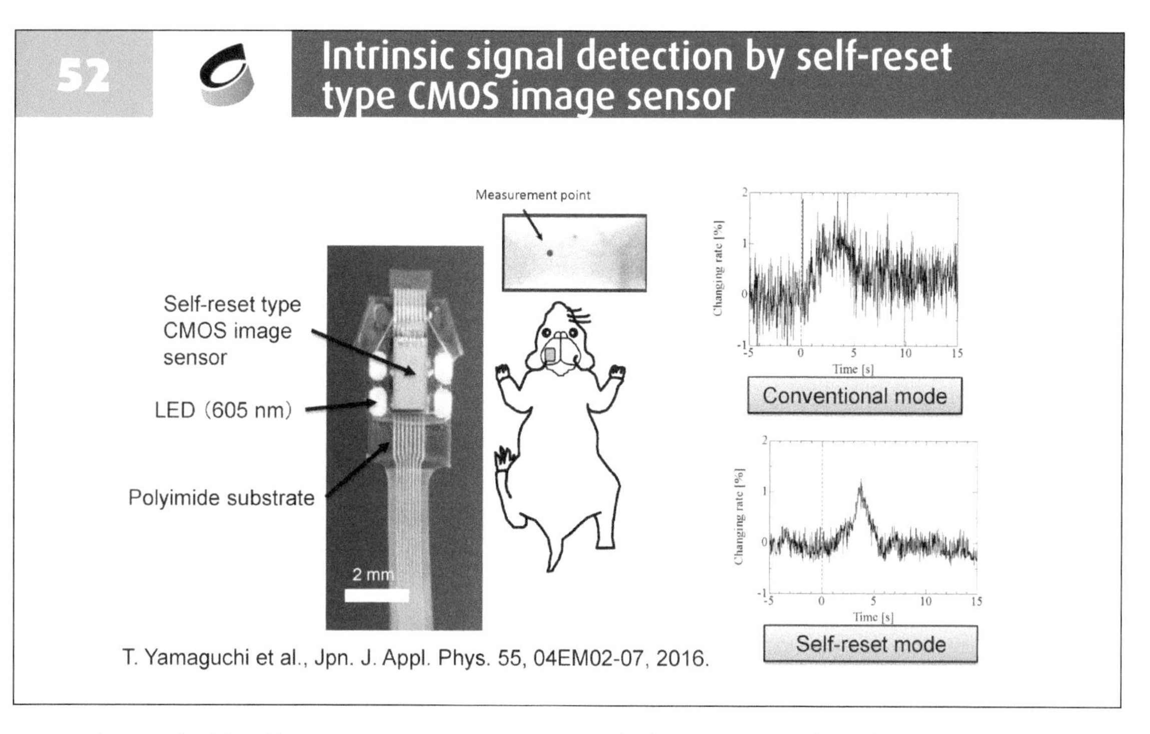

We then apply this self-reset type image senor to measure the neural signals of a trained mouse. This is the transient outputs of the measured signal, with the conventional mode on top, and the self-reset mode at the bottom. We can clearly see the increase in SNR by using the self-reset mode.

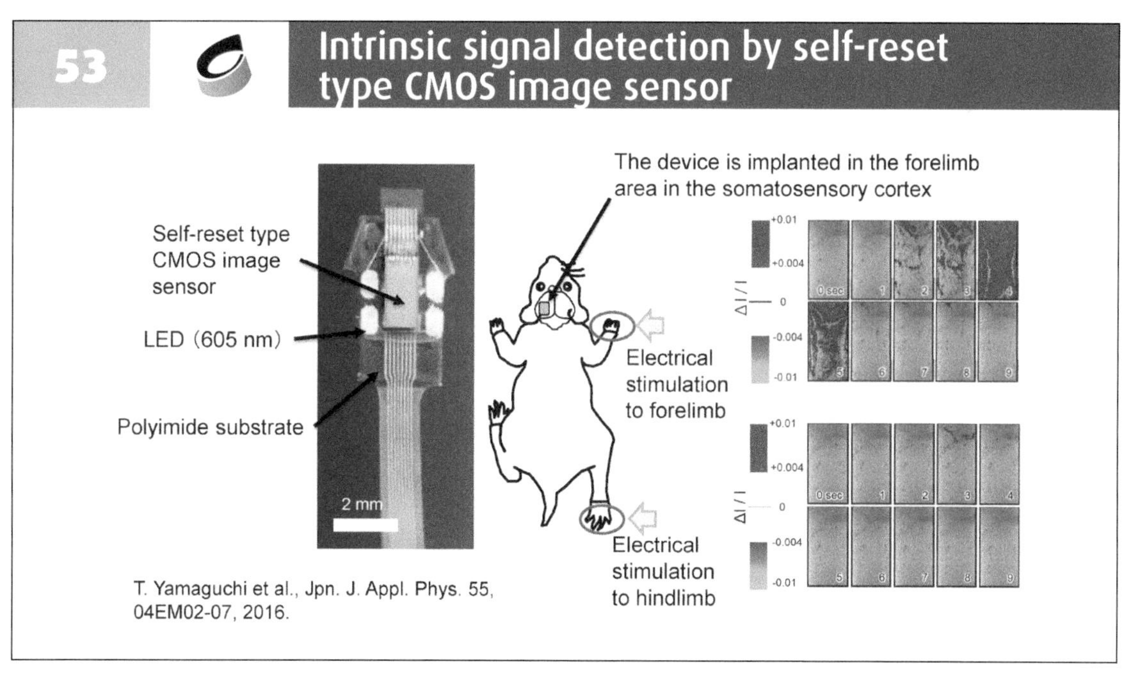

Here we measure the response of electrical stimulation through the forelimb and hindlimb by measuring the response of the brain blood. According to the stimulation, we can observe the response in the forelimb region in the brain but not in the hindlimb region. So by using this self-reset type cmos image sensor, we can obtain clearly the stimulation response.

54  Outline

- **Image sensor technologies**
  - Trend of CMOS image sensors
  - Fundamental characteristics of CMOS image sensors
  - Classification of image sensors for biomedical applications
- **Fluorescence detection**
  - Introduction
  - Voltage sensitive dye: Measurement of cell activities through fluorescence with high speed CMOS image sensor
  - FLIM (fluorescence lifetime imaging microscopy): Measurement of fluorescence lifetime with ultrafast CMOS image sensor
- **Retinal prosthesis**
  - Introduction
  - Implantable stimulators in subretinal space
- **Summary**

INTRODUCTION

55  Overview of retinal prosthesis

- Enables blind patients to partially regain their vision.
- Retinitis pigmentosa (RP), Age-related macular degeneration (AMD)*
  - □ Photoreceptor cells are degenerated
  - □ No remedies
  - □ Most of retinal cells are alive

⇒Electrical stimulation to the rest of retinal cells causes visual sensation (phosphene)

Principle of retinal prosthetic device

Implanted Chip
Pigment epithelium
Bipolar cell
Ganglion cell
Optic nerve
Amacrine cell
0.3-0.4mm
Photoreceptor
Horizontal cell

*AMD is the largest portion of blind diseases in US.

Retinal prosthesis offers to restore the visual function for the patient. This is the structure of the retina composed of layers of neurons. This is the photoreceptor region like photodiode. The photoreceptor produces neural signals analogues to photocurrents in photodiode. The output signal is transferred and processed through these neurons, and finally reaches the ganglion cells. This output ganglion cells are connected to the brain through the optic nerve. If the photoreceptors malfunctions due to some disease, then one can suffer from blindness. Some disease such as retinitis pigmentosa, the photoreceptor cells are degenerated, but the other neural cells, such as the output ganglion cells, are still intact. This means that the photodiode is malfunctioned, but the other transistors are still functioning. If you introduce some electrical stimulation to the retinal neural cells, then the ganglion cells can produce the output, and this output is transmitted to the brain which enables one to see again. This electrical stimulation pattern corresponds to the output image that the blind patient can see using this method. So the retinal prosthesis device basically provides electrical stimulation to the retinal neural cells.

56

Electrical stimulation of cells

Return Electrode
Stimulus electrode
Cells
V_{re}
V_{se}
Interstitial fluid (Electrolyte)
Electric double layer: C_{ed}
C_{ed}
V_{se}
Cells
R_s
V_{re}
Return Electrode
Z_F
Stimulus electrode

Cells are evoked through electrolyte. Electrodes do not directly contact cells. → Many cells are evoked simultaneously when stimulated.

Current
0
I_a
I_c
Time
t_c t_i t_a
t_f

- Under unstable biological condition, constant current injection is preferable.
- Biphasic pulse is necessary to keep charge neutrality or to avoid electrolysis.

When retinal cells are evoked by stimulus current, electrical double layer is formed because interstitial fluid is electrolyte or contains much ions. This electric double layer acts as a capacitor.

It should be noted that stimulus electrodes do not directly contact cells; cells are evoked through electrolyte. This means that many cells are simultaneously evoked by the stimulation.

This electrical double layer capacitance is charged by stimulation and finally causes electrolysis due to the increase of the electrode potential by the capacitance charge.

To avoid this, biphasic pulse is used to keep charge neutrality.

In addition, under unstable biological conditions, constant current injection is preferable.

Consequently, biphasic current pulse is used to evoke retinal cells stably and safely.

57

Signal path and implementation

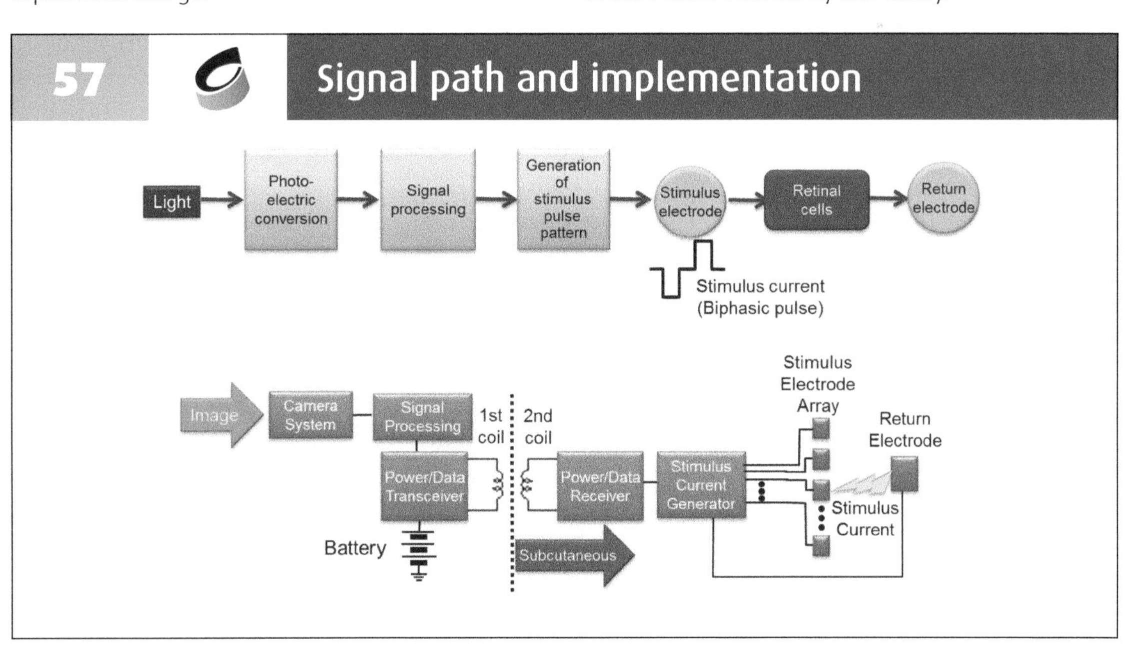

This slide shows the signal path and implementation of retinal prosthesis. It is composed of a camera system, followed by signal processing, and then the transceiver with two coupling coil for power and data transmission. Inside the body, the stimulation current generator is implemented for stimulating the neural cells.

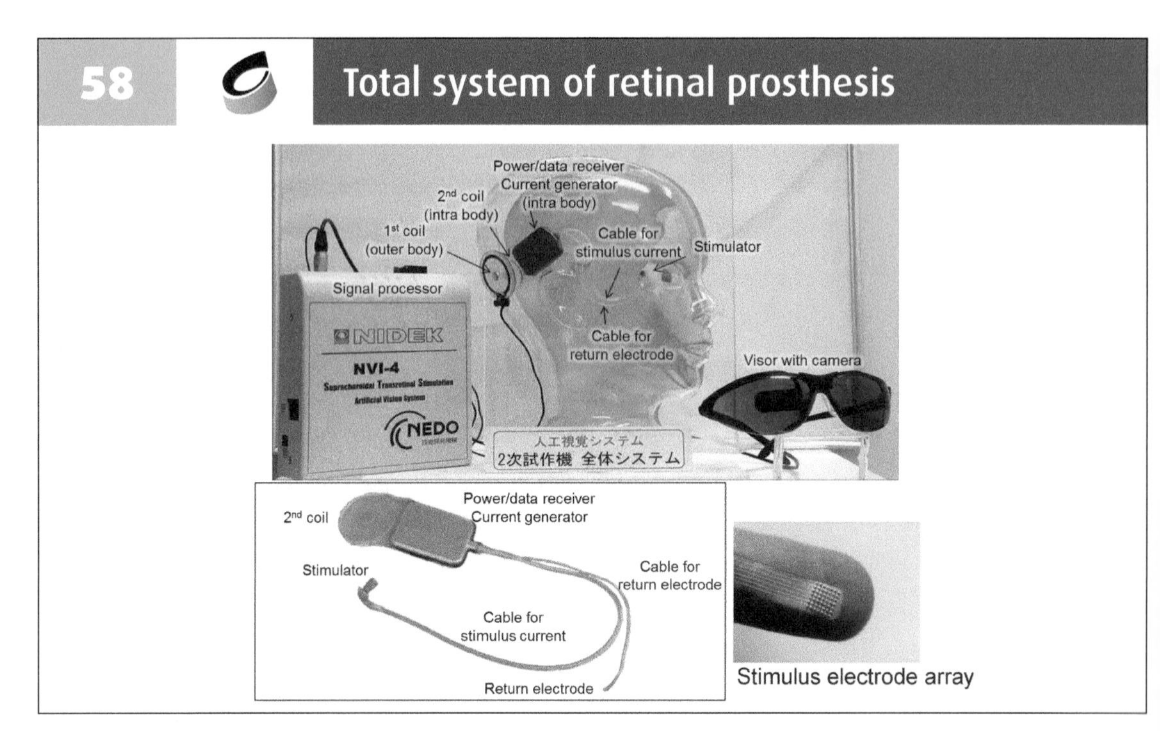

This shows an example for retinal prosthesis. You can see the outside and inside coils.

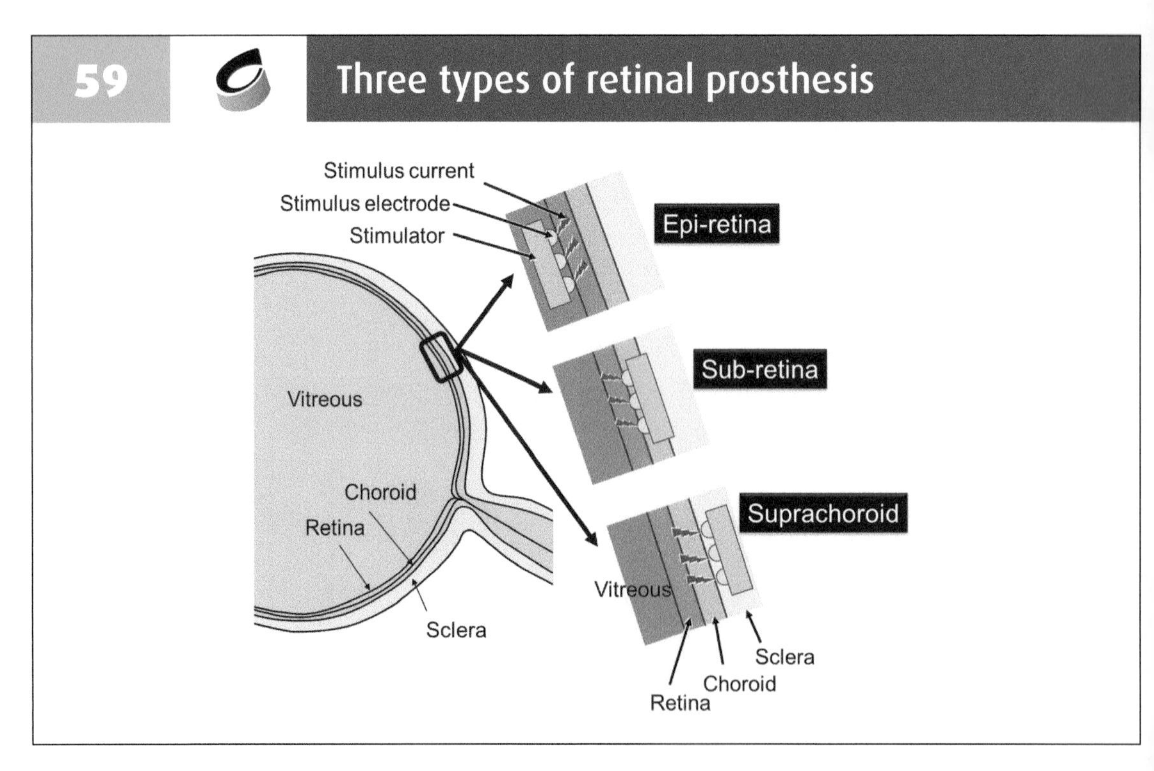

Retinal prosthesis is classified into three types in a viewpoint of the implantation space; epi-retina, sub-retina and suprachoroid.

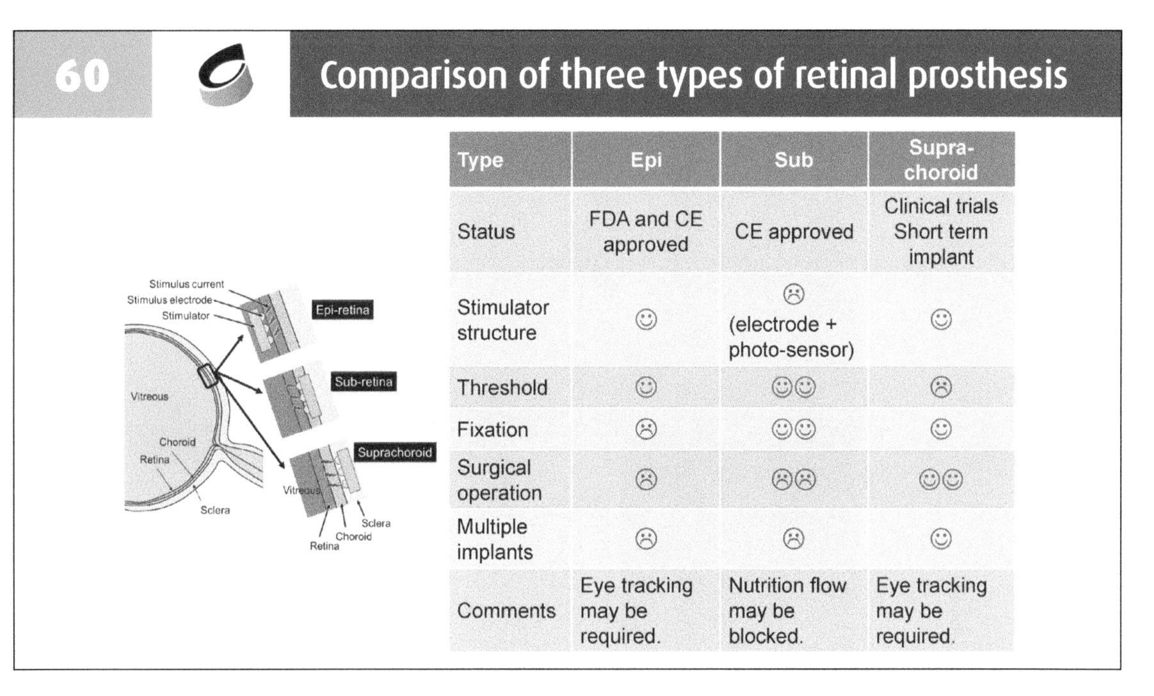

Type	Epi	Sub	Supra-choroid
Status	FDA and CE approved	CE approved	Clinical trials Short term implant
Stimulator structure	☺	☹ (electrode + photo-sensor)	☺
Threshold	☺	☺☺	☹
Fixation	☹	☺☺	☺
Surgical operation	☹	☹☹	☺☺
Multiple implants	☹	☹	☺
Comments	Eye tracking may be required.	Nutrition flow may be blocked.	Eye tracking may be required.

This table shows comparison among three types of retinal prosthesis.

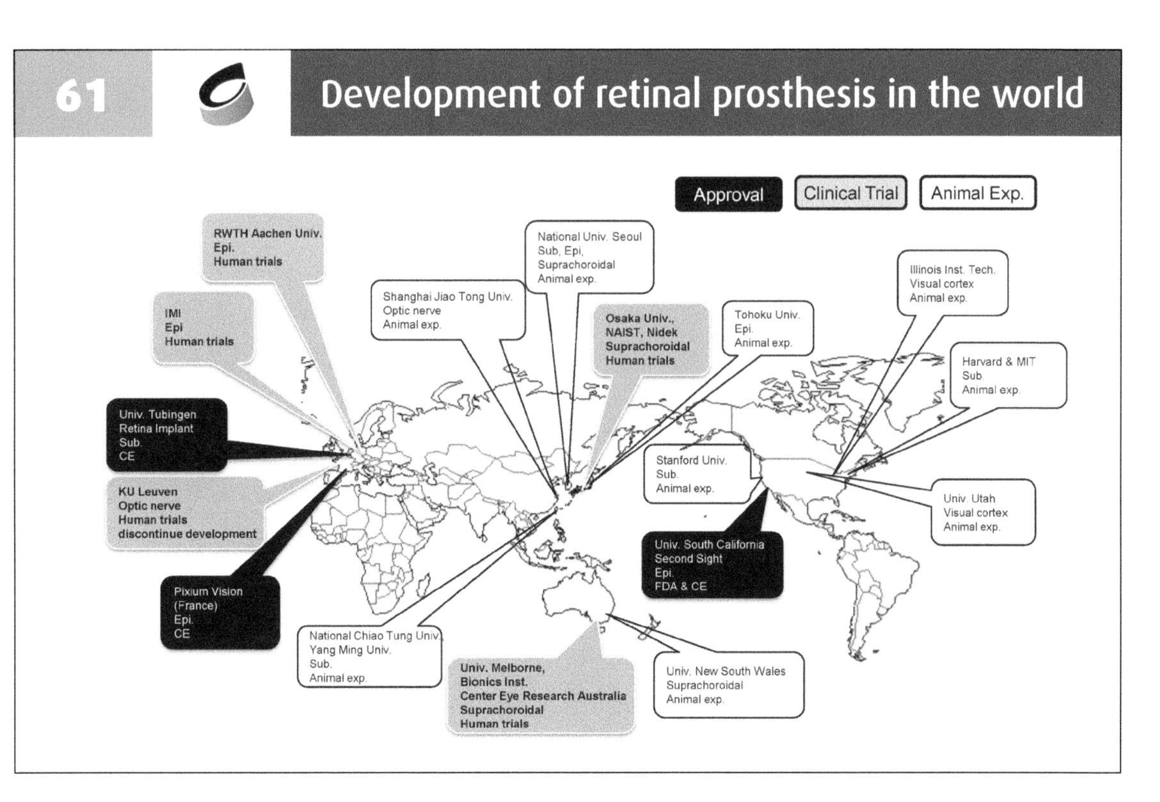

In the present state, three retinal prosthesis companies in different places have already established and gained approval from the government.

62 Epiretinal implantation

Second Sight (USA): FDA & CE approved

A B C D

Standard VCF

Scrambled VCF

RT EA ON

Stimulator: 60-electrode

L. da Cruz L, et al., Br. J. Ophthalmol. 97(5):632-6, 2013.

This shows one of the examples for such commercialized retinal prosthesis devices made by Second Sight in the US. In this case, the camera is outside the body. However, you can also install the camera system inside the eye. In that case, both the photo sensing and stimulation elements should be integrated into one device.

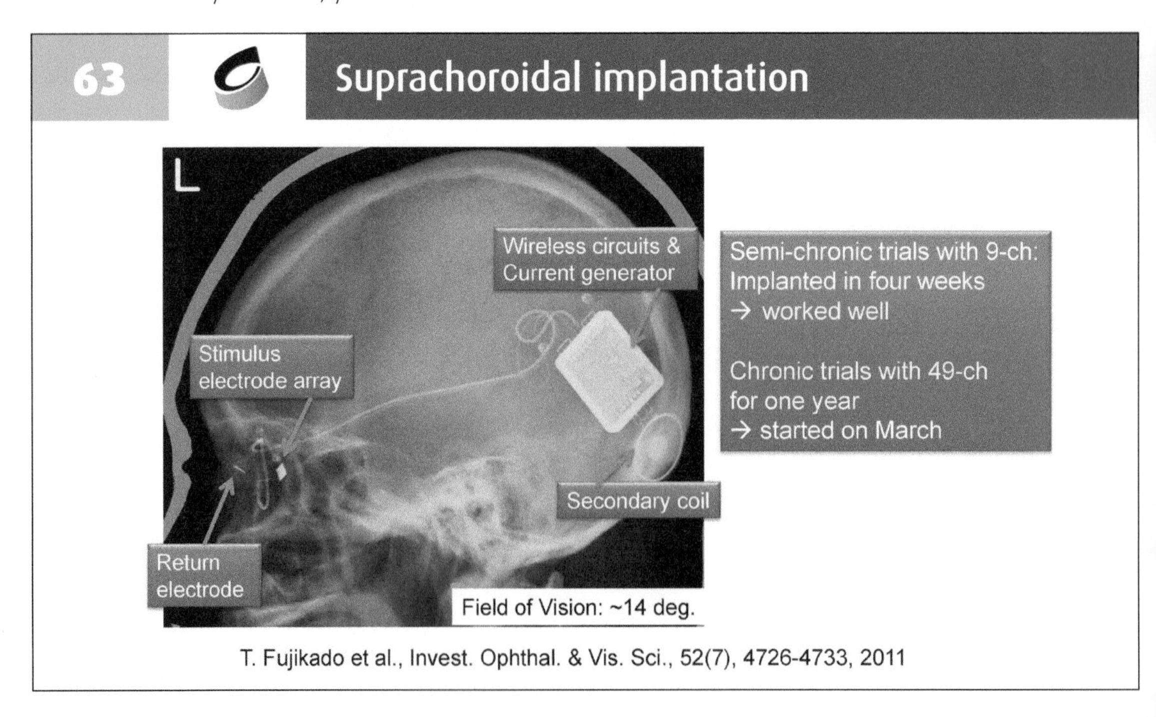

Osaka University Medical School and Nidek have developed STS type retinal prosthesis and successfully executed chronic trials with 49-ch devices for one year.

IMAGE SENSORS FOR RETINAL PROSTHESIS

64 Imaging system in retinal prosthesis

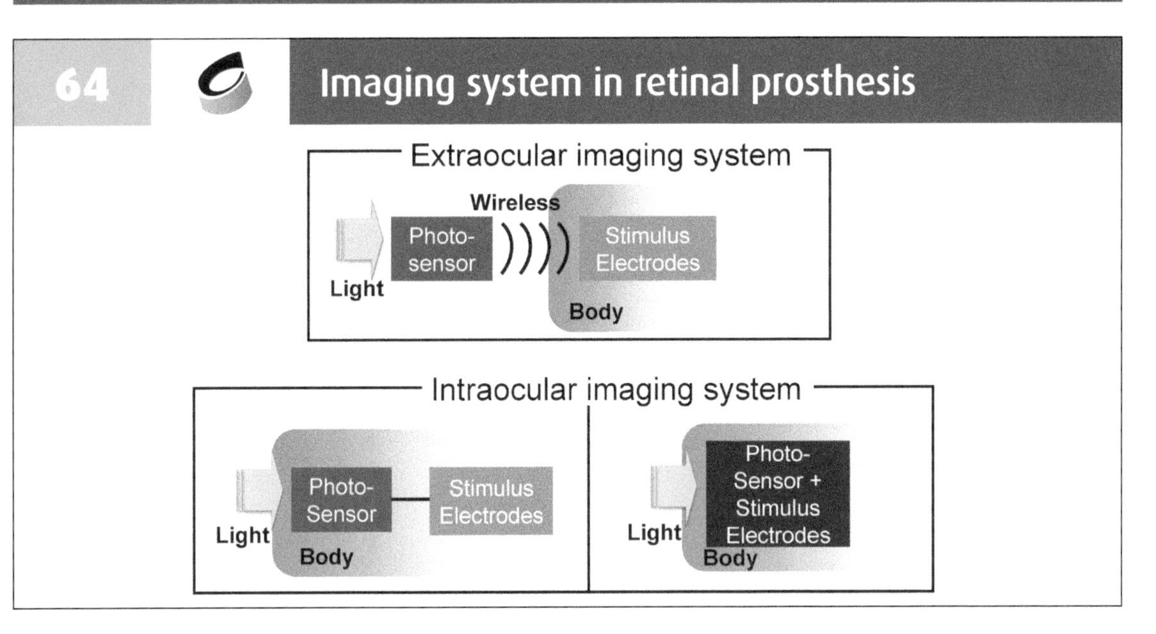

There are two types of imaging systems for retinal prosthesis; extraocular and intraocular imaging systems. In the extraocular imaging system, a photosensor is located outside a body.

The imaging data is transferred by a wireless device. In intraocular imaging system, a photosensor is implanted in the body. The system is usually used in a sub retinal implantation.

In this system, a wireless imaging data transfer device is not required so that the system becomes simple. In the system, a photosensor can be integrated with stimulus electrode in one device, which has a similar function of a photoreceptor cell.

65 Retinal prosthesis with light sensing (Intraocular imaging system)

Configuration		Stimulus place	Affiliation
Implanted micro camera		Optic nerve	C-Sight
Micro photodiode Array (Solar cell mode)	-	Sub.	Optobionics
	NIR-converted image	Sub.	Stanford U.
	Image: HMD Power: NIR	Sub.	National Chaio Tung U.
CMOS sensor integrated with electrodes	Log sensor	Sub.	Retina Implant/ Tubingen U.
	3D integration chip	Epi.	Tohoku U.
	APS-Based Photo-sensor	Sub./ STS	Osaka U. /NAIST/ Nidek

This table summarize the types of retinal prosthesis devices using light-sensing.

66 Intraocular imaging system
Integration of photosensor and electrode

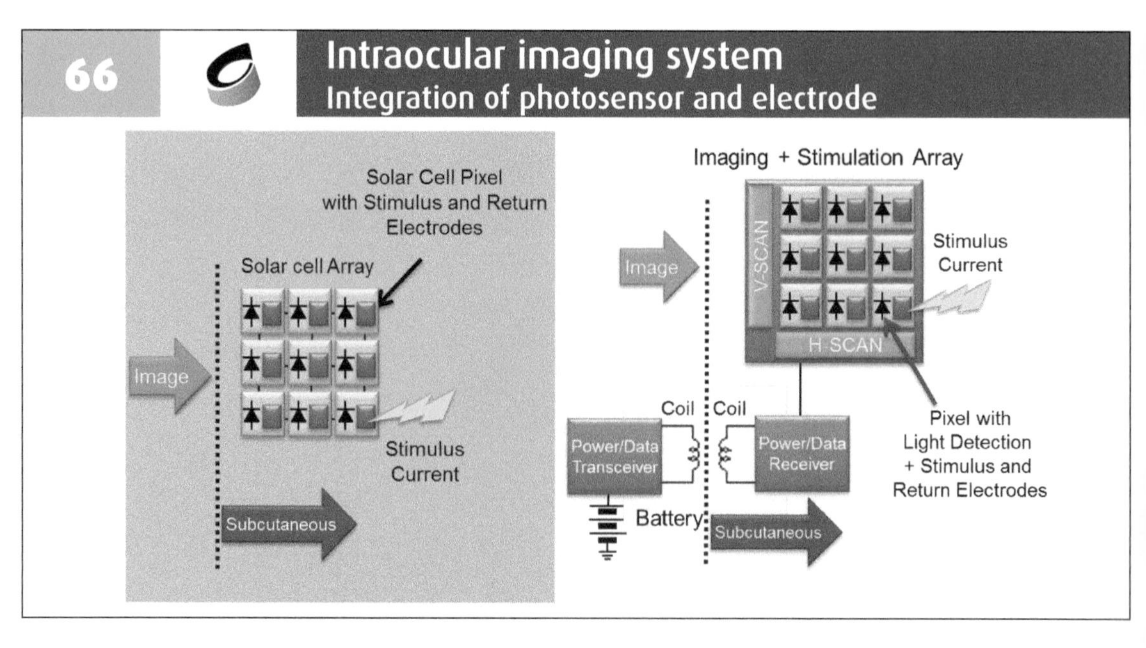

For an intraocular imaging system, it is comprised of a photodiode plus electrode, i.e. each pixel has a photodiode and the corresponding electrical stimulation circuitry. For an array of pixel, when the light enters a photodiode, a proportional stimulation current is produced.

67 Micro solar array for retinal prosthesis

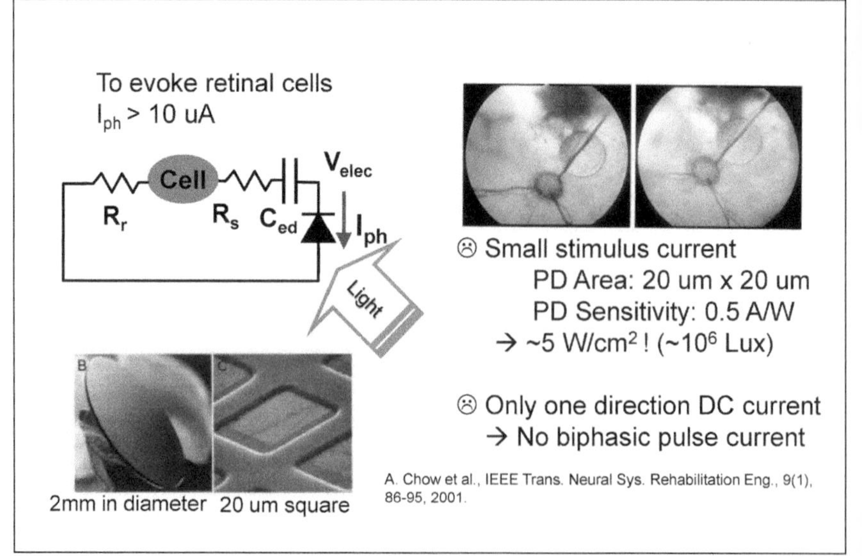

The retinal prosthesis with micro solar array was developed. The most important advantage of the device is that it works only by implanting the device with no electric power delivery, thus a surgical operation of the implantation of the device is very easy.

It, however, finally failed to be commercialized in market. The reason is that a solar cell cannot produce enough stimulus current to evoke retinal cells.

Generally to evoke retinal cells, the amount of stimulus current must be larger than 10 micro amperes. To obtain this current value, about 100,000 lux is required for a PD with the area of 20 micrometer square.

This is the brightness level in outdoor during midsummer! In addition, a solar cell can only produce one direction DC current, not biphasic pulse current. Thus, only a micro solar array device is not suitable to retinal prosthesis.

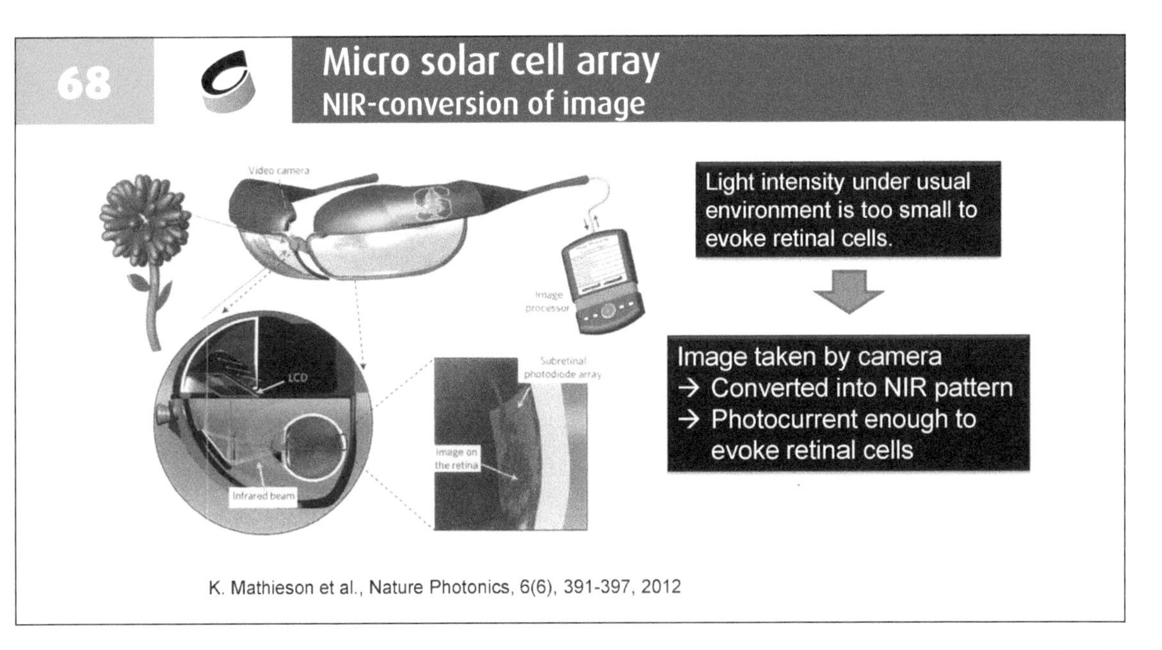

To overcome the limitation by using a solar cell array in retinal prosthesis, Palanker's group in Stanford University has developed a system where image is converted into near infrared (NIR) light pattern and enhanced its intensity enough to evoke retinal cells. Although this system requires an extraocular device that converts visible image into NIR one and projects it into the retina, surgical operation is still simple and easy.

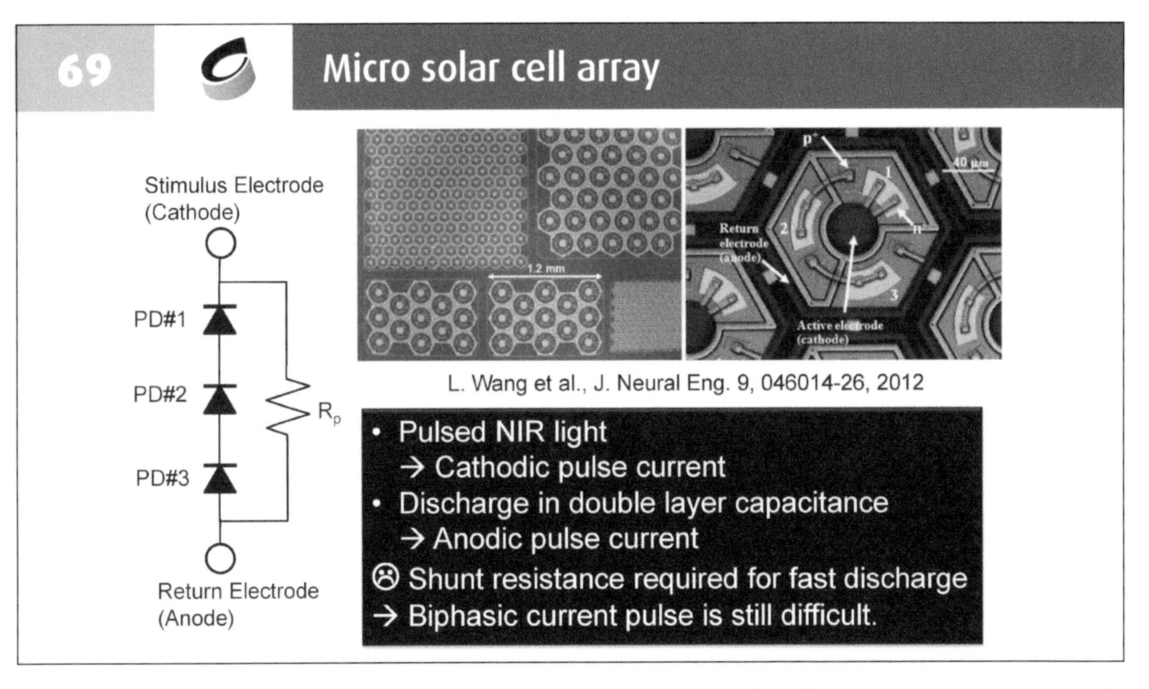

This slides shows the implementation of the subretinal device with micro solar cell array developed by Palanker's group.

To produce biphasic current, the discharge in double layer capacitance is used. In addition, to produce enough voltage, three PDs are serially connected.

To discharge quickly, a shunt resistance is required as shown in the figure.

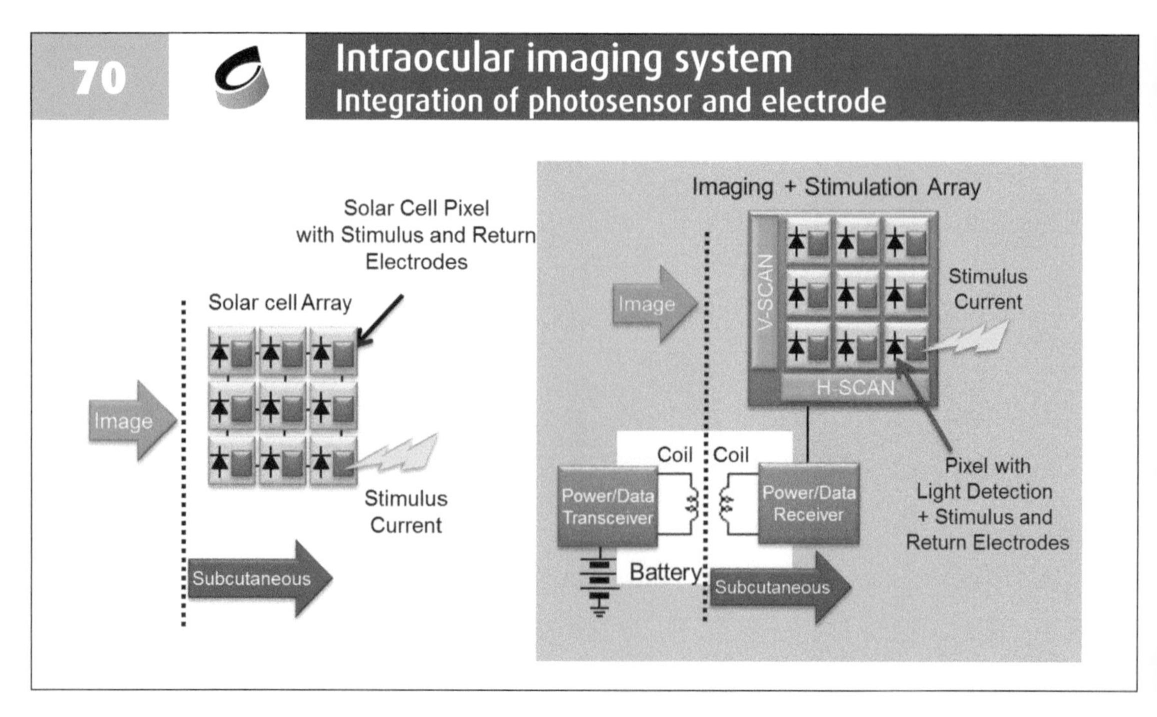

Next, another intraocular imaging system is introduced. The system requires an electrical power supply.

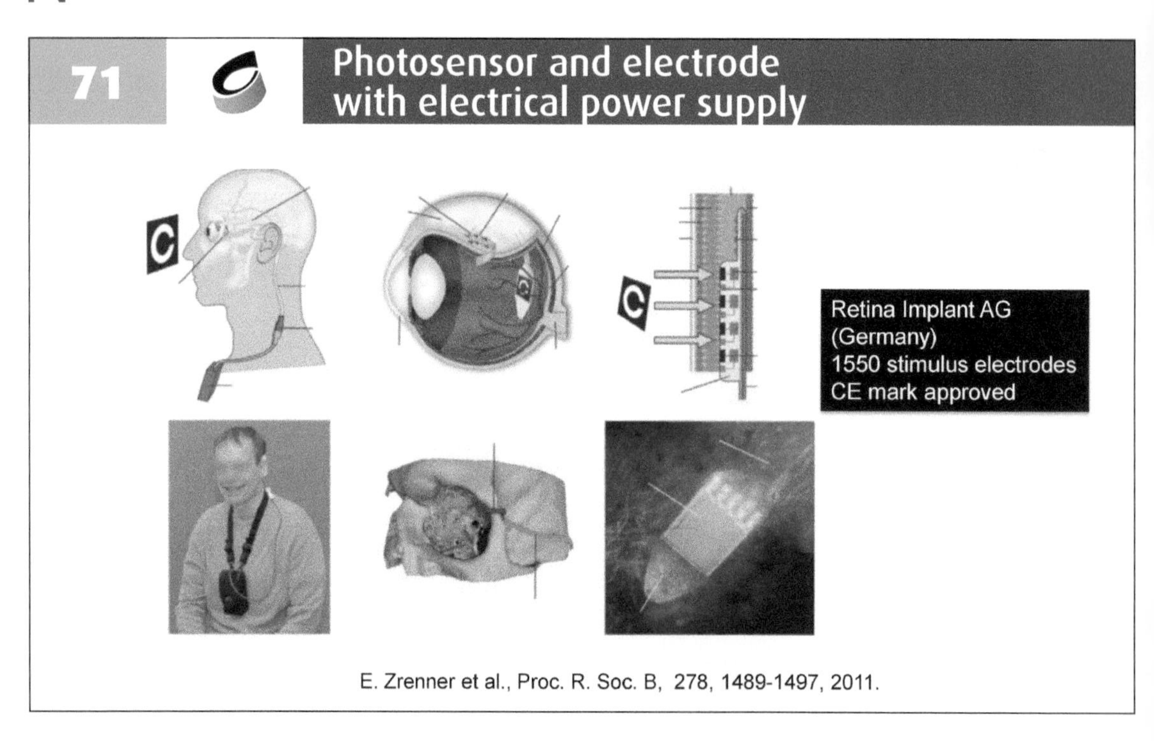

The subreinal implantation device developed by Retinal Implant ATG in Germany has been approved in Europe.

It has 1550 stimulus electrodes.

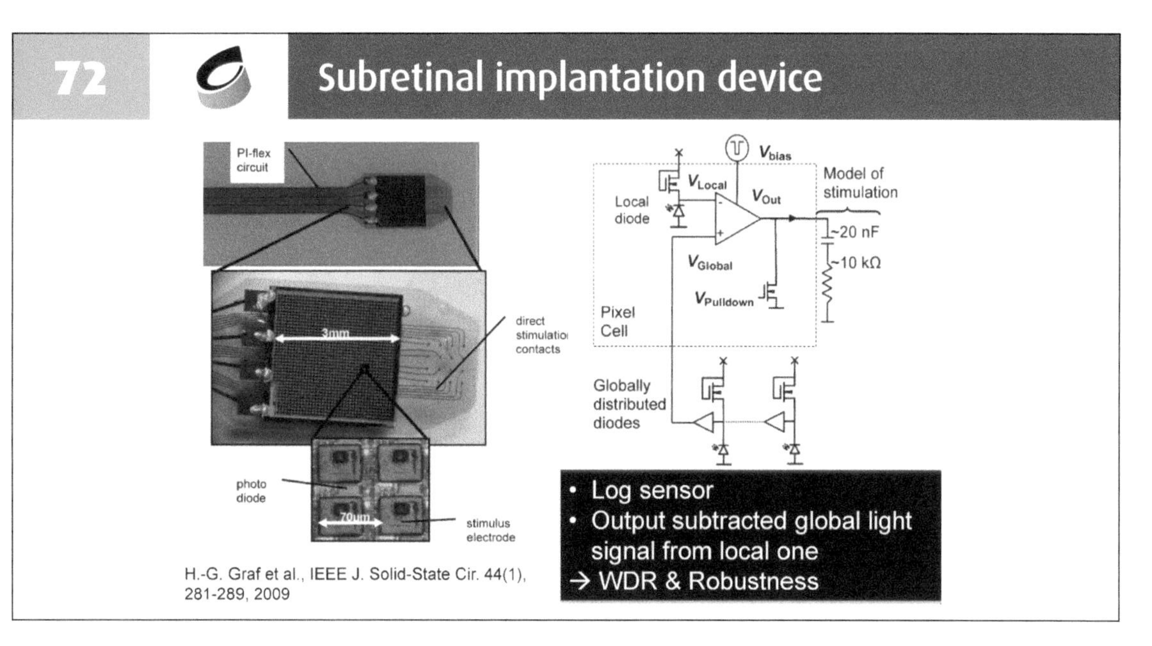

This slide shows the subretinal implantation device developed by Retina Implant in Germany, with the photodiodes, electrodes and some circuits here. These photodiodes here work as the global sensor, and the difference between the local/global intensity is used to generate the output stimulation current. These globally distributed diodes output can eliminate the problem of intensity fluctuations due to the outside environment. One drawback of this design is the large chip size, which is up to 3 mm. As you know, the eye ball is round, and the large flat implantable device can produce some problem as it is in direct contact with the retina.

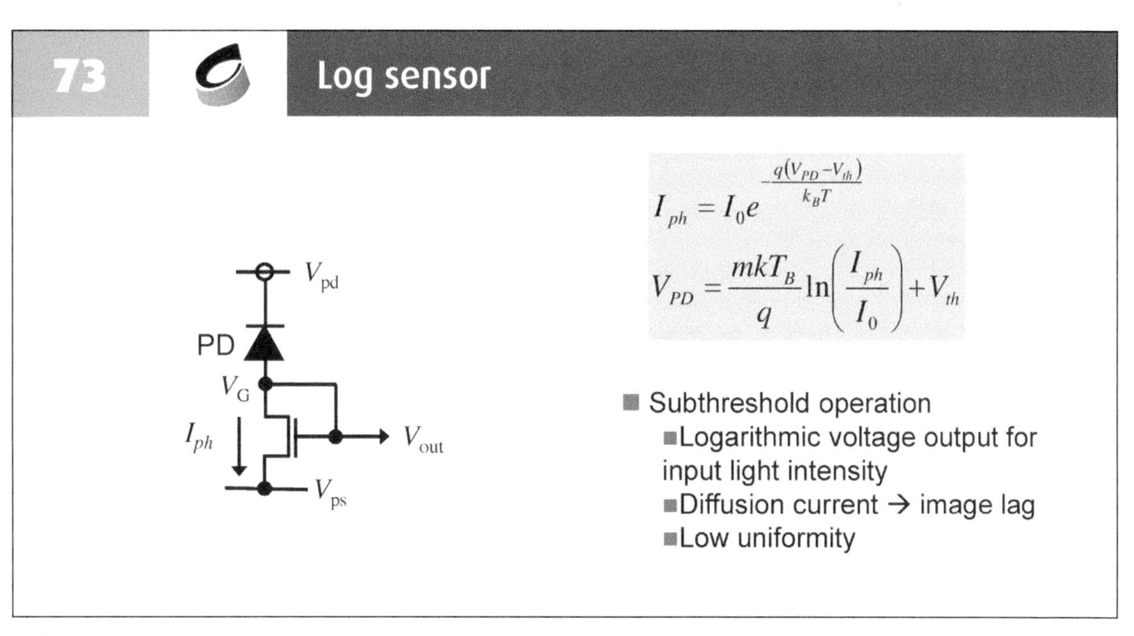

This slides shows a log sensor where output signal voltage increases in a logarithmic manner depending as input light intensity increases.

Such logarithmic output characteristics is useful for wide dynamic range image sensor but image lag and low uniformity are drawbacks in the sensor due to subthreshold operation.

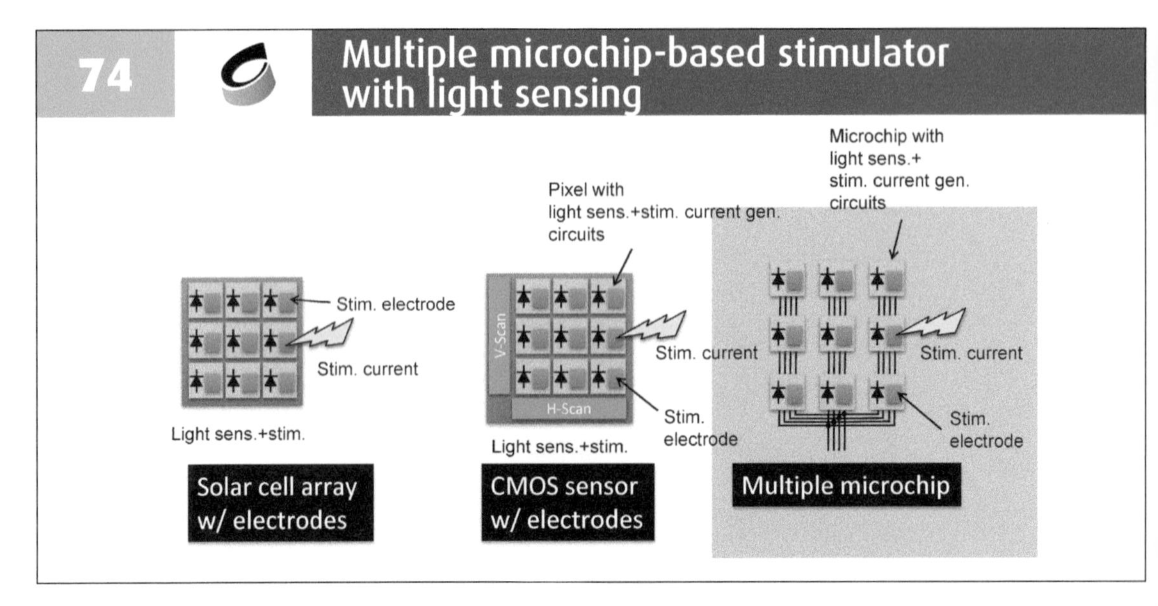

Finally, multiple microchip based simulator with light sensing is introduced.

In this architecture, each microchip contains a photosensor, a stimulus electrodes and circuits to control stimulation current by input digital data.

The advantage of the architecture is small number of wiring and bendability. The number of wires are four or five irrespective to the number of microchips, so that it can achieve many number of electrodes such as 1000 stimulus electrodes in one system in a bendable manner.

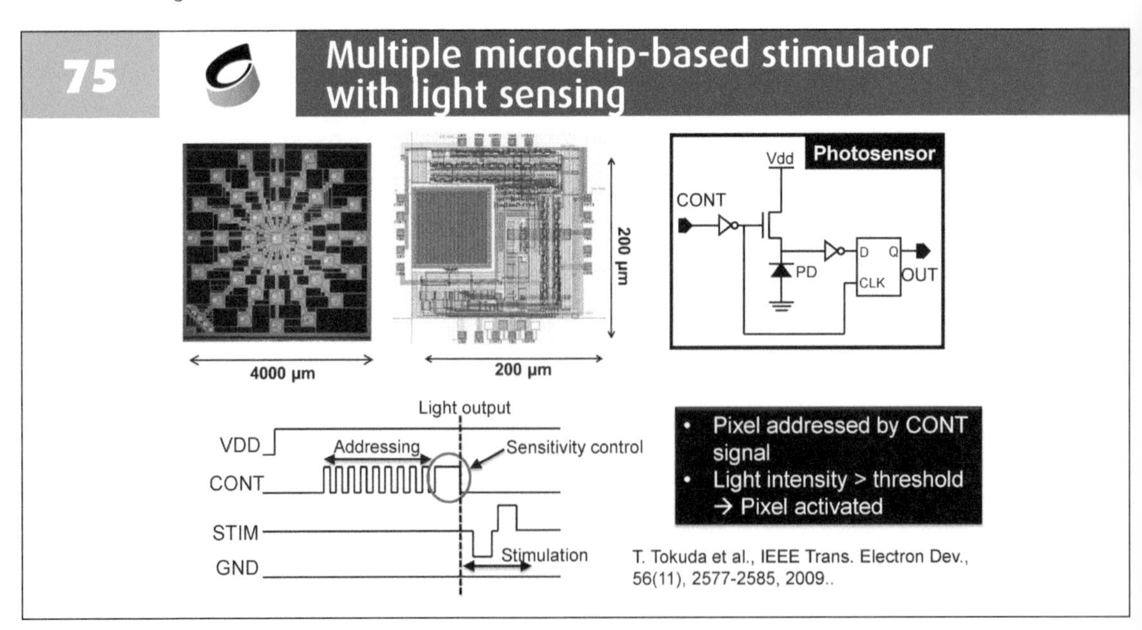

This slide shows the details of the microchip. The microchip employs a photosensor where the sensitivity is controlled by a control pulse width as shown in the timing chart.

When the anode voltage value of the PD is larger than the threshold value, the pixel is activated and simulation current is injected. The amount of the stimulation current can be determined by the digital data in "CONT".

The size of the microchip is 200 µm x 200 µm. The mother chip consists of the number of microchips placed along radial direction as shown in the figure. The size of the mother chip is 4mm x 4 mm.

The final fabrication is shown later.

76 Multiple microchip-based stimulator with light sensing

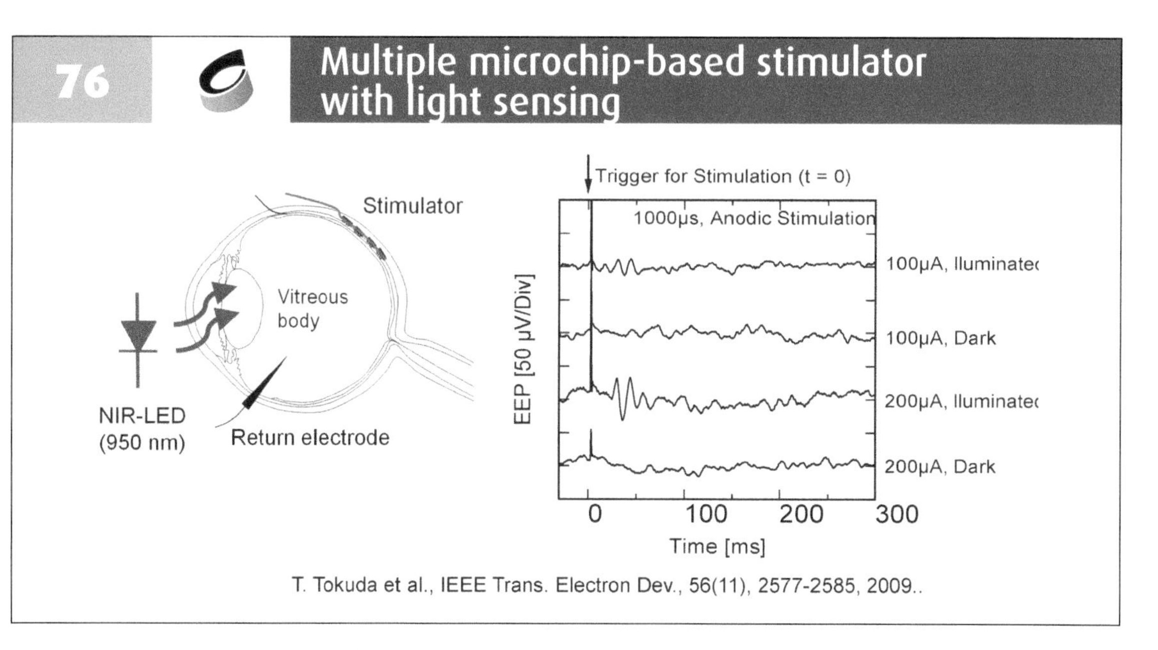

T. Tokuda et al., IEEE Trans. Electron Dev., 56(11), 2577-2585, 2009..

The microchip is validated by electro-evoked potential (EEP) signal when it is implanted in a rabit's eye ball and illuminated by NIR light.

The left figure shows the experimental setup and the right figure shows the experimental results of EEP.

Only when light is illuminated, EEP signal appears. It is noted that NIR light cannot evoke photoreceptor cells and only activates the photosensor on the microchip.

77 Experimental result of microchip with light sensing and current generator

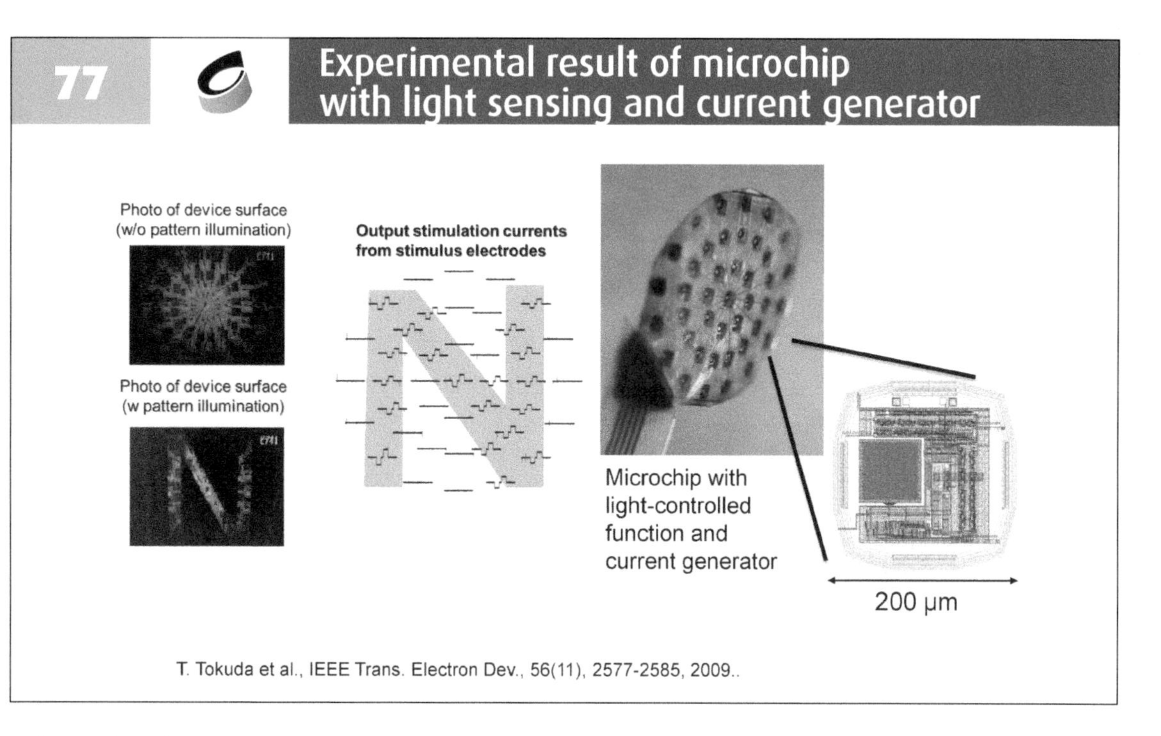

T. Tokuda et al., IEEE Trans. Electron Dev., 56(11), 2577-2585, 2009..

The mother chip is illuminated by the light pattern "N" as shown in the left figure and it produces stimulus current pattern according to the light pattern.

The right figure shows the fabricated device where each microchip connected with thin wires. The whole device can be bent.

So this is the summary of my talk. I first introduced some basics of CMOS image sensor. Next, I showed fluorescence detection examples using CMOS image sensor. Finally, I briefly introduced the application of CMOS image sensor for retinal prosthesis applications.

CHAPTER 6

Ultra-low Power/ Energy Efficient High Accuracy CMOS Temperature Sensors

Man-Kay Law

University of Macau, Macao, China

Ultra-low power/energy efficient wireless sensing microsystems plays an important role in the upcoming internet of things (IoT) era, with emerging applications including medical, surveillance, and environmental monitoring and many more. Being one of the most common measurement parameters, high performance CMOS temperature sensors become an important design element of such microsystems. This chapter covers the fundamental knowledge and operating principles of CMOS temperature sensors. The main sources of inaccuracy will be identified, followed by possible circuit/system level solutions to achieve high power/energy efficiencies. Case studies of state-of-the-art CMOS temperature sensors will also be introduced.

1 

Temperature Sensing in Our Daily Life

Why is temperature sensing important? In fact, temperature is one of the major physical parameters that we experience every day. You will be able to find all kinds of temperature sensors in many of our daily necessities, including various home appliances, automotive, medical supplies, food, and of course your PCs, laptops, and even your mobile phones. Local temperature measurements can be useful to boost the system performance and improve the user experience.

2 

Emerging Sensing/Biomedical Applications

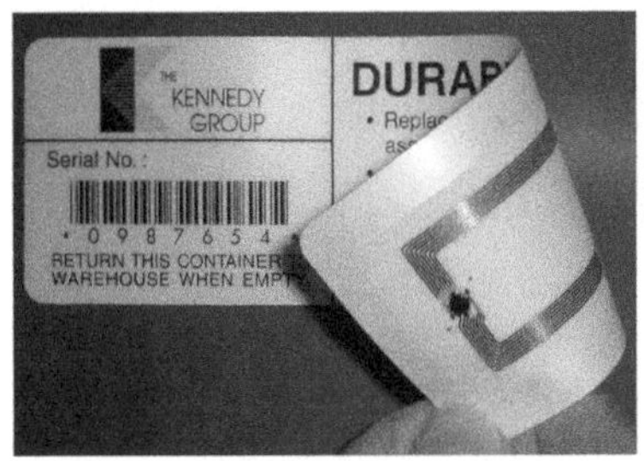

- Wireless sensor networks/RFID[1,2,3]
- Implantable devices[4,5,6]
- Temperature compensation for various sensors[7,8,9]

Small battery/battery-less
1. **Ultra-low power (µW~nW)**
2. **High accuracy w/ less calibration efforts**
3. **Good supply rejection**

Nowadays, there are many emerging sensing or biomedical applications for internet of things (IoT) such as wireless sensor network, RFID and implantable devices. Also, temperature compensation for other kinds of sensors is widely used due to the intrinsic temperature dependence is many commercially available sensors. Typically, these kind of systems are usually miniaturized for reduced cost, and equipped with different sensors to perform different application specific sensing tasks. As you can see here, these kinds of systems are usually very small in size. As a result, their energy storage size (e.g. battery) is generally limited, and some (e.g. passive RFID) maybe even battery-less for further production/maintenance cost reduction as well as application specific considerations, such as the potential leakage of hazardous battery chemicals for implantable devices. Consequently, it is challenging to achieve ultra-low power which can be in the order of µW to even nW, while achieving high accuracy without elaborate calibration efforts. Apart from that, good supply immunity is also necessary to ensure accurate sensing under significant supply noise (e.g. RFID).

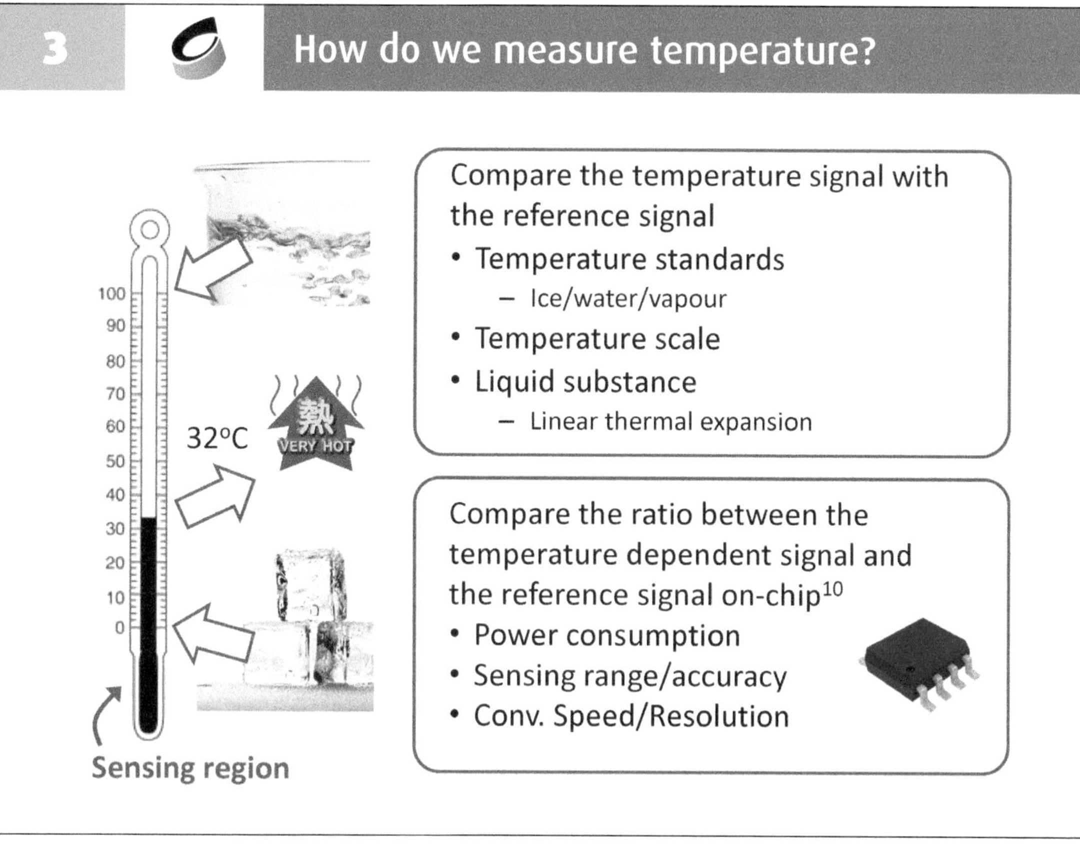

So, how do we measure temperature? Let me first introduce to you some basics in temperature sensing. Here shows a simple traditional liquid-glass thermometer. We can obtain the current temperature by reading the mark on the glass tube which coincide with the volume of the liquid inside. You may notice several basic ingredients here required for temperature measurement. First, we need the two reference points, which are inferred by the volume of the liquid at 0 and 100°C when ice melts and water boils. Scientifically, it should be more accurate by using the triple point of water, but the basic definition should suffice in our discussion. Then we need an accurate temperature scale. This can be achieved by simple linear interpolation if the liquid exhibits a linear thermal expansion characteristics. Notice that the liquid inside the bulb here shows the sensing region. For CMOS temperature sensors, what we need is to translate this liquid-glass thermometer requirements into electrical parameters. Essentially, what we need is to compare the ratio between a temperature dependent signal and a reference signal, both of which should be generated on chip. The design metrics include power consumption, sensing range and accuracy, conversion speed as well as resolution.

4  Specifications

- Power consumption
  - Tens to hundreds of nW ~ few mW
- Sensing Range/Accuracy/ Resolution
  - Military:
    -55 to 125°C/±0.1°C/0.01°C
  - Clinical:
    25 to 45°C/±0.1°C/0.01°C
  - Deep freeze/frozen storage:
    -30 to -10°C/±0.5~1°C/0.05°C
  - Pharmaceutical:
    0 to 10°C/±0.5°C/0.1°C
- Conversion speed
  - Few sample/s (typical)

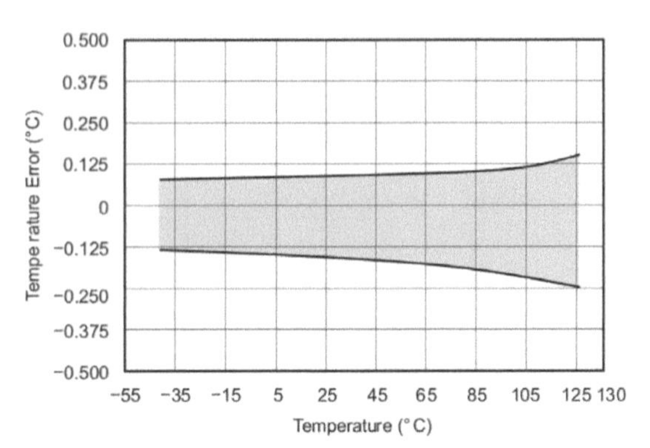

Datasheet: TMP275 from TI

These specifications can be different depending on the target application. For the emerging battery-powered as well as battery-less systems, power consumption is essentially one of the major performance metric, with a power budget in the order of few micro-watt or even less. The sensing range and accuracy can vary from -55 to 125°C in the military range with ±0.1°C accuracy in typical high-end applications, from -25 to 0°C with ±0.5°C for cold-chain and food monitoring, and even from 25 to 45°C with ±0.3°C in clinical applications. Typically, the conversion speed in most applications is not as stringent, say in order of few samples per second should be enough. Even if your circuit can sample very fast, the thermal time constant for the package can be a limiting factor. Typically, one obvious exception is for microprocessor applications, as we are sensing the substrate temperature and can change very fast. But this is not the focus of this talk. For the resolution, it should be higher than the accuracy requirement, and is limited by the thermal noise. As you can observe here, accuracy is also an important parameter for CMOS temperature sensor. Similar to other circuit designs, CMOS temperature sensors suffer from both the systematic error and random error. Systematic error, such as the linearity of the temperature signal, can be corrected by circuit techniques, but not so for random error. This ultimately limits the sensing accuracy.

5  CMOS Temperature Sensors

As illustrated here, CMOS temperature sensors generally requires an analog frontend, a sensor interface which is generally a custom designed ADC, and a digital backend for control, calibration and data processing purposes. Similar to the conventional liquid-glass thermometer, the analog frontend consists of temperature sensitive devices to generate the temperature dependent signal, as well as a reference signal. Notice that electrical signals of any forms (e.g. voltage, current etc.) can be used for this purpose. The sensor interface then digitizes and compares these two signals, and the resultant digital output D_{out} will then be interpreted as the instantaneous temperature through the digital processing block.

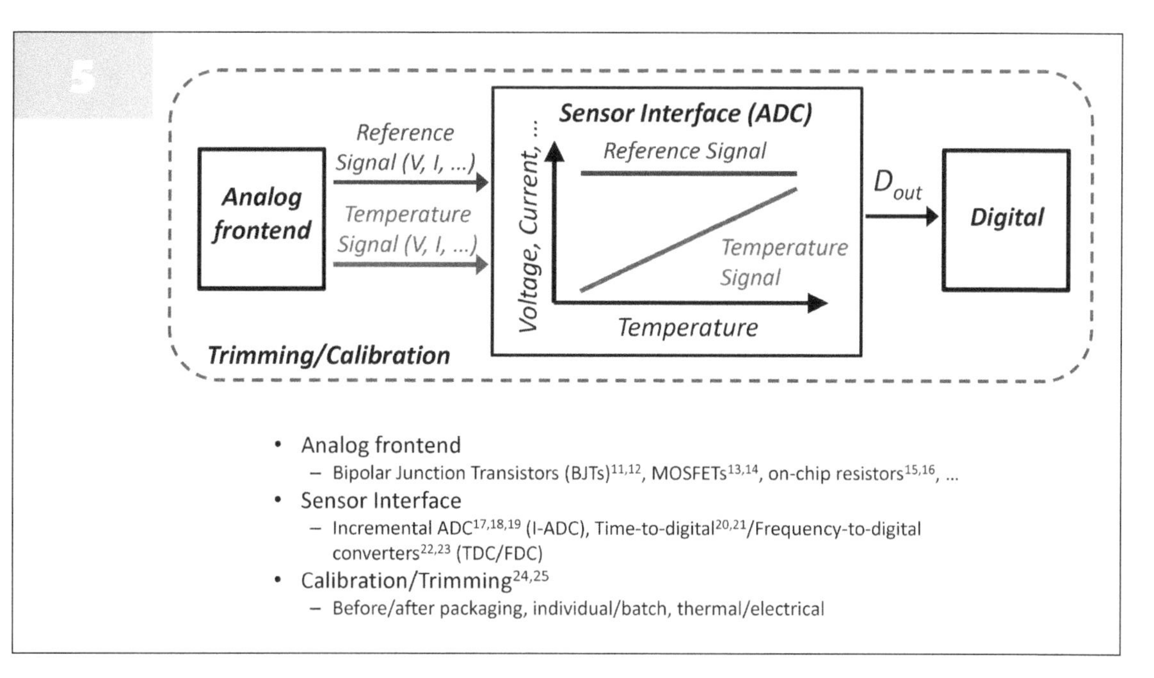

So what kind of devices can be used for temperature sensing? In fact, silicon is a semi-conducting material, and most of its electrical properties are temperature dependent. As a result, different devices available in a typical CMOS process can be utilized for temperature sensing purpose, such as BJTs, MOSFETs, on-chip resistors etc. In most cases, bipolar transistors (BJTs) and MOSFETs are selected due to their good process tolerance in generating the temperature dependent signal. BJT-based temperature sensors exhibit higher accuracy, and MOSFET-based ones are capable of low voltage low power operations. Recently, highly energy efficient CMOS temperature sensors using on-chip resistors are demonstrated. In this talk, we will mainly talk about BJTs and MOSFETs for temperature sensing.

In terms of the sensor interface, the incremental ADC is widely used due to its high resolution property, but generally consumes higher power to achieve the required integrator gain and bandwidth. For ultra-low power alternatives, time-to-digital converter and frequency-to-digital converter (TDC/FDC) are potential candidates but generally have limited resolution. The choice depends on the application specific requirements, but it should suffice to state for now that the sensor interface should be selected based on the analog frontend for optimal performance. We will discussion more when we cover the design examples later in this talk.

Without trimming/calibration, CMOS temperature sensors can only achieve limited accuracy (in the order of few degree Celsius) mainly due to the process spread and packaging stress. To characterize the temperature sensor performance, we need to calibrate the sensor output to a known reference sensor. After calibration, we can then bin different devices into different classes with different accuracies, or trim the device by adjusting some physical parameters of the sensor to bring its measured value closer to the reference at known temperatures. This calibration/trimming process can take place in the wafer level or after packaging. We can also trim devices individually. Or, we can extract the variation in device parameters within a production batch using N random samples by exploiting the strong intra-batch correlation. This can significantly reduce the calibration time/cost at the expense of reduced accuracy. Traditionally, we calibrate CMOS temperature sensors thermally, i.e. the sensor is required to thermally settle at a well-defined temperature, requiring long calibration time as the packaged time constants are typically in the order of seconds. This limits the accuracy of commercial sensors to about 0.5°C with a 1 second per point calibration time. Recently, electrical calibration method is proposed to significantly reduce the calibration time, and we will cover this briefly at the end of this talk.

6  BJT for Temperature Sensing[26,27]

- Vertical/Lateral PNP/NPN BJT
- For a BJT, the collector current I_C is dependent on the base-emitter voltage V_{BE}

$$I_C = I_S \cdot \left(e^{\frac{V_{BE}}{kT/q}} - 1\right) \approx I_S \cdot e^{\frac{V_{BE}}{kT/q}}$$

- The saturation current I_S exhibits strong temperature dependence

$$I_S = A_E C T^{\eta} \cdot e^{-\frac{V_{g0}}{kT/q}}$$

- A_E = emitter area, C is a process dependent term, $\eta \approx 4$, V_{g0} is the extrapolated bandgap voltage of silicon

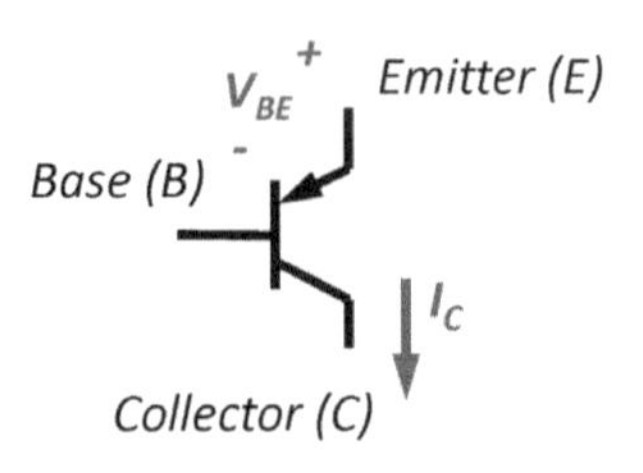

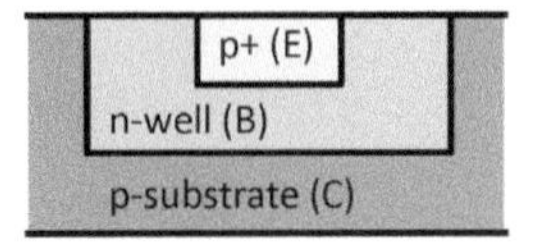

So, let's move to the discussion of BJT for temperature sensing. I believe most of you should have some experience in semiconductor device physics, and I'll just illustrate something basic here. Here shows a PNP BJT, which is a 3-terminal device. For a typical p-substrate process, a vertical PNP BJT can be easily fabricated by forming the p+ emitter inside a n-well base. The substrate then becomes the collector. If you are using a deep-nwell process, NPN BJT can be available. It can have a higher current gain than a PNP BJT in the same process, and can be an advantage considering the exceeding low gain for PNP BJT in advanced processes (less than 5). Also, unlike the PNP BJT where the collectors are all connected together through the p-substrate, NPN BJT can have all of their 3 terminals available, which can be good for more circuit configurations. However, it suffer more from packaging stress, so we should take this also into account. Of course, lateral BJTs are also possible, but they are not often considered for CMOS temperature sensor implementations due to their non-ideal BJT characteristics as a result of parasitic substrate BJTs.

In our discussion here, we'll just use the vertical PNP BJT as an example. For simplicity, the parameters here are all positively defined. As you should know, the collector current I_C is having an exponential relationship with the base-emitter voltage V_{BE}, as illustrated in this equation. Generally speaking, we can design I_C to be much larger than I_S, and we can approximate the I_C-V_{BE} relationship and remove the second term, with I_C equals to the saturation current I_S multiplied by the exponent term of V_{BE}. Here, q is the electron charge and k is the Boltzmann constant. T represents the absolute temperature in Kelvin. The saturation current is dependent on the emitter area A_E, a process dependent term C, a temperature dependent term η which is as large as 4 in a typical process, and another exponent term which is dependent on the extrapolated bandgap voltage of silicon V_{g0}. Notice that I_S is process dependent as well as a strong function of temperature, and we need to take these into account for CMOS temperature sensor designs.

7

V_{BE} and ΔV_{BE} Characteristics

- For $I_C >> I_S$

$$V_{BE} = \frac{kT}{q} ln \frac{I_C}{I_S}$$

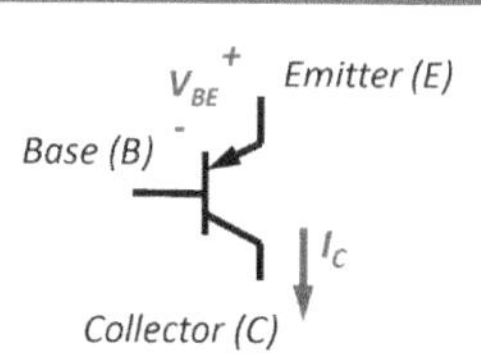

- V_{BE} is complementary-to-absolute-temperature (CTAT) and non-linear, as governed by I_S, and is therefore process dependent

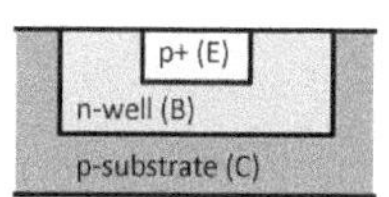

- The I_S term can be eliminated by measuring the difference of two collector currents

$$\Delta V_{BE} = V_{BE2} - V_{BE1} = \frac{kT}{q} ln \frac{I_{C2}}{I_{C1}}$$

As discussed before, we can design I_C to be much larger than I_S. As a result, V_{BE} can simply be expressed as $(kT/q)\cdot\ln(I_C/I_S)$. As the first glance, it may appear that V_{BE} is linearly dependent on temperature due to the (kT/q) term. However, due to the strong temperature dependence of I_S, V_{BE} is in fact having a complementary-to-absolute-temperature (CTAT) characteristic, meaning that it decreases as temperature increases, of approximately -2mV/°C. It is also process dependent due to I_S. It is interesting that if we measure the difference of two V_{BE}, we can essentially cancel out the I_S dependent term. Consequently, ΔV_{BE} equal to $(kT/q)\cdot\ln(I_{C2}/I_{C1})$, and is free from process variation. It is also highly linear to temperature if we can generate an accurate I_{C2}/I_{C1} ratio, which is possible in integrated circuit designs.

8

V_{BE} and ΔV_{BE} Characteristics

- By controlling the collector current ratio (p) and emitter-area ratio (r)

$$\Delta V_{BE} = V_{BE2} - V_{BE1} = \frac{kT}{q} ln(k \cdot r)$$

- With $k = 5$ and $r = 1$, ΔV_{BE} ~ 126 µV/°C
- ΔV_{BE} is process tolerant with highly linearity and proportional-to-absolute-temperature (PTAT)

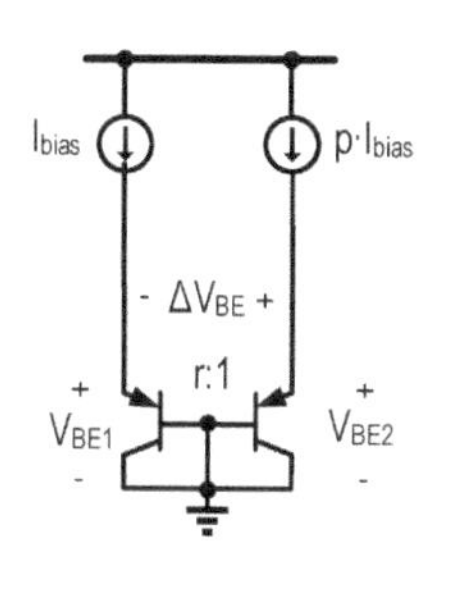

So what we can do is to bias two BJTs simultaneously at two different collector current levels, say with a ratio of p, for extracting ΔV_{BE}. We can also have another degree of freedom by using BJTs with a different emitter area ratio r, so we can get ΔV_{BE} to be equal to $(kT/q)\cdot\ln(p\cdot r)$. Typically, ΔV_{BE} is in the order of 100 µV/°C. As an example, with p = 5 and r = 1, ΔV_{BE} is approximately 126 µV/°C. This process independent characteristic with high temperature linearity in ΔV_{BE} makes BJT an excellent candidate for temperature sensing applications. Here, we need to ensure accurate ratios n and r. Generally speaking, a matching accuracy of up to 10-bit should be possible with careful analog layout. If a higher accuracy is required, circuit techniques such as dynamic element matching using unit current sources and BJTs should be considered, and we will discuss more in this talk.

9

Basic Operation Principle

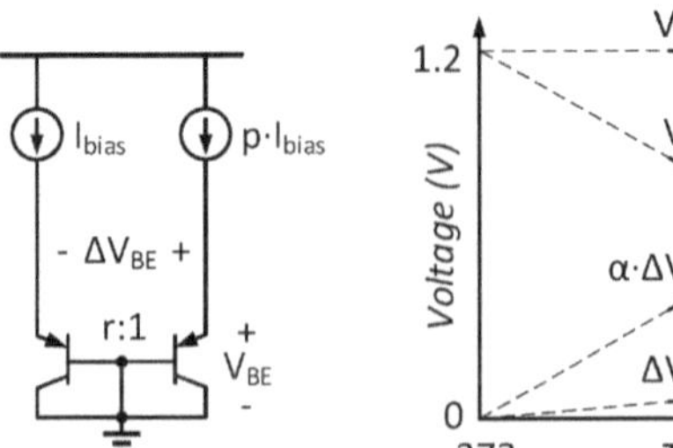

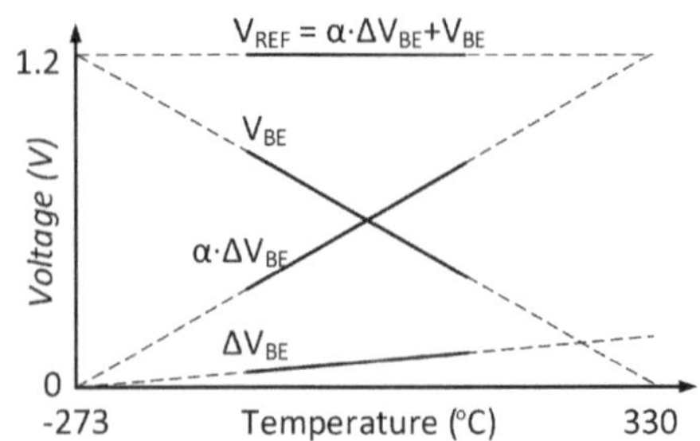

- **Two BJTs biased at different current densities for generating**
  - ΔV_{BE} : Proportional-to-absolute-temperature (PTAT)
  - V_{BE} : Complementary-to-absolute-temperature (CTAT)
- **V_{REF} generated by linear combination of αV_{BE} and ΔV_{BE}**

Here shows the basic operation principle by biasing two BJTs with different collector currents with a current ratio p to obtain both the PTAT signal ΔV_{BE}, and the CTAT signal V_{BE}. Here, we first assume that the BJT current gain is large so $I_E \approx I_C$. Remember that V_{BE} is having a temperature coefficient of roughly -2mV/°C. If we multiply ΔV_{BE} with a gain α, we can combine with V_{BE} to generate a temperature independent voltage V_{REF}. As an example, α is approximately 16 with p = 5 and r = 1. Up to now, we already have the basic ingredients for temperature measurement, a temperature dependent signal $\alpha \cdot \Delta V_{BE}$, and a reference signal V_{BE}.

10

Basic Operation Principle

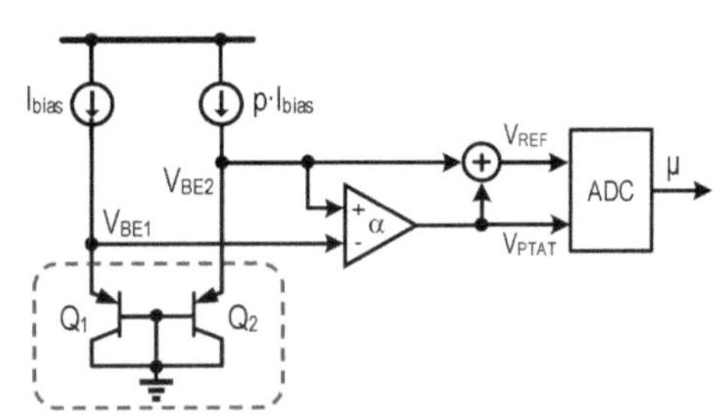

- **Current mirror ratio (*k*), BJT emitter ratio (*r*) and amplifier gain (*α*) generate** $V_{PTAT} = \alpha \cdot \frac{kT}{q} ln(p \cdot r)$
- **Ratiometric temperature readout:** $\mu = \frac{\alpha \cdot \Delta V_{BE}}{V_{BE} + \alpha \cdot \Delta V_{BE}}$
- Ratio μ varies from 0 t0 1 roughly from -273 to 300° --> 16-bit

So, here shows a generic concept diagram for a CMOS temperature sensor using BJT as the temperature sensing element. As discussed before, we can generate the PTAT voltage using circuit parameters α, p and r. The sensor readout provides the appropriate V_{REF} and V_{PTAT} signals for generating the ratiometric temperature measurement μ, which varies from 0 to 1 from roughly -273 to 330°C. As a result, a resolution of 0.01°C requires a 16-bit ADC. So everything looks very good here. So what can go wrong? First, we need to notice that the BJT characteristics is in fact not ideal.

11 

Non-linearity in V_{BE}

- For practical reasons, I_C is generally made proportional to some power of T

$$I_C \propto T^m = I_C(T_r)\left(\frac{T}{T_r}\right)^m \qquad I_C = A_E C T^{\eta} e^{\frac{V_{BE}-V_{g0}}{kT/q}}$$

- Express V_{BE} as the sum of a constant term, a proportional to T term, and higher-order terms[28]

$$V_{BE} = V_{BE0} - \lambda T + c(T)$$

$$V_{BE0} = V_{g0} + (\eta - m)\frac{kT_r}{q} \qquad \lambda = \frac{V_{BE0} - V_{BE}(T_r)}{T_r}$$

$$c(T) = (\eta - m)\frac{k}{q}\left(T - T_r - T \cdot \ln\frac{T}{T_r}\right)$$

From our previous discussion, the CTAT characteristics of V_{BE} of roughly -2mV/°C is in fact an approximation. Here, we'll spend some time to discuss about the non-linearity in V_{BE}. For practical reasons, notice that I_C is generally generated to be proportional to some power of T, and we define I_C to be proportional to T to the power m. Recall that for a BJT, I_C is exponentially dependent on V_{BE}. For easy analysis, we can express V_{BE} as the sum of a constant term, a proportional to T term, and higher-order terms. Here, V_{BE0} denotes the V_{BE} at absolute 0. Essentially, the linear coefficient λ determines the dominant temperature behavior of V_{BE}, while the term c(T) shows the non-linearity in V_{BE}.

12 

Non-linearity in V_{BE}

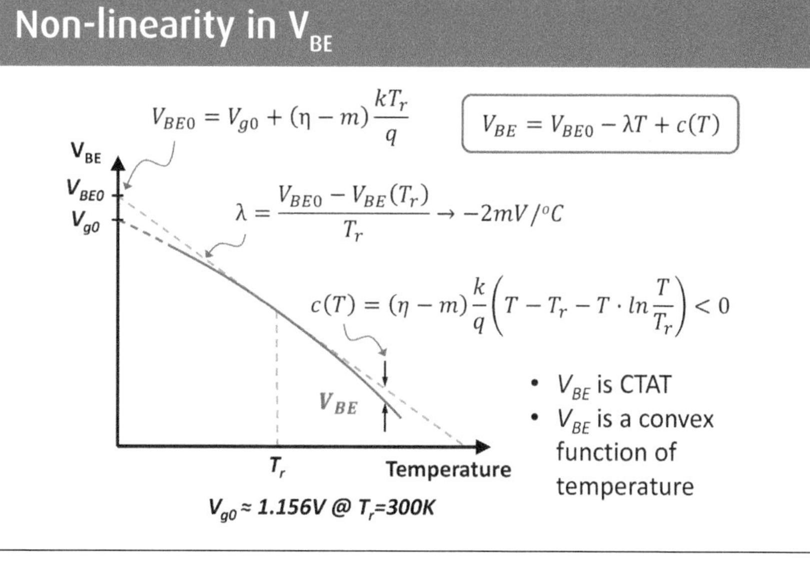

It should be easier to obtain essential design insights by plotting V_{BE} against temperature as illustrated here. Notice that V_{g0} is the extrapolated bandgap voltage of silicon and is process independent at 0K. V_{BE0}, however, varies depending on the calibration temperature T_r, the biasing condition, and process. The temperature dependency of roughly -2mV/°C is defined as the tangent of V_{BE} at T_r. As discussed, the linearity of V_{BE} is mainly determined by c(T). For I_C to be generated on-chip, it generally exhibits values of m equals -1, 0 and 1, which can be generated through a resistor with V_{BE}/R, V_{REF}/R and V_{PTAT}/R, respectively. As η is approximately 4 in a typical process. It can be proved that c(T) is positive for the entire temperature range. As a result, V_{BE} is a convex function of temperature. This makes the generated reference voltage V_{REF} using $V_{BE}+\alpha\cdot\Delta V_{BE}$ slightly temperature dependent. As a rough calculation, with a V_{BE} slope of -2mV/°C, a c(T) of 1 mV can already contribute to 0.5°C error. Obviously, the value of c(T) depends on the temperature range, and is typically in the order of few mVs.

13  I_S Process Spread

- Mainly impacts I_S
  - Base doping
  - Geometry
- Process spread in I_S
  → process spread in V_{BE}

$$\delta V_{BE} = -\frac{kT}{q}\cdot\frac{\delta I_S}{I_S}$$

- V_{BE} spread is PTAT[29]

  → can be PTAT trimmed (e.g. by adjusting I_C)

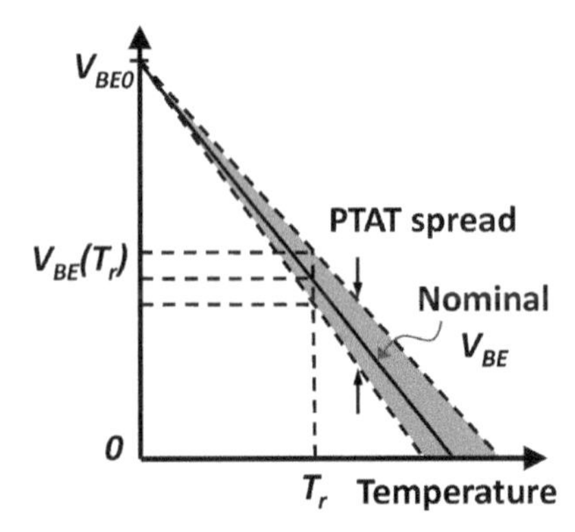

Another major error is contributed by the BJT process spread due to variation in base doping concentration and geometry size. This impacts I_S and in turn V_{BE}. If we differentiate V_{BE} with respect to I_S, we can get this equation, where V_{BE} changes as a result of process variation in I_S. Notice that δI_S is much smaller than I_S. As you can observe here, the change in V_{BE} is in fact PTAT provided that $\delta I_S/I_S$ is temperature independent. This assumption should be quite accurate as far as δI_S experience similar variation to I_S due to process. In practice, this residue error is demonstrated to be very small. Consequently, the variation of V_{BE} due to process increases as a function of temperature. As illustrated in the figure, this spread only contribute to the change in V_{BE} slope and has only 1 degree of freedom. This means that we only need to adjust 1 parameter, e.g. by adjusting I_C, to correct this V_{BE} spread back to its nominal value. Notice that this is a very special kind of spread, and not all the spreads in CMOS temperature sensors exhibit the same behavior.

At this point, I would like to add one comment, as you may have noticed that a diode can also be used for temperature sensing. In fact, you are correct, but generally a diode is not good for accurate sensing. The short answer is that its current-voltage relationship depends on a non-Ideality factor, which dependents on the recombination current and is therefore temperature sensitive. This makes the PTAT voltage sensitive to process spread.

14  Current gain in BJT

For vertical PNP BJT, the collector is always tied to the substrate and is therefore not available. As a result, we mainly bias the BJT through the emitter. Notice that I_C and I_E is related by the $\beta_F/(\beta_F+1)$ factor, and β_F is a strong function of temperature with a temperature exponent XTB generally >2. This can result in a non-PTAT error which cannot be completely trimmed out in V_{BE}. As a result, a large β_F is generally desired to eliminate this error source. In advanced CMOS processes, this β_F can be <5 or even lower, as a result, beta-compensation is usually implemented to compensate for this.

In terms of ΔV_{BE}, we need to ensure accurate 1:n ratio for biasing, so we need to make sure that the β_F for the two BJTs should have the same current gain when they are biased at two different current levels. Fortunately, a constant β_F region often exists due to the wide base width of vertical PNP BJTs. So we may need to extract such flat β_F region for generating ΔV_{BE}.

In terms of process spread, notice that as $\delta\alpha_F/\alpha_F$ is in fact temperature dependent as shown, it can lead to non-PTAT errors that cannot be easily trimmed.

14

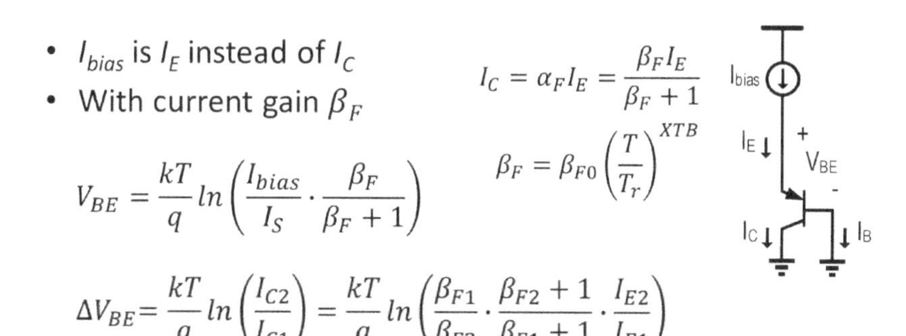

15 

Biasing for BJT

- $I_C >> I_S$

$$I_C = I_S\left(e^{\frac{V_{BE}}{kT/q}} - 1\right)$$

$$\Delta V_{BE} = \frac{kT}{q} ln\left(\frac{nI_{C1} + I_S}{I_{C1} + I_S}\right)$$

- Constant β_F biasing

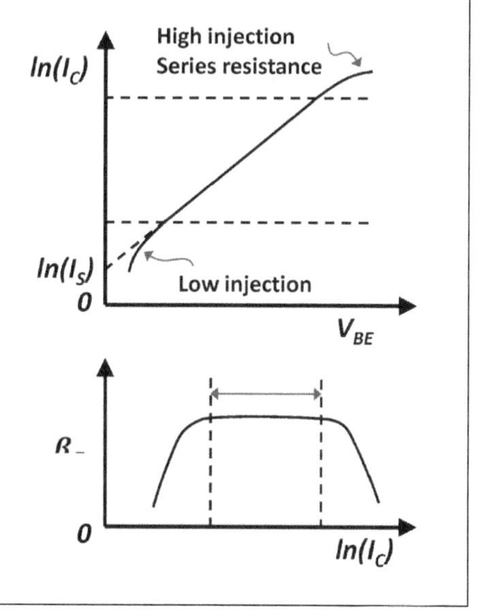

How can be choose for the biasing current for BJTs? For high power/energy efficiency, we would like to reduce the biasing current while minimizing the V_{BE} non-linearity, i.e. use a smaller I_C. This essentially avoids the possibility to biasing the BJT in the high injection region where the error due to series resistance can be significant. The first thing we need to consider is that I_C should be much larger than I_S. If not, we cannot omit the constant term in the equation. This leads to an error term in ΔV_{BE} and directly affect the sensing accuracy. We also need to choose an I_C which can avoid the low injection region that can break the exponential I_C-V_{BE} relationship. Notice that as I_S is highly non-linear with temperature, normally I_C should be determined at the high temperature end of the required sensing range. Another consideration is the constant β_F biasing as described before to ensure the 1:p ratio for generating ΔV_{BE}. A small current ratio 1:p can be good for limiting the required constant β_F region. This is a direct tradeoff with the magnitude of ΔV_{BE}.

16  MOSFET Characteristics

- Subthreshold current I_D is dependent on the gate-source voltage V_{GS}[31]

Drain, Gate, I_D, +, V_{GS} -, Source

$$I_D = \mu C_{OX}\left(\frac{W}{L}\right)\left(\frac{kT}{q}\right)^2 e^{\frac{V_{GS}-V_{th}}{nkT/q}}\left(1 - e^{-\frac{V_{DS}}{kT/q}}\right)$$

- For $V_{DS} >> 3kT/q$

$$I_D \approx \mu C_{OX}\left(\frac{W}{L}\right)\left(\frac{kT}{q}\right)^2 e^{\frac{V_{GS}-V_{th}}{nkT/q}}$$

Now, we move to the discussion of MOSFET. For a NMOS operating in the subthreshold region, i.e. $V_{GS} < V_{th}$, its drain current I_D is defined by this equation. For V_{DS} >> 3kT/q, the last term become negligible.

17  MOSFET Characteristics

- μ = mobility, W/L is sizing, V_{th} = threshold voltage, n is a process dependent term

Drain, Gate, I_D, +, V_{GS} -, Source

$$V_{GS} \approx V_{th} + \frac{nkT}{q} ln \frac{I_D}{\mu C_{OX}\left(\frac{W}{L}\right)\left(\frac{kT}{q}\right)^2}$$

$$\begin{cases} V_{th} = V_{th}(T_r) + \alpha_{V_{th}}(T - T_r), & \alpha_{V_{th}} \approx (-1, -4)\ \mathrm{mV/^oC} \\ \mu \approx \mu(T_r)\left(\frac{T}{T_r}\right)^{\alpha_\mu}, & \alpha_\mu \approx (1.2, 2.4) \end{cases}$$

- With two ideal MOSFET with $(W/L)_1/(W/L)_2$ and I_{D2}/I_{D1} ratio equals r and p

$$\Delta V_{GS} = \frac{nkT}{q} ln(p \cdot r)$$

We can rewrite the previous equation into this form to better illustrate the temperature dependency of V_{GS}. In this equation, there are mainly two temperature dependent terms, the threshold voltage V_{th}, and the mobility μ. V_{th} is a CTAT signal with a slope α_{Vth} equals to -1 to -4 mV/°C, while μ is a non-linear function of temperature with a temperature exponent α_μ ranging from 1.2 to 2.4 in typical processes. By sizing the W/L and drain current ratios of r and p, we can generate ΔV_{GS} similar to ΔV_{BE}. As a result, using MOSFET to generate both the temperature and reference signals should be possible.

But why MOSFET is not commonly used in temperature sensor designs?

18  Process Spread in V_{GS}

- Impact by μ, V_{th}, C_{OX} and W/L
- Process in V_{GS}

$$\delta V_{GS} = \delta V_{th} - \frac{nkT}{q}\left(\frac{\delta\mu}{\mu} + \frac{\delta C_{OX}}{C_{OX}} + \frac{\delta(W/L)}{(W/L)}\right)$$

- V_{GS} spread requires two- or more-point calibrations for higher accuracy[32]

The major reason is due to the increased process spread. The process spread in V_{GS} is mainly dependent on mobility, threshold voltage, oxide thickness, and geometry size. As illustrated using this equation, both V_{th} and μ can essentially contribute to non-PTAT spread, which cannot be compensated using simple trimming techniques. As a result, MOSFET-based temperature sensors generally require two- or more-point calibrations in order to achieve a better accuracy. The linearity is also an issue and curvature compensation is often required.

19  Biasing for MOSFET[33]

- V_{DS} >> 3kT/q, otherwise, both V_{GS} and ΔV_{GS} become a high-order function of temperature → increased non-linearity
- I_D is larger than leakage at high temperature

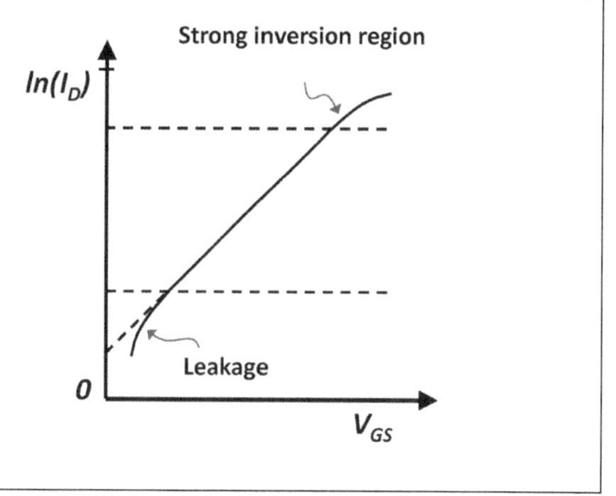

For MOSFET biasing, one major condition for ultra-low power applications is that the biasing current should be high enough to ensure that V_{DS} is much larger than $3V_{th}$ over the required sensing range. Otherwise, both V_{GS} and ΔV_{GS} become a high-order function of temperature, i.e. the non-linearity increases which can limit the sensing accuracy. Also, we need to make sure that the subthreshold current I_D is larger than the source/drain junction leakage at high temperature.

Up till now, we have covered the basics of both BJT and MOSFET for temperature sensing. If high accuracy is required, we would prefer BJT due to its intrinsic property to cancel PTAT spread using simple trimming. However, if some level of sensing accuracy can be sacrificed for more power/energy efficiencies, MOSFET can be a good candidate and a lower supply voltage can be used. Yet, the non-PTAT error is still an issue and extra calibration effort may be needed.

20 

Error Sources

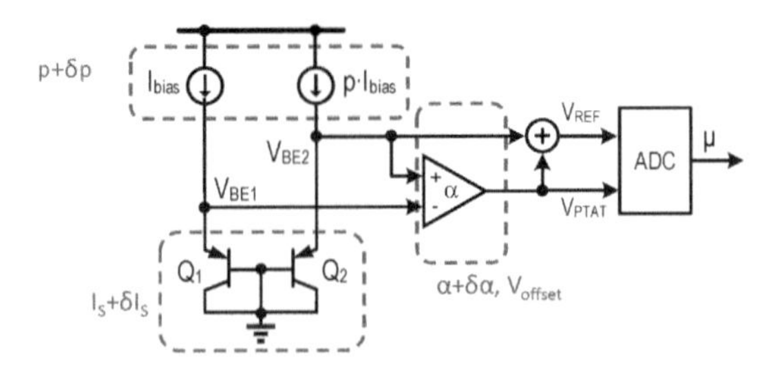

- Offset, gain error, mismatch and process spread
  - Dynamic element matching
  - Auto-zero/chopping
  - Trimming/calibration
  - ...

So how can we readout the temperature and reference signal? In the case of BJT, the sensing core generates both the V_{BE} and ΔV_{BE} signals. These two signals are then readout using a readout stage and quantized using an ADC.

And what can limit the sensing accuracy? Here shows different errors sources that can appear in CMOS temperature sensors. We have already covered the process spread error in I_S for BJT devices. Apart from that, we have offset error V_{offset}, gain error **δα**, current source mismatch error **δp** etc. All of these can contribute to error in the final temperature output. What we can do is to exploit the dynamic techniques such as chopping, auto-zeroing, dynamic element matching to reduce most of the noise sources, except from the process spread in I_S which can be tackled using trimming. As discussed before, PTAT errors can be trimmed at one temperature point, and we would like to avoid non-PTAT errors which require more elaborate trimming procedures.

21 

Design Examples

- Ultra-low Power High Accuracy CMOS Smart Temperature Sensor for Clinical Use[34]
- Embedded Temperature Sensor with Process Spread Compensation[35]
- Ultra-low Power Embedded Temperature Sensor with MOSFET[21]
- Voltage Calibration with On-chip Heater[36]

Up till now, we have introduced the basics of CMOS temperature sensors. I'm going to discuss with you on a few design examples. The first one is an ultra-low power CMOS temperature sensor targeting for clinical applications. It employs some curvature correction techniques, and dynamic element matching for generating accurate capacitor and current source ratios. In the second example, we'll move to an embedded temperature sensor in a passive RFID, and discuss about how to exploit curvature corrections and achieve better process spread immunity in the system level under a very limited power budget. The third example is an ultra-low power CMOS temperature sensor using MOSFET as the sensing element. Finally, we will talk about calibration, and outline a recently proposed voltage calibration technique using on-chip heaters to significantly reduce the calibration time.

22

Clinical Temperature Sensor

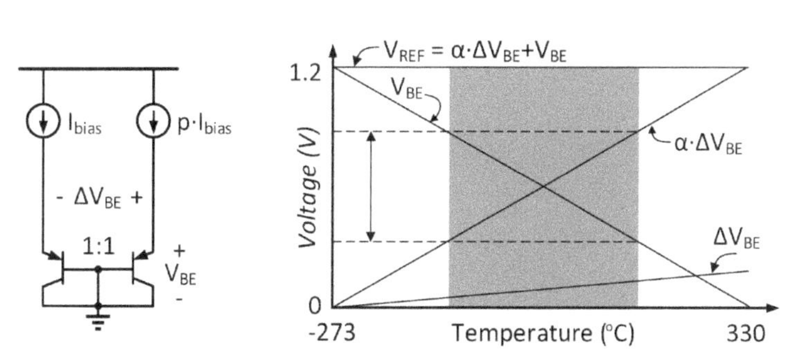

- Input range depends on temperature range
  - Input range under-utilization for applications requiring limited sensing range (e.g. human body temperature monitoring)

Let us first recap the basic operations for CMOS temperature sensors. Due to the high accuracy requirement for the target clinical application, we choose to use BJT as the sensing element. As illustrated in this diagram, conventional BJT-based CMOS temperature sensors normally occupies only a portion of the available input range even for an extended sensing range. As an example, for a temperature sensor operating in the military range from -55 to 125°C, only ~1/3 of the available input range will be utilized. This leads to an increased ADC resolution requirement, which can in turn result in increased readout power consumption for achieving the required sensing accuracy.

23

Clinical Temperature Sensor

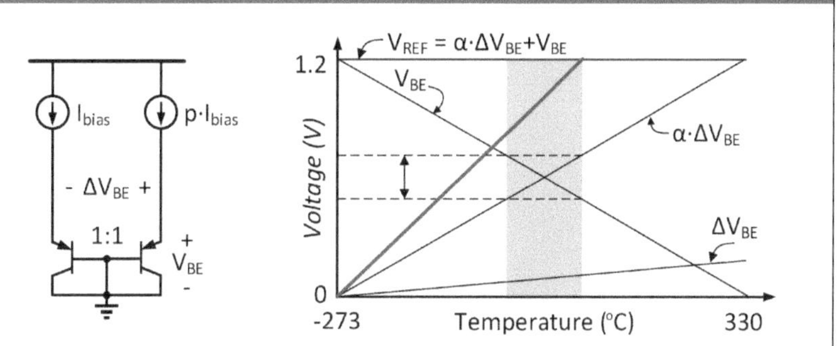

- Input range depends on temperature range
  - Input range under-utilization for applications requiring limited sensing range (e.g. human body temperature monitoring)
- Increase temperature signal to relax ADC resolution requirement

In many emerging applications where only a narrow sensing range is necessary, such as from 25 to 45°C for the clinical application, the input range under-utilization issue is becoming more significant, specifically <4%. If we can increase the temperature signal within the sensing range, we can indeed relax the ADC resolution requirement proportionally, leading to significant power reduction. Unfortunately, if we just increase the ΔV_{BE} signal, as shown by the red line here, the improvement can be very limited as most of the available input range is in fact wasted by the large signal offset.

24

Clinical Temperature Sensor

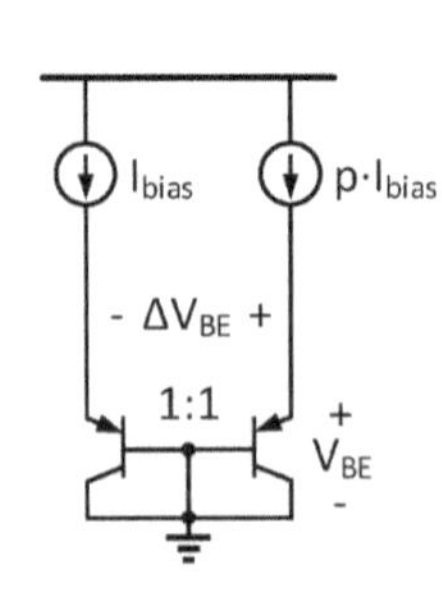

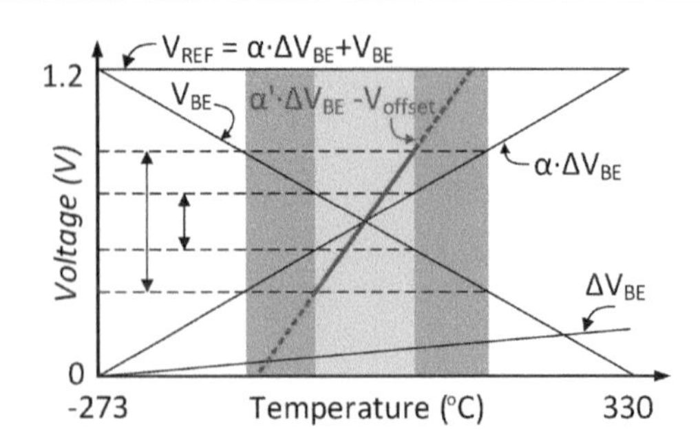

- **Input range depends on temperature range**
  - Input range under-utilization for applications requiring limited sensing range (e.g. human body temperature monitoring)
- **Increase temperature signal to relax ADC resolution requirement**
- **Accurate gain stage (α') and offset (V_{offset})**

So ideally, what we want is to increase the temperature signal, while generating an input offset so that we can optimally cover the input range for reduced power budget, as denoted by α' and V_{offset} here. It should be noted that these two terms should be highly accurate so as not to sacrifice the sensing accuracy. So how can we generate the required α' and V_{offset}?

25

Operation Principles

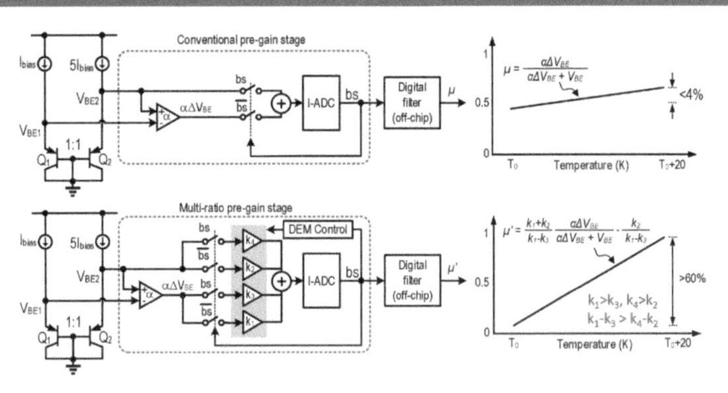

- **Dynamic element matching for accurate gain/offset generation using $k_{1\text{-}4}$**
- **Improve ADC input range utilization from < 4% to > 60%**
- **Relax ADC resolution to 12-bit for 0.01°C resolution**

Here shows the solution to the above stated problem. Instead of using a fixed pre-gain stage α followed by an I-ADC, we can integrate a multi-ratio pre-gain stage $k_{1\text{-}4}$ to amplify the temperature signal. As a result, the input range utilization can be increased from <4% to >60%.

During temperature signal conversions, ΔV_{BE} and V_{BE} are amplified using the multi-ratio pre-gain stage according to the bitstream bs. The gain ratios (including α and $k_{1\text{-}4}$) are implemented by the first stage integrator using capacitors to balance the temperature coefficients of V_{BE} and ΔV_{BE}. Based on charge balancing, the integrator is charged with gain ratios αk_1, k_2 when bs=0, and discharged with gain ratios αk_3, k_4 when bs=1, respectively. As a result, we can generate a new ratio μ' which is composed of an extra gain term $(k_1+k_2)/(k_1-k_3)$, while providing an offset $k_2/(k_1-k_3)$. This can relax the I-ADC requirement as only a moderate resolution is required (e.g. 12-bit for 0.01°C). Here the $k_{1\text{-}4}$ values can be implemented by using capacitor ratios as in conventional switched-capacitor charge balancing solutions. Notice that $k_{1\text{-}4}$ cannot be arbitrary, and $k_1 > k_3$, $k_4 > k_2$ and $k_1-k_3 = k_4 - k_2$ should be enforced to ensure proper operations.

26  Block-based DWA

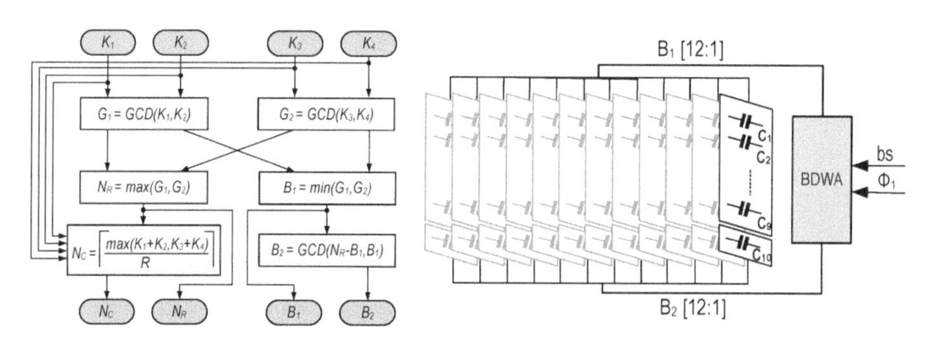

- For $\alpha = 18$, $k_1 = 6$, $k_2 = 9$, $k_3 = 5$, $k_4 = 10$
  - Requires 117 unit capacitors in conventional DWA
- Blocked-based DWA
  - Group maximum number of unit capacitors into blocks while providing the flexibility offered by DWA
  - Control wiring reduce by 4.8x

To ensure a high gain, a large ratio is required. The required design values are: α = 18, k_1 = 6, k_2 = 9, k_3 = 5, k_4 = 10. To implement all these ratios needs a total capacitors of 117 by using the conventional data weighted averaging (DWA) technique, which is a commonly used to achieve dynamic element matching. Despite the fact that the conventional DWA can achieve an accurate gain value with the presence of capacitor mismatch, the implementation complexity can lead to significant design overhead. To resolve this problem, block-based data weighted averaging (BDWA) can be used to achieve the required multi-ratios with high accuracy while significantly reducing the routing cost. As illustrated in this figure, the control resources are allocated by grouping a maximum number of unit capacitors into blocks while still providing the flexibility offered by the DWA. With the two ratios (K_1, K_2) and (K_3, K_4), first their corresponding greatest common divisor (GCD) is found. The required number of rows N_R is assigned to be the maximum of G_1 and G_2. The number of columns N_C can then be selected as the smallest integer which can realize the required gain ratios. As a result, a total number of elements required for BDWA is N_R x N_C. In this work, the BDWA algorithm results in two block control allocations B_1 and B_2 to realize dynamic block-level matching, and the control wiring can be significantly reduced from 117 to 24, corresponding to a 4.8x improvement.

27  Block-based DWA

This slide illustrates how the block-based DWA works. When bs=0, a total of 9 + 108 = 117 unit capacitors are selected to realize the gain ratios αk_1 and k_2, leaving 3 capacitors unused. This corresponds to a complete capacitor element utilization in 52 cycles. Similarly, when bs=1, a 10 + 90 = 100 capacitors are selected for αk_3 and k_4 in each cycle, and complete capacitor selection can be accomplished in 12 cycles.

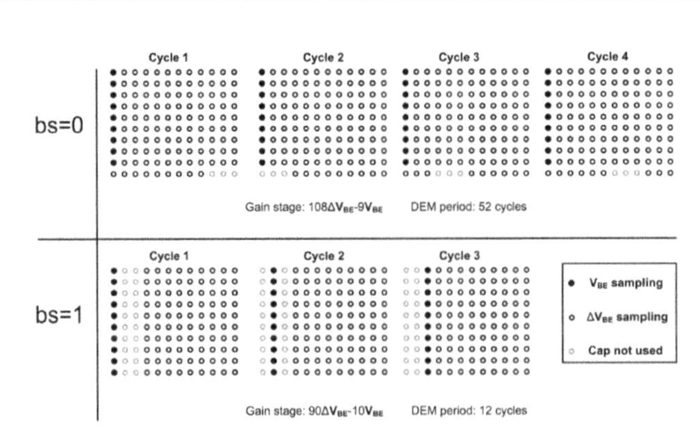

- bs=0: 9 + 108 = 117 capacitors selected, remaining 3 capacitors → Complete selection in 52 cycles
- bs=1: 10 + 90 = 100 capacitors selected, remaining 20 capacitors → Complete selection in 12 cycles

28

Block-based DWA

- Achieve >12-bit resolution of -74 dB in <80 cycles, with a capacitor mismatch of 4%

This slide shows the simulated performance comparison between DWA and BDWA with $K_{1\text{-}4}$ equal to 108, 9, 90 and 10, respectively, assuming a capacitor mismatch of 1% and 4%. The capacitor array size for the DWA is 117 while that for the BDWA is 120 (N_R=10 and N_C=12). B_1 and B_2 are set to 9 and 1, respectively. It can be observed that the achieved error level for K_3/K_4 is very close to that achieved by DWA. For K_1/K_2, due to the presence of a residue gain error in BDWA, slightly more number of conversion cycles are required when compared to DWA. It can also be observed that the ratio error quickly converges to a level which is well below the required 74dB to achieve 12-bit resolution even with a mismatch error as large as 4%.

29 

System architecture

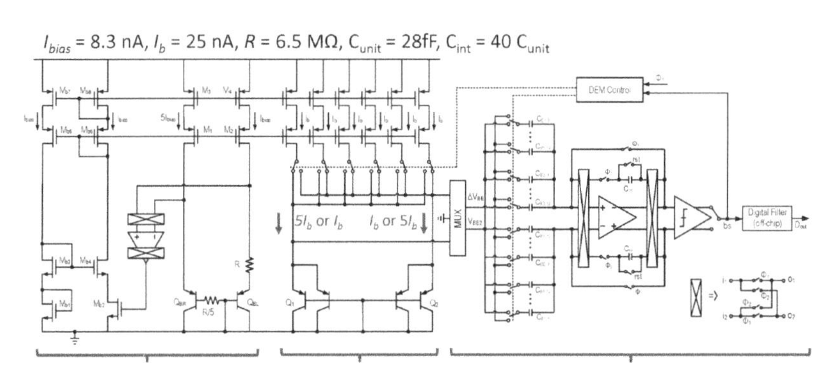

- Bias generation + sensor core + incremental ADC
- BDWA with 120 unit capacitors ($C_{B1,1\text{-}12}$, $C_{B2,1\text{-}12}$)
- Chopping to eliminate amplifier offsets
- Unit current sources switch between $Q_{1,2}$ in alternative cycles
- Beta compensation with R and R/5

Here shows the simplified schematic of the complete temperature sensor implementation. It is composed of an analog frontend followed by a first-order I-ADC. The analog frontend consists of a bias current generator and a bipolar core. The bias current I_{bias} is PTAT to enhance the linearity of V_{BE} and V_{REF}, improving the sensing accuracy. In order to achieve ultra-low power consumption while fulfilling the sensing accuracy and readout speed requirements, the values for I_{bias}, I_b and R are designed to be 8.3 nA, 25 nA and 6.5 MΩ, respectively. The Opamp is adaptive self-biased to increase the DC gain while drawing only 85 nA at 37°C from a 1V supply. Accurate I_{bias} generation is ensured by chopping the opamp offset as well as beta compensation using R and R/5. Two V_{BE} signals (V_{BE1} and V_{BE2}) are derived using 6 identical current branches from the bias current generator with DEM to minimize the mismatch error, with I_b = 25nA for complete V_{BE} settling using a sampling clock f_{clk} of 8 kHz. To enhance the output impedance and mirror the current to the branches more accurately, cascoding has been used. As the targeted temperature sensing range is from 25°C to 45°C, the corresponding V_{BE} is expected to be close to 620 to 500 mV, respectively. As a result, a supply voltage of 1 V for the analog blocks would suffice for proper circuit operations. The proposed multi-ratio pre-gain stage is implemented using a fully differential integrator, and a unit capacitor array of 120 elements, with C_{unit} = 28fF and C_{int} = 40 C_{unit}. The capacitor blocks $C_{B1,i}$ and $C_{B2,i}$ are designed according to the proposed BDWA algorithm.

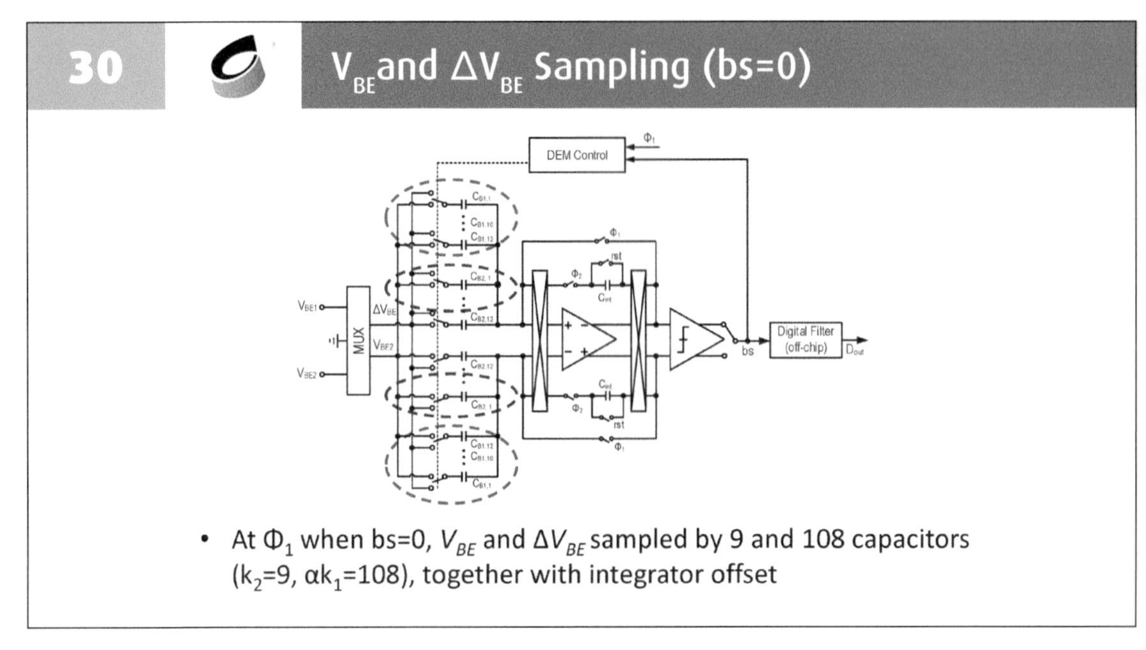

So how does the proposed circuit work? As in conventional incremental ADCs, we perform charge balancing by integrating either one combination of V_{BE} and ΔV_{BE}, or another based on the output bitstream. If bitstream bs=0, we will integrate $k_1 \cdot \alpha\Delta V_{BE} - k_2 \cdot V_{BE}$. Both V_{BE} and ΔV_{BE} can be selected by controlling the switches. As $k_2 = 9$, we select 9 of the C_{B2} block, $C_{B2,1\text{-}9}$ in this case, which contains 9 unit capacitors connected in parallel, and we connect the opamp in unity gain. So V_{BE} will be stored in the sampling capacitors. Note that the opamp offset will also be sampled at this phase.

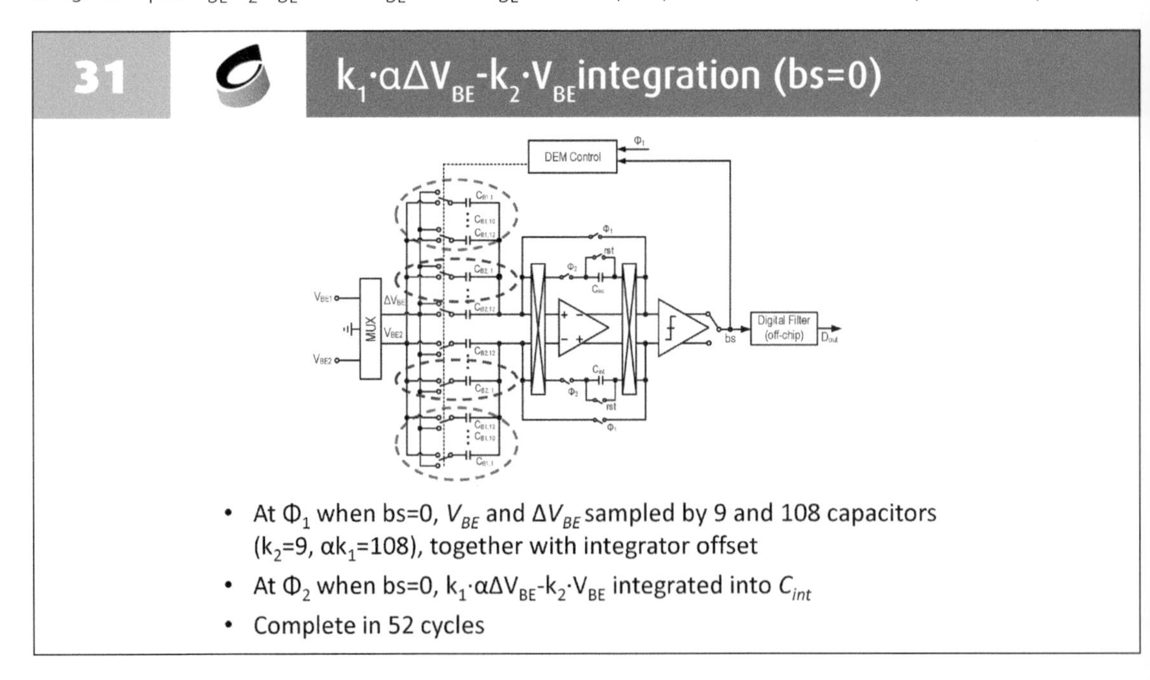

In the next phase, we connect ΔV_{BE} across the input terminals. The stored V_{BE} in the 9 $C_{B2,1\text{-}9}$ capacitors together with ΔV_{BE} using108 unit capacitors, e.g. $C_{B1,1\text{-}12}$, will be integrated across C_{int}, with a charge package proportional to $k_1 \cdot \alpha\Delta V_{BE} - k_2 \cdot V_{BE}$. Based on our previous analysis, the use of every unit capacitors can be completed in 52 cycles.

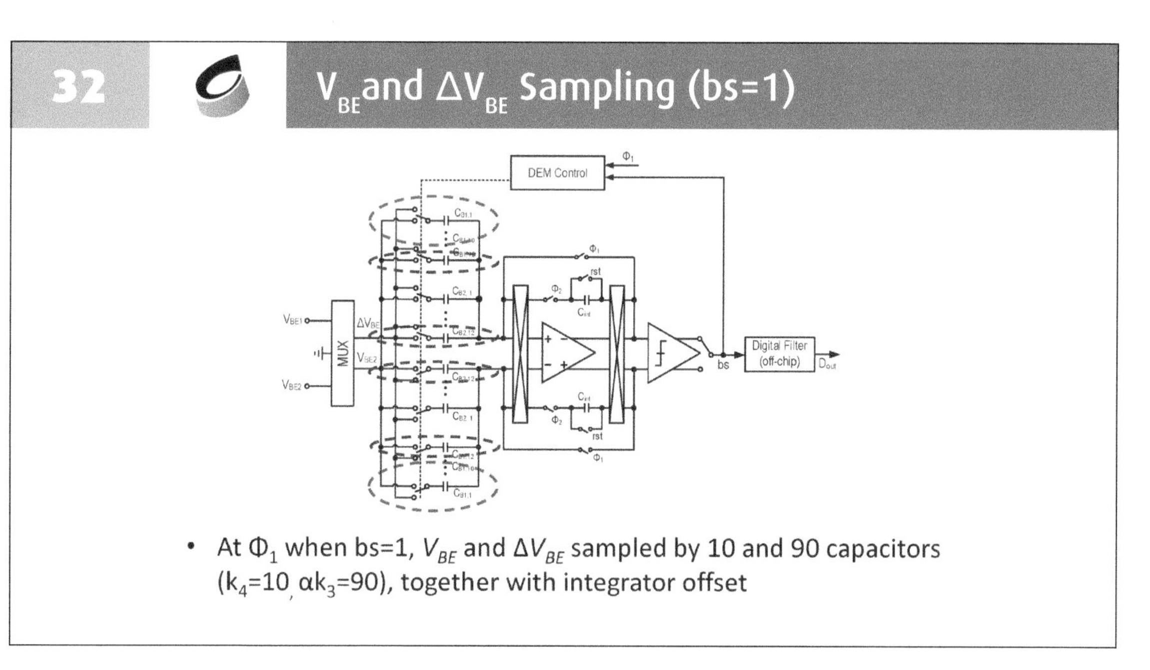

For bs = 1, as $\alpha{\cdot}k_3 = 90$, we select 90 of the C_{B1} block, $C_{B1,1\text{-}10}$ in this case, which contains 90 unit capacitors connected in parallel. Similar to the previous case, we connect the opamp in unity gain, ΔV_{BE} will be stored in the sampling capacitors.

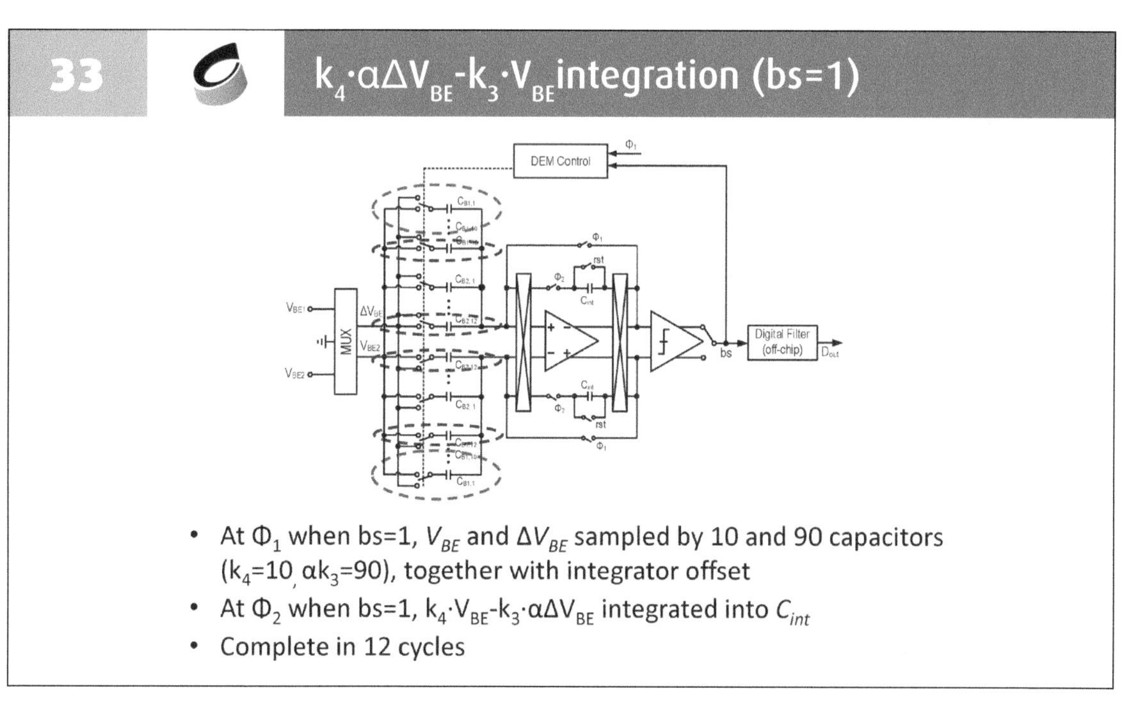

In the next phase, we connect V_{BE} across 10 unit capacitors, e.g. $C_{B1,12}$ and $C_{B2,\ 12}$. As a result, both the stored ΔV_{BE} in the 90 capacitors will be integrated together with V_{BE} with 10 unit capacitors across C_{int}. The charge package is proportional to $k_4{\cdot}V_{BE}\text{-}k_3{\cdot}\alpha\Delta V_{BE}$. In this mode, complete capacitor cycling can be achieved in 12 cycles.

34  Experimental Results

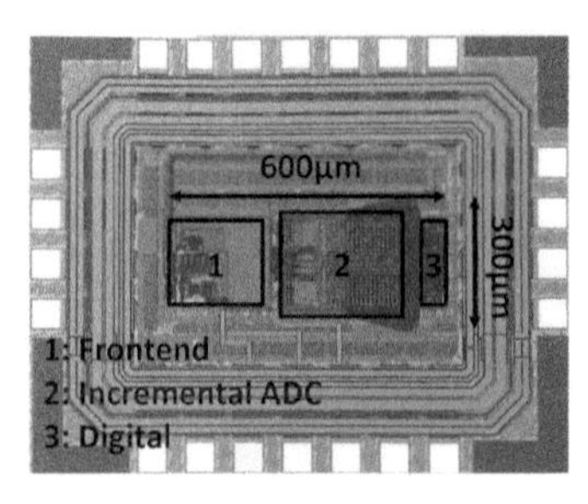

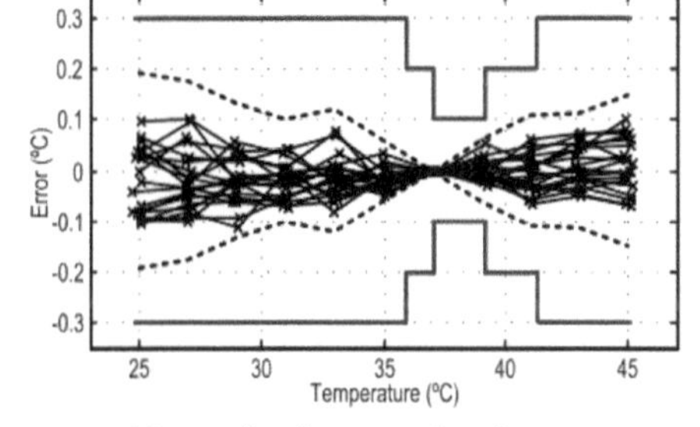

- Area: 0.18 mm² in 0.18-μm CMOS
- Power: 0.9 μW (Analog)
  0.2 μW (Digital)
- Resolution: 0.01 °C

- 20 samples from one batch
- Calibration and trimming at 37°C
- Inaccuracy (±3σ) < 0.2 °C
- Fulfill human body temperature monitoring requirements

This smart temperature sensor is implemented in a standard 0.18μm CMOS process, occupying an active area of 0.198 mm². With the analog and digital supply set to 1V and 1.8V respectively, the sensor dissipates a measured power of 1.1μW at 37°C (0.9 μW analog, 0.2μW digital). The achieved resolution is 0.01°C at 12-bit. The measured temperature error from 25 to 45°C with 20 sample chips after one-point digital calibration at 37°C is as shown. It can be observed that an inaccuracy of ±0.1°C (3σ) is obtained from 37 to 39°C (±0.2°Cfrom 25 to 45°C). The maximum error tolerance for human body temperature monitoring as discussed in the Standard Specification for Electronic Thermometer for Intermittent Determination of Patient Temperature is also indicated using the blue bold line. It can be concluded that the temperature sensor achieves an accuracy which is well within the specification, making it suitable for human body temperature monitoring applications.

35  Embedded Temperature Sensor

Now let's move to our second example. This work is an embedded temperature sensor design inside a passive RFID tag targeting for cold chain, medicines and health commodities storage and environmental monitoring applications. For cost considerations, it is a battery-less system and all the system power is wirelessly delivered using RF. As you may expect, the tag available power is very limited, in the order of few μW. Due to the very limited power budget, the embedded sensor should dissipate ultra-low power. To achieve this, system level co-optimization is used, and both the system clock and the existing bandgap reference are utilized for temperature sensing. The lack of a dedicated sensor frontend can significantly limit the sensing accuracy. Apart from that, the use of the system clock can result in sub-optimal performance as a result of error propagation. But you get much improvement in the system efficiency, and this is the tradeoff you need to pay. This work demonstrate systematically how different error sources can be optimized to achieve improved sensor performance in such a constrained system.

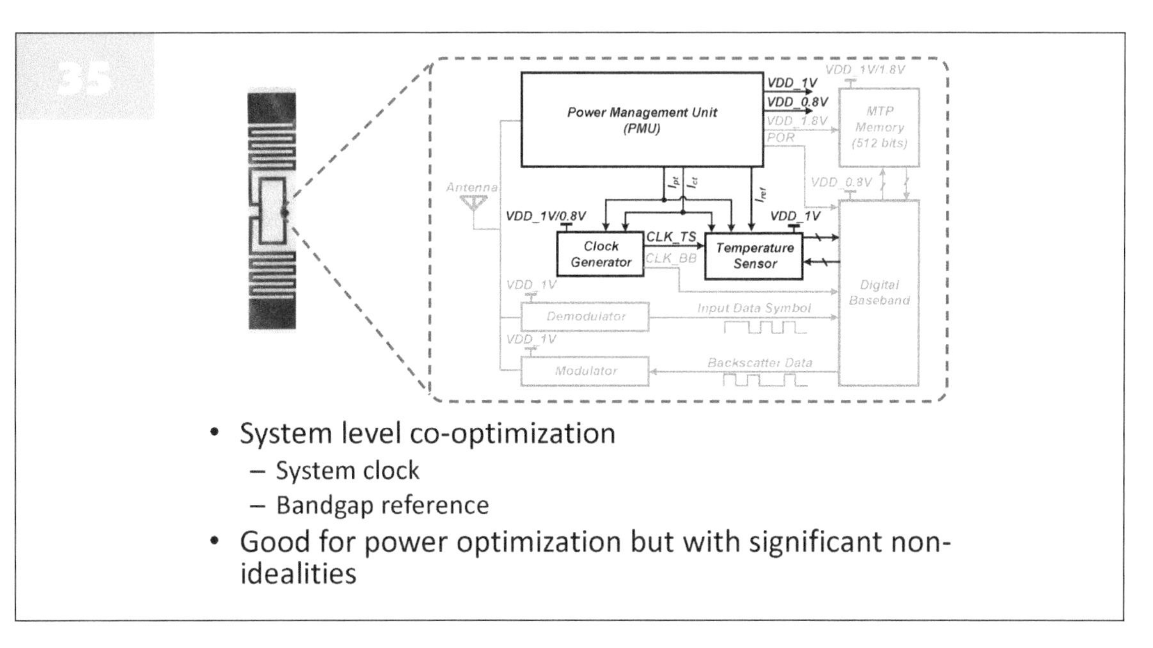

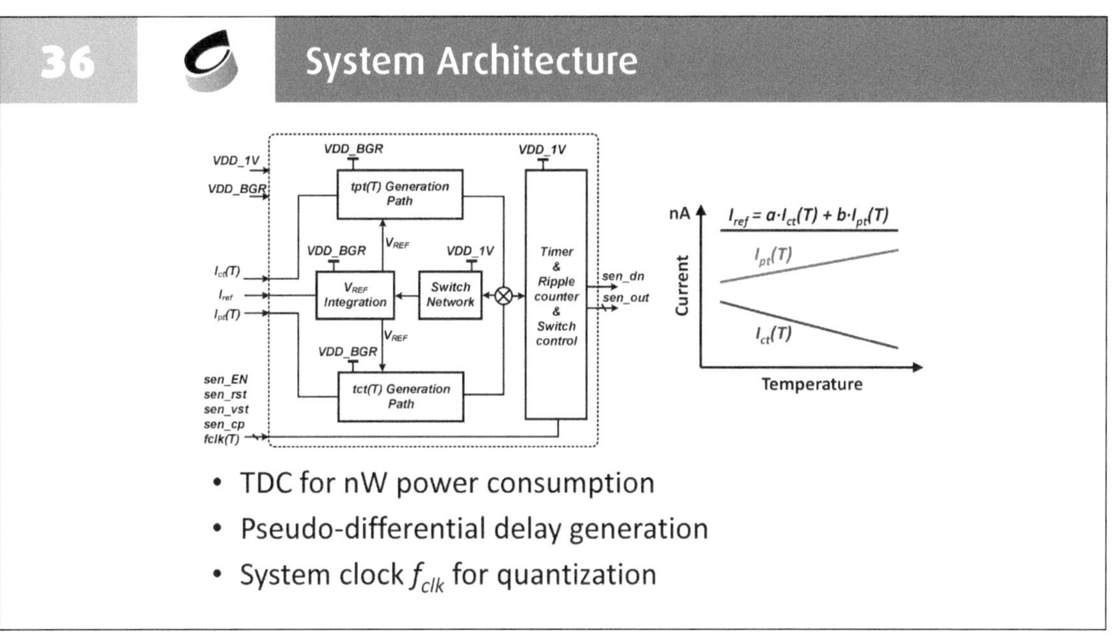

This slide shows the block diagram of the embedded sensor and its interfaces with other system sub-blocks. The embedded sensor is powered by a low-dropout regulator (LDO) output. A shared bandgap reference serves as the temperature sensor front-end and the system clock f_{clk} is used by the sensor for quantization purpose. The critical control signals are from the tag baseband. As a common supply is used to power up different building blocks, noise coupling is inevitable and the sensor should have a good supply rejection in order to ensure its sensing accuracy. In this work, the bandgap reference generates all the I_{pt}, I_{ct} and I_{ref} signals in the nA levels for temperature sensing as illustrated in the figure on the right. This current level is chosen to provide enough SNR as a result of leakage especially at high temperature. As mentioned before, this sensor utilizes the system bandgap reference to generate the sensing signals instead of constructing a dedicated sensing frontend for reduced power consumption. This, however, leads to the penalties of reduced SNR and degraded signal linearity.

37 Sensor Frontend

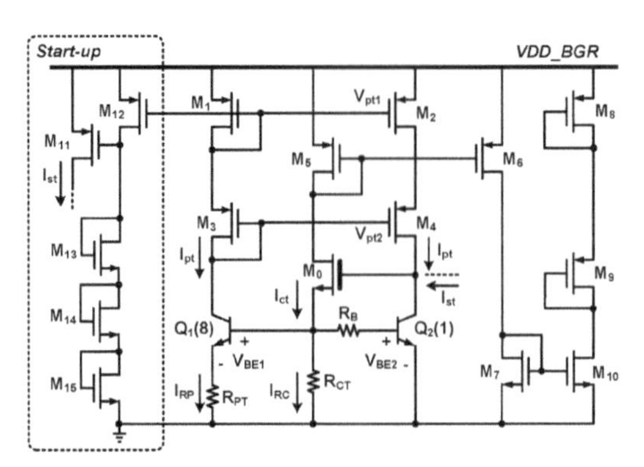

- Ultra-low power consumption
  - BJT sizing ratio and current mirror ratio of 8:1 and 1:1
  - Current reuse to for base currents of $Q_{1,2}$
- Process spread error and non-linearity in I_{pt} and I_{ct}
- No dedicated BJTs with dynamic element matching

This slide shows the bandgap reference circuit, which also serves as the sensor frontend. Here, two matched vertical NPN transistors serve as the sensing devices, and a PTAT current is generated by ΔV_{BE} and R_{PT}. Here the BJT size ratio and the current mirror ratio are 8:1 and 1:1, respectively. A current mirror ratio of 1:1 is selected for reduced power consumption and alleviate the error due to β variation. Here, we reuse the current from I_{ct} to generate the base currents for Q1 and Q2 for power saving. The native device M_0 serves as a low threshold voltage buffer. We also cascode M_{1-4} to improve the supply rejection.

Up till now, you may have noticed that this BGR is in fact designed mainly for power saving instead of accurate temperature sensing. We cannot use a high gain amplifier to ensure the drain current matching for M_1 and M_2, dynamic element matching technique also cannot be used as the biasing currents are also used by other building blocks. The base currents can also exhibit temperature dependencies which can affect the linearity of I_{ct}. It looks everything is not optimal for temperature sensing, but that's life and this is also the exact reason to have you as the engineer to solve the problem.

So exactly what we can do? In this work, low cost is very important. At least we have BJTs here so we can trim out some PTAT errors. Also, we can improve the systematic error such as the linearity of the signal to avoid multi-point calibration. Let's first take a look at what's going on. As we cannot use dynamic element matching techniques here, there'll be mismatch in the drain current. The spread of this error can be defined using this equation. Current mirrors tend to have some level of temperature dependencies especially if the ratio is large. However, as we choose to use a current mirror ratio of 1:1, the spread in current ratio should be weakly temperature dependent, and the resultant error can be considered as a PTAT spread. How about the BJT sizing spread? The ratio of 8:1 can suffer from process and can be represented as an I_S spread, which is PTAT and can be trimmed out relatively easily. At the first glance, the situation does not seem as bad, and we'll revisit the discussion of the sensor frontend noise source later.

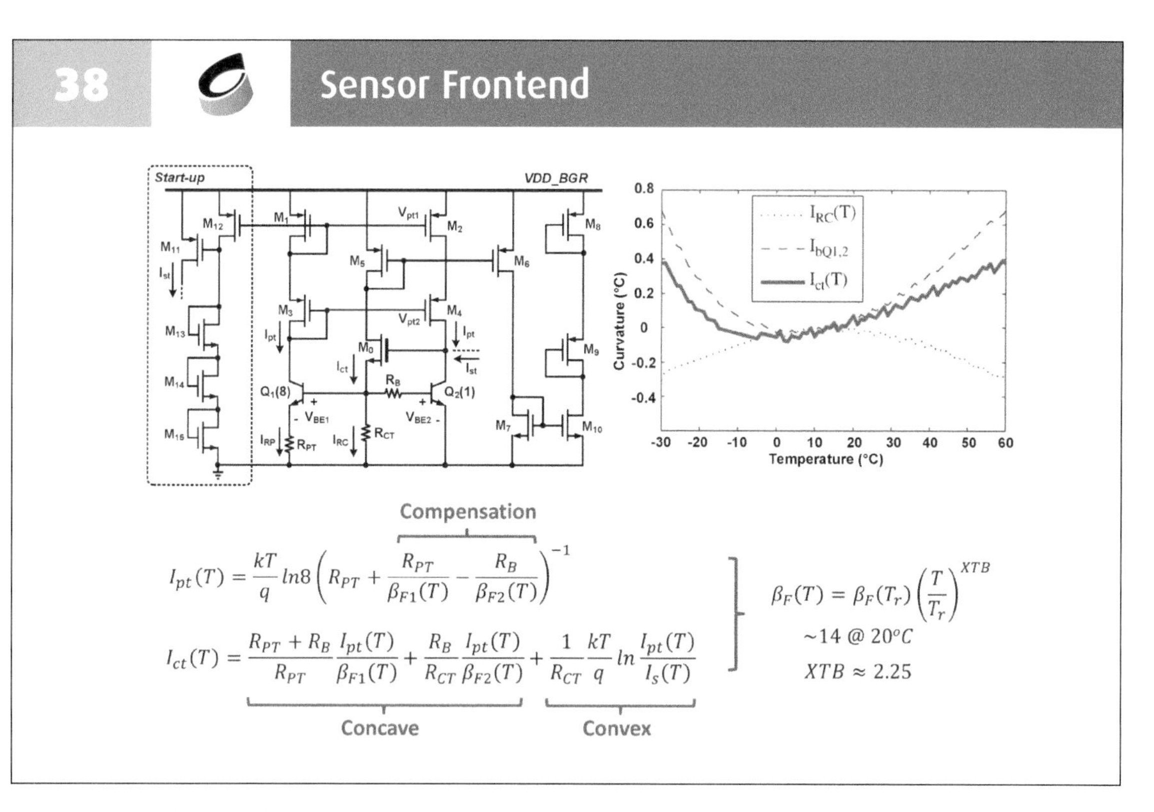

How about the non-linearity in I_{pt} and I_{ct}? We first study I_{pt}. We can observe that I_{pt} should be the sum of I_{RP} and the base current of Q_1. As the nominal β is only ~14, we also need to take care of this base current as it cannot be considered small, and its temperature dependencies can affect the linearity of I_{pt}. What we can do is to implement a β compensation resistor R_B at the base of Q_2. With R_B, I_{PT} can be expressed as the $(kT/q)\cdot\ln(8)$ term here, which is the nominal ΔV_{BE}, divided by $(R_{PT} + R_{PT}/\beta_{F1} - R_B/\beta_{F2})$. Here, notice that the second and third term mainly contributes to the non-linearity of I_{pt}, and the fact that they subtract each other can effectively reduce this error term. In theory, β_{F1} and β_{F2} are temperature dependent with XTB to be roughly 2.25 in the selected process. For further power saving, we have chosen a low biasing current which is out of the constant β range. This results in a mismatch between β_{F1} and β_{F2}, and R_{PT} is designed to be slightly larger than R_B to compensate for this effect. However, mismatch exists between R_{PT} and R_B. This results in a non-PTAT error and is roughly ±0.1°C with ±10% mismatch, so this is still tolerable.

In terms of I_{ct}, it is the sum of the base current of $Q_{1,2}$, which is denoted by the first term, and the current passing through R_{CT}, as expressed by the second and third term here, respectively. Notice that due to the strong positive temperature dependency of β, the first two terms in this equation is concave. While the third term is convex as described earlier. As a result, their sum should improve the overall I_{ct} linearity. Simulation result is provided in this plot here. It can be observed that the temperature dependency well-matched our expectation, and the sensing signals I_{pt} and I_{ct} should be good enough for our purpose.

39

Readout Principles

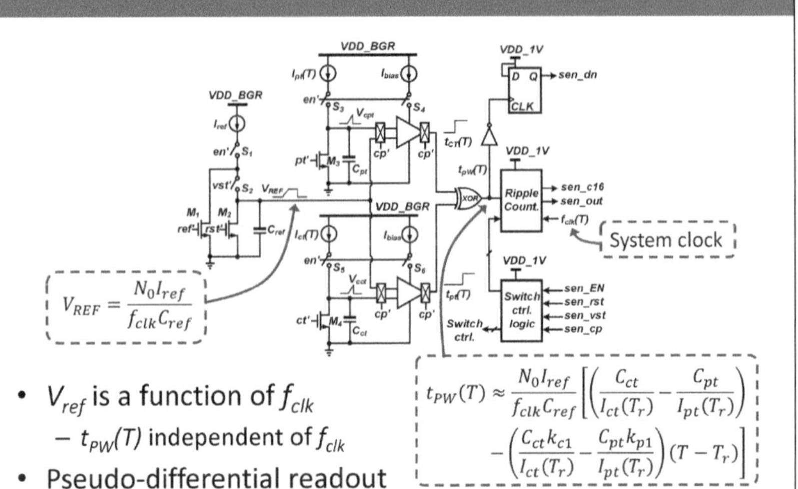

In this work, a TDC readout is chosen for its ultra-low power consumption. The two delay paths integrates I_{pt} and I_{ct} up to the reference level V_{REF} using C_{pt} and C_{ct}, and generate a temperature dependent pulse t_{PW}. The pulse width is then quantized by a ripple counter using a clock signal.

As discussed before, the system clock is reused for temperature signal quantization. However, due to the significant process spread, the variation in the quantization levels can lead to significant sensing error. To solve this problem, we include a third integration path C_{ref}, which is charged up to V_{REF} using I_{ref} in N_0 clock cycles. To improve the linearity of t_{PW}, we can minimize the second-order non-linear terms by optimizing the design parameters. The resultant t_{PW} can be expressed as this equation. The N_0/f_{clk} term can be cancelled out after quantization using the system clock. By doing so, the spread in clock frequency can be compensated to the first-order. Apart from that, due to the pseudo differential readout architecture, the supply rejection of t_{PW} can also be improved.

40

Noise Sources

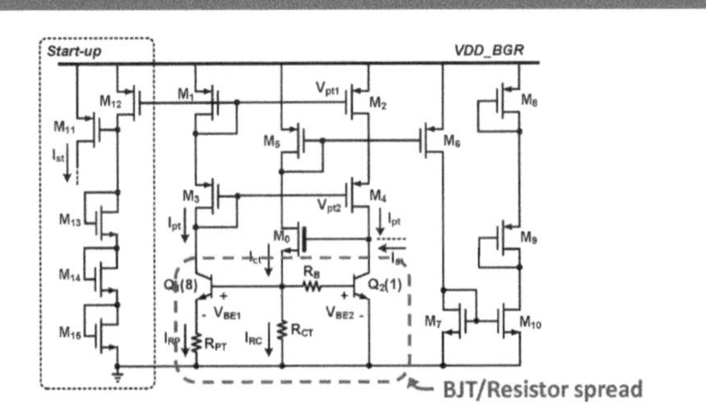

• BJT/Resistor leads to PTAT spread

$$\Delta e_1 \approx \frac{N_0(\delta_{I_S} + \delta_R)}{2} \left(\frac{1}{I_{ct}(T_r) R_{CT}} + \frac{1}{I_{pt}(T_r) R_{PT}} \right) \frac{kT}{q}$$

Here, we discuss on the sensor frontend noise sources on BJT and resistor spread. As the BJT sizing spread depends on I_S, its induced error is therefore PTAT. For the spread of resistors, it can lead to variation in I_{ct} and I_{pt} due to V_{BE}. The combination error is described in this equation. It can be observed that if $\delta R/R$ is constant within the target sensing range, the overall BJT/resistor spread can be considered to be PTAT, and such errors can be trimmed out after fabrication.

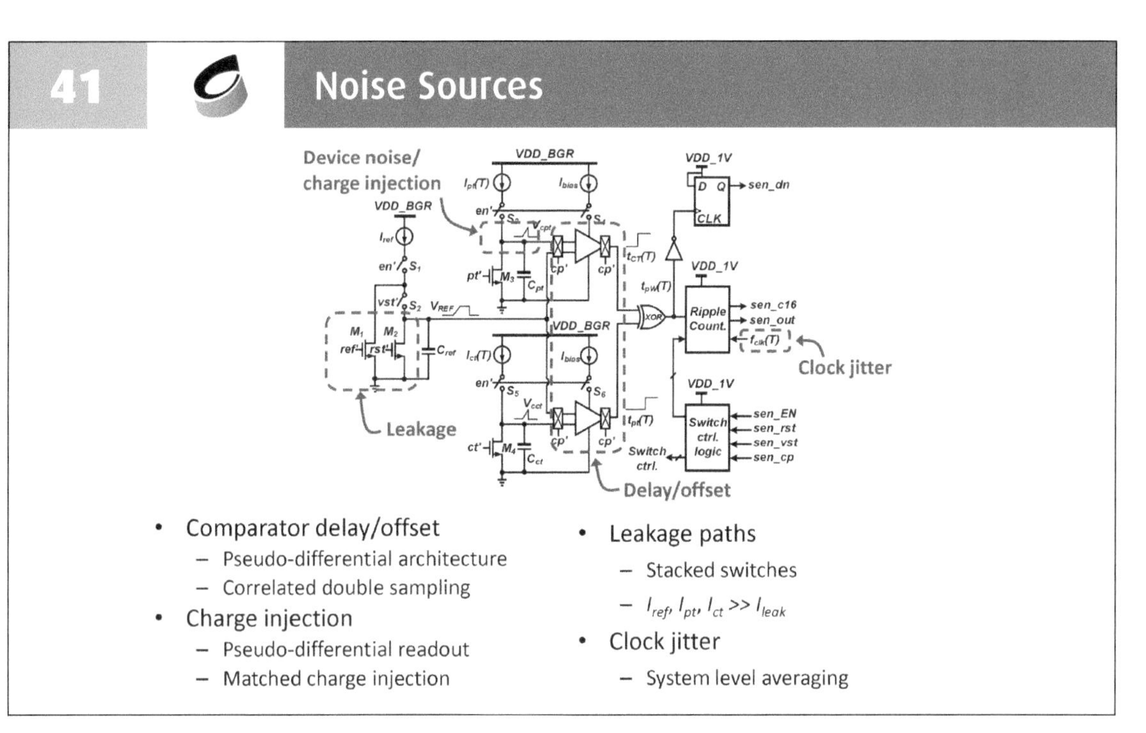

For the readout noise sources, the comparator generates non-negligible delay due to the low power consumption with limited voltage gain. This error can be cancelled by using the pseudo-differential architecture. In terms of its offset, a time-domain offset cancellation scheme similar to CDS by reading out two consecutive two t_{PW} signals is employed. This can minimize the comparator offset to a negligible level.

In terms of charge injection of switches, such errors are mainly coming from: 1) at the start of each sensing cycle where all capacitors are reset; 2) when the current sources are switched in for integration; and 3) when I_{ref} is disconnected from C_{ref}. Here, we size C_{ref} to be twice the size of C_{pt} and C_{ct}, and M_2 is also twice the size of $M_{3,4}$ so as to maintain the same common mode voltage levels when integration starts. The resultant common mode noise can be eliminated using the pseudo-differential readout.

Another noise source is the leakage paths especially from off switches. To reduce this subthreshold leakage, $M_{1\text{-}4}$ and S_2 are all implemented with stacked transistors. I_{pt}, I_{ct} and I_{ref} are all designed to be much larger than the leakage currents. By doing so, the error due to leakage becomes negligible, with a maximum error of 0.01°C.

In terms of thermal noise, the dominant noise sources is mainly coming from the system clock in the form of jitter, which can be significant due to the limited power consumed by the clock building block. As jitter is a form of random noise, we can eliminate this error by using system level averaging to tradeoff accuracy with conversion time.

42 

Experimental Results

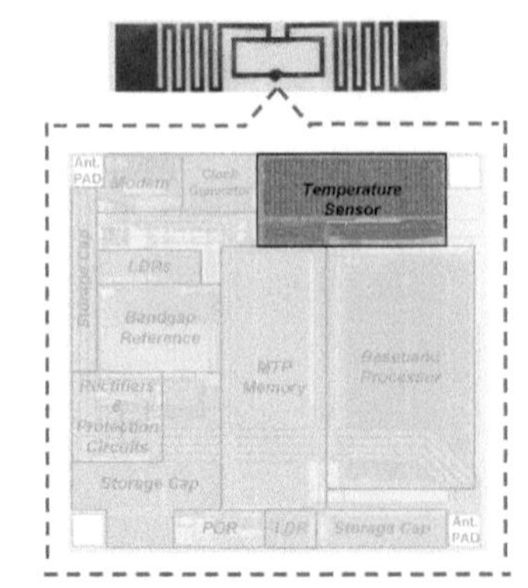

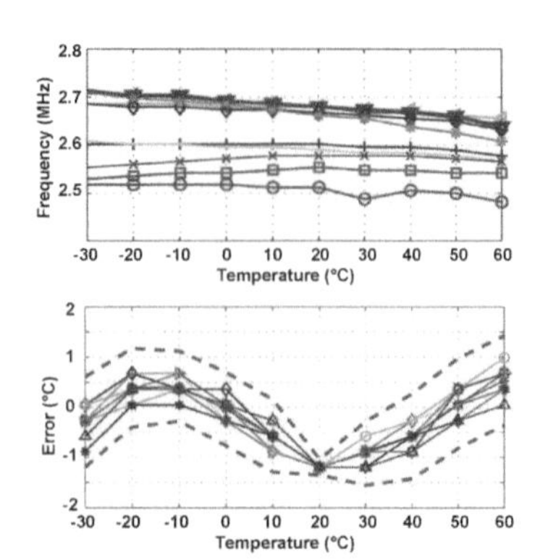

- Area: 0.13 mm² in 0.18-µm CMOS
- Power: 0.35 µW
- Resolution: 0.3 °C

- 12 samples from one batch
- Calibration and trimming at 20°C
- Inaccuracy (3σ) < ±1.5 °C

This slide shows the chip micrograph, with the temperature sensor within the complete RFID system as shown. The area is 0.13 mm^2 in the 0.18-um CMOS technology, with a power consumption of 350nW at a resolution of 0.3°C. The top figure on the right shows the measured clock frequency spread. As you can see here, the clock frequency varies from 2.5 to 2.7 MHz with a CTAT dependency. By using the noisy system clock, the sensor can still achieve a ±1.5°C (3σ) accuracy using one-point trim at 20°C from 12 samples in one batch. As expected, the sensing error is curvature limited due to the non-linear nature of I_{ct} and I_{pt}, but this performance is already acceptable in many power-limited applications where the accuracy requirement is not as important.

43 

Subthreshold MOS-based Temperature Sensor[21]

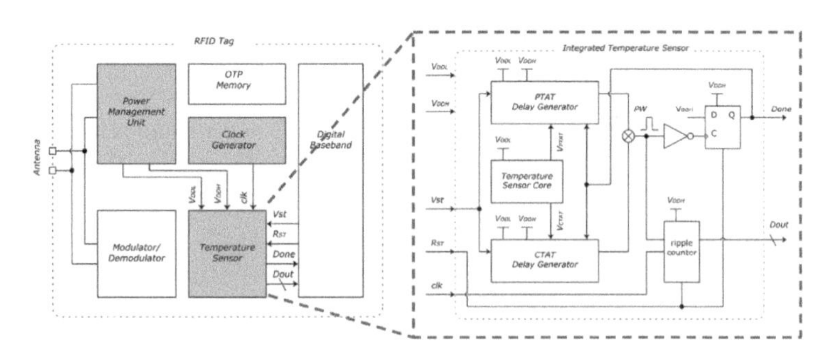

- Building block reuse
  - Existing regulated sensor supply from PMU
  - Existing system clock for quantization
- Low supply voltage (0.5V) for further power reduction

Here shows the third example, which is MOSFET-based CMOS temperature sensor embedded in a passive RFID tag. For improved system efficiency, we reused both the existing regulated supply from the power management unit and the existing system clock as the quantization clock. For this implementation, there is a temperature sensing core, a PTAT and a CTAT delay generator, and a time-to-digital converter. We implemented the sensor core so that it is working in the subthreshold region and can operate under a 0.5V supply for further power reduction.

44

Sensor Frontend

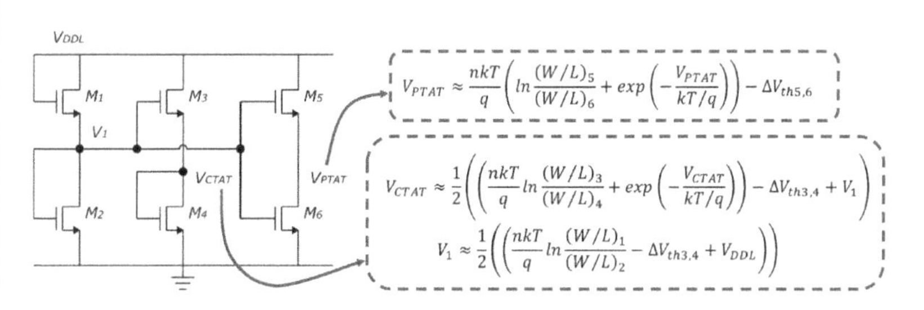

$$V_{PTAT} \approx \frac{nkT}{q}\left(ln\frac{(W/L)_5}{(W/L)_6} + exp\left(-\frac{V_{PTAT}}{kT/q}\right)\right) - \Delta V_{th5,6}$$

$$V_{CTAT} \approx \frac{1}{2}\left(\left(\frac{nkT}{q}ln\frac{(W/L)_3}{(W/L)_4} + exp\left(-\frac{V_{CTAT}}{kT/q}\right)\right) - \Delta V_{th3,4} + V_1\right)$$

$$V_1 \approx \frac{1}{2}\left(\left(\frac{nkT}{q}ln\frac{(W/L)_1}{(W/L)_2} - \Delta V_{th3,4} + V_{DDL}\right)\right)$$

- The lambert-W function G(.) introduced (ΔV_{th} ignored):

$$V_{PTAT} \approx \frac{kT}{q}\left(n \cdot ln\frac{(W/L)_5}{(W/L)_6} + G\left(n \cdot exp\left(-n \cdot ln\frac{(W/L)_5}{(W/L)_6}\right)\right)\right)$$

PTAT controlled by size of $M_{5\text{-}6}$

$$V_{CTAT} \approx \frac{V_{DDL}}{4} + \frac{kT}{q}\left(\frac{n \cdot ln(K)}{4} + G\left(\frac{n}{2} \cdot exp\left(-\frac{1}{4}\left(\frac{V_{DDL}}{kT/q} + n \cdot ln(K)\right)\right)\right)\right)$$

$$K = \left(\frac{(W/L)_3}{(W/L)_4}\right)^2\left(\frac{(W/L)_1}{(W/L)_2}\right)$$

CTAT controlled by size of $M_{1\text{-}4}$

This slide shows the temperature sensor core design. It is composed of 6 transistors operating in the subthreshold region. If we equate the current through M_5 and M_6 to be equal, V_{PTAT} can be expressed using this equation. Similarly, V_{CTAT} can be expressed using these equations. Notice from this V_{CTAT} equations that they are recursive functions, which be solved numerically but lack intuition. Instead, we first ignore the body effect term for now and use the lambert-W function to solve for both V_{PTAT} and V_{CTAT}. It can be observed that V_{PTAT} can be controlled by the sizing of $M_{5,6}$. Similarly, VCTAT can be controlled by the sizing of $M_{1\text{-}4}$. Notice that V_{CTAT} is in fact dependent on the supply voltage, and its effect will be explained later.

45

Sensor Frontend

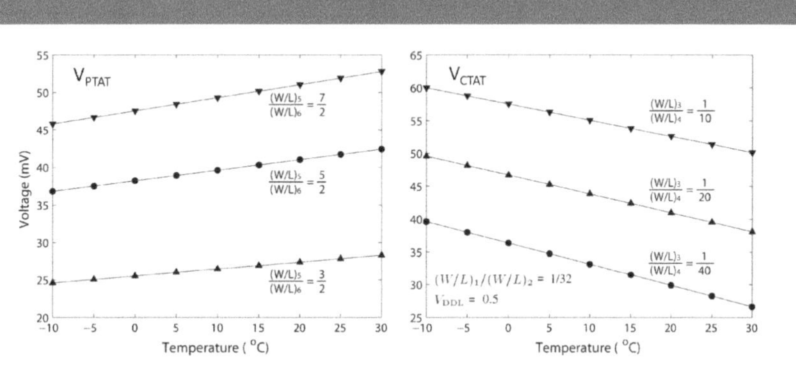

- Sensitivity can be adjusted using different transistor ratios
- Linearity can still be preserved

This slide shows the simulation results of V_{PTAT} and V_{CTAT} using different transistor sizing as shown. It can be concluded that the sensitivity can be adjusted by using different transistor ratios, and the linearity can still be preserved.

46 

Sensor Frontend

- ΔV_{th} terms posess PTAT or CTAT behavior

$$\Delta V_{th}(T) = K_{T2} \cdot V_{bseff}\left(\frac{T}{T_0} - 1\right)$$

- The overall output linearity of V_{PTAT} and V_{CTAT} can still be preserved with body-effect

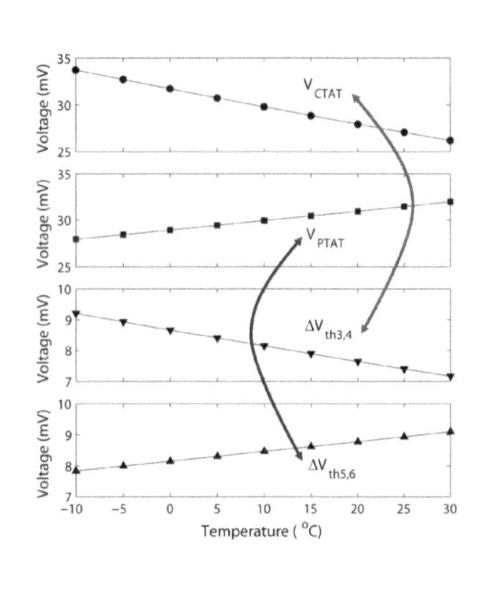

Now we reconsider the V_{PTAT} and V_{CTAT} voltage with body effect. In the BSIM model, the difference in V_{th} can be expressed using this equation. Notice that this term V_{bs} corresponds to the V_{PTAT} and V_{CTAT} signal, and the corresponding body effect term should also exhibit linear to temperature relationship, as verified through simulation. It can be observed that the corresponding body effect terms also exhibit similar linear to temperature relationship to V_{PTAT} and V_{CTAT}, meaning that even with body effect, the output linearity of V_{PTAT} and V_{CTAT} can still be preserved.

47 

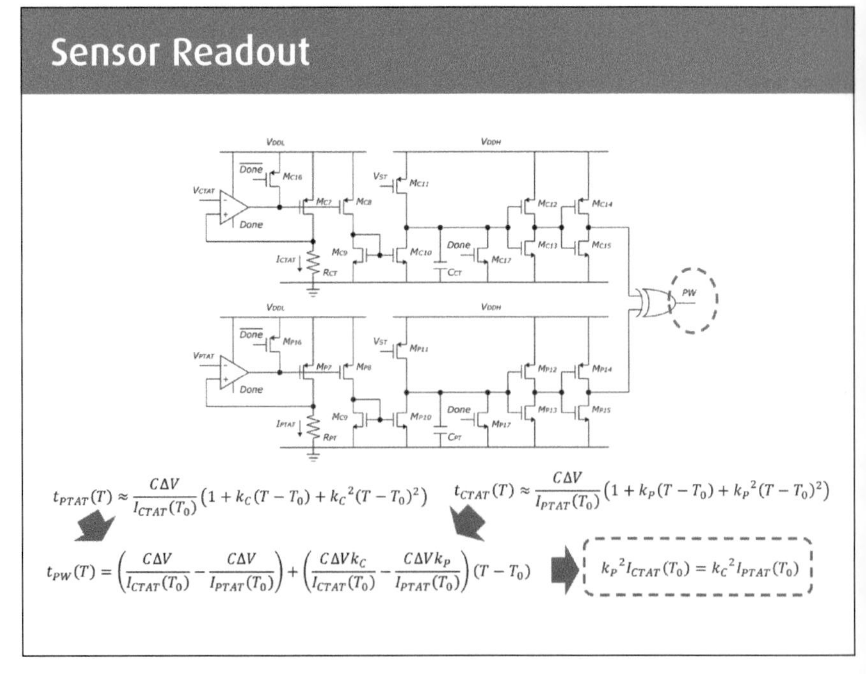

This slide shows the implementation of the differential delay generators. Using the V_{PTAT} and V_{CTAT} signals from the sensor core, the corresponding PTAT and CTAT delays can be represented by these equations. Notice that they are in general non-linear. However, the linearity of the pulse width at the output of the XOR gate can still be preserved if this condition is fulfilled, which can be achieved in the design stage.

48 

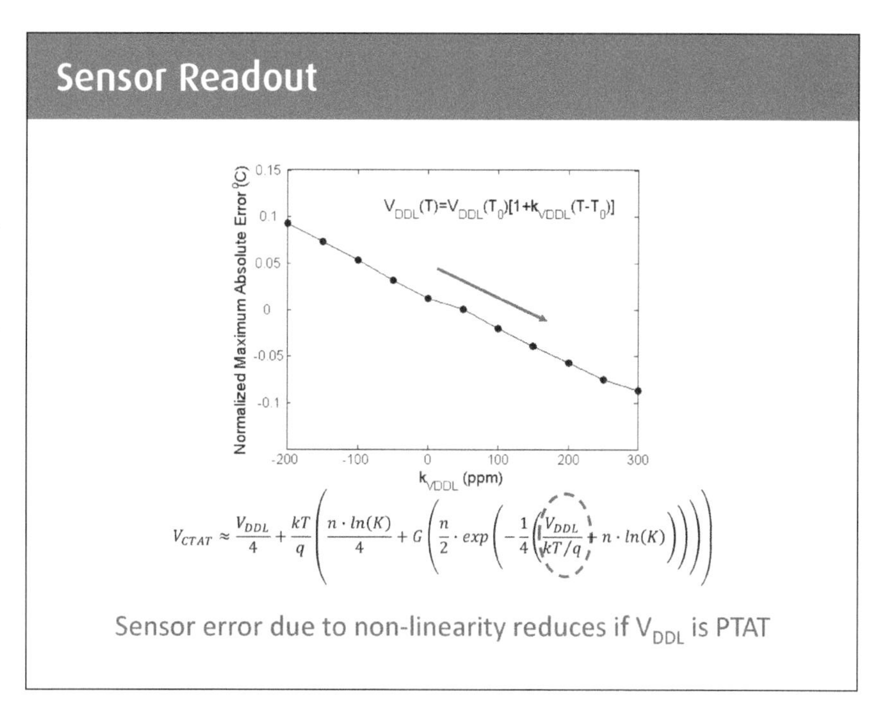

This slide shows the effect of V_{CTAT} with respect to the supply voltage. Due to the effect of the thermal voltage, the linearity of V_{CTAT} is expected to improve if the supply voltage is PTAT. As shown in the above figure, a reduction in sensing error is obtained when the supply voltage is PTAT, and this is what we designed for the on-chip supply.

49 

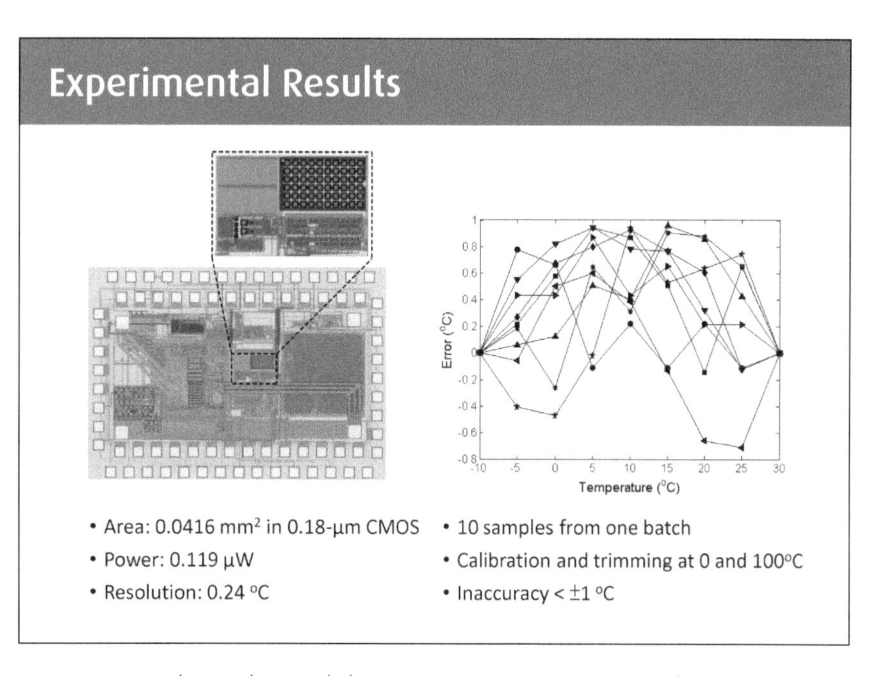

The sensor is fabricated using the TSMC 0.18um 1P6M process, and is embedded in the RFID tag system. The sensor occupies an area of 0.0416mm². This slide shows the measurement result of the sensor. We averaged 30 samples so as to reduce the sensor error, and the result is shown here. It can be seen that a sensing inaccuracy of +1/-0.8°C is achieved over 9 measured samples, and the power consumption is merely 119nW with a resolution of 0.24°C.

50  Trimming/Calibration

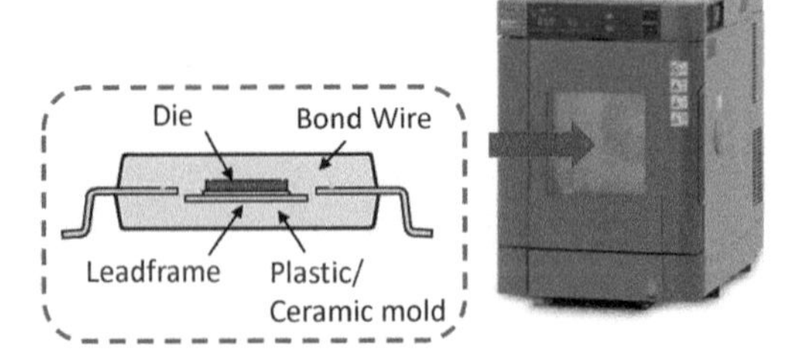

- Wafer-level calibration
  - Cannot compensate for the effects due to packaging stress
- Calibrate after packaging
  - Extended calibration time due to thermal settling
  - Thermal time constant for different packaging technologies
  - Packaging stress induces non-PTAT error[37,38]

Now we briefly discuss on trimming and calibration. As discussed, CMOS temperature sensor generally require calibration and trimming to achieve high accuracy, and the process spread of I_S can already lead to temperature sensing error of up to few °C. Wafer level calibration is a cost efficient solution to improve the sensor accuracy. We can also assume that the entire wafer is more or less at the same temperature. However, this kind of calibration is only good if the temperature is measured at the die level. When you package your die individually using plastic or ceramic packaging, stress is induced and the BJT characteristic changes. It should be clear that wafer level calibration is not useful for such kind of error, which can result in a sensing accuracy of about 0.5°C. A better solution is to calibrate the sensor after packaging so as to tackle such packaging stress induced error. However, each package can have different thermal time constant, which can be in the order of few seconds. This kind of thermal calibration leads to an extended calibration time required for thermal settling. What's more, the packaging stress can lead to non-PTAT errors, meaning that you need to calibrate using more than one temperature. This can lead to hours of calibration effort which can significantly increase your part cost.

51  Heater-Assisted Voltage Calibration[36]

Recently, a heater-assisted voltage calibration scheme is proposed to significantly improve the calibration time. It exploits an electrical calibration method instead of a thermal one. As a result, there is no need to wait for the package thermal settling. The basic idea is to measure ΔV_{BE} and compare with an accurate external voltage V_{ext}. As ΔV_{BE} is process independent, you should be able to tell the die temperature electrically. You can then trim your device by adjusting say I_C to obtain the required digital output that represents the die temperature. As discussed before, packaging stress can induce non-PTAT errors, and it requires more than one-point calibration. Instead of using an external heat source, an on-chip heater can be placed on top of the sensing BJTs so that the die temperature can be increased instantaneously. There is no need to wait for thermal settling, and the value of ΔV_{BE} can be extracted to trim the device again at a different temperature. Of course, circuit errors still exist and dynamic element matching technique should be used.

During calibration, what you need to do is to drive the heater with an external current to heat up the die. The calibration time required as shown for the traditional thermal calibration and the heater-assisted voltage calibration shows a significant calibration time improvement from 2 hours to 0.5s while achieving similar accuracy of ±0.1°C (ceramic) and ±0.2°C (plastic).

51

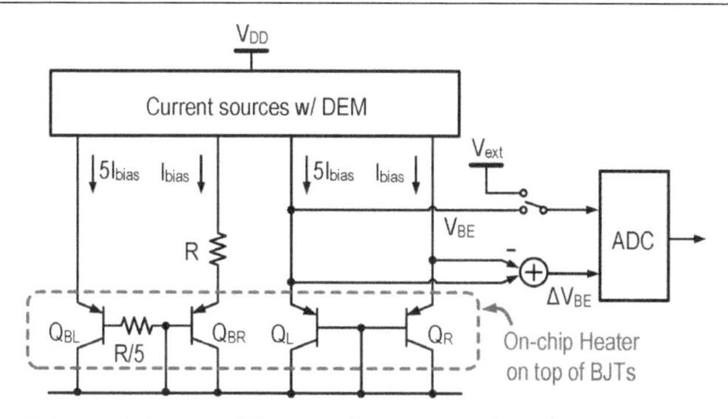

- Compare ΔV_{BE} with a calibrated external voltage V_{ext}
  - ΔV_{BE} is process tolerant
  - Fast calibration due to electrical measurement (no need to wait for thermal settling)
- On-chip heater on top of BJT
  - Fast increase of die temperature
  - Two-point trim to calibrate non-PTAT error due to packaging stress
- Significant calibration time improvement from 2 hours to 0.5s with ±0.1°C (ceramic) and ±0.2°C (plastic) accuracy

52 

Conclusions

- **Sensing Devices (BJT and MOSFET) with design considerations**
- **Application specific/System level optimization possible for improved power/energy efficiency and accuracy**

In this talk, we have introduced the use of different sensing devices, especially BJT and MOSFET, for CMOS temperature sensor implementations. We have covered both of their advantages and disadvantages for fulfilling different application specific requirements. Apart from that, various ways to achieve application specific and system level optimization are also outlined to achieve improved power/energy efficiency and accuracy for CMOS temperature sensor implementations.

53 

Acknowledgements

- All the members involved in the RFID project
- MYRG of University of Macau/Macao Science and Technology Development Fund (FDCT)

Thank you for your attention.

Here, I would like to thank all the members involved in the RFID project, as well as the Macau FDCT and RC of University of Macau for financial support. Thank you for your attention.

54 

References

1. N. Cho, S.-J. Song, S. Kim, S. Kim, H.-J. Yoo, "A 5.1-μW UHF RFID tag chip integrated with sensors for wireless environmental monitoring," *ESSCIRC*, Sep. 2005, pp. 289-282.
2. F. Kocer, M. P. Flynn, "An RF-powered, wireless CMOS temperature sensor," *IEEE Sensors J.*, vol. 6, no. 3, pp. 557-564, Mar. 2006.
3. A. Ramos, D. Girbau, A. Lazaro, A. Collado, A. Georgiadis, "Solar-Powered Wireless Temperature Sensor Based on UWB RFID With Self-Calibration," *IEEE Sensors J.*, vol. 15, no. 7, pp. 3764-3772, Jul. 2015.
4. K. Stangel, S. Kolnsberg, D. Hammerschmidt, B. J. Hosticka, H. K. Trieu, W. Mokwa, "A programmable intraocular CMOS pressure sensor system implant," *IEEE J. Solid-State Circuits*, vol. 36, no. 7, pp. 1094-1100, Jul. 2001.
5. M. A. Ghanad, M. M. Green, C. Dehollain, "A 30 μW Remotely Powered Local Temperature Monitoring Implantable System," *IEEE Trans. Biomed. Circuits Systs.*, vol. 11, no. 1, pp. 54-63, Jan. 2017.
6. Y. C. Shin, T. Shen, B. P. Otis, "A 2.3μW Wireless Intraocular Pressure/Temperature Monitor," IEEE J. Solid-State Circuits, vol. 46, no. 11, pp. 2592-2601, Nov. 2011.
7. F. Sebastiano, L. J. Breems, K. A. A. Makinwa, "A 65-nm CMOS Temperature-Compensated Mobility-Based Frequency Reference for Wireless Sensor Networks," *IEEE J. Solid-State Circuits*, vol. 46, no. 7, pp. 1544-1552, Jul. 2011.
8. U. Kang, K. D. Wise, "A high-speed capacitive humidity sensor with on-chip thermal reset," *IEEE Trans. Electron Devices*, vol. 47, no. 4, pp. 702-710, Apr. 2000.
9. K. Sundaresan, P. E. Allen, F. Ayazi, "Process and temperature compensation in a 7-MHz CMOS clock oscillator," *IEEE J. Solid-State Circuits*, vol. 41, no. 2, pp. 433-442, Feb. 2006.
10. T. Verster, "P-n junction as an ultralinear calculable thermometer," *Electronics Letters*, vol. 4, no. 9, May 1968, pp. 175–176.

11. A. Bakker, J. H. Huijsing, "Micropower CMOS smart temperature sensor with digital output," *IEEE J. Solid-State Circuits*, vol. 31, no. 7, pp. 933-937, Jul. 1996.
12. F. Sebastiano, L. J. Breems, K. A. A. Makinwa, S. Drago, D. M. W. Leenaerts, B. Nauta, "A 1.2-V 10-µW NPN-Based Temperature Sensor in 65-nm CMOS With an Inaccuracy of 0.2oC (3σ) From -70oC to 125oC," *IEEE J. Solid-State Circuits*, vol. 45, no. 12, Dec. 2010, pp. 2591-2601.
13. I. M. Filanovsky and W. Lee, "Two temperature sensors with signal-conditioning amplifiers realized in BiCMOS technology," *Sensors and Actuators*, vol. 77, pp. 45-53, Sep. 1999.
14. G. C. M. Meijer, "Thermal sensors based on transistors," *Sensors and Actuators*, vol. 10, pp. 103-125, Sep. 1986.
15. S. Pan, Y. Leo, S. H. Shalmany, K. A. A. Makinwa, "A Resistor-Based Temperature Sensor with a 0.13pJ · K^2 Resolution FOM," *ISSCC*, Feb. 2017, pp. 158-159.
16. C.-H. Weng, C.-K. Wu, T.-H. Lin, "A CMOS Thermistor-Embedded Continuous-Time Delta-Sigma Temperature Sensor With a Resolution FoM of 0.65 pJ°C^2," *IEEE. J. Solid-State Circuits*, vol. 50, no. 11, pp. 2491-2500, Nov. 2015.
17. M. A. P. Pertijs, A. Niderkorn, X. Ma, B. McKillop, A. Bakker, J. H. Huijsing, "A CMOS smart temperature sensor with a 3σ inaccuracy of ±0.5°C from -50°C to 120°C," *IEEE J. Solid-State Circuits*, vol. 40, no. 2, pp. 454-461, Feb. 2005.
18. M. A. P. Pertijs, K. A. A. Makinwa, and J. H. Huijsing, "A CMOS temperature sensor with a 3σ inaccuracy of ±0.1°C from −55°C to 125°C," *ISSCC*, Feb. 2005, pp. 238–239.
19. M. Tuthill, "A switched-current, switched-capacitor temperature sensor in 0.6-µm CMOS," *IEEE J. Solid-State circuits*, vol. 33, no. 7, pp. 1117-1122, Jul. 1998.
20. M. K. Law, A. Bermak, "A 405-nW CMOS temperature sensor based on linear MOS operation," *IEEE Trans. Circuits Systs. II: Exp Briefs*, vol. 56, no. 12, pp. 891-895, Dec. 2009.
21. M. K. Law, A. Bermak, H. C. Luong, "A Sub-µW Embedded CMOS Temperature Sensor for RFID Food Monitoring," *IEEE J. Solid-State Circuits*, vol. 45, no. 6, pp. 1246-1255, Jun. 2010.
22. K. Ueno, T. Asai, Y. Amemiya, "Low-power temperature-to-frequency converter consisting of subthreshold CMOS circuits for integrated smart temperature sensor," *Sensors and Actuators A: Physical*, vol. 165, pp. 142.137, Jan. 2011.
23. S. Jeong, Z. Foo, Y. Lee, J.-Y. Sim, D. Blaauw, D. Sylvester, "A Fully-Integrated 71 nW CMOS Temperature Sensor for Low Power Wireless Sensor Nodes," *IEEE J. Solid-State Circuits*, vol. 49, no. 8, pp. 1682-1693, Aug. 2014.
24. A. L. Aita, M. A. P. Pertijs, K. A. A. Makinwa, "A CMOS smart temperature sensor with a batch-calibrated inaccuracy of ±0.25°C (3σ) from −70°C to 130°C," *ISSCC*, Feb. 2009, pp. 342-343.

25. K. Souri, Y. Chae, K. A. A. Makinwa, "A CMOS Temperature Sensor With a Voltage-Calibrated Inaccuracy of ±0.15°C (3σ) From – 55°C to 125°C," *IEEE J. Solid-State Circuits*, vol. 48, no. 1, pp. 292-301, Jan. 2013.

26. D. F. Hilbiber, "A new semiconductor voltage standard," *ISSCC*, Feb. 1964, pp. 32–33.

27. R. J. Widlar, "New developments in IC voltage regulators," *IEEE J. Solid-State Circuits*, vol. 6, no. 1, pp. 2–7, Feb. 1971.

28. G. C. M. Meijer, "Integrated circuits and components for bandgap references and temperature transducers," Ph.D. dissertation, Delft University of Technology, Delft, The Netherlands, Mar. 1982.

29. G. C. M. Meijer, G. Wang, F. Fruett, "Temperature Sensors and Voltage References Implemented in CMOS Technology," *IEEE Sensors J.*, vol. 1, no. 3, pp. 225-234, Oct. 2001.

30. M. A. P. Pertijs, Johan H. Huijsing, *Precision Temperature Sensors in CMOS Technology*, The Netherlands: Springer, 2006.

31. K. Roy, S. Mukhopadhyay, H. Mahmoodi-Meimand, "Leakage current mechanisms and leakage reduction techniques in deep-submicrometerCMOS circuits," in *Proceedings of the IEEE*, vol. 91, no. 2, Feb. 2003, pp. 305-327.

32. S. Hwang, J. Koo, K. Kim, H. Lee, C. Kim, "A 0.008 mm^2 500 μW469 kS/s Frequency-to-Digital Converter Based CMOS Temperature Sensor With Process Variation Compensation," *IEEE Trans. Circuits Systs. I: Reg. Papers*, vol. 60, no. 9, pp. 2241-2248, Sep. 2013.

33. I. M. Filanovsky, S. T. Lim, "Temperature Sensor Applications of Diode-Connected MOS Transistors," *ISCAS*, May 2002, pp. 149-152.

34. M. K. Law, S. Lu, T. Wu, A. Bermak, P.-I. Mak, R. P. Martins, "A 1.1 μW CMOS Smart Temperature Sensor With an Inaccuracy of ±0.2°C (3σ) for Clinical Temperature Monitoring," IEEE Sensor J., vol. 16, no. 8, pp. 2272-2281, Feb. 2016.

35. B. Wang, M. K. Law, A. Bermak, "A Passive RFID Tag Embedded Temperature Sensor With Improved Process Spreads Immunity for a -30°C to 60°C Sensing Range," *IEEE Trans. Circuits Systs. I: Reg. Papers*, vol. 61, no. 2, pp. 337-346, Feb. 2014.

36. J. F. Creemer, "The effect of mechanical stress on bipolar transistor characteristics," Ph.D. dissertation, Delft University of Technology, Delft, The Netherlands, Jan. 2002.

37. F. Fruett, G. C. M. Meijer, and A. Bakker, "Minimization of the mechanical-stress-induced inaccuracy in bandgap voltage references," *IEEE J. of Solid-State Circuits*, vol. 38, no. 7, pp. 1288–1291, July 2003.

38. B. Yousefzadeh, K. A. A. Makinwa, "A BJT-Based Temperature Sensor with a Packaging-Robust Inaccuracy of ±0.3°C (3σ) from -55oC to +125oC After Heater-Assisted Voltage Calibration," *ISSCC*, Feb. 2017, pp. 162-163.

CHAPTER 7

Millimeter-Wave CMOS Power Amplifiers

Dixian Zhao[1]
and Patrick Reynaert[2]

1. Southeast University, China

2. KU Leuven, Belgium

The rapid growth of mobile data and the use of smart phones are making unprecedented challenges for wireless service providers to overcome a global bandwidth shortage. Millimeter-wave (mm-Wave) technology is widely considered as one of the key technologies that will continue to serve the consumer demand for increased wireless data capacity. Meanwhile, the advanced CMOS can now well operate in mm-Wave bands, permitting the integration of a full transceiver in a low-cost, high-yield technology. However, the design of mm-Wave transceivers in advanced CMOS still poses many challenges at device, circuit and architecture levels. In addition to generic difficulties, such as high-frequency operation and low active gain, mm-Wave designers must deal with issues like low breakdown voltage, high interconnect loss, unwanted mutual coupling, poor device matching, inaccurate PDK high-frequency models, strict design rules, long EM-simulation time, etc. At transmitter side, all these critical issues limit the output power and efficiency, prolong the design time and make it difficult to guarantee the success of tape-out.

This chapter focuses on realizing CMOS mm-Wave power amplifiers (PAs) towards more output power, higher efficiency and broader bandwidth. All the aforementioned design challenges will be described. Novel design techniques at mm-Wave will be presented, followed by prior-art PA and TX designs in advanced CMOS technologies.

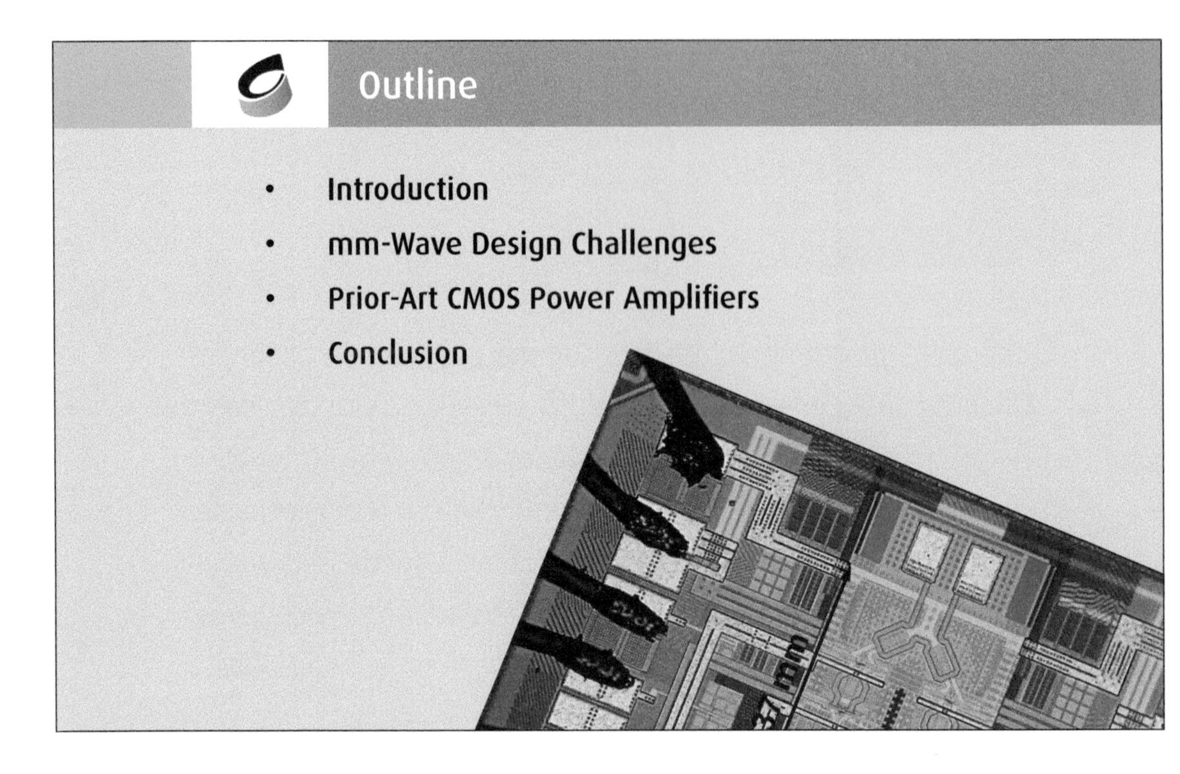

1  Outline

- **Introduction**
- mm-Wave Design Challenges
- Prior-Art CMOS Power Amplifiers
- Conclusion

To begin with, We will briefly introduce why mm-Wave techniques are important and how they will benefit us. It will be followed by a detailed analysis of the design challenges of making circuits at mm-Wave. After that, several prior-art mm-Wave PA designs will be described.

2  Global Mobile Data Traffic

Let's look at the measured and expected data traffic from 2012 to 2022. Thanks to the daily booming mobile services, the mobile data traffic grows rapidly. In 2013, the data traffic is first time reach more than one ExaByte every month. We can also see the mobile data traffic generated by smartphones are more than the sum of other devices. By the end of 2022, the mobile data traffic is expected to have 8-fold increase.

So how can we get there?

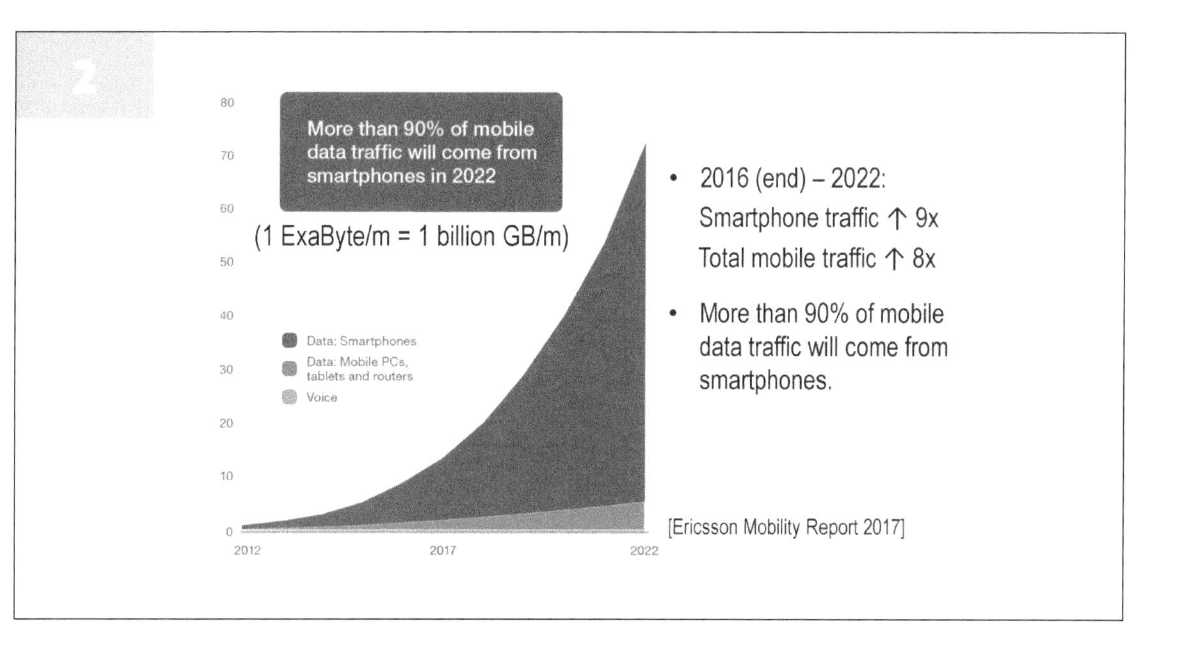

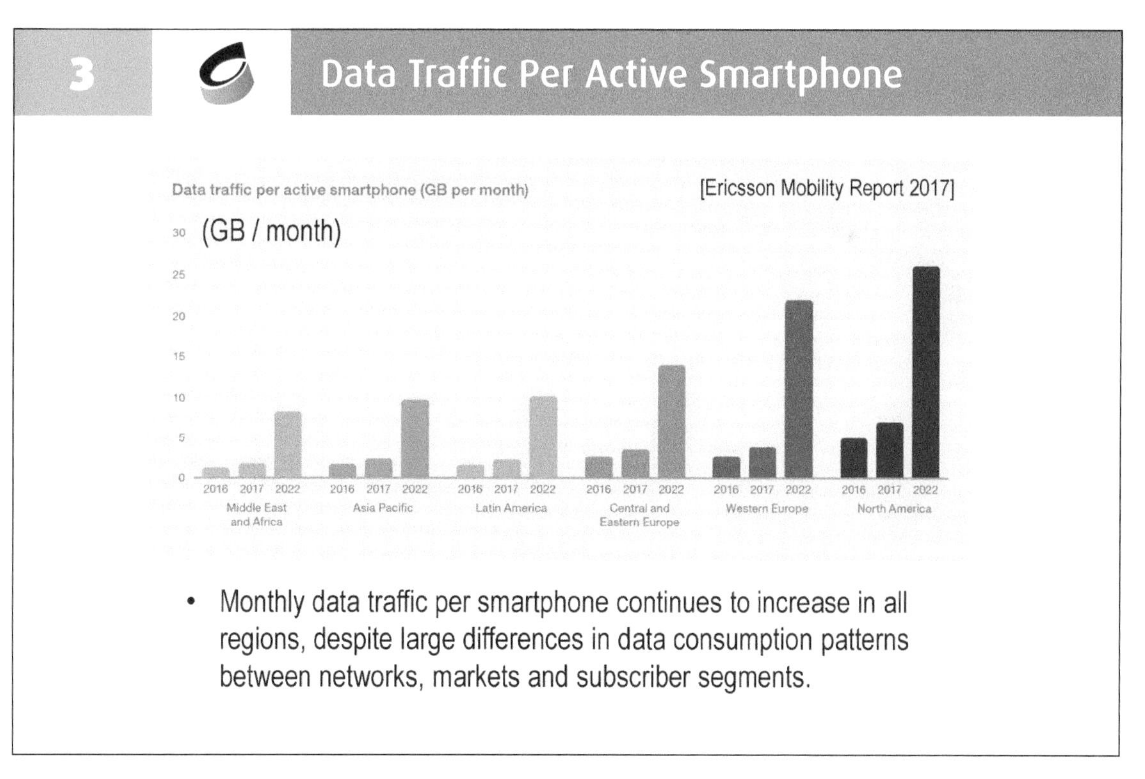

Monthly data traffic per smartphone continues to increase in all regions, despite large differences in data consumption patterns between networks, markets and subscriber segments. North America has the highest usage and traffic which is almost twice as high as Western Europe by the end of 2017. In 2022, North America will still be the region with the highest monthly usage at 26 GB, but other regions will be catching up.

Well, how can we reach such a fast increase in the consumed data traffic?

4 Microwave & mm-Wave Frequencies

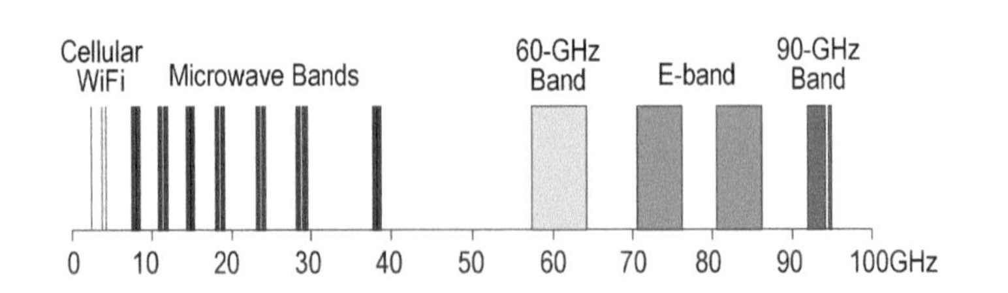

- **Microwave & mm-Wave Frequencies → MORE BW**
  - Communication: MORE data rate (~Gb/s)
  - Radar Sensing & imaging: BETTER resolution (~cm)

Due to the limited spectrum resource at low-GHz frequency, 3G and 4G networks are about to reach the theoretical limit. To have more data, we need more bandwidth. The total bandwidth available for all the cellular and WiFi standards is less than 1 GHz. In comparison, there are much more bandwidth available at microwave and millimeter-wave. For example, there are 9GHz bandwidth available at 60 and 10GHz in total at E-band. For the 5G bands around 26GHz and 39GHz, there are also multi-GHz bandwidth available. Moving to higher frequency is the way to go for our next mobile generation.

In addition, with more bandwidth, the millimeter-wave radar is able to achieve cm resolution.

5 Outline

- Introduction
- **mm-Wave Design Challenges**
- Prior-Art CMOS Power Amplifiers
- Conclusion

Obviously, millimeter-wave technology is great and important to us. However, we know that there is no free lunch. What are the design challenges for mm-Wave ICs?

6 Design Challenges at mm-Wave

The first challenge is the transistor has low power gain at mm-Wave. In the left figure, we can see the maximum power gain G_{MAX} drops at higher frequency. In addition, transistor has bad reverse isolation at mm-Wave and it is still not unconditional stable at 60GHz and E-band.

In the figure, we can also see the S_{21} is much lower than G_{MAX} when terminated by 50Ohm. That means you have to use passives to tune the circuit and make S_{21} approach G_{MAX} in the frequency of

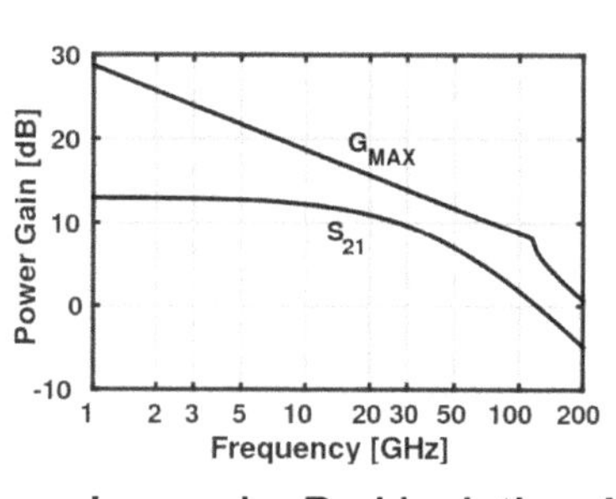

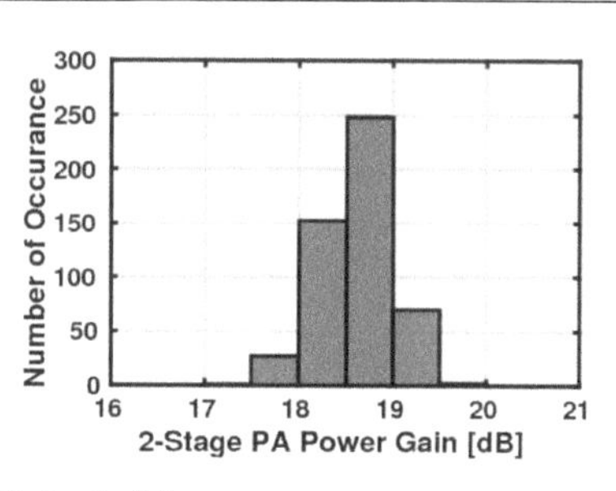

- **Low gain, Bad isolation, Not stable...**
- **Poor passives, Long EM-simulations...**
- **Sensitive to PVT, Mismatches, Strict DRC rules ...**

Small Design Margin, Long Design Cycle

interest. You can never get there because passive performance is usually poor at mm-Wave. Besides, you have to do many time-consuming EM-simulations to characterize the passive at mm-Wave.

mm-Wave circuits are also sensitive to process variations and device mismatch. In the right figure, we show the Monte Carlo simulation of a 2-stage PA that we designed. You can see the PA gain varies by more than 2 dB under corners and device mismatches.

When designing circuits in nano-meter CMOS process, you also have to deal with very strict DRC rules.

All these issues results in small design margin and long design cycle.

7 

More Challenges on mm-Wave PA (1)

- **Trade-off: Efficiency vs. Linearity**
  - peak PAE near P_{SAT}
  - 3-6dB back-off due to high PAPR
  - low P_{AVG} & low PAE_{AVG}

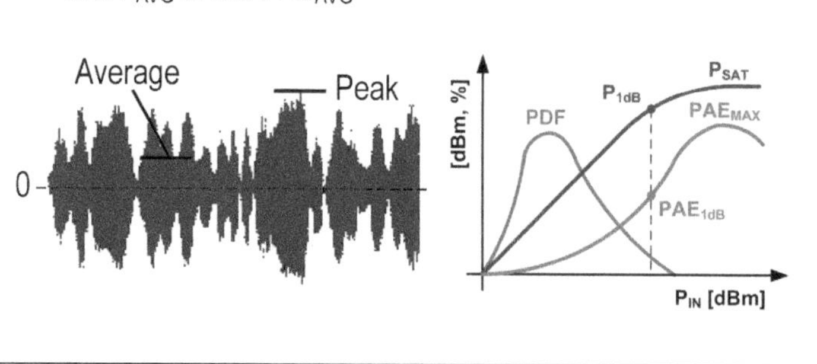

Things go even worse for the PA, which is one of the most challenging building block at mm-Wave.

The left figure shows the time domain signal that a PA needs to process. The average signal is usually much lower than its peak signal. We call this ratio as peak-to-average power ratio (PAPR). From the right figure, we can see the famous PA trade-off. The PA has the highest efficiency near its saturated output power. But to have sufficient linearity, the actual average output power is normally 3-6 dB lower than the 1-dB compression point. As a result, the PA normally has low average efficiency and low output power when transmitting modulated signals.

8 

More Challenges on mm-Wave PA (2)

- **Single-stage PA: low P_{OUT}**
  - fast transistor → low breakdown voltage
  - loss from long interconnects
  - large transistor → small transformer/inductor (low Q)

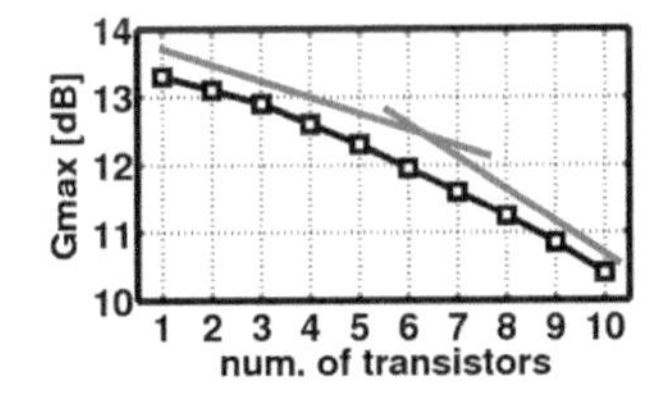

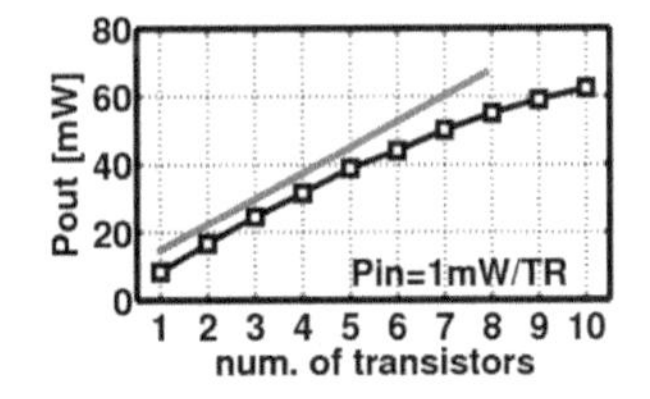

The other big issue is it is difficult to generate high output power at mm-Wave. At mm-Wave, you need fast transistors but fast transistor has low breakdown voltage which limits the output swing. And the worse thing is you cannot enhance the power by increasing the transistor size as the loss of the long interconnects at mm-Wave is high.

From the figure, you can see the gain drops fast with the increase of the normalized transistor size and the increase in output power will also saturate.

Apart from that, when you use large transistor, you have to use small transformer and inductor for the tuning while small transformer and inductor have reduced Q-factor. This makes thing even worse.

9 

CMOS Technology

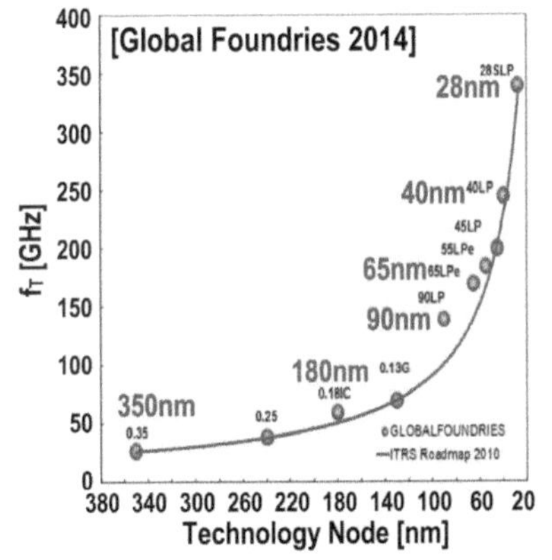

- **Moore's law → Integration**
- **Planar device → Yield**
- **Low cost in large volume**
- **Speed: f_T , f_{MAX} ↑↑**
- **Thick top metals**
- **Substrate: 10-15 Ω·cm**

All the PAs that will be discussed in the following slides are implemented in CMOS. Then Why is CMOS?

Golden Moore predicted that the number of the transistors in a chip doubles every 2 years. CMOS technology has been the key to take the challenge of Moore's law. As a result, within a fixed chip area, more transistors can be integrated and more functions can be realized in CMOS. It is also a planar technology which has high yield. More integration and improved yield makes CMOS a low-cost technology in large production volume.

In the left figure, we can see along with the transistor scaling, the speed of the transistor also

CMOS Technology

increase. This makes it possible to design mm-Wave circuits in CMOS. The 1st generation of 60-GHz circuits uses 90nm CMOS where f_T is about 120GHz. Thanks to scaling, transistor becomes faster and faster. Today, CMOS has become the mainstream technology for mm-Wave circuits.

We would like to emphasize that CMOS technology has also been improved for RF design. Now we have several thick top metals and substrate resistivity is much higher than before. About 10 years ago, this number is 10-100 times lower. So thanks to the integration, we RF people do make an influence on the technology development.

10

Outline

- Introduction
- mm-Wave Design Challenges
- **Prior-Art CMOS Power Amplifiers**
- Conclusion

With the challenges in mind, let's see how we can tackle them in CMOS technology.

11

Objective

- **Realize mm-Wave PAs towards**
  - –higher efficiency → low power
  - –more output power → far distance
  - –broader bandwidth → high data rate
  - –compact area in CMOS → low cost

In the remaining slides, we will focus on realizing mm-Wave PAs in CMOS technology towards higher efficiency for low power consumption, more output power to reach far distance, broader bandwidth for high data rate. All the designs have to be realized within a compact chip area to reduce the cost. Many useful techniques will be discussed to achieve these goals.

12  I. CMOS 60-GHz Dual-Mode PA

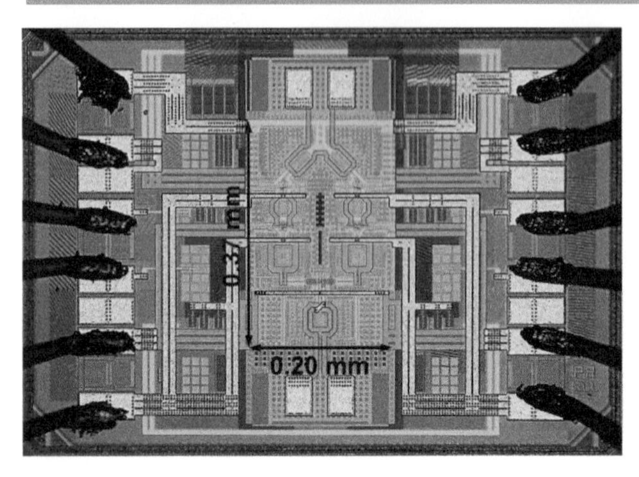

- D. Zhao et al., "A 60-GHz Dual-Mode Class AB Power Amplifier in 40-nm CMOS," IEEE Journal of Solid-State Circuits (JSSC), vol. 48, no. 10, pp. 2323-2337, 2013.
- D. Zhao et al., "A 60 GHz Dual-Mode Power Amplifier with 17.4 dBm Output Power and 29.3% PAE in 40-nm CMOS," in Proceedings of ESSCIRC, 38, pp. 337-340, Bordeaux, 2012.

Let's move on to the first design. It is about a CMOS 60-GHz dual-mode power amplifier. By dual-mode, it means the PA has two modes: high output power mode or low output power mode.

It has been published in ESSCIRC 2012 and JSSC 2013.

13  Device: Transistor Layout

We start our optimization from the transistor layout. We know that device parasitics and long interconnects will limit the PA output power and efficiency at mmWave frequencies. To minimize these effects, we propose a transistor layout. The key idea is we try to minimize the overlap between gate and drain and the overlap between gate and source to reduce the parasitic capacitance. In addition, the gate and source are connected out from both sides to reduce the parasitic resistance. After the layout, we extract all the transistor parameters. We found important parasitics, C_{GS} and C_{GD} are only slightly increased. The value of C_{DS} is almost doubled. But since C_{DS} can be tuned out by matching circuits. It has much less effects. So these is also trade-off in layout design. The simulated maximum unilateral gain is reduced by 2.5 dB and maximum stable gain is reduced by only 1.3 dB. To give you a number to compare, the conventional layout reduce the MSG by about 3dB, which shows the benefits of our proposed layout.

Another key advantage of the proposed layout is it can be easily duplicated for high output power. As shown in the right figure, five transistor cells are connected seamlessly through wide metal tracks or thick metal lines. I put the complete layout in EM simulator and found out the unilateral gain is only reduced by 0.7 dB. How good is 0.7dB? It corresponds to if you put 0.5 **Ω** lump resistor and 1.5 pH inductor in series with the source node to degenerate the transistor.

The two-dimentional layout technique also eases the connections to the input and output matching circuits as the drain and gate terminals appear at two ends and it is good for power amplification.

13

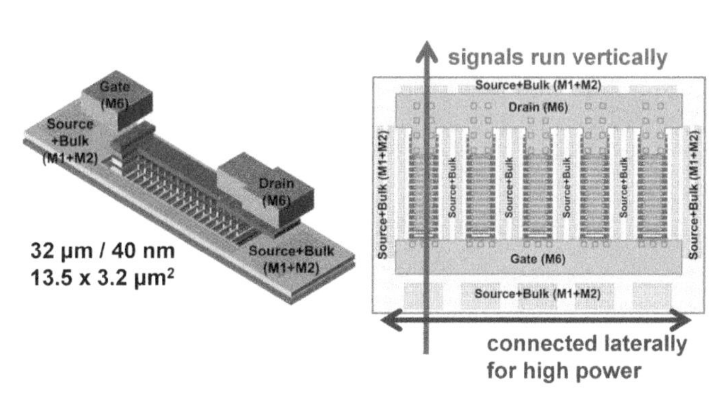

- $C_{GS}\uparrow < 20\%$, $C_{GD}\uparrow < 10\%$, $C_{DS(B)}\uparrow 2x$
- Easy to duplicate, easy to route signal lines.

14 

Amplifier: Neutralized CS

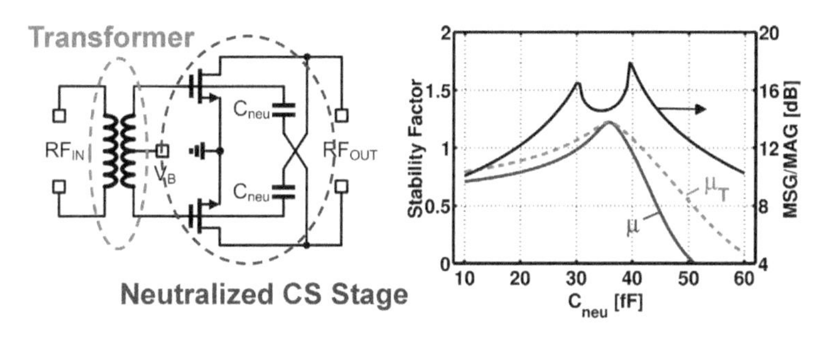

- Neutralization improves stability, gain & S_{12}.
- $\pm$20% C_{neu}: unconditional stable w/ transformer .

Neutralization technique is also used to compensate the capacitance C_{GD}. We can see by proper choosing the value of C_{neu}, both the power gain and stability can be improved. We can also see over-compensating the C_{GD} is not good which will further degrade the stability.

To encounter the process variations, such as C_{neu} deviates from its desired value, additional margin should be provided to ensure stability. You can add extra series or shunt resistance to further stabilize the transistor, but that compromises the power gain or complicates the layout design. Here, we consider that the on-chip passives are lossy, so the stability of amplifier stage can be improved if the inherent loss resistance of the passives is included in the simulation during the design of the single amplifier stage. In the figure, we see by including the loss resistance of the transformer, the amplifier stage is unconditional stable even when C_{neu} changes by plus minus 20%.

This active and passive co-design technique simplifies the design procedure and optimizes the power gain and stability simultaneously.

15 

Dummy-Prefilled Transformer

- **Strict DRC rules in nm-CMOS**
  - design for manufacture (DFM)
- **Cautious metal filling**
- **Negligible impact with 10% density**
  - Q↓ < 5%, ~L & ~k_m

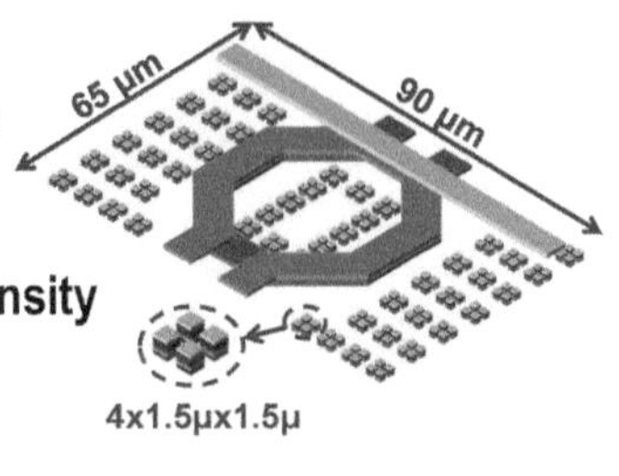

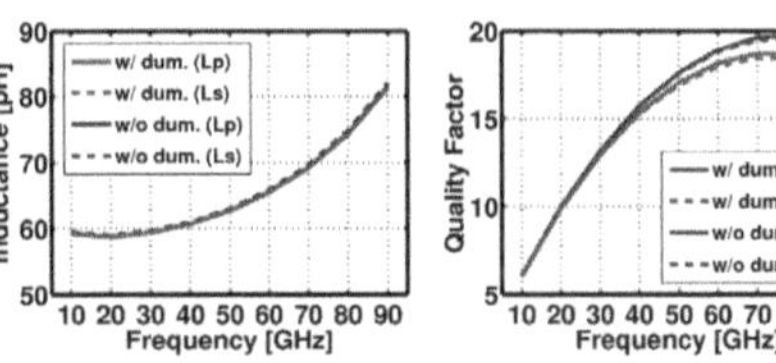

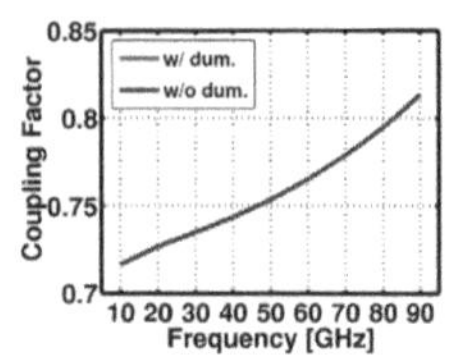

Another important concern is metal dummies have to be filled inside the transformer to fulfill the design rule. These dummies have relatively big influence on mmWave passive components. The actual influence very much depends on how much dummies you put inside and around the transformer and how big those dummies are. To relax the influence, the minimum amount of metal dummies are prefilled manually and carefully inside and around the transformer. Due to our cautious dummy filling, the metal dummies have negligible impact on the self-inductance and the coupling factor of the transformer. The Q-factor is reduced by less than 5%.

16 

Dual-mode Power Amplifier

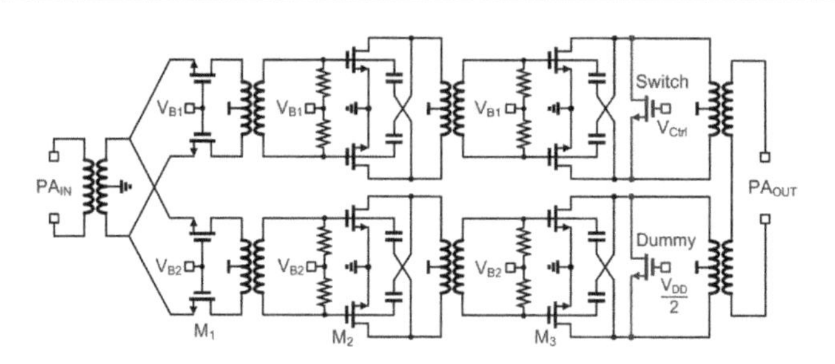

- **3-stage Class AB power amplifier.**
- **WiGig & IEEE802.15.3c: low power mode.**

This slide shows the schematic of the dual-mode PA with a power combiner at the output. Neutralization technique is used in the output stage and driver stage while the input stage is realized by common-gate amplifier to simplify the input matching. Matching networks are designed based on the load-pull simulations. Moreover, the pad capacitance and output transmission line transfer the 50Ω load to the desired impedance required by the transformer to maximize its efficiency. The superior power gain of our optimized amplifier stage provides extra design margin. It allows the PA to be biased in Class AB where the extra gain can be traded for back-off efficiency.

To further reduce the power consumption, the PA can be switched to LP mode by tuning off one unit PA and saves about 50% DC power. The low power mode is also defined in WiGig and IEEE802.15.3c standards, which can support relatively low data rate or shorter range communication. We introduce a switch to short the output of the non-operating PA, which help improve the combiner efficiency in LP mode, as explained in the next slide.

17 

Low-Power Mode

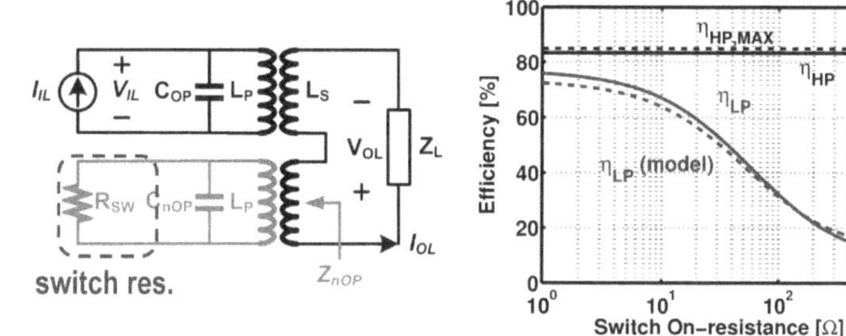

- **Higher loss in output combiner**
  - $k_{m,LP} = k_m/2^{0.5}$, load changed, voltage division
- **MOS switch ↑η_{LP} from 10% to 76% in low-power mode**
- **R_{ON} = 10 Ω → 67%**

Here I show a simple model of PA output stage in the low-power mode. Upper side is the operating PA and downer side is the non-operating PA. The combiner efficiency in LP mode is reduced due to three reasons. (1) The effective coupling factor is reduced as only one of the primary coils delivers power to the load. (2) The load impedance seen by the operating transformer is changed, which deviates from optimum load impedance. (3) Some portion of the power delivered by the operating PA is lost due to the voltage division between the load impedance and the impedance seen at the secondary side of the non-operating transformer. The latter two loss mechanisms can be alleviated by shorting the output stage of the non-operating PA using a switch transistor. In the right figure, it shows efficiency increases from 12% to 76% by lowering the on-resistance of the switch. To avoid large parasitic capacitance introduced by the switch, an R_{ON} of 10 Ω is used in the design, which gives 67% efficiency. The simulated results of the combiner circuit and the simplified model are in good agreement.

We can also see the combiner efficiency in HP mode is very close to the maximum power transfer efficiency, which is above 80%.

18 

60-GHz Class AB Dual-Mode PA

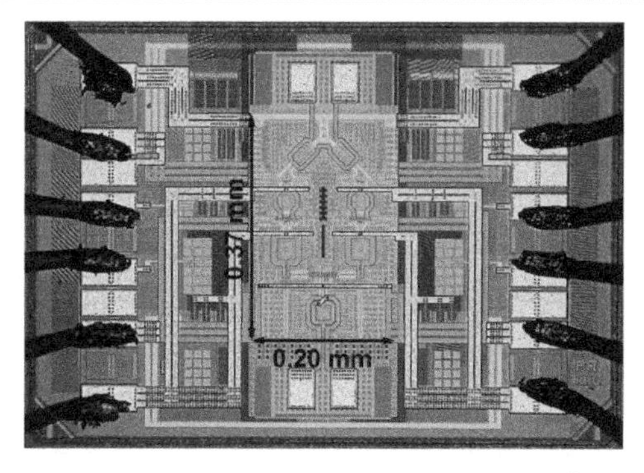

- **40-nm CMOS LP, core area: 0.074mm^2**
- **WiGig & IEEE802.15.3c: low-power mode**
- **1st application of dual-mode technique at mm-Wave**

The 60-GHz Class AB dual-mode power amplifier is implemented in 40-nm CMOS LP technology and the core area is less than 0.1mm^2. This work is the first reported dual-mode PA at millimeter-wave frequencies.

19

S-parameter Measurement

Here we show the S-parameter measurements. The small-signal gain has a peak value of about 24 dB at 63 GHz in HP mode. The gain difference between two modes is small, which is not easy to achieve if you refer to some dual-mode PA designs at low-GHz frequency. Compared with the simulation, the measured S_{21} is about 1 dB lower and there is a slight shift in frequency. In the right figure, the measured S_{11} is better than -20 dB in LP mode while it is about -10 dB in HP mode. It is because we optimize the input matching for LP mode to balance the performance. The S_{12} is better than -40 dB for both modes in the 60-GHz band. The design is unconditional stable.

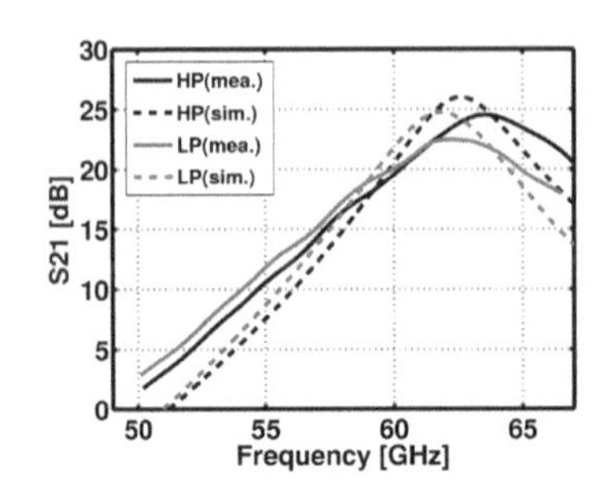

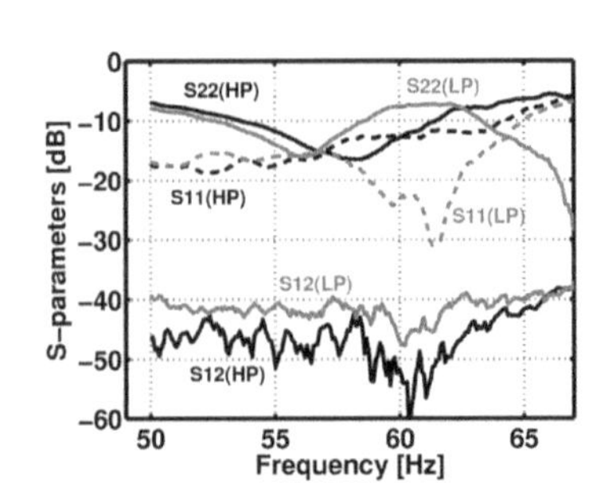

- S_{12} < -40 dB, unconditional stable.

20

Class AB

Here we compared measured PA performance with different bias settings. With reduced bias voltage, the DC power consumption also drops while there is little change in 1dB compressed power. So we can achieve better back-off efficiency if we operate the PA in Class AB.

We also noticed the gain decreases by about 7dB with reduced bias voltage. I would like to mention that the gain cannot be further reduced because a 60-GHz mixer also consumes quite some power and its linear efficiency is usually less than 2%. To have an overall low-power transmitter, the PA gain has to be larger than 15 dB.

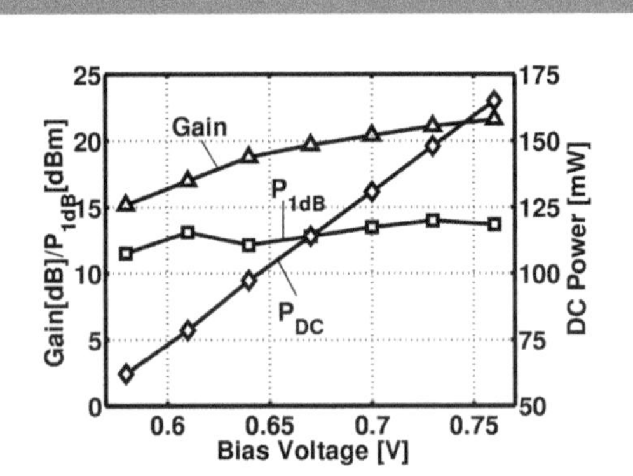

- Class AB: low P_{DC} and high back-off PAE.
- Gain > 15 dB for an efficient TX (η_{Mixer} < 2%).

21

The PA in LP mode is measured by switching off one unit PA. The figure shows when the switch at the output of the non-operating PA is open, both the gain and P_{OUT} are reduced by about 5 dB. So the switch greatly improves the combiner efficiency in low power mode. The PA in LP mode achieves a gain of 21.6 and P_{1dB} of 9.5 dBm.

Low Power Mode

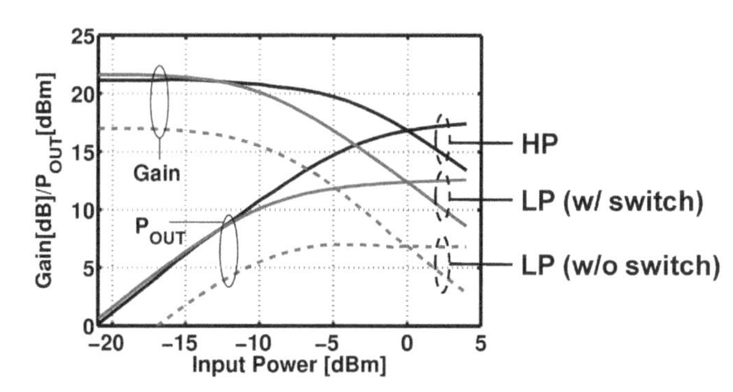

- Switch improves the Gain & P_{OUT} > 5 dB.
- Gain: 21.6 dB, P_{1dB}: 9.5 dBm (61 GHz).

22

In this slide, I show the measured PAE to demonstrate the benefit of Class AB and dual-mode operations. We can see with Class AB, the PAE at P_{1dB} is improved by about 6%. At 6-dB back-off, it is improved by 7% and nearly 3 times better. When the PA in low power mode, it also saves 50% dc power.

Measured PAEs

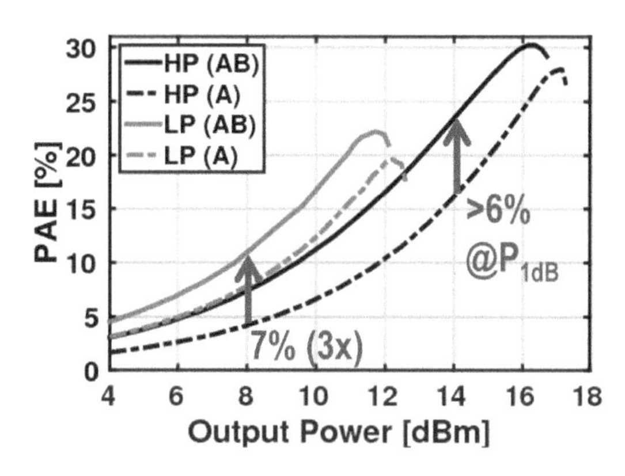

- Enhance back-off η (~3x) & save 50% dc power.

23  Performance Summary

	Gain [dB]	P_{SAT} [dBm]	P_{1dB} [dBm]	PAE_{MAX} [%]	PAE_{1dB} [%]	$P_{DC,1dB}$ [mW]
Class A (HPM)	21.2	17.4	14.0	29.0	16.3	153
Class AB (HPM)	17.0	17.0	13.8	30.3	21.6	106
Class A (LPM)	21.6	12.6	9.5	19.6	11.1	80
Class AB (LPM)	16.8	12.1	9.1	22.2	14.1	56

V_{DD} = 1V; active area = 0.074 mm^2.

Here I summarize our PA performance in different operating modes. While achieving similar output power, the PA in Class AB has about 5% better efficiency at P_{1dB} as it consumes less DC power. When the PA is switched to low-power mode, it saves about 50% power consumption.

24  Comparison

	Tech. [nm]	V_{DD} [V]	P_{SAT} [dBm]	P_{1dB} [dBm]	PAE_{MAX} [%]	PAE_{1dB} [%]
JSSC 13	40	1.0	17.0	13.8	**30.3**	**21.6**
ISSCC13	40	1.1	10.2	8.0	22.5	**16.0**
ISSCC12	40	1.1	10.5	10.2	10.8	10.4
RFIC10	45	2.0	14.5	11.2	14.4	7.0
ESSCIRC09	65 (**SOI**)	1.8	14.5	12.7	**25.7**	**15.0**
RFIC10	65	1.0	10.6	6.8	18.0	7.7
ISSCC11	65	1.0	18.6	15.0	15.1	6.8
ISSCC10	90	1.2	19.9	18.2	14.2	12.8
JSSC09	90	1.2	12.5	10.2	19.3	11.0
RFIC09	90	1.0	11.6	10.1	11.5	10.0

We compare the PA performance in the high-power mode with the state-of-the-arts in different technology nodes. Our PA operates at 1V. It achieves 17dBm output power. It has about 5% higher PAE than the best reported PAE which is implemented in SOI technology. The PAE at 1dB compression point of our design is also 5% better than other designs.

Summary: 60-GHz Dual-Mode PA

Design techniques		60-GHz Class AB PA
Optimal transistor layout		30% PAE_{MAX}, 20% PAE_{1dB}
T./Amp. co-design		17 dBm P_{SAT}, 14 dBm P_{1dB}
T. optimized for η_{MAX}		>17 dB Gain for eff. TX
Dummy prefilled T.		1st dual-mode PA
Switch for η_{LP}		Long-term reliability (1V)

Let's summarize the work of the 60-GHz dual-mode PA. In this work, we propose several design techniques to optimize the transistor layout, amplifier stage, transformer and output combiner, which leads to a highly efficient PA with relatively high output power. It has 20% PAE and 14dBm linear output power. It achieves sufficient gain, which is needed for a high efficient transmitter. It also realizes the 1st dual-mode PA at 60 GHz for low-power applications. The long-term reliability of this PA is also validated.

26

II. CMOS 60-GHz Outphasing Transmitter

- D. Zhao, S. Kulkarni, P. Reynaert, "A 60-GHz outphasing transmitter in 40-nm CMOS," IEEE Journal of Solid-State Circuits (JSSC), vol. 47, no. 12, pp. 3172-3183, 2012.
- D. Zhao, S. Kulkarni, P. Reynaert, "A 60GHz Outphasing Transmitter in 40nm CMOS with 15.6dBm Output Power," in IEEE International Solid-State Circuits Conference (ISSCC), Digest Tech. Papers, Feb. 2012, pp. 170-171.

The 2nd design is about a 60-GHz outphasing transmitter. It has been published in ISSCC 2012 and JSSC 2012.

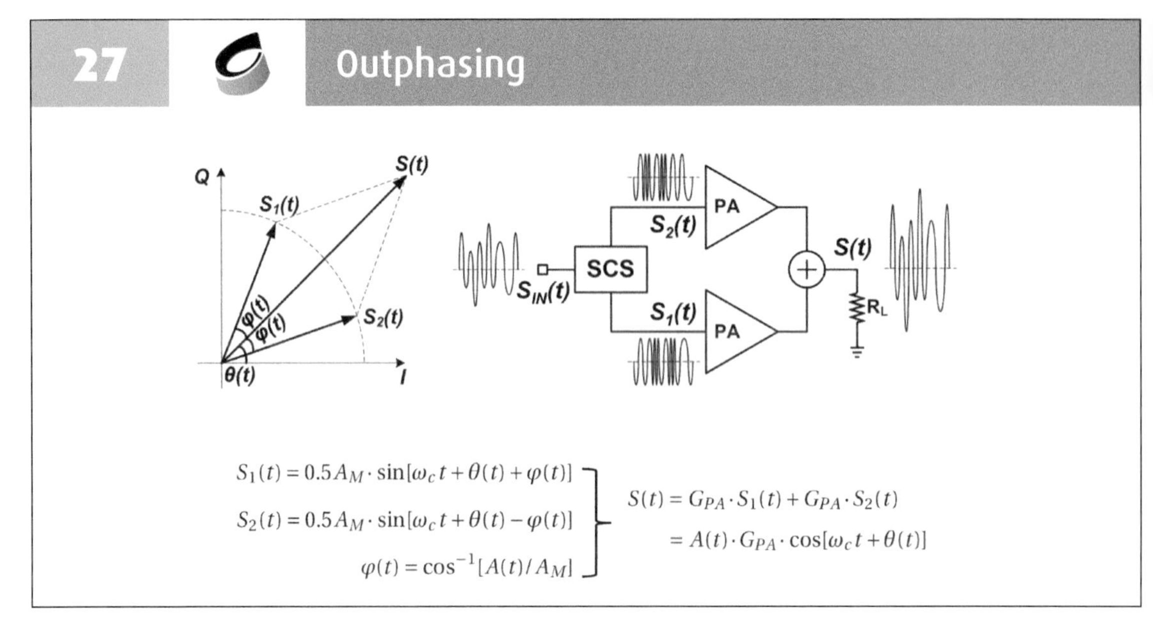

What is outphasing? This slide explains the basic principle of the outphasing PA.

In an outphasing system, any arbitrary amplitude and phase modulated input signal $S_{IN(t)}$ can be separated into two constant-envelope signals, $S_{1(t)}$ and $S_{2(t)}$ by an signal component separator S_{CS}. SIN has both amplitude modulation and phase modulation. S_1 and S_2 have constant envelope, we call them outphasing signals, represented by these two equations with the same constant amplitude. They will be amplified by the PA and combined at the output. After combining, the amplitude modulation can be reconstructed.

It is more clear if you look at the figures on the left. G_{PA} is the gain of the PA. S_1 and S_2 are constant-envelope phase-modulated signals. $\theta(t)$ is the original phase modulation. $\phi(t)$ is the outphasing angle. $\phi(t)$ equals inverse cosine of A(t) over AM. So the phase of S_1 is $\theta(t)+\phi(t)$ and the phase of S_2 is $\theta(t)-\phi(t)$. After vector-summing these two signals, the amplitude modulation can be restored. With smaller $\phi(t)$, you get larger amplitude. With larger $\phi(t)$, you get smaller amplitude. So the amplitude modulation is determined by the outphasing angle.

As there is no amplitude variation in the signals S_1 and S_2, nonlinear and highly efficient PA can be used. This gives outphasing PA or transmitter huge benefits over other transmitter modulation technique.

28  Transmitter modulation

Here we compare the outphasing with other commonly used transmitter modulation techniques.

On the left side is the conventional I/Q modulator, which usually has low efficiency because both I and Q paths have variable envelope.

The polar transmitter in the middle makes the situation better. Its phase path has constant envelope and it improves the overall efficiency by doing the supply modulation.

Our work, the outphasing transmitter, is shown on the right side. As I mentioned, $S_{1(t)}$ and $S_{2(t)}$ are two constant-envelope signals. So the PAs in outphasing system can always operate at peak output power and peak efficiency and at the same time, sufficient linearity can be maintained. This attribute of outphasing makes it quite suitable for mm-Wave applications.

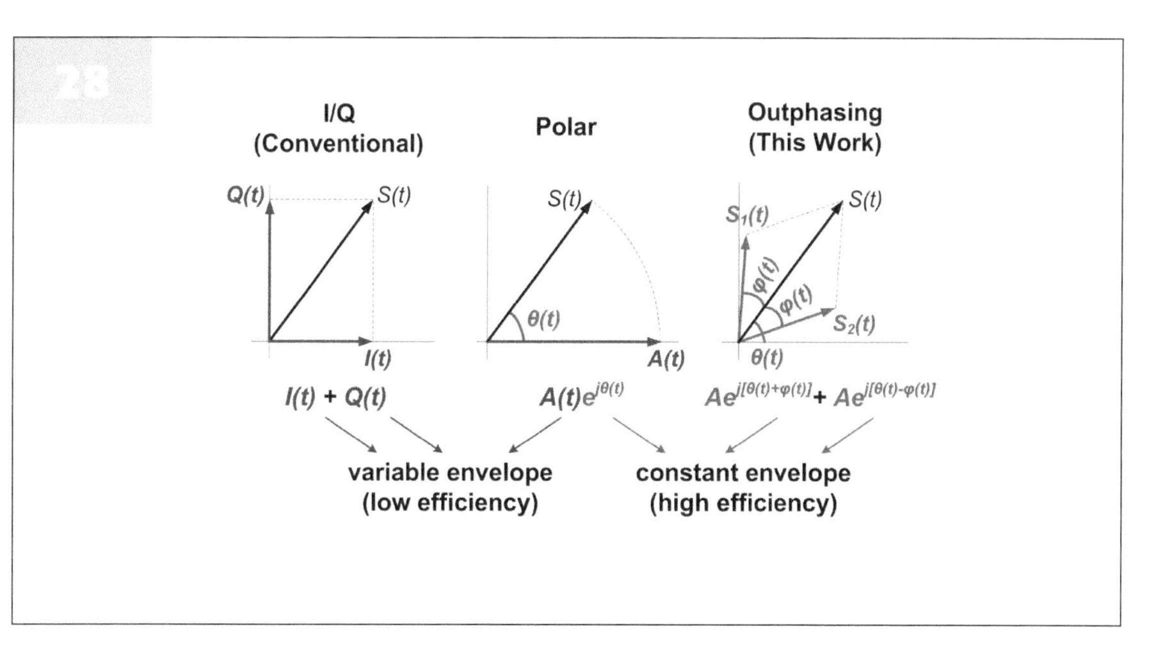

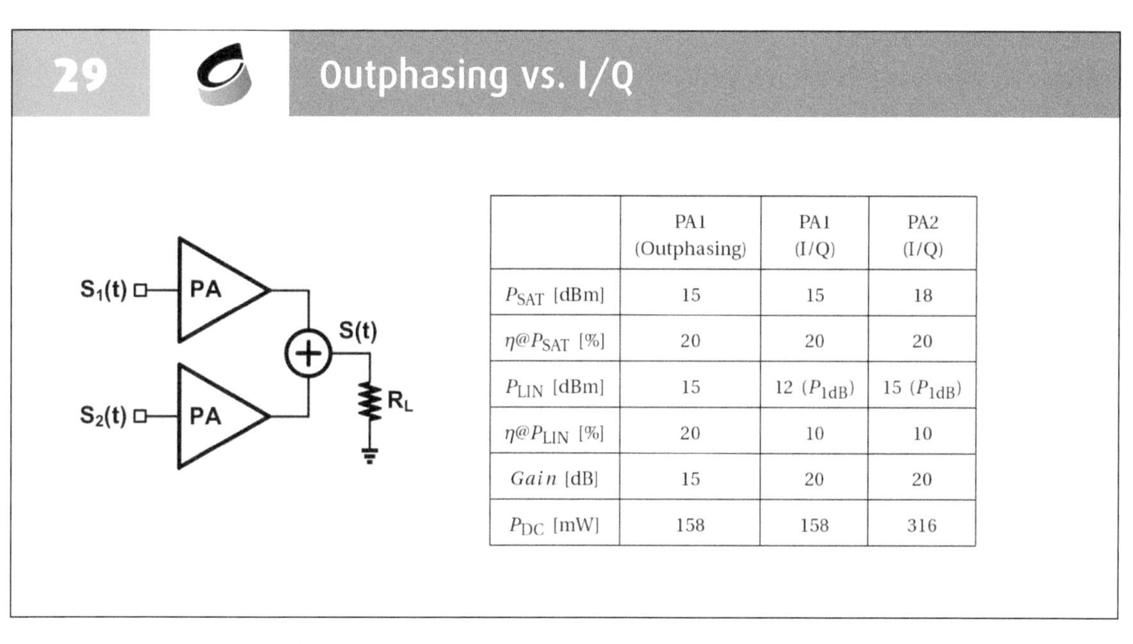

	PA1 (Outphasing)	PA1 (I/Q)	PA2 (I/Q)
P_{SAT} [dBm]	15	15	18
$\eta@P_{SAT}$ [%]	20	20	20
P_{LIN} [dBm]	15	12 (P_{1dB})	15 (P_{1dB})
$\eta@P_{LIN}$ [%]	20	10	10
Gain [dB]	15	20	20
P_{DC} [mW]	158	158	316

So the outphasing TX has linear output power and efficiency benefits at mm-wave frequencies. To prove our concept, I give an example here. Let's assume we have one power combining PA. As the power combining PA has two unit PAs with an output power combiner, it can work both as a outphasing PA and a conventional I/Q PA. This power combining PA can be realized by PA1 or PA2. For the PA1, it can work both in outphasing and I/Q. PA1 has a P_{SAT} of 15dBm and PA2 18dBm. They both have 20% efficiency. The outphasing PA has 5-dB lower in gain because it operates in saturation. PA1 consumes half the power consumption. All the numbers listed in this table are based on the state-of-the-art published PAs.

So we compare three cases: PA1 with 15dBm output power in outphasing, PA1 with 15dBm output power in I/Q and PA2 with 18dBm output power in I/Q.

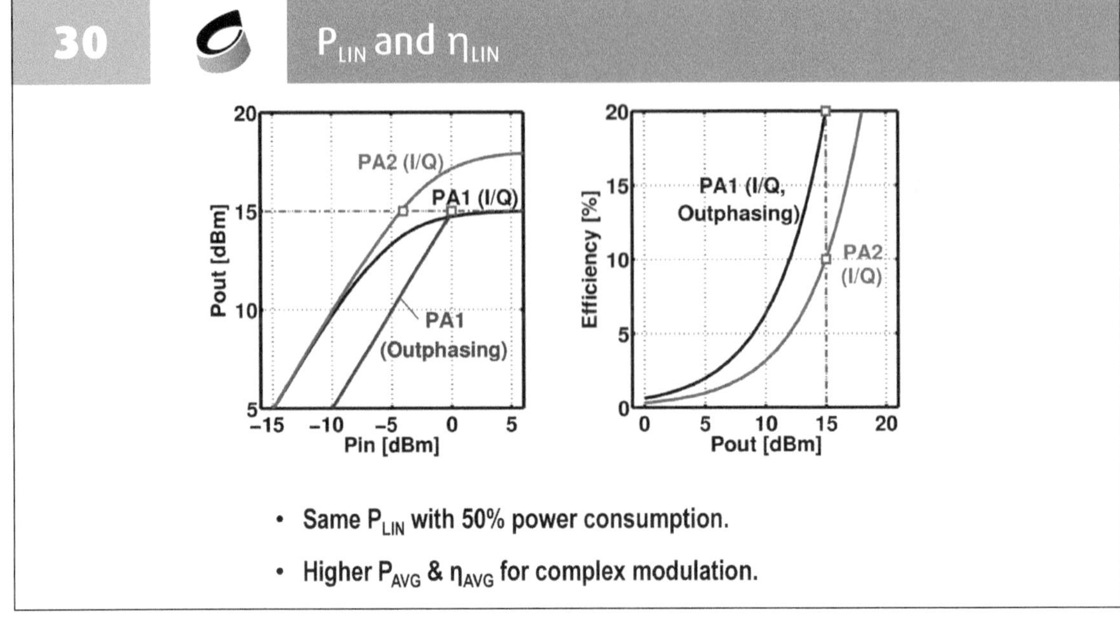

We plot the results of these three PAs. PA2 has 3-dB higher output power and it also has 3-dB higher P1dB than PA1. We define the max. linear output power of the I/Q PA is its P_{1dB}. So for the PA2, it is 15 dBm. For the PA1, it is 12 dBm. The PA1 in outphasing mode operates linearly up to the maximum output power, so the maximum linear output power of the outphasing PA is also 15 dBm. So the outphasing PA achieved the same linear output power but only consumes 50% power consumption.

On the right plot, we plot the efficiency. PA1 and PA2 have the same peak efficiency. The PA1 in outphasing and I/Q modes has the same efficiency. In a few slides, I will show the outphasing PA actually has better back-off efficiency. Here we assume they have the same efficiency. We can see to have the same linear output power of 15dBm. The outphasing PA has 10% more efficiency.

Because of this, the outphasing TX has much better back-off efficiency and such efficiency enhancement is not attainable by other techniques at mm-wave frequencies.

As a conclusion, the ouphasing TX can achieve higher linear output power with higher efficiency, which means when transmitting modulated signals, outphasing achieves much better average output power and efficiency. This is one of the major reasons we propose outphasing for mm-wave applications.

31  Load Modulation

Another benefit of outphasing is it has better back-off efficiency, which is caused by the effect called "load modulation". It can be explained by this figure.

The saturated PA output stage is modeled by a voltage source with an output impedance of R_O. The parasitic capacitance C_P of the output stage is resonated out by the magnetizing inductance L_M of the ideal coupled transformer, which does not affect the analysis.

The two PAs are coupled through transformer. Due to the interaction between two PAs, the load impedances seen by each PA vary with the outphasing angle, which are given by the two equations in this slide.

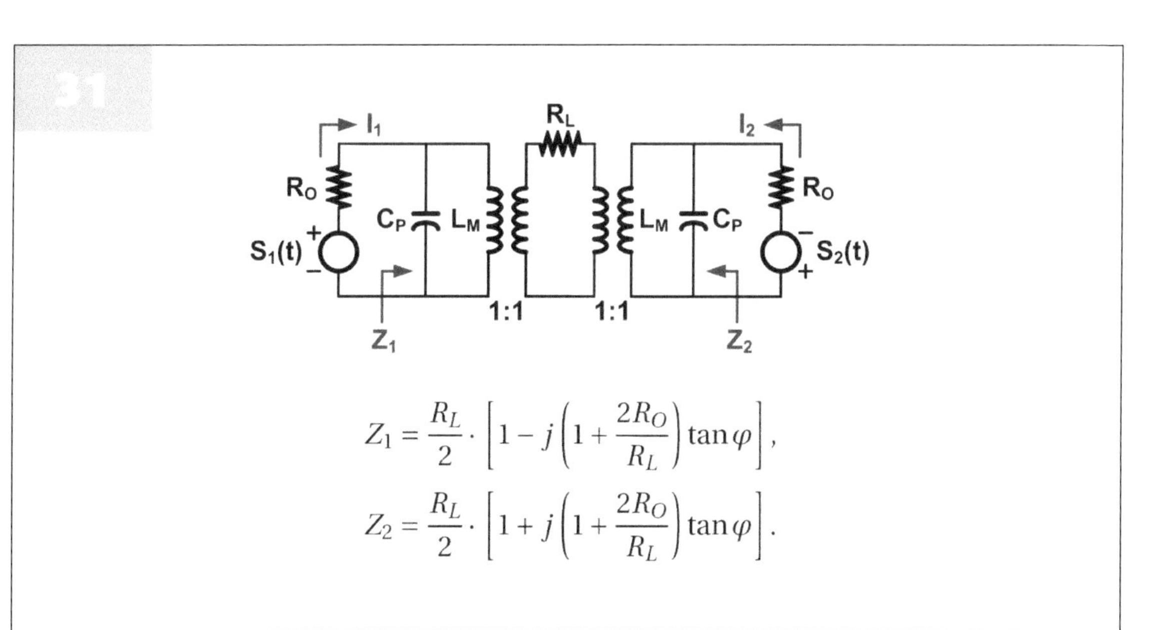

$$Z_1 = \frac{R_L}{2} \cdot \left[1 - j\left(1 + \frac{2R_O}{R_L}\right)\tan\varphi\right],$$

$$Z_2 = \frac{R_L}{2} \cdot \left[1 + j\left(1 + \frac{2R_O}{R_L}\right)\tan\varphi\right].$$

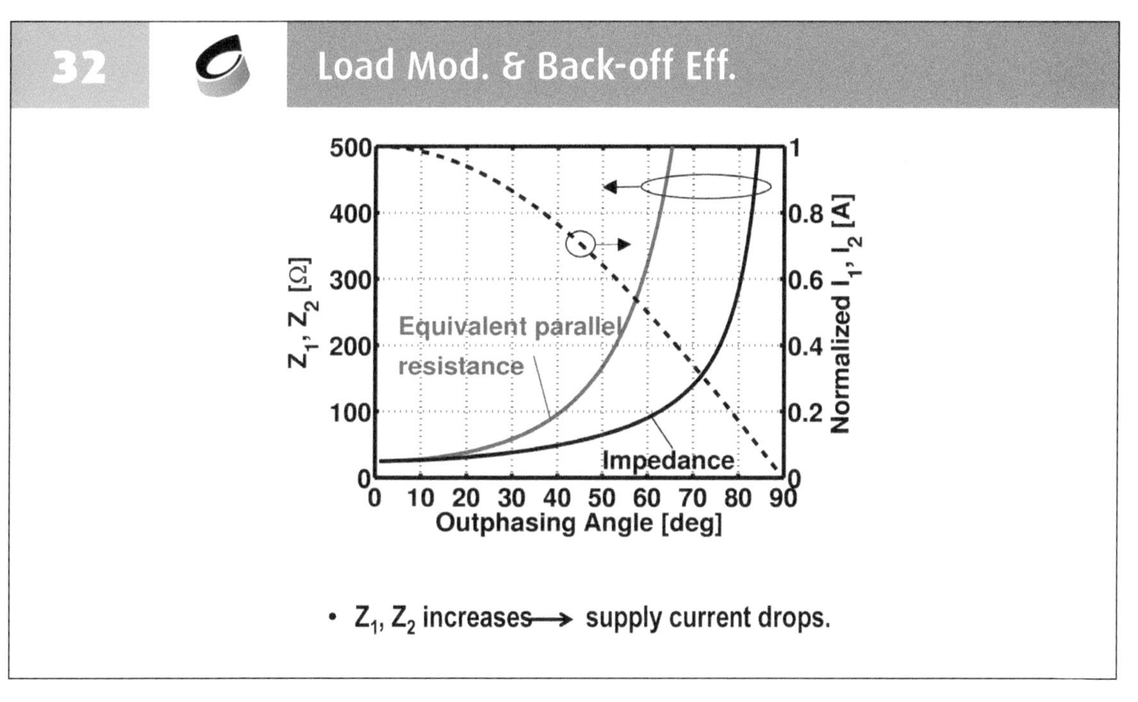

If you plot the impedance, you can see it increases at larger outphasing angle, which means for the same voltage swing, you need less supply current. I also plot the supply current in this figure, you can see it drops at large outphasing angle. This load modulation effect reduces power consumption at back-off, giving the outphasing PA back-off efficiency benefit.

33

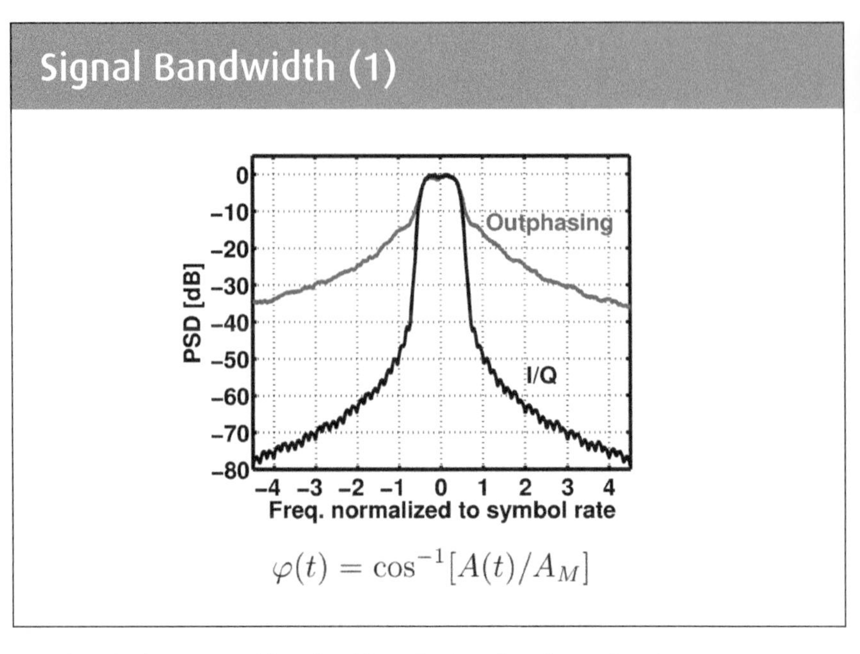

In order to realize the outphasing TX, we need to do a bit more thing. This slide shows the spectrum of the outphasing signal before the combiner. The spectrum of the outphasing signals expands. This is mainly due to the nonlinear inverse cosine operation involved when generating outphasing signals.

Therefore, it is instructive to identify the required bandwidth in the baseband signal paths.

34

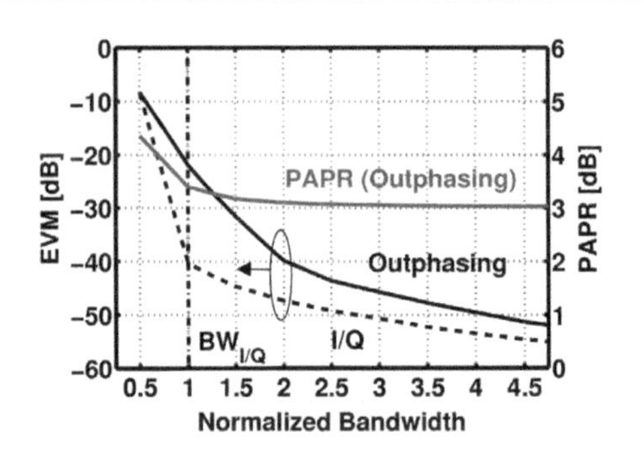

- 2 X BW$_{I/Q}$ for outphasing
- Trade linearity for bandwidth
- No power penalty

We build some system models for both I/Q TX and outphasing TX to investigate this issue. The normalized bandwidth required by I/Q TX is 1. If you look at the figure, when the outphasing TX has a bandwidth of 2, it can achieve the same EVM as I/Q TX, -40 dB. This means the circuit needs to process the outphasing signal, especially the analog baseband circuit, need a bandwidth of 2.

You can also see at the bandwidth of 2, the PAPR of the outphasing signal flattens out to 3 dB. Note that the signal we process is the pulse-shaped16QAM. The PAPR of pulse-shaped16QAM in our case is 7.4dB. So the linearity requirement of the analog baseband circuits is relaxed.

The outphasing circuit trade linearity for the bandwidth. Considering the power consumption of analog baseband circuits is in proportion to bandwidth and in exponential proportion to linearity. The analog baseband circuit for both outphasing and I/Q transmitters end up with comparable power consumption in analog baseband.

On the other hand, the bandwidth of the mm-wave front-end circuits is not a problem since the carrier frequency is at 60GHz.

35

Gain and phase mismatches of the outphasing TX will deteriorate the EVM and cause spectral regrowth. We assume we have some amplitude and phase mismatches in one path. After the combining, you can find at the output, besides the desired signal, you also have some other components, which is related to the outphasing angle.

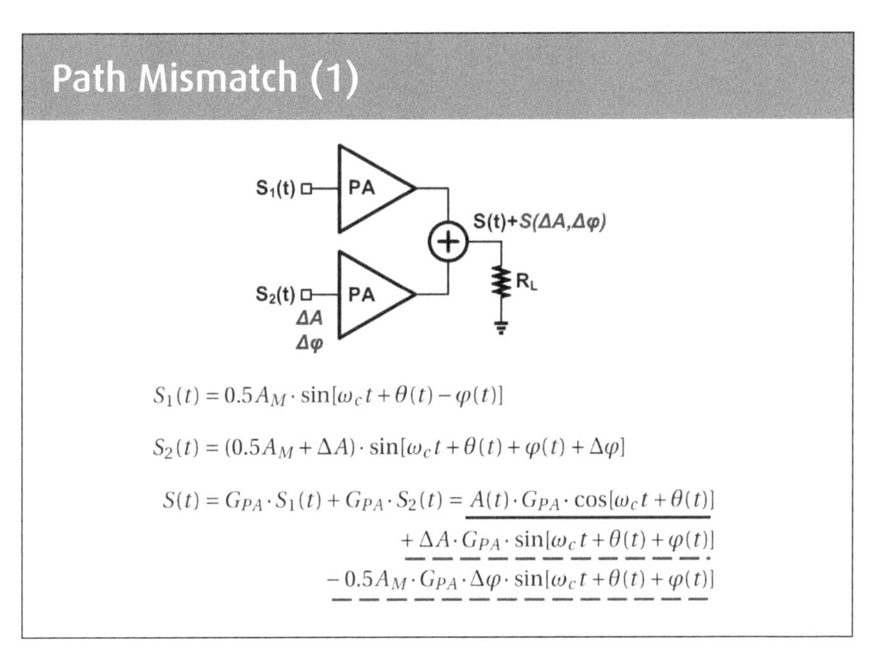

$$S_1(t) = 0.5A_M \cdot \sin[\omega_c t + \theta(t) - \varphi(t)]$$

$$S_2(t) = (0.5A_M + \Delta A) \cdot \sin[\omega_c t + \theta(t) + \varphi(t) + \Delta\varphi]$$

$$S(t) = G_{PA} \cdot S_1(t) + G_{PA} \cdot S_2(t) = A(t) \cdot G_{PA} \cdot \cos[\omega_c t + \theta(t)] + \Delta A \cdot G_{PA} \cdot \sin[\omega_c t + \theta(t) + \varphi(t)] - 0.5A_M \cdot G_{PA} \cdot \Delta\varphi \cdot \sin[\omega_c t + \theta(t) + \varphi(t)]$$

36

I plot the effect of phase mismatch between two outphasing paths. For the integrated outphasing TX, gain mismatch in the two outphasing pahts can be tolerated because both PA output stages are placed closely and driven into saturation. In the left plot, it shows that the EVM of the 16QAM degrades to about -20 dB with a phase mismatch of 6 degree. In the right plot, it shows the spectrum is worsened by 20 dB.

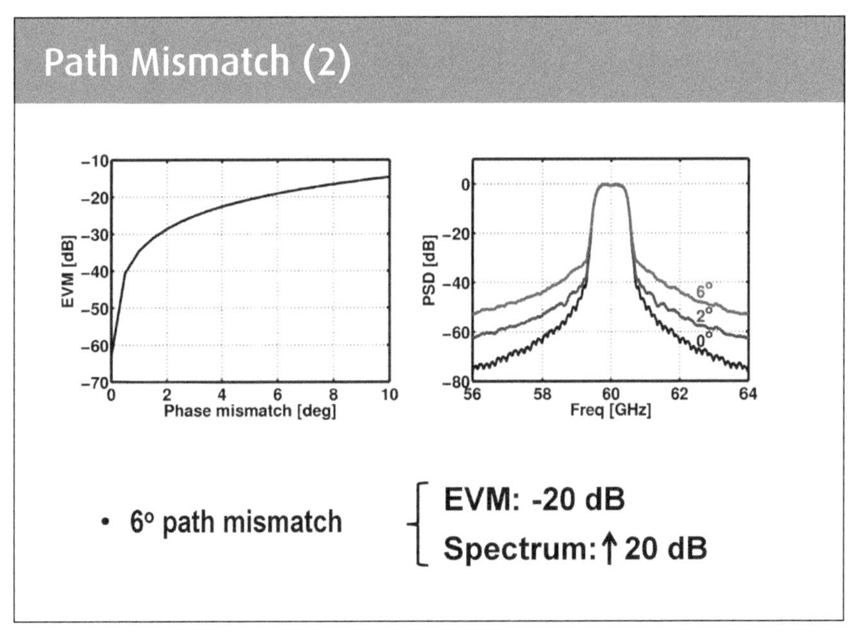

To reduce these mismatch effect, we try to do symmetrical floorplan and layout during the design. The floor plan of the whole TX will be discussed later on. In the measurement, we found the impact of the mismatches is small.

37

Outphasing: Pros & Cons

- **Outphasing TX provides higher P_{LIN} and η_{LIN}, suitable for mm-wave applications.**
- **Load modulation improves back-off efficiency.**
- **Trade linearity for bandwidth ($2BW_{IQ}$).**
- **Path mismatch degrades performance, requiring symmetrical floor plan and layout**

A short summary about what we have discussed:

· The outphasing PA operates linearly up to the maximum output power, provides higher P_{LIN} and η_{LIN}, and as a result, it shows much better average output power and efficiency for modulated signal. It is suitable for mm-wave applications because such benefit is not easy to get with other techniques.

· The outphasing TX needs to double the processing bandwidth but requires less linearity. So the analog baseband of the outphasing and I/Q TXs consumes similar power but the outphasing TX has quite some benefits at its mm-wave front-end.

· The load modulation can help improve the back-off efficiency.

· Path mismatch degrades performance, requiring symmetrical floor plan and layout.

38

Outphasing TX

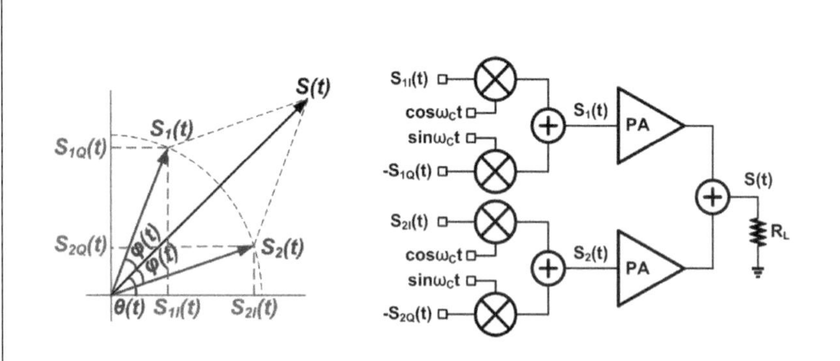

$$S_1(t) = S_{1I}(t)\cos(\omega_C t) - S_{1Q}(t)\sin(\omega_C t)$$

$$S_2(t) = S_{2I}(t)\cos(\omega_C t) - S_{2Q}(t)\sin(\omega_C t)$$

How to generate two constant-envelope outphasing signals $S_{1(t)}$ and $S_{2(t)}$ at PA inputs? A conventional way is that we generate the I/Q signals which constitute S_1 and S_2 at baseband. Then, by simply using quadrature upconverion, we can generate S_1 and S_2 at RF frequency. These four signals can be generated based on original I/Q signals by doing some signal processing or simply using a look-up table.

39

System Diagram

- **Outphasing**
- **Conventional I/Q**

This is the system diagram of the outphasing transmitter.

The outphasing TX requires 4 baseband inputs: S_{1I} (t), S_{1Q} (t), and S_{2I} (t), S_{2Q} (t), which are in blue in the figure. The four baseband signals are upconverted by the I/Q mixer to generate the desired outphasing signals ($S_{1(t)}$ and $S_{2(t)}$) at PA inputs. The two PAs only need to amplify constant-envelope signals, which guarantee the high efficiency. At the output, the power combiner sums the amplified signals and reconstructs the amplitude modulation.

On the other hand, this TX can also operate in conventional I/Q up-conversion mode if we apply identical S_{1I} (t), S_{2I} (t) and also identical S_{1Q} (t), S_{2Q}(t).

To show the benefit of the outphasing, the measurement results of these two modes, the outphasing mode and the conventional mode, will be compared.

40

60-GHz Outphasing Transmitter

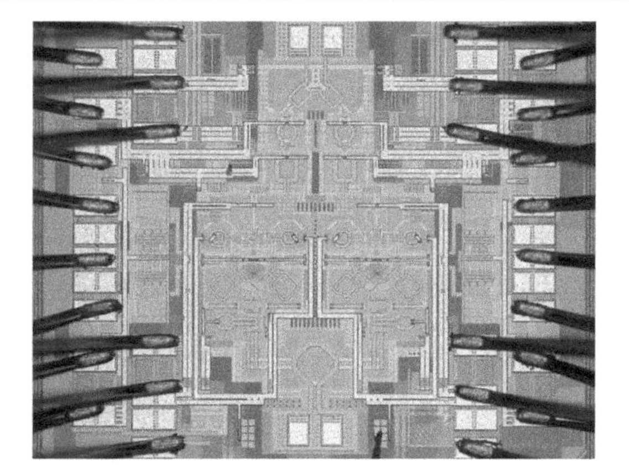

- **40-nm CMOS LP, 0.96 mm^2 (core: 0.33 mm^2)**
- **1st outphasing TX at mm-Wave**

This work is implemented in 40-nm CMOS LP. The chip area is less than 1 mm^2 with the core area of 0.33mm^2. It is the first reported outphasing TX at mm-Wave frequencies.

41

Chip Micrograph (RF)

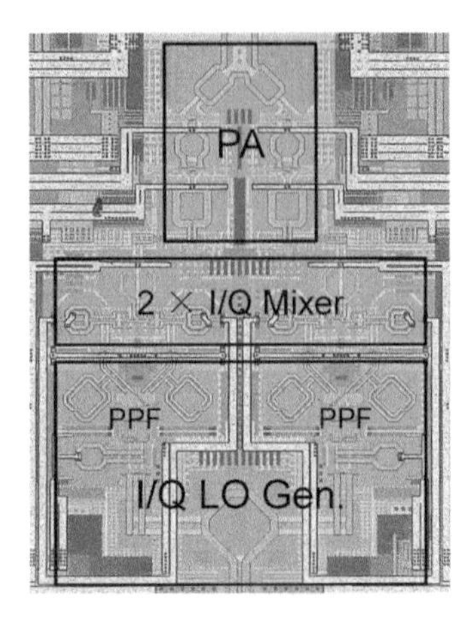

- **Symmetry**
  - **BB, LO & RF paths**
  - **1.5° phase error**
- **DRC clean**
  - **TF, TL & L...**
- **Active area**
  - **0.33 mm²**

Let's zoom in to the RF part of the transmitter.

We can clearly see the PA, two I/Q mixers, PPF and the LO distribution networks in the chip photo. The complete layout floorplan looks very symmetrical. Indeed, for the outphasing transmitter, the matching between the two outphasing signal paths is extremely important to outphasing system. Therefore, in the layout, we minimized the asymmetries in baseband path, in I/Q LO distributions and also between two RF signal paths. And due to these layout techniques, we don't need to apply any mismatch compensation or predistortion for this transmitter during the measurement.

Besides, we can also see the transformers and inductors are extensively used in the RF building blocks. In the design, all the inductors, transformers and transmission lines are filled with metal dummies to fulfill all the design rules. The performance degradation due to these dummies is less than 2% according to field simulation.

The active area of the TX is only 0.33mm².

42

Measurements

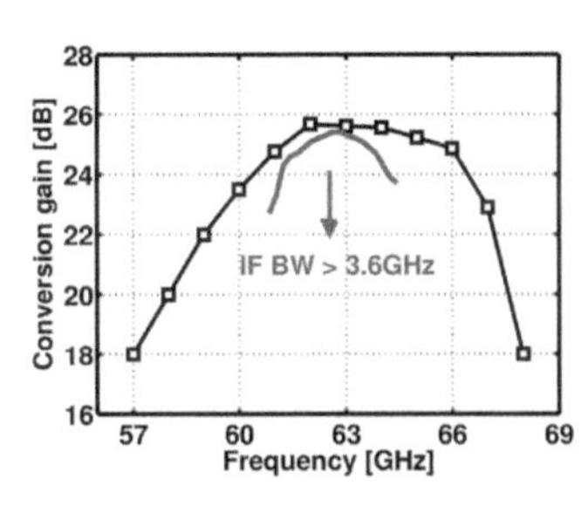

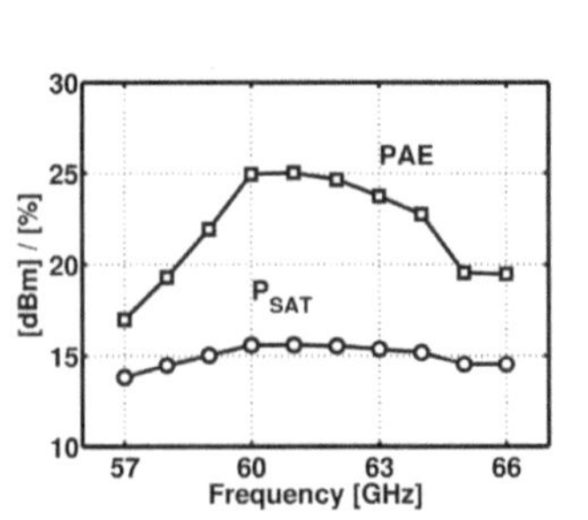

- **Conversion gain: 26dB**
- **RF & IF bandwidth: 7.5 and >3.6 GHz**
- **P_{SAT}: 13.8 – 15.6 dBm, PAE: 17 – 25%.**

Here we show the measurement results. The TX has a conversion gain of 26dB and the BW is more than 7GHz. The IF BW is more than 3.6GHz and sufficient for outphasing operation.

From 57 to 66 GHz, the P_{SAT} varies between 13.8 and 15.6 dBm while the PAE of the PA is higher than 17% with a peak value of 25% at 60GHz.

In fact, we measured several chips and the performance is quite consistent. The P_{SAT} of 15.6 dBm is lowest number we got. The peak output power of the TX is around 16 dBm.

43

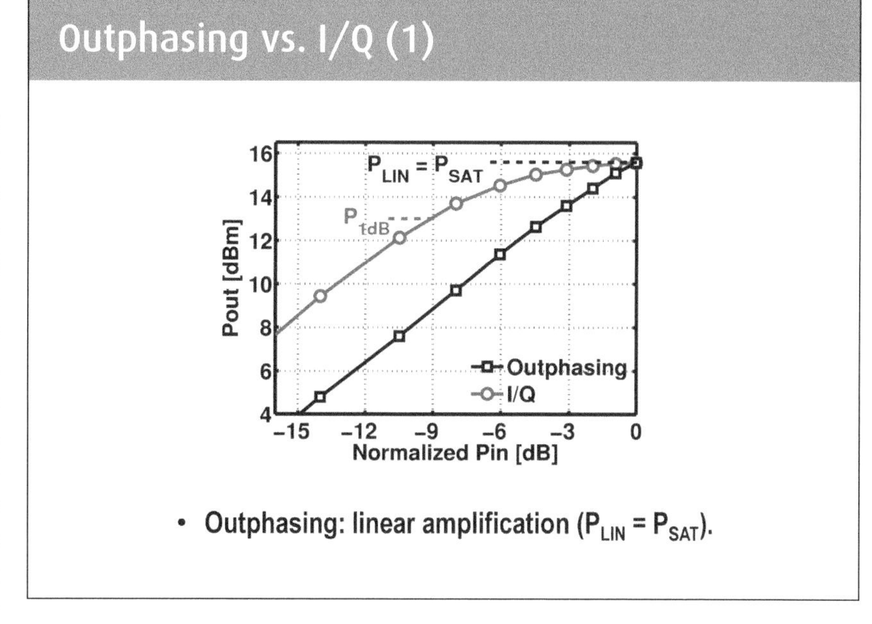

To show the benefits of outphasing transmitter, we operate the transmitter in two modes: the outphasing mode and I/Q mode, and compare the results in the slide. The left graph shows the measured output power. The TX in conventional mode starts to saturate at P_{1dB} while the outphasing TX is able to operate linearly up to the maximum output power. Because of its ultra linear property, the outphasing TX does not need to operate at back-off when processing the modulated signals, which guarantee the high efficiency.

44

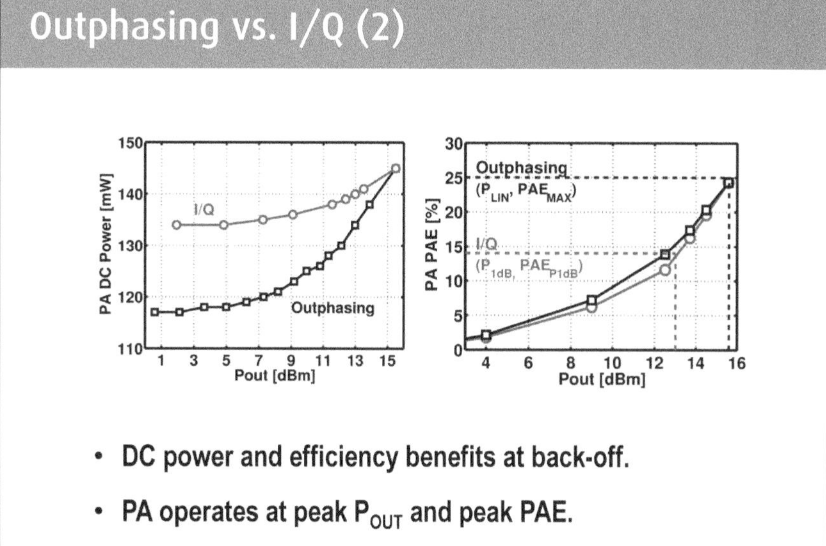

The complete outphasing PA consumes about 13% less DC power at back-off because of the load modulation effect discussed before. Note that the consumed current cannot be further reduced at large outphasing angle since the increase in the PA load impedance is limited by the loss of the transformer.

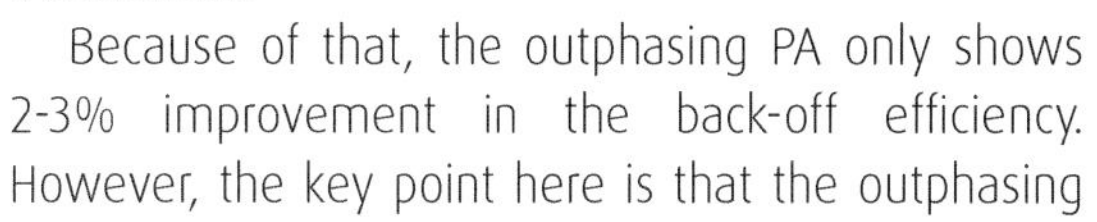

Because of that, the outphasing PA only shows 2-3% improvement in the back-off efficiency. However, the key point here is that the outphasing PA can operate with P_{SAT} and peak PAE, leading to higher average output power and better average efficiency with complex modulation.

45

Outphasing vs. I/Q (3)

Here we do another comparison and we show the DC power consumption versus output power. We compared the measured outphasing PA with a simulated conventional PA with the same amount of linear output power. If assume the max linear power of a conventional PA is its P_{1dB}, the outphasing PA is able to achieve the same linear output power with only 50% power consumption.

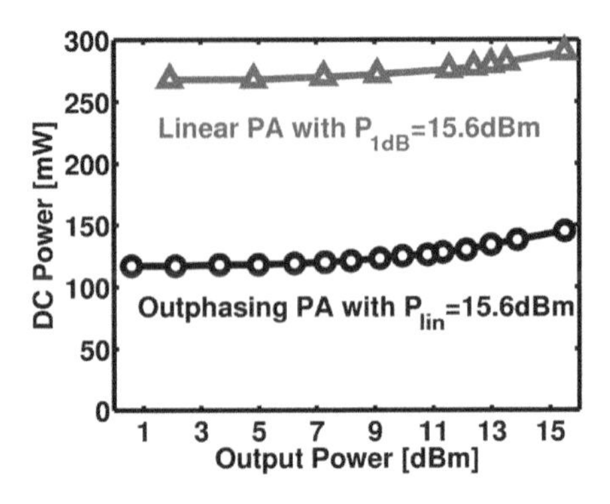

- Outphasing: 50% P_{DC} for the same P_{LIN}.

46

Constellation & EVM (16QAM)

The benefits of the outphasing transmitter are further confirmed by EVM measurements. We did the comparison by again operating this TX in outphasing and IQ modes. The figure on the left shows the 500Mb/s 16QAM signals. The figure in the middle plots the average output power versus EVM and the figure on the right plots the PA average efficiency versus EVM.

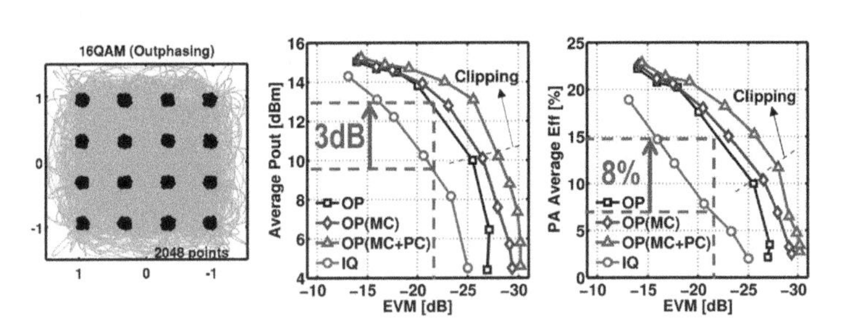

- Data rate = 500 Mb/s, PAPR = 7.4 dB (RCF, α=0.35).
- 12.5 dBm P_{AVG} & 15% η_{AVG} @ -22dB EVM.
- Mismatch Comp. & Phase Corr. ↑1.6 dB P_{AVG} & 4% η_{AVG}.

It shows at the EVM of -22dB the transmitter in outphasing mode, the black curve, achieves 12.5dBm P_{AVG} and 15% average PA efficiency. It performs two times better than the conventional IQ mode, the grey curve.

Later on, we also applied some predistortion to future improve the performance.

47 

Summary: Outphasing PA

- **Outphasing facilitates system-circuit co-design.**
- **Highly efficient mm-wave TX solution.**
  - P_{LIN}= P_{SAT} (15.6 dBm)
  - constant envelope and high efficiency (25%)
  - higher P_{AVG} and better $_{AVG}$
- **Outphasing trades amplitude linearity for phase (time) resolution, well matching the transistor scaling.**

A short summary on the outphasing power amplifier:

1) Outphasing facilitates system-circuit co-design. Outphasing is actually a system & circuit co-design concept. To have better overall performance, you not only do things at circuit level, you also do optimizations at system level. Why do we need system and circuit co-design at mm-wave frequencies? Because at mm-wave, both transistor and passive performance are poor. You can not have sufficiently good performance by just doing circuit optimization.

2) The proposed outphasing TX is ultra linear. It only deals with constant envelope and therefore is highly efficient. It has better average output power and average efficiency when transmitting the complex digital modulated signals.

3) Last but not the least, outphasing trades amplitude linearity for phase (time) resolution, which well matches the trend of transistor scaling.

48 

III. CMOS Broadband Power-Combining PA

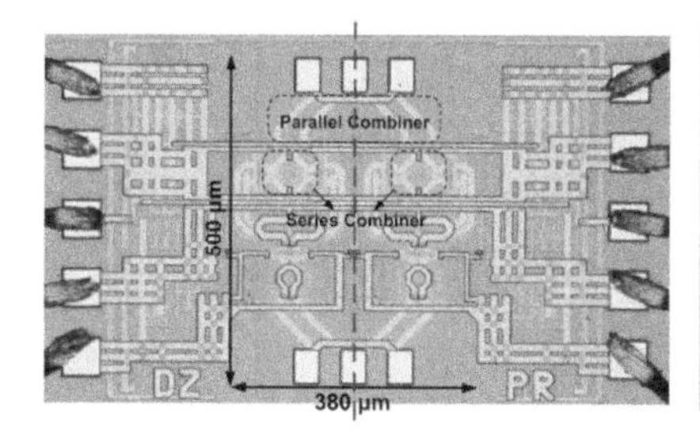

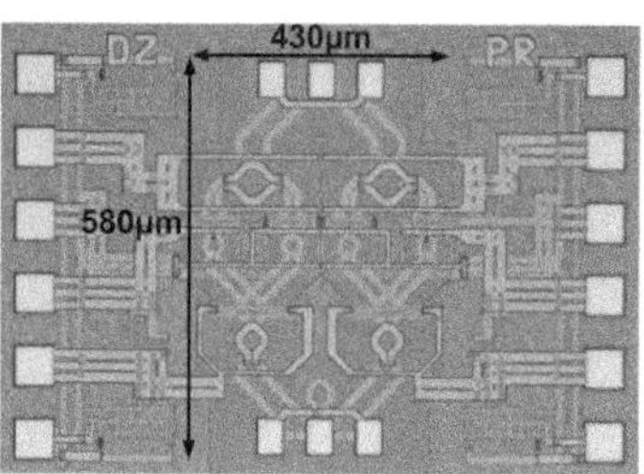

- D. Zhao et al., "A 40-nm CMOS E-band 4-Way Power Amplifier with Neutralized Bootstrapped Cascode Amplifier and Optimum Passive Circuits," IEEE Trans. Microwave Theory and Techniques, vol. 63, no. 12, pp. 4083-4089, Dec. 2015.
- D. Zhao et al., "An E-band Power Amplifier with Broadband Parallel-Series Power Combiner in 40-nm CMOS," IEEE Trans. Microwave Theory and Techniques, vol. 63, no. 2, pp. 683-690, Feb. 2015.
- D. Zhao et al., "A 0.9V 20.9dBm 22.3% E-band Power Amplifier with Broadband Parallel-Series Power Combiner in 40nm CMOS," in IEEE International Solid-State Circuits Conference (ISSCC), Digest Tech. Papers, Feb. 2014, pp. 248-249.

Finally, we will discuss two E-band PAs. The focus will be on the broadband power combining techniques.

49  E-band Technology

- **Allocation: 71-76 & 81-86 GHz (2× 5GHz BW).**
- **Licensed band: regulatory protection.**
- **Propagation characteristics**
  - low atmospheric attenuation (0.2dB/km)
  - rain attenuation (10dB/km for heavy rain)
  - unaffected by other deteriorations (e.g., fog, dust...)

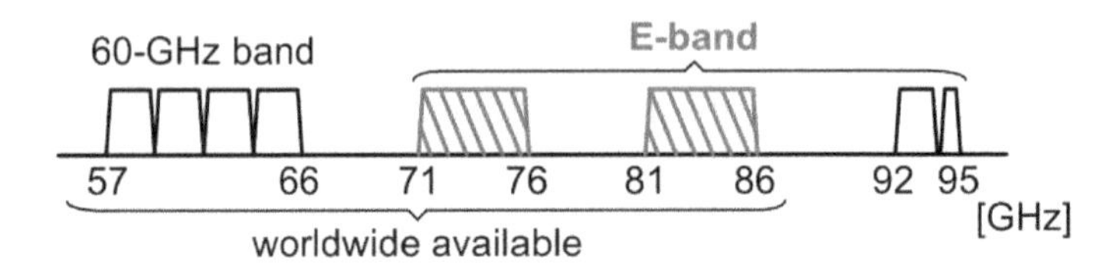

Let's first look at some properties of E-band. For E-band high data rate communication, it usually refers to two sub-bands: 71-76 GHz and 81-86 GHz. Like 60GHz, these two sub-bands are also available worldwide. There are 5GHz bandwidth available in each sub-band.

Different from 60-GHz, E-band needs licensing. So there will be no interference or there is guaranteed interference during transmission, which ensures robust communication.

E-band has much lower attenuation in the air compared to 60-GHz, only 0.2db/km. This makes it suitable for long-range communication.

Rain attenuation is a potential problem for all the outdoor mm-Wave applications. It needs to be considered when planning the link. For E-band, studies show only a few hours over the whole year will be affected by the heavy rain in most countries and areas for km range communication.

Another important feature of E-band is the transmission will not be affected by most other particles in the air, such as fog and dust (mist, sand...).

50  E-band Applications

All the properties of E-band make it suitable for long-range high-speed point-to-point communications, which enable lots of applications, such as mobile backhaul and enterprise connection. It can also be used to extend the fiber optic network.

Compared to fiber, E-band wireless system gives much lower overall cost and super easy to install, which results in fast paybacks.

The figures in the slide show the typical E-band application. E-band can be used to setup the connection between core network and backhaul network and also the link between backhaul and the access networks. To extend the communication range, a very directional antenna has to be used. The antenna gain is normally larger than 45dBi. To derive the PA specification, we assume some numbers in the link. The communication range is 2km. We will make use of the complete 5GHz bandwidth. Noise figure at receiver side is about 10dB.

50

- Mobile backhaul
- Enterprise connections
- Fiber extension / replacement

→ low cost, easy to install, fast paybacks...

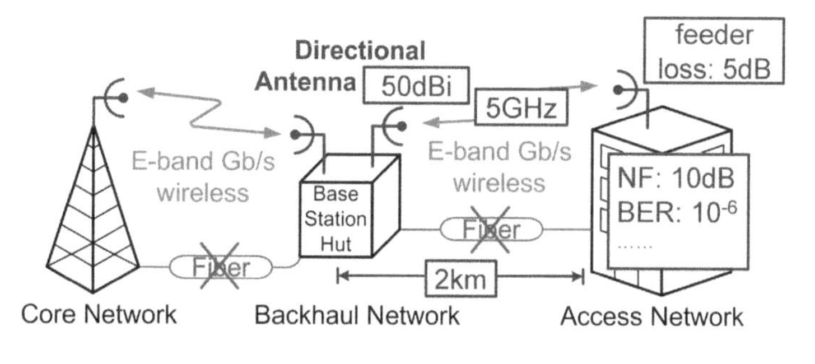

51

E-band PA Requirements

- **Output power = 20dBm (2km)**
  - 64QAM: max. data rate 30Gb/s.
  - QPSK: max. data rate 10Gb/s (heavy rain).
- **Large signal -1dB BW = 15GHz**
  - FDD: 10-GHz spacing between TX/RX.
  - constant P_{OUT} across 71-76 & 81-86 GHz.
- **Small signal -3dB BW = 15GHz**
  - broadband communication in each sub-band.
- **High efficiency**

Based on the link calculation, we have the design requirements for our E-band PA.

An output power of 20dBm is required to reach a communication range of 2km. In good weather condition, 64QAM can be achieved with a theoretical max. data rate of 30Gb/s. Under heavy rain, the system can switch to QPSK while still making the link with 10Gb/s data rate.

E-band system is based on frequency division duplexing (FDD). So it would be good if the E-band system can operate in both 70 and 80 GHz bands. As a result, the transmitter can operate in one of the sub-bands and the receiver works in the other band. In that case, there will be 10GHz spacing between transmitter and receiver which make the system design a lot easier. This also requires the PA operates in both 70 and 80 GHz bands. To make sure that the PA can deliver similar output power in both bands and the same communication range can be reached, we need about 15GHz -1dB large-signal bandwidth.

For broadband communication, it also requires about 15GHz -3dB small signal bandwidth. So in each 5GHz sub-band, the gain variation will be about 1dB.

Of course, high efficiency is always preferred to have a low power system.

52 

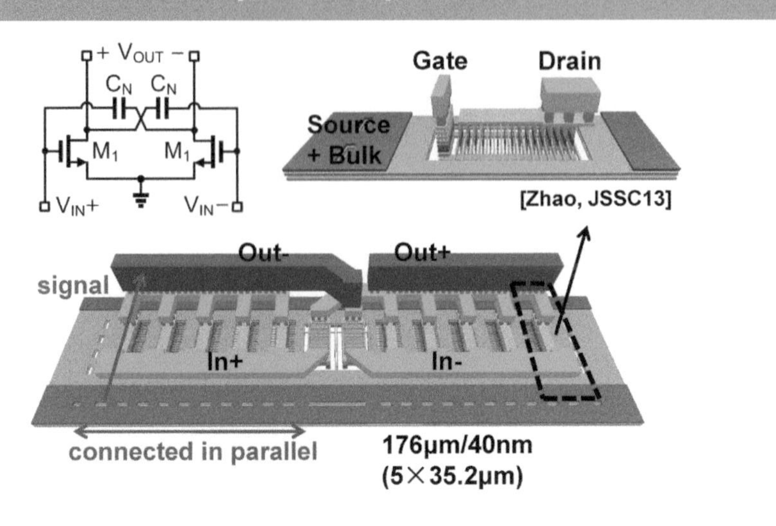

We start our discussion with the mm-Wave amplifier design.

Neutralization technique is commonly used at mm-Wave frequencies to improve the stability and gain of the amplifier. It is also used in this design.

The transistor layout has a big impact on the mm-wave designs, especially for the transistors with large size. The device parasitics and long interconnects from the transistor layout actually determines the gain and output power that can be eventually achieved by the amplifier.

Here we show the layout of the neutralized differential amplifier at the bottom of the slide. The transistors are at two sides and the neutralization capacitors are sitting at the center. The input and output of the amplifier can be directly connected to the matching circuits.

To minimize device parasitics, we propose a power transistor layout for mm-Wave PA in our previous work, which is used as a unit transistor. Compared to conventional layout, this layout has less layout-related parasitics.

The key advantage of the proposed layout is it can be easily duplicated for high output power. As shown in the figure, five such unit transistors are connected through wide metal tracks or top thick metal lines. With this layout, it proves that the impact of long interconnects can be minimized.

To see the benefits and limitation of the layout, we simulate the amplifier performance with different number of the unit transistors being connected in parallel.

53  vs. number of unit transistors

Here we show the simulated output capacitance, maximum gain, output power and required neutralization capacitor versus the number of the transistors.

We can see in the top left figure the output capacitance of the amplifier increases proportionally with transistor number.

Top right figure, the amplifier maximum gain is reduced by less than 1dB when the transistor number equals 5. It corresponds to a transistor width of 176$_{\mu m}$. This clearly shows the advantage of the proposed layout.

The amplifier gain drops faster and decreases by about 3dB when the transistor number further increases due to the long interconnects.

To simulate the output power, we scaled the input power according to the transistor size. We can also see the increase in the output power starts to saturate with large transistor size. The neutralization capacitor also levels when the transistor number increase. Note that Cneu sees more interconnect effects than the output capacitance as it is a lumped capacitor.

So at mmWave, the gain and output power you

53

vs. number of unit transistors

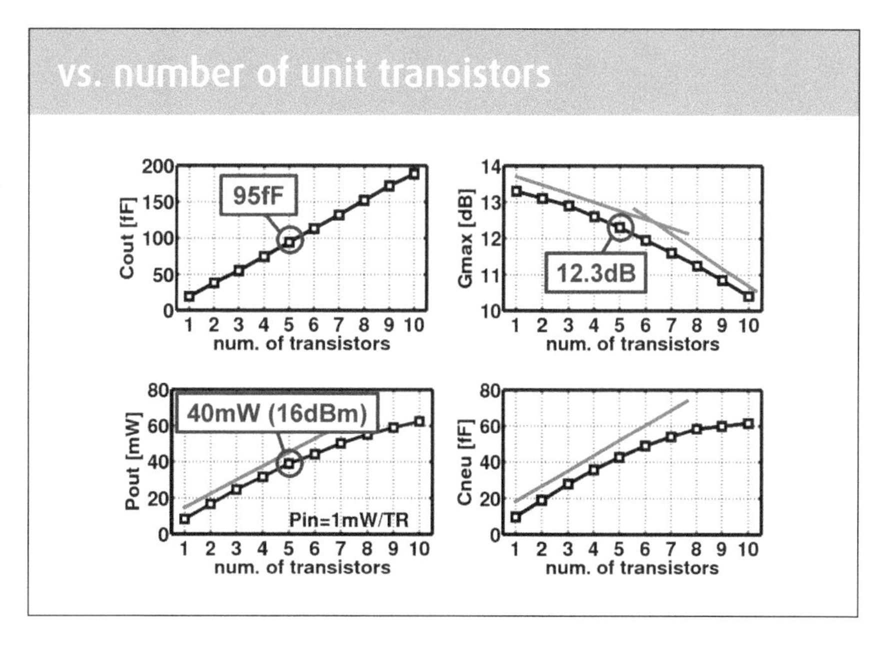

can achieve will be limited by the transistor size.

In this case, we choose transistor number equals 5 while the long interconnects still do not severely affect the performance. The output capacitance is about 95fF and the output power it can deliver is about 16dBm.

54

Transformer Efficiency

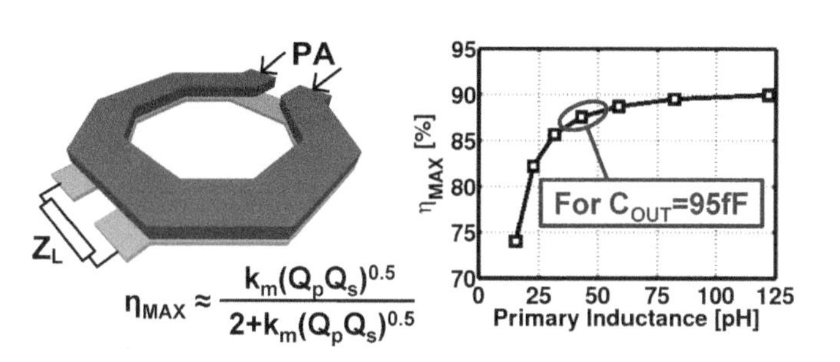

$$\eta_{MAX} \approx \frac{k_m(Q_pQ_s)^{0.5}}{2+k_m(Q_pQ_s)^{0.5}}$$

- Large transistor size → small transformer → low η.
- Transistor size: 176μm/40nm (5×35.2μm) → 16dBm.
- 4-way power combiner for more than 20dBm P_{OUT}.

Transformers are extensively used in mm-Wave designs. Transformer power transfer efficiency depends on the quality factors of the primary and secondary coils and coupling factor between them. The transformer performance also has an influence on the transistor size we pick.

Just like large transistor having less gain, transformer design also has similar trade-offs. To absorb more capacitance of the transistor, the diameter of the transformer has to be decreased to reduce its self-inductance. However, with smaller diameter, we can also see the maximum power transfer efficiency is reduced from 90% to about 75%. This is because the Q and coupling factor are reduced with smaller diameter. So we choose an amplifier with about 95fF output capacitance while the transformer still maintains high efficiency.

So for a single stage amplifier, the output power, gain and efficiency will be limited by both transistor parasitics, long interconnects and passive matching circuits. Considering the limitations from actives and passives, in this design, for each differential amplifier stage, we choose transistor size of 176$_{\mu m}$. The amplifiers give about 16dBm output power. To further improve the output power and efficiency, power combining technique is needed.

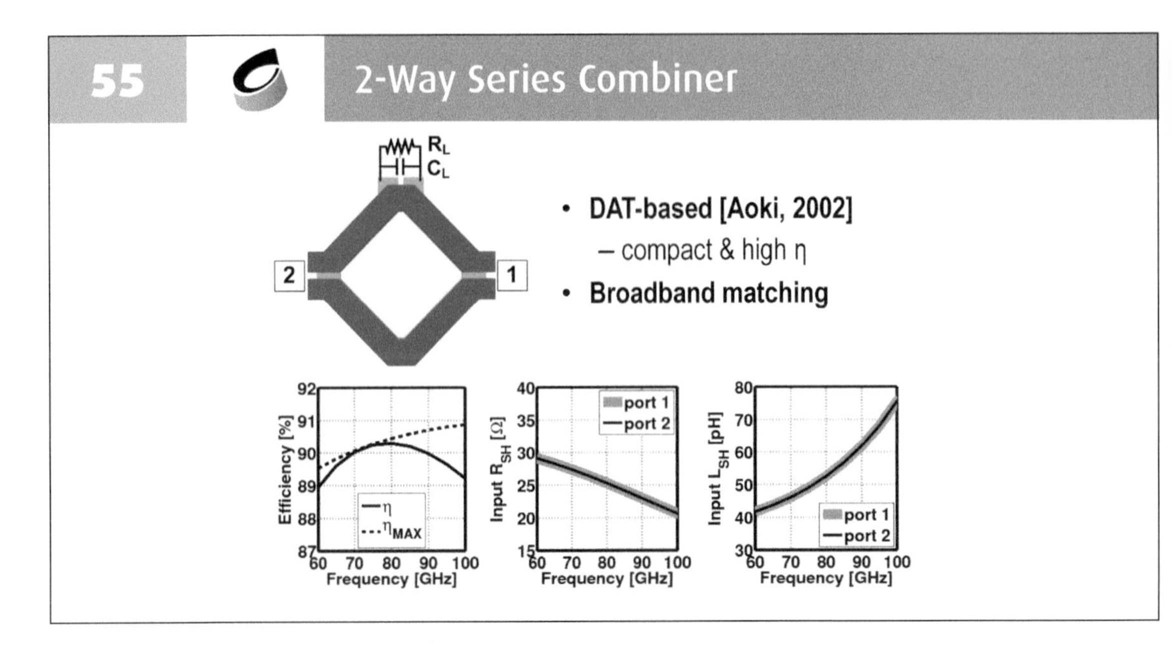

To develop a broadband efficient 4-way combiner, let's first look at the DAT-based 2-way series combiner. This structure is the most often used combiner at mm-Wave frequency and it does have many advantages. Compared with the series combiner with separate primary windings, the DAT-based series combiner can achieve high efficiency as it has bigger coils. From the figure, we can see if we tune the combiner at 80GHz, it can achieve high efficiency over wide bandwidth.

This series combiner can also achieve very broadband power matching. Here we plot the equivalent shunt resistance and inductance seen from the two input ports of the combiner. With 50Ohm load and tuning capacitor, we can see input shunt resistance maintains around 25Ohm over broad bandwidth. In comparison, the input shunt inductance has more variation. So if we can dynamically tune the load capacitance C_L, the power matching bandwidth of this series combiner can be further improved.

56  N-Way Series Combiner Challenges

Well, 2-way series combiner is still not enough to achieve more than 20dBm output power. So we need to combine more PAs. However, at mm-Wave frequencies, the N-way DAT-based series combiner generally has 3 problems for further improving the output power.

First, for a series combiner, when the number of combining paths increases, the input impedance at each port also increases. For a 4-way combiner with 50Ohm load, the input impedance reduces to 12.5Ohm. This means it needs very large transistor, but we know that large transistor has less gain. That will make the PA less efficient.

Second, as the dimension of a mm-Wave transformer is small, the interwinding parasitic capacitance between the primary and second coils has big impact on the input impedance seen at each port.

From the simulation, we can see when we tune the 4-way combiner at 80-GHz, the impedance seen at 4 ports are different from the theoretical value. The shunt resistance seen at port 1 & 2 are nearly two times larger than the value seen at port 3 & 4. This effect limits further improving the output power by combining more PAs using series combiner.

Another important limitation of N-way series combiner is it is very difficult to evenly distribute the input signals of the PAs at 4 corners. It usually results

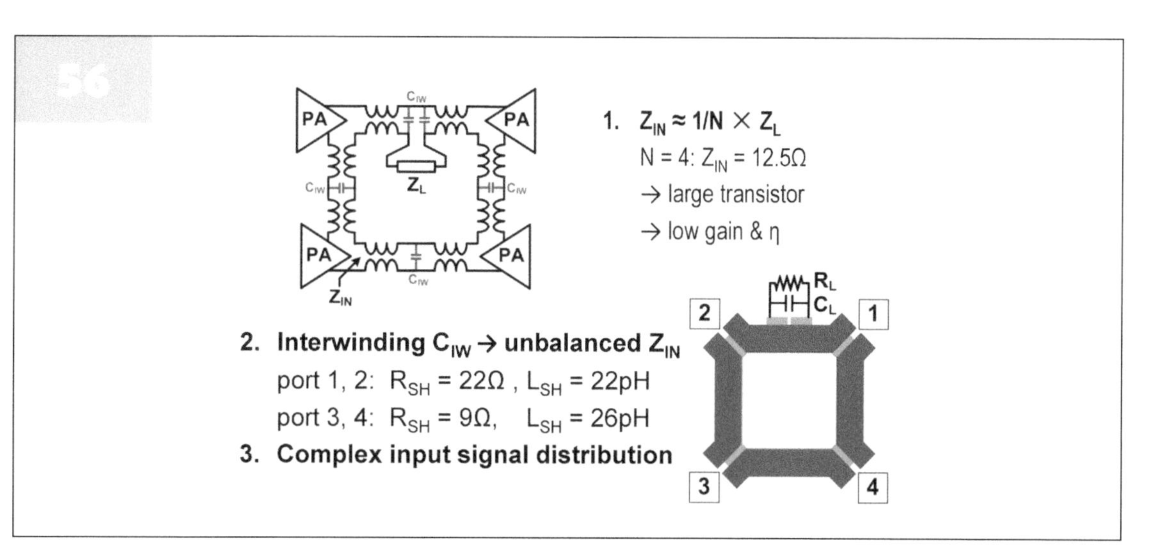

in large chip area and high loss in the interstage matching network.

Therefore, not like RF PAs at low-GHz frequencies, N-way series combiner is not a solution at mm-wave frequencies to achieve high output power.

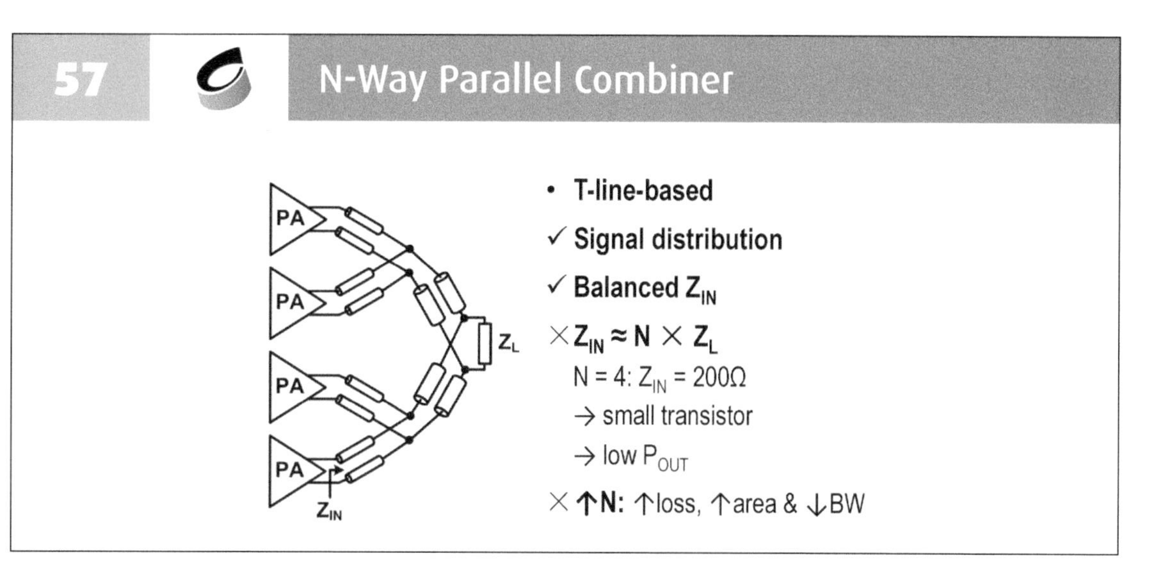

Now let's look at the parallel combiner.

Due to the high operating frequency, if you want to combine PAs with large transistor size, only T-line-based parallel combiner is feasible. The good thing about T-line-based combiner is the length of these transmission line can be made use of to distribute the signals. And if you use a "tree-like" distribution topology, its input impedance is always balanced.

Different from the series combiner, if you increases the combining paths, the input impedance of the parallel combiner is increased. For a 4-way combiner with 50Ohm load, its input impedance increases to 200Ohm. This means the unit PA will be realized with small transistor, which also results in low output power from each unit PA.

Of course, you can use transmission line to do some impedance transformation, but using N-way parallel ocmbiner usually results in large area, high overall loss and lower bandwidth.

So the series combiner and parallel combiners have their own advantages and disadvantages.

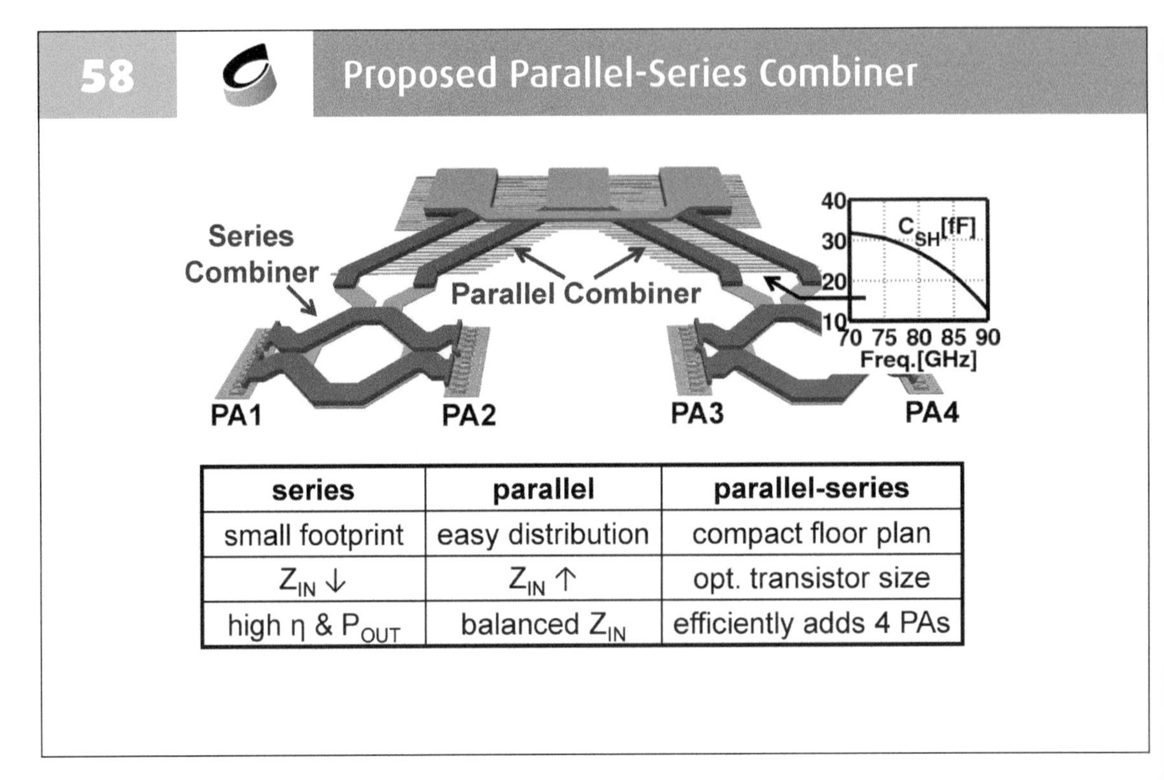

series	parallel	parallel-series
small footprint	easy distribution	compact floor plan
Z_{IN} ↓	Z_{IN} ↑	opt. transistor size
high η & P_{OUT}	balanced Z_{IN}	efficiently adds 4 PAs

In this design, we proposed a parallel-series combiner and try to get benefits from both structures.

In the slide, we see that we have 2 DAT-based series combiner. Each one sums the output power of 2 PAs. The output signals from left and right sides are further routed by the slow-wave transmission lines and combined in parallel at the output RF pads. Floating metals are also placed under the RF pads to minimize its loss.

The proposed parallel-series combiner actually avoids the drawbacks of the series and parallel structures and take advantage of both structures.

First, the parallel-series combiner makes use of the small footprint of series combiner and easy signal distribution of the parallel combiner. With parallel-series combiner, you can make very compact floor plan and it is also easy to distribute the input signals to each PA.

Second, with the parallel-series combiner, the input impedance of the combiner can maintain at a certain level where optimum transistor size can be used that give high gain and relatively high output power.

Third, with the parallel-series combiner, the input impedance is balanced so you can combine 4 PAs efficiently to achieve high output power.

One additional feature of our combiner is due to the impedance transformation of the transmission line, the equivalent shunt capacitance seen by series combiner is reduced at higher frequencies. This actually help us to achieve broadband matching.

Let's look at some simulation results of this combiner.

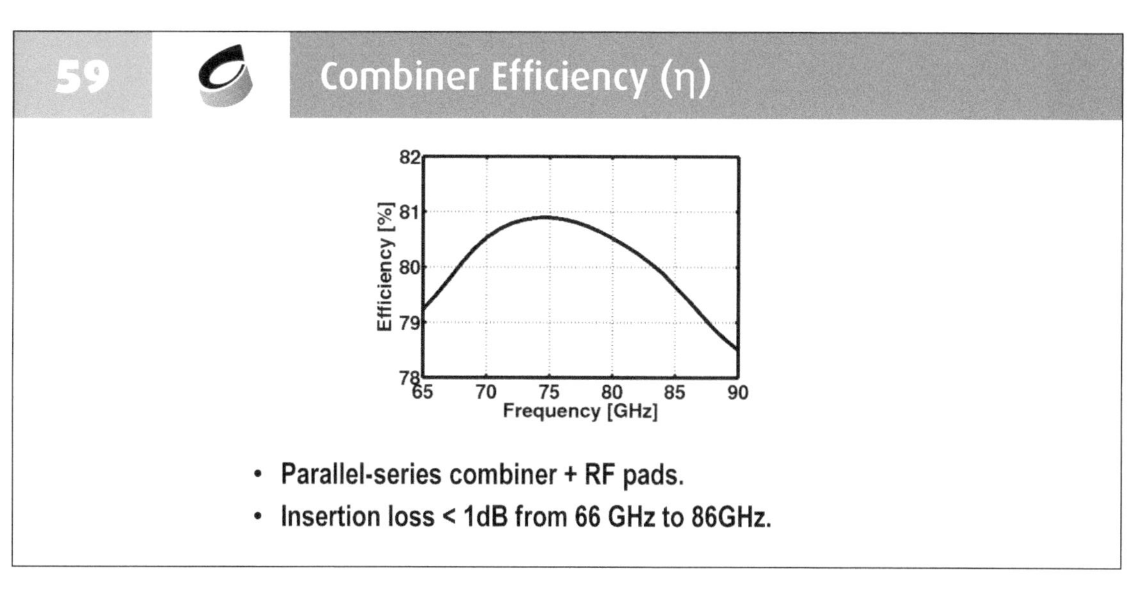

The simulated efficiency of the complete combiner is above 79% from 65 to 86 GHz. It also includes the losses of output RF pads. This number corresponds to an insertion loss of less than 1dB.

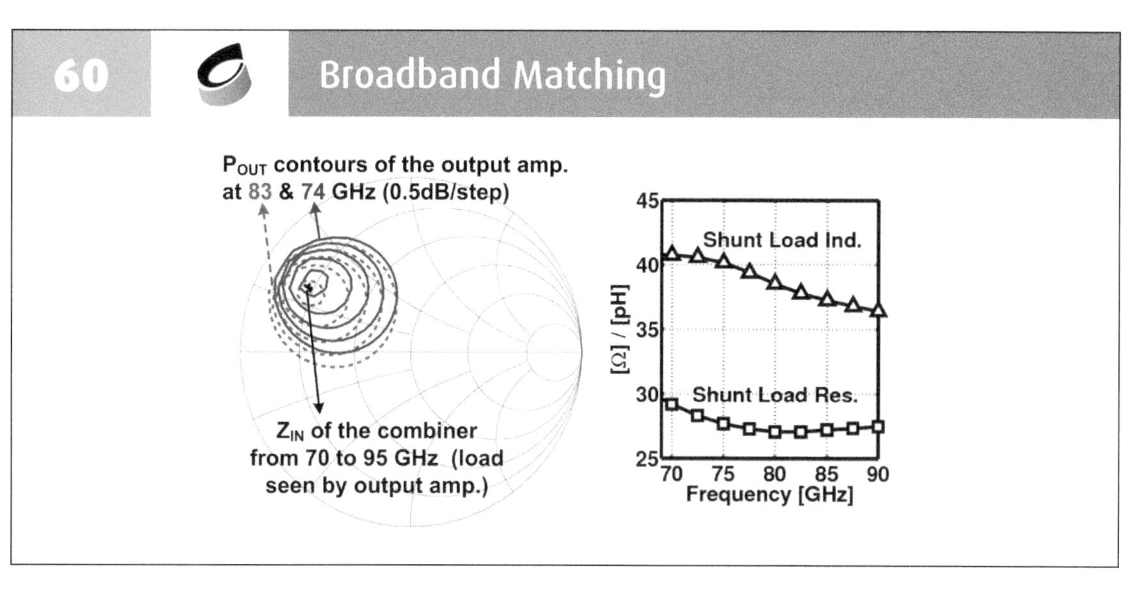

The proposed combiner also achieves broadband matching.

The Smith chart on the left shows the output power contours of the unit PAs at 83 & 74GHz. The black dots in the center of the contours are actually the input impedance of the combiner from 70 to 95GHz, which is also the load impedance seen by each unit PA.

The figure on the right shows the equivalent shunt impedance of this load impedance. We can see the load resistance maintains around 28Ohm which is the optimum load of the PA while the load inductance decreases at higher frequencies. These two effects are desired for broadband matching.

Therefore, with the proposed parallel-series combiner, the 4 unit PAs can nearly deliver constant power in the complete E-band.

61 Complete PA Circuit

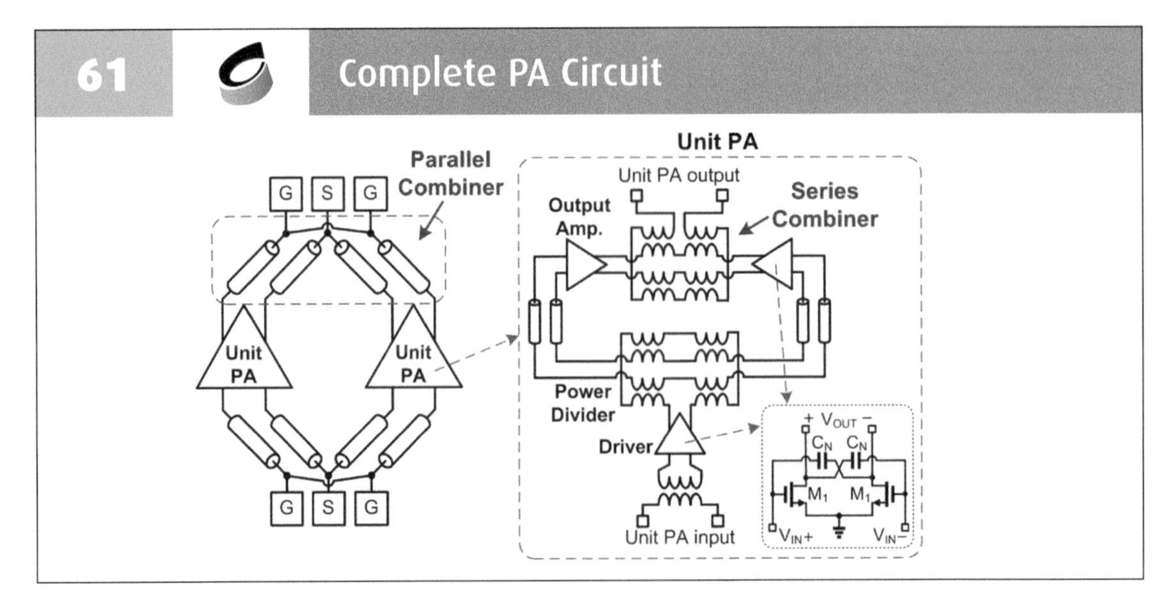

Here we show the complete E-band PA schematic. It has two unit PAs with T-line based parallel combiner at the output. At input, we also have transmission line based divider. It can help lower the input impedance by a factor of 2 and simplify the input matching.

Each unit PA consists of two-stage amplifier: one driver stage and two output stages. The driver stage drives two output stages through a transformer-based divider and differential transmission lines. Due to the input impedance of the output stage and also the relatively complex structure of the interstage network. The driver stage cannot see an optimum load impedance over wide bandwidth. In this design, the size the driver amplifier is sufficiently large to make sure the linear power can be delivered by the driver even when the output amplifier is in saturation over the complete E-band. So here we trade the power consumption of the driver stage with the broadband linearity of the complete PA.

And due to the optimization of the amplifier stage and passive network, the PAE of the complete chip is still above 20%.

The two output stages are combined by the transformer-based series combiner as we already discussed. Both the driver stage and the output stages are realized by neutralized common-source amplifier.

We have discussed the optimization of amplifier stage and broadband 4-way parallel-series combiner. Next, let's look at the remaining design issues.

62  Mutual Coupling

Mutual coupling between passives is another problem to solve. We made very compact floor plan in this work. It helps to reduce the interconnect loss and also the chip area. However, it also results in unwanted mutual coupling between passive components which will affect the performance in most cases. In the slide, we can see the distance between the series combiner and the inter-stage power divider is less than $35_{\mu m}$.

In this design, we minimize its effect or say make use of the unwanted mutual coupling by playing with the polarity of the signal. In the slide, we can see there are two ways to connect the power divider and the output stages, referred as interconnects A and B. By swapping the two lines, the polarity of coupling between the power combiner and power divider is also changed. Simulations predict the complete PA achieves 1.7dB higher gain with interconnect A and it even outperforms the case when no coupling exists.

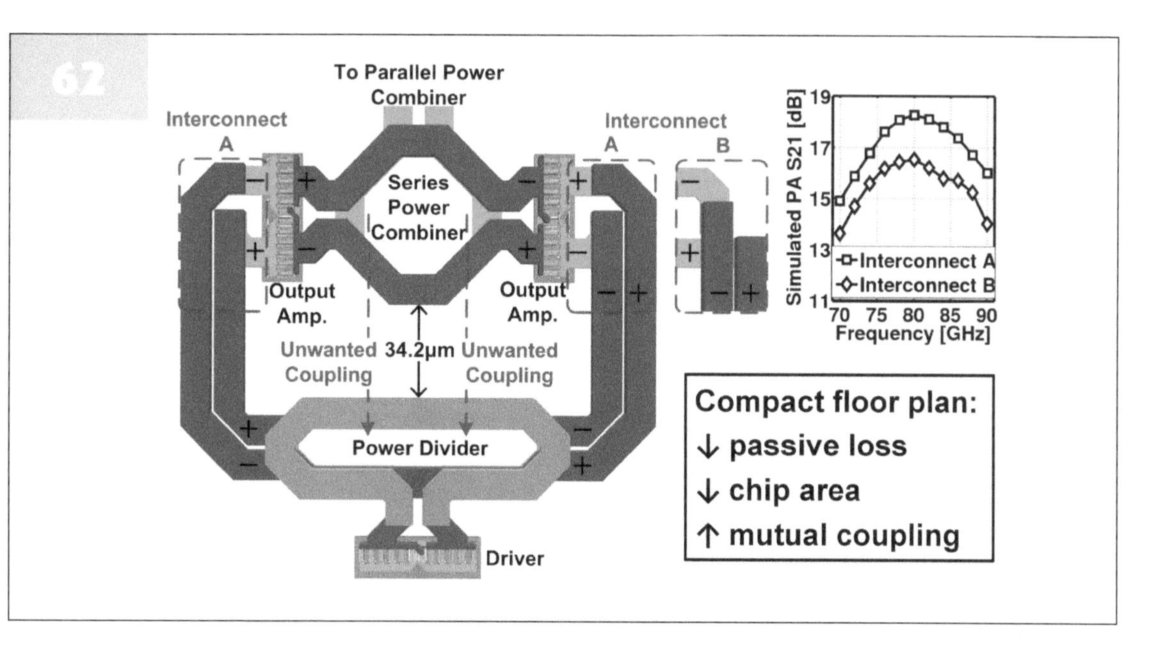

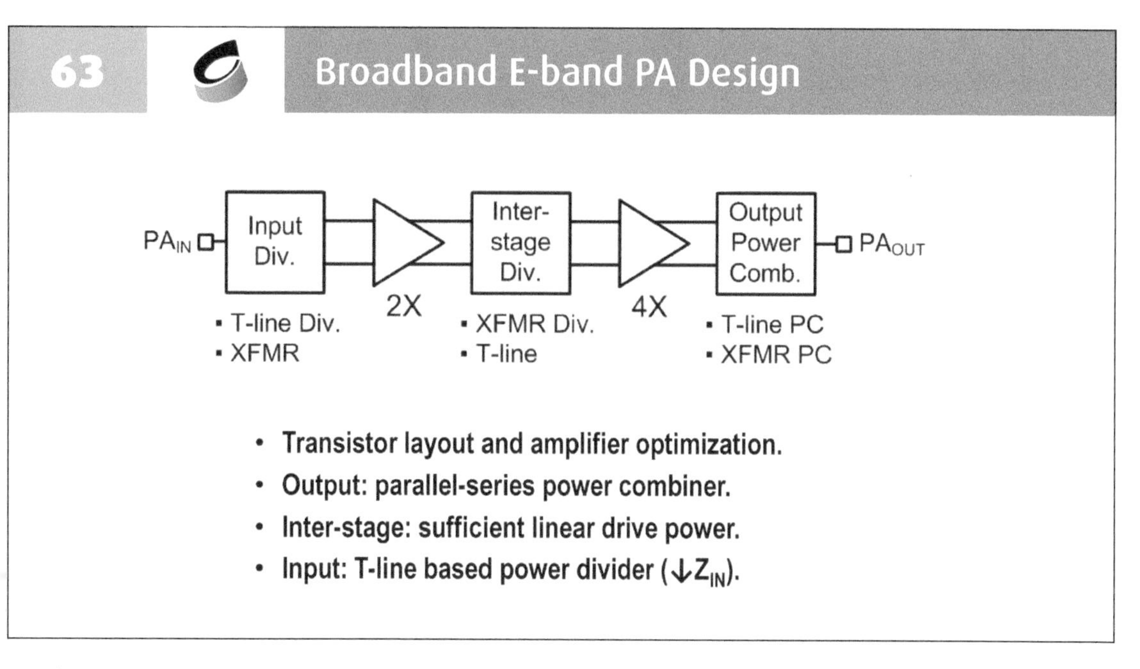

Let's summarize the broadband PA design.

We start with transistor and amplifier stage optimization to ensure each unit PA achieves high gain and delivers relatively high output power.

At output, we proposed the parallel-series power combiner which achieves high efficiency with broadband matching.

We optimize the interstage matching circuits and make sure sufficient linear power can be delivered by the driver stage to broaden the large signal bandwidth.

At input, the transmission-line based power divider also lowers the input impedance and simplify the input matching.

64

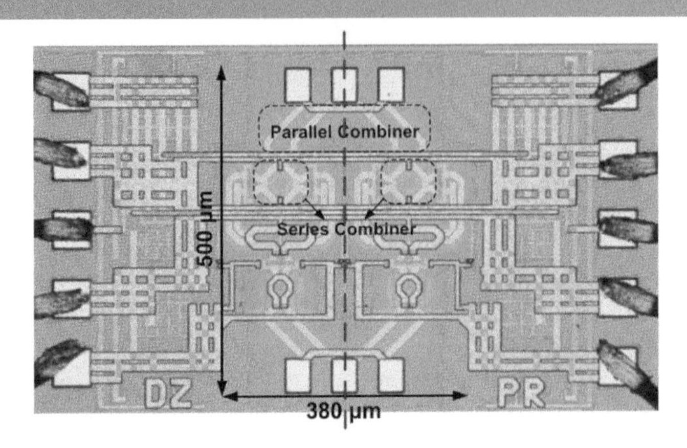

We showed the chip micrograph in this slide. The chip is fabricated in 40nm CMOS. The area is less than 0.2mm^2 including RF pads. Compared with other mm-Wave PA that combines 4 differential or 8 single-ended amplifiers, the area of our PA is extremely small.

In the figure, you can clearly see the parallel combiner and two series combiners at the output.

We can see that two parallel-combined unit PAs are perfectly symmetrical against the central dashed line. This gives us additional advantage as the magnetic couplings of the two unit PAs to the previous stages will be cancelled. So if this PA is integrated in a transmitter, the problems like VCO pulling can be alleviated.

65

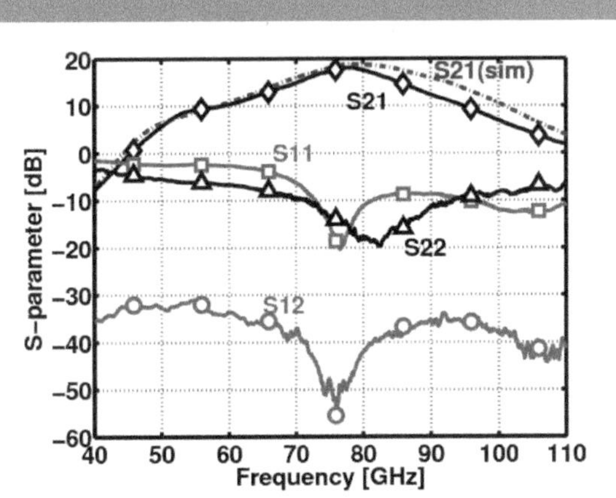

This slide shows the measured s-parameters.

The 2-stage PA achieves a peak S_{21} of 18dB with more than 15GHz small-signal BW. In simulation, we have similar peak gain but we achieved even wider bandwidth.

We also achieved broadband input and output matching. You can see the measured S_{11} and S_{22} are smaller than -8 and -10 from 71 to 86 GHz.

With only 2-stage neutralized amplifier, we also achieved good reverse isolation. The S_{12} is smaller than -37dB from 71 to 86 GHz.

We measured the PA from low-GHz frequency up to 110 GHz. The PA is unconditional stable in the measured frequency with the µ-factor always larger than 1.

66

From the phase response of measured S_{21}, we can derive the group delay performance of the PA. We can see the delay variation from 71-86 GHz is less than 23ps. The variation in each sub-band is less than 13ps.

There is very good correspondence between simulation and measurement.

Measured Group Delay

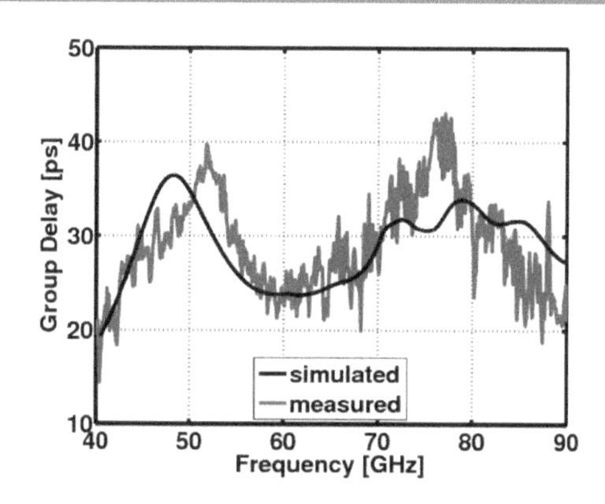

- Delay variation of 22.6ps from 71 to 86 GHz.
- Delay variation < 13ps in 71–76 & 81-86GHz bands.

67

This slide shows the large-signal measured results. At 0.9V supply voltage, the PA consumes about 375mW at DC.

At 80GHz, the PA achieves about 17.5dB gain. It has more than 17dBm 1dB compressed power and more than 20dBm saturated power. The peak efficiency is about 22%.

From the figure, you can see even at saturation, this PA still has about 12dB gain.

Measured P_{OUT} & PAE

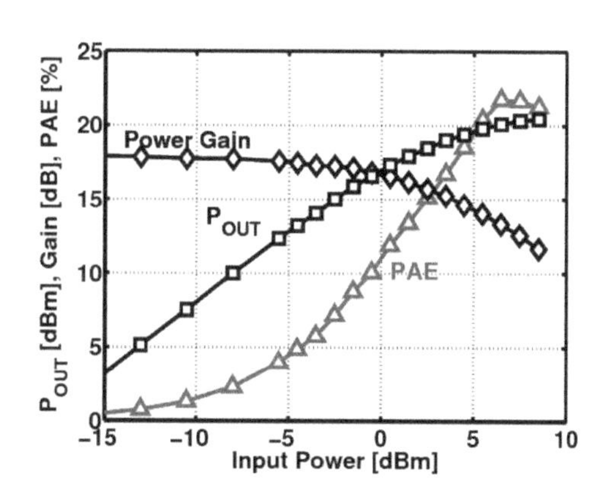

- dc power: 375 mW @ 0.9 V with no RF input.
- 17.4 dBm P_{1dB}, 20.4dBm P_{SAT} & 22% PAE_{MAX}@80GHz.

68

P_{SAT}, P_{1dB}, PAE vs. Frequency

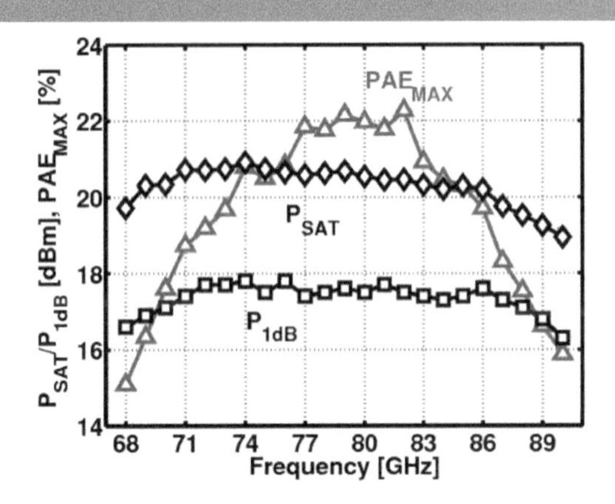

- P_{1dB} varies only 0.5 dB (71-87 GHz).
- Nearly uniform performance across E-band.

Wideband performance is important to E-band PA. Here we showed measured P_{SAT}, P_{1dB} and peak-PAE versus frequency. We can see the saturated output power is above 20dBm over the complete E-band. The peak efficiency is also quite constant, around 20%.

From 71 to 87 GHz, the 1dB compressed power is only varied by 0.5dB.

Thanks to the design techniques we proposed and applied in this work, this PA achieved nearly uniform performance in the complete E-band.

69

Modulated Signal Measurement

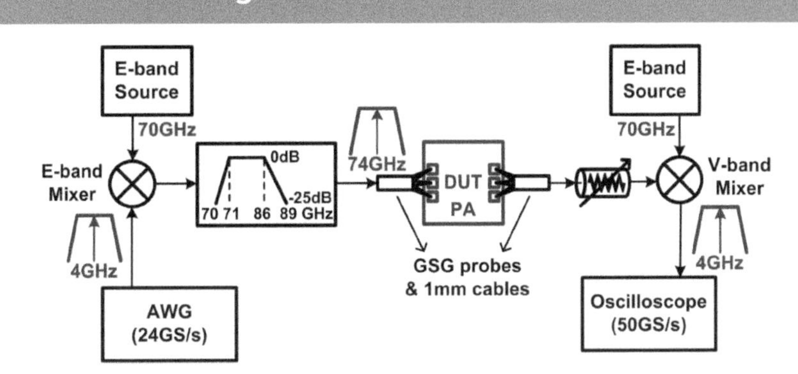

- SSB up-converter = balanced Mixer + BPF.
- Measurement performed w/ & w/o DUT (PA).

To further characterize the PA, we also made the modulated signal measurement with the help of some off-chip E-band components.

High data rate input data are generated by an arbitrary waveform generator. They are up-converted by an E-band balanced mixer and being filtered to generate the SSB signals. The signal is amplified by our PA. At output side, an attenuator is inserted to make sure the down-conversion mixer will not distort the signals. The down-converted signals are then sampled by the oscilloscope and further analyzed to obtain the EVM of the data.

Note that the complete measurement is also performed without the DUT PA where the output of the filter is directly connected to the attenuator at the output side. The attenuation level is adjusted to make sure the down-conversion mixer has the same input signal power.

70

QPSK & 16QAM

Modulation	QPSK	16QAM
PAPR	3.9 dB	6.5 dB
Constellation [2048 points]	EVM=-14dB	EVM=-18dB
Data Rate	5 Gb/s	2 Gb/s
EVM (w/ DUT)	-14 dB	-18 dB
EVM (w/o DUT)	-16 dB	-18.5 dB
Average P_{OUT}	13.0 dBm	12.5 dBm
dc Power	395 mW	390 mW

Here we shows the measured constellations and EVMs of QPSK and 16QAM modulations.

With QPSK, we are able to achieve 5Gb/s data rate with EVM of -14dB. For the 16QAM, 2Gb/s data rate with -18dB EVM was achieved.

The measured EVMs with DUT are actually very close to the EVMs that directly measured from the set-up. We found that the measured EVM is generally limited by the input data generated by the AWG, the frequency response of the off-chip component and spurs in the oscilloscope.

71

Comparison

	This Work	ISSCC12	TMTT13	ISSCC06	JSSC12	TMTT13
Technology	**40nm CMOS**	65nm CMOS	65nm CMOS	130nm SiGe	130nm SiGe	180nm SiGe
V_{DD} [V]	**0.9**	1.0	2	-2.5/0.8	2.5	4
Freq. [GHz]	**70.3-85.5**	79	77	85	84	83
Max. S_{21} [dB]	**18.1**	24.2	20.9	9	27	25
BW_{-3dB} [GHz]	**15.2**	10	N/A	>18‡	8	9.6
P_{SAT} [dBm]	**20.9**	19.3 #	15.8	21	18	14.7
PAE_{MAX} [%]	**22.3**	19.2 #	15.2	4 (η_{Drain})	9	8.1
P_{1dB} [dBm]	**17.8**	16.4 #	13	N/A	16	12.5
P_{1dB} variation [dB (GHz)]	**0.5 (71-87)**	N/A	2 (75-82)	N/A	2 (75-90)	1 (75-88)
Area [mm²]	**0.19**	0.855	0.21	2.4†	0.68†	0.34
Path combined	**4-way diff.**	8-way	2-way diff.	4-way	2-way diff.	2-way
Topology	**2-stage CS**	4-s. CS	2-s. CA	DA / CA	3-s. CB	2-s. CA

The loss of the on-chip output balun was de-embedded from the measured P_{SAT}, P_{1dB} and PAE_{MAX}.
‡ The gain is only shown from 72 to 90 GHz. † Include the area of DC pads.

OK, let's compare our work with other silicon-based PAs.

This design is made in 40nm CMOS. It operates at lowest supply voltage, 0.9V. It achieves 18dB gain with only 2-stage common-source amplifier.

Our design achieves 15GHz small-signal bandwidth. It is only lower than a distributed amplifier topology.

We have more than 20dBm output power with highest peak-PAE.

We also achieves highest 1-dB compressed power in the table. P_{1dB} only varies by 0.5dB from 71-87GHz. It means both 70 & 80GHz bands can be covered by our PA with nearly uniform performance.

We use 4-way differential combining. For a similar combining topology, the area of this design is very small. It is less than 0.2 mm² including the input and output RF pads.

72 

4-way E-band PA with NBCA topology

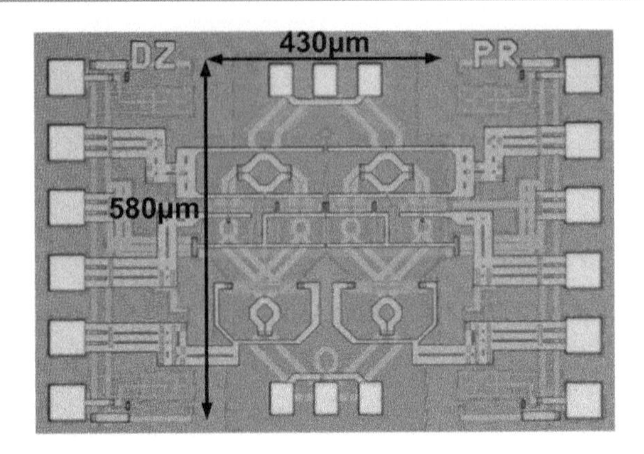

- **40-nm CMOS GP, 0.25 mm²**
- **Neutralized bootstrapped cascode amplifier (NBCA)**
- **Optimal Passive Circuits**

The 2nd E-band PA is based on parallel-series power combiner as well. In addition to that, the neutralized bootstrapped cascade amplifier topology is proposed.

73 

Neutralized Bootstrapped Cascode Amplifier

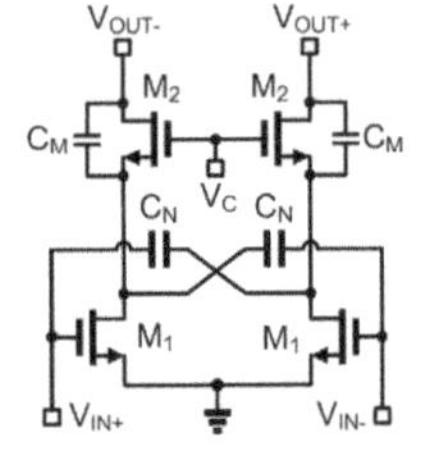

- **"Neutralized"**
  - C_N compensates C_{GD} of M_1
- **"Boostrapped"**
  - $(1\text{-}A_{M2})C_M \rightarrow$ Inductor
- **"Cascode"**
  - M_1 & M_2 stacking for high V_{OUT}

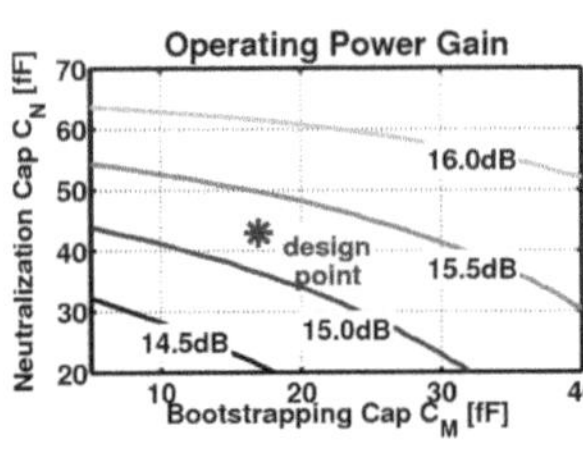

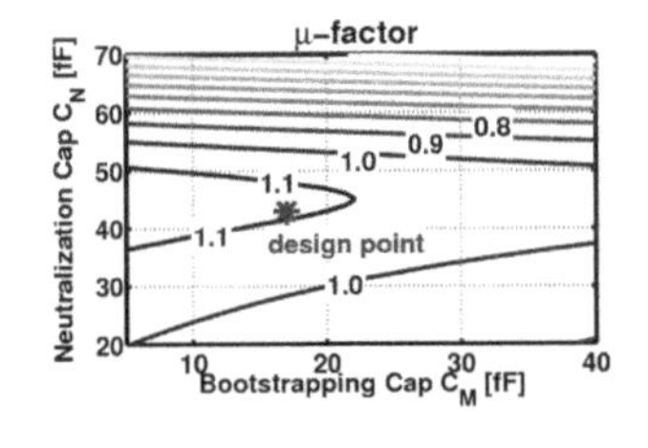

In this slide, we show the schematic of the neutralized bootstrapped cascode amplifier.

In the schematic, C_N cancels the effect of C_{GD} of M_1 and therefore neutralize the common-source stage for high gain and improved stability.

C_M which is across the drain and source of M_2 introduces a negative capacitance at the intermediate node. Its effect is similar to miller effect but the voltage gain of M_2 (A_{M2}) is positive so the capacitance become negative. The negative capacitor behaves as an inductor to tune out the parasitic capacitance at this intermediate node to further boost the gain.

And by stacking two transistors, the output voltage swing can be extended which further improves the output power.

Figures below show the power gain and stability factor as functions of C_M and C_N. In the design, optimum values for C_M and C_N are chosen to enhance the amplifier gain while maintaining the stability simultaneously.

74 

Optimal Passive Circuits

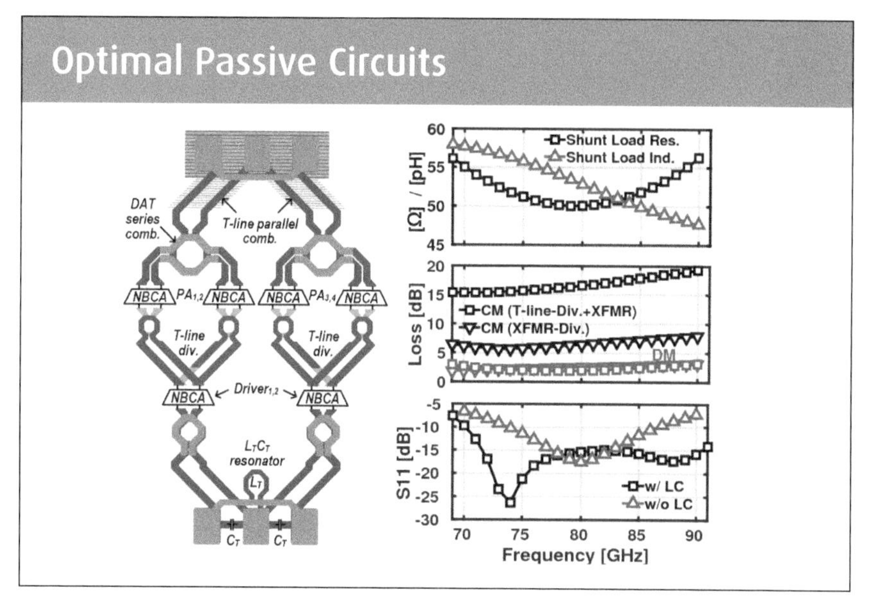

This slide shows the optimization of output, interstage and input passive circuits of the PA.

At output, the parallel-series combining structure is again used to sum up the output power from 4 PAs. We re-optimized it for the new amplifier stage and you can see very broadband impedance matching is also achieved.

At interstage, instead of using transformer-based divider as our previous design, we employ a transmission-line-based divider which is a more balanced structure. After the divider, we use a transformer couples the differential signals to the amplifier and rejects the common-mode signal. In the simulation, you can see in differential mode, both designs have similar insertion loss but the current design has much better common-mode rejection.

At input, we introduced an LC resonator. The LC tank resonates at about 78 GHz and it appears inductive at lower frequencies and capacitive at higher ones, which helps to achieve broadband matching. In the figure, the simulated S_{11} is smaller than -15dB from 72 to 90 GHz with the LC resonator.

75 

Comparison

	This Work	ISSCC14	ISSCC12	ISSCC13	TMTT07
Technology	40nm CMOS	40nm CMOS	65nm CMOS	130nm SiGe	130nm SiGe
Freq. [GHz]	73	78.5	79	42	60
Gain [dB]	25.3	18.1	24.2	18.5	20
Gain/Stage [dB]	**12.65**	9.05	6.05	6.17	10
P_{1dB} [dBm]	18.9	17.8	16.4 #	N/A	N/A
P_{SAT} [dBm]	**22.6**	20.9	19.3 #	28.4	23
PAE_{MAX} [%]	19.3	22.3	19.2 #	10	6.3
Core Area [mm^2]	0.25	0.19	0.855	4.7 †	2 †
P_{SAT}/ Core Area [mW/mm^2]	**728**	647	99.5. #	150	100
FOM	**98**	90.5	94.3 #	89.4	86.6
Power Combiner	**4-way diff. parallel-series**	**4-way diff. parallel-series**	8-way	**16-way T-line-parallel**	**4-way diff. DAT-series**
PA Topology	NBCA	NCS	CS	CE	CASC

The loss of the on-chip output balun was de-embedded from the measured P_{SAT}, P_{1dB} and PAE_{MAX}.

Again, let's compare. Thanks to the NBCA topology we proposed, we achieved highest reported gain/stage at mm-Wave. This design also achieves the highest output power in CMOS at 60GHz and above.

Thanks to the proposed parallel-series power combiner, we achieved much better power density in the PAs we designed. You can see compared to some excellent designs with high output power at 42GHz and 60GHz, the P_{SAT}/ core area of our design is much higher. The work in ISSCC13 use T-line parallel combiner and the work in TMTT07 use dat-series combiner, which clearly shows the advantage of our proposed parallel-series power combining structure.

76 

Summary: E-band PA

- **E-band for long-haul high-speed PTP**
- **mm-Wave PA customized for E-band**
  - optimized transistor layout & transistor size
  - broadband parallel-series combiner
  - compact floor plan
- **Nearly uniform performance across E-band**
  - 18dB gain with 15 GHz BW-3dB
  - P1dB= 17.55±0.25 dBm, PAEMAX= 20.5±1.8%
- **Neutralized Bootstrapped Cascode Amplifier for high gain and high output power**

Summary: **(1)** E-band has very low atmospheric loss and is suitable for long-haul high-speed point-to-point communications.

(2) In this work, we proposed a broadband PA that is suitable for E-band application. In this design, we optimize the transistor layout and choose a proper transistor size to boost the performance of the single-stage amplifier. We propose a broadband parallel-series combiner that has low insertion loss and provides wideband optimum impedance to the PA. We also optimize the driver stage and interstage network to ensure the broadband performance. The complete design has very compact floor plan that minimize the interconnect loss and also the silicon area.

(3) The proposed PA achieves nearly uniform performance in both 71-76 & 81-86 GHz bands. It has 18dB gain with more than 15GHz bandwidth. The 1-dB compressed power varies less than 0.5dB. The peak-PAE is about 20% in the complete E-band.

(4) Neutralized Bootstrapped Cascode Amplifier for high gain and high output power is also presented.

77 

Conclusions

- **Millimeter-wave technology is one of the key enablers for high-speed 5G, broadband satellite, radar sensing/ imaging.**
- **Advanced CMOS technologies catch mm-Wave frequencies and it has the advantages of high integration and low cost in large volume.**
- **Prior-art CMOS PAs have been realized that achieve high output power, high efficiency and broad bandwidth.**

From this chapter, we draw the following conclusions: **(1)** millimeter-wave technology is one of the key enablers for high-speed 5G, broadband satellite, radar sensing/imaging; **(2)** advanced CMOS technologies catch mm-Wave frequencies and it has the advantages of high integration and low cost in large volume; **(3)** prior-art CMOS PAs have been realized that achieve high output power, high efficiency and broad bandwidth.

78  Reference

1. D. Zhao, P. Reynaert, "A 60-GHz Dual-Mode Class AB Power Amplifier in 40-nm CMOS," *IEEE Journal of Solid-State Circuits (JSSC)*, vol. 48, no. 10, pp. 2323-2337, 2013.
2. D. Zhao, S. Kulkarni, P. Reynaert, "A 60 GHz Dual-Mode Power Amplifier with 17.4 dBm Output Power and 29.3% PAE in 40-nm CMOS," in *Proceedings of ESSCIRC*, 38, pp. 337-340, Bordeaux, 2012.
3. D. Zhao, S. Kulkarni, P. Reynaert, "A 60-GHz outphasing transmitter in 40-nm CMOS," *IEEE Journal of Solid-State Circuits (JSSC)*, vol. 47, no. 12, pp. 3172-3183, 2012.
4. D. Zhao, S. Kulkarni, P. Reynaert, "A 60GHz Outphasing Transmitter in 40nm CMOS with 15.6dBm Output Power," in *IEEE International Solid-State Circuits Conference (ISSCC), Digest Tech. Papers*, Feb. 2012, pp. 170-171.
5. D. Zhao, P. Reynaert, "A 40-nm CMOS E-band 4-Way Power Amplifier with Neutralized Bootstrapped Cascode Amplifier and Optimum Passive Circuits," IEEE Trans. Microwave Theory and Techniques, vol. 63, no. 12, pp. 4083-4089, Dec. 2015.
6. D. Zhao, P. Reynaert, "An E-band Power Amplifier with Broadband Parallel-Series Power Combiner in 40-nm CMOS," *IEEE Trans. Microwave Theory and Techniques*, vol. 63, no. 2, pp. 683-690, Feb. 2015.
7. D. Zhao, P. Reynaert, "A 0.9V 20.9dBm 22.3% E-band Power Amplifier with Broadband Parallel-Series Power Combiner in 40nm CMOS," in *IEEE International Solid-State Circuits Conference (ISSCC), Digest Tech. Papers*, Feb. 2014, pp. 248-249.
8. Y. Zhang, D. Zhao, P. Reynaert, "A Flip-Chip Packaging Design With Waveguide Output on Single-Layer Alumina Board for E-Band Applications," *IEEE Trans. Microwave Theory and Techniques*, vol. 64, no. 4, pp. 1255-1264, Apr. 2016.
9. E. Kaymaksut, D. Zhao, P. Reynaert, "Transformer-Based Doherty Power Amplifiers for mm-Wave Applications in 40-nm CMOS," *IEEE Trans. Microwave Theory and Techniques*, vol. 63, no. 4, pp. 1186-1192, 2015.
10. Y. Zhang, D. Zhao, P. Reynaert, "Millimeter-wave packaging on alumina board for E-band CMOS power amplifiers," *IEEE Topical Conference on Power Amplifiers for Wireless and Radio Applications (PAWR)*, 2015.
11. E. Kaymaksut, D. Zhao, P. Reynaert, "E-band Transformer-based Doherty Power Amplifier in 40 nm CMOS," in *Proceedings of RFIC*, 2014.
12. D. Zhao, P. Reynaert, "A 3 Gb/s 64-QAM E-band Direct-Conversion Transmitter in 40-nm CMOS," in *Proceedings of A-SSCC*, Nov. 2014.
13. D. Zhao, P. Reynaert, " A 40-nm CMOS E-band Transmitter with Compact and Symmetrical Layout Floor-Plans," *IEEE Journal of Solid-State Circuits (JSSC)*, vol. 50, no. 11, pp. 2560-2571, 2015.

Above are the references given which have been used throughout the chapter. Many figures have been adopted from these references. Readers can refer to the corresponding paper for more detailed information.

CHAPTER 8

Ultra-Low Power Zigbee/BLE Transmitter for IoT Applications

Jun Yin

University of Macau, Macao, China

Ultra-low-power (ULP) short-range radios are the key enabler of a wide variety of Internet-of-Things (IoT) products such as wearable/implantable healthcare monitoring. The small physical dimension of those products also significantly limits the size of the battery, pressuring the power budgets of the radios at both architectural and circuit levels. The transceiver efficiency has been significantly reduced in the past few years, thanks to both CMOS technology down scaling and the development of the new circuit techniques for power reduction. In this chapter, the design of ultra-low power transmitters for Zigbee/BLE standard will be discussed from system architecture to circuit techniques. In addition, a 2.4-GHz transmitter using a function-reuse Class-F DCO-PA achieving 22.6% system efficiency at 6-dBm P_{out} will be presented as a case study.

1 **Outline**

- **Introduction**
- **Transmitter (TX) Architecture**
- **Function Reuse Class-F DCO-PA**
- **All-Digital PLL for Frequency Modulation**
- **Measurement Results**

In this talk, first I will introduce you the background and motivation. Then I will discuss the transmitter architecture of the Zigbee/BLE transmitter. After that, I will present the function reuse class-F DCO-PA that helps reduce the transmitter power consumption and chip area. Also, the all-digital PLL to stabilize the DCO-PA frequency during the data transmission will be studied. Finally, the measurement results will be discussed and compared with the state-of-the-art designs.

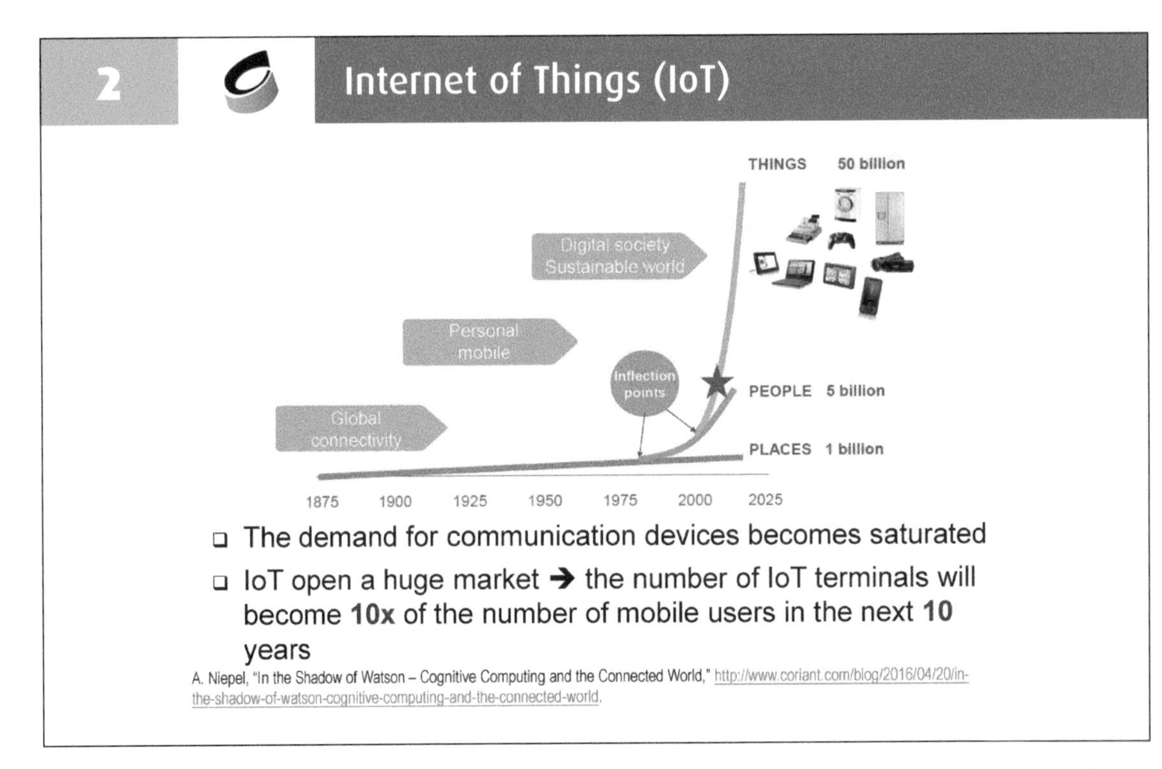

The demand for communication devices today becomes saturated since the number of users is limited by the population of the world. Now around 5 billion people over the world use the wireless communication services. On the other hand, the number of articles over the world is much larger. Connecting all these articles together and linking them to the cloud will create a huge market for the electronic devices and semiconductor chips. People forecast that the number of IoT terminals will exceed 50 billion in the next 10 years, which is ten times of the mobile users today.

3

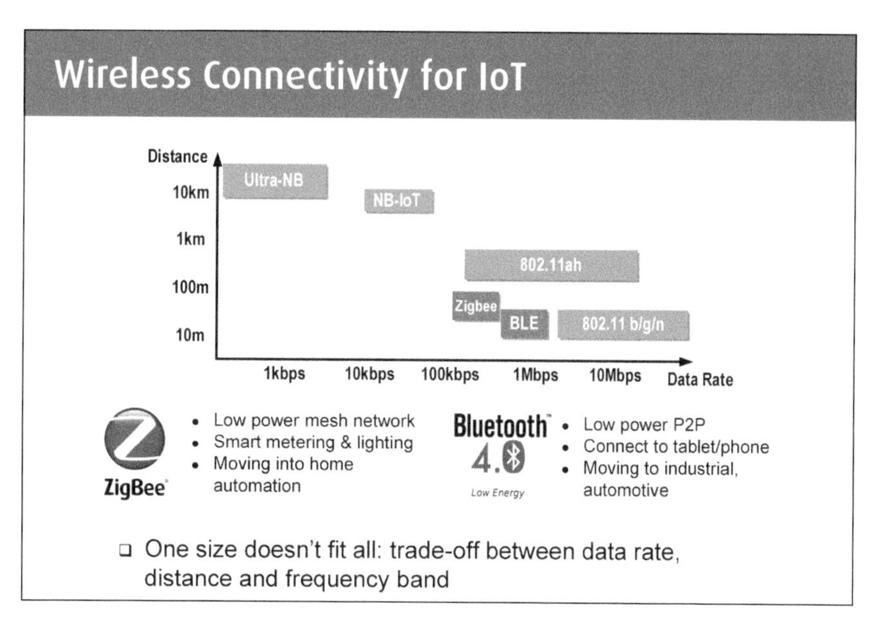

When we talk about the wireless connectivity for IoT applications, we do not specify only a single standard. Actually, there are many standards to fit the different requirements of IoT scenarios. For example, the ultra-narrow-band (ultra-NB) standards such as SigFox and LoRa, target at the applications requiring low data rate and long distance of longer than 10 kilometers while the conventional WiFi standards such as 802.11b/g/n aims at the applications requiring high data rate but with a short distance of only several tens of meters. We can see the data rate and communication distance is significantly traded off. My focus here will be on the two standards, e.g. Zigbee and Bluetooth low energy (BLE). Their data rates and distances are in the middle between the Ultra-NB and WiFi standards. For the Zigbee standard, one most important feature is that the low-power devices can support mesh network. Its application mainly focuses on the home automation such as smart metering and lighting. The BLE standard targets for connecting device only from point to point. It is compatible with the smart phones and tablets. Its application is also moving towards the industrial control and automotive.

4

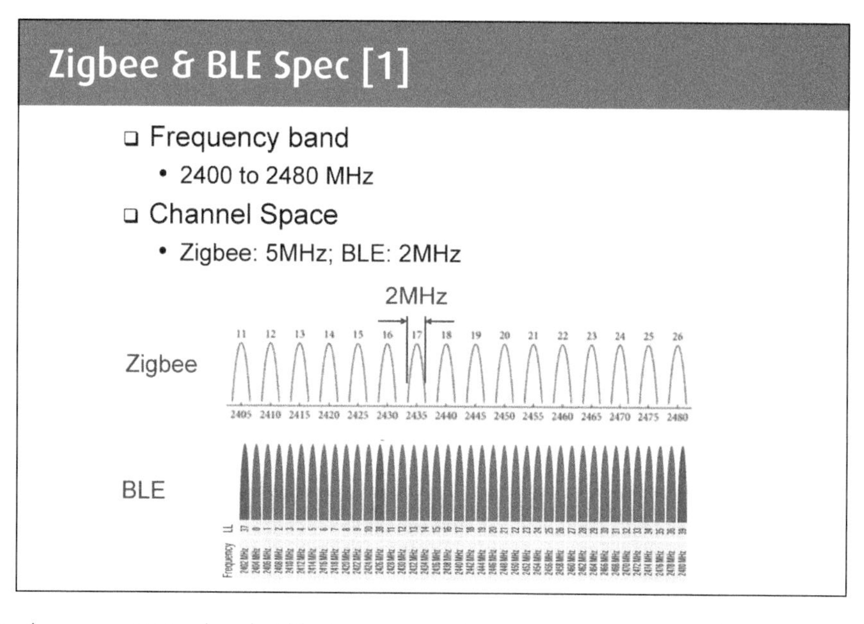

The Zigbee and BLE standards share some similarities. Their frequency bands are both located between 2400 and 2480MHz covering 80MHz bandwidth. The channel space of Zigbee is 5MHz but the signal energy is mainly concentrated in 2MHz bandwidth. For BLE, the channel space is only 2MHz, which can have more channels within the same 80MHz bandwidth. More channels mean that the standard can support more users at the same time.

5 

Zigbee & BLE Spec [2]

- Data-Rate
  - Zigbee: 250kbps
  - BLE 4.0: 1Mbps
  - BLE 5.0: 2Mbps
- Modulation Scheme: frequency modulation (FM)

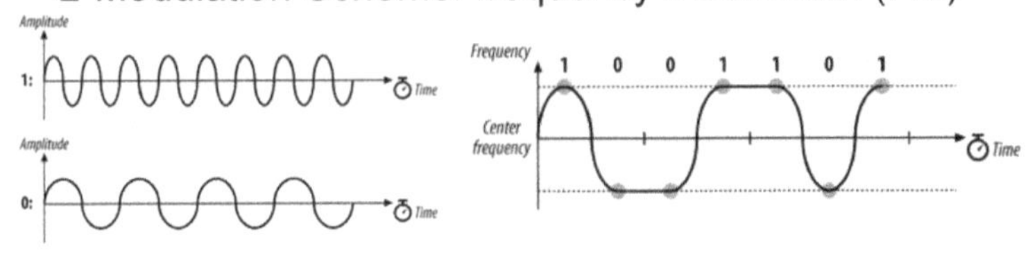

The data rate of the Zigbee is 250 kbit per second while that of the BLE is higher, e.g. 1 Mbit and 2 Mbit per second for BLE4.0 and 5.0 respectively. These two standards use similar modulation schemes, i.e. frequency modulation.(FM) The simple modulation scheme can save the power consumption. The working principle of the FM is like this. We can use a higher freuency to represent the data 1 and a lower frequency to represent the data 0 as shown in the bottom left figure. The bottom right figure shows the relationship between frequency and time when transmitting data.

6 

Basic TX Architecture for FM

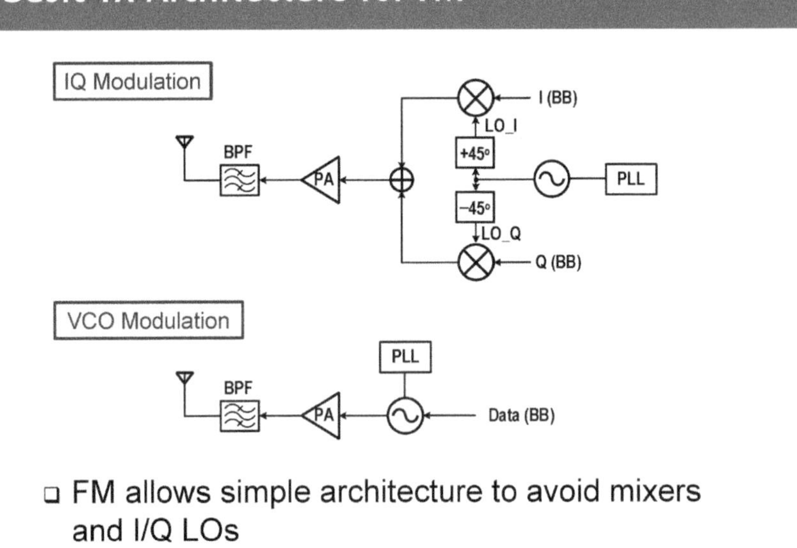

In the conventional transmitter architecture as shown in the top figure, I/Q LOs which have 90-degree phase shift are required for the upconversion of the baseband signal. The upconverted signal is then amplified by the power amplifier (PA) and sent out through the antenna. Here, the generation of I/Q LOs and the mixers for upconversion consume large power. Since the FM scheme for Zigbee and BLE has a constant envelope, one possible way to reduce the complexity of the transmitter is to directly control the VCO frequency for FM modulation as shown in the bottom figure. The direct VCO modulation scheme avoids the use of power hungry mixers and I/Q LOs and becomes popular in the low-power FM transmitter design.

7

BLE & ZigbeeTX [1]

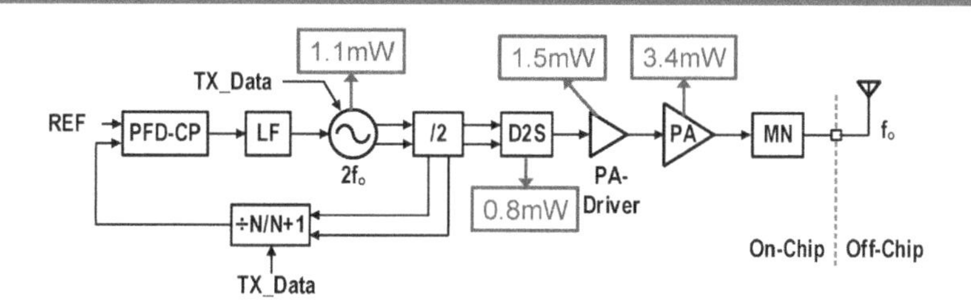

- Separate VCO and PA
- VCO operates at $2f_0$:
  - Reduce the pulling from PA to VCO when PA starts up
- Low system efficiency of ~9.9% at P_{out} = 0dBm

J. Prummel *et al.*, "A 10 mW Bluetooth low-energy transceiver with on-chip matching," *IEEE J. Solid-State Circuits*, vol. 50, Dec. 2015.

This slide shows the typical transmitter architecture using VCO modulation scheme. The VCO operates at twice of the carrier frequency to reduce the pulling from PA to VCO when PA starts up. The PA consumes 3.4mW and is the most power-hungry blocks. However, the VCO plus the differential-to-single-ended (D2S) converter and PA driver consumes another 3.4mW, which lowers the system efficiency to only ~9.9% when delivering an output power of 0dBm. How can we further reduce the power consumption and improve the system efficiency?

8

BLE & ZigbeeTX [2]

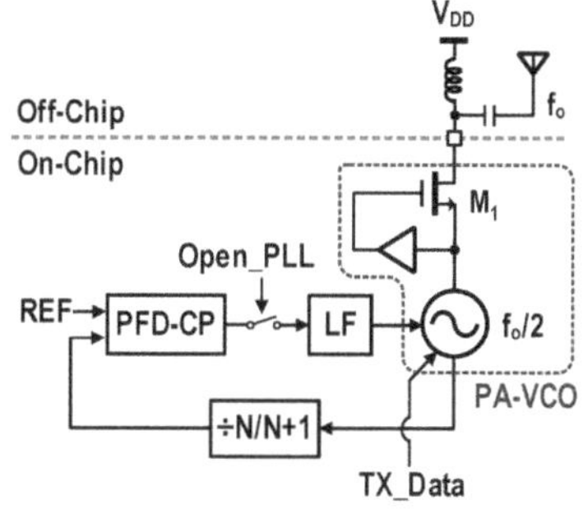

- Current-reuse PA-VCO
- Reduced voltage headroom ➔ Limited maximum P_{out} of −1dBm
- Need off-chip passives (1 inductor and 1 capacitor)

C. Li and A. Liscidini, "Class-C PA-VCO cell for FSK and GFSK transmitters," *IEEE J. Solid-State Circuits*, vol. 51, Jul. 2016.

People proposes the architecture that stacks the PA on top of the VCO so that the current can be reused between the PA and VCO. Here, the VCO is differential and operating at half of the carrier frequency to reduce its power consumption. The transistor M_1 serves as a common-source amplifier which extracts and amplifies the second harmonic of the oscillation signal from the common-mode node of the VCO. The positive feedback from the source to the drain nodes of M_1 is utilized to enhance the PA gain. Although the current-reuse helps save the power consumption, the VCO will reduce the voltage headroom of the PA, which limits the maximum output swing of the PA. For a standard supply of 1V, the reported maximum output power is −1dBm. This work also requires one off-chip inductor as the RF choke and one off-chip capacitor to prevent the DC current from flowing into the antenna. The off-chip components will increase both the cost and form factor of the final product.

9

Proposed TX

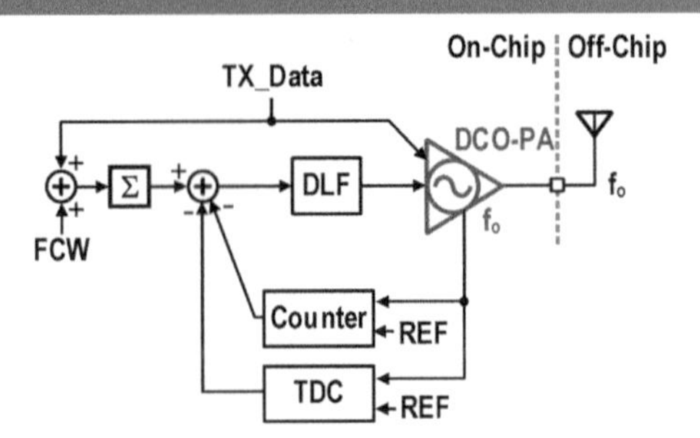

- Function-reuse DCO-PA to absorb the power of DCO and PA-driver → Save power w/o voltage headroom limit
- A 6-port transformer to combine the resonant tank of DCO and matching network (MN) of the PA → Save chip area

X. Peng *et al.*, "A 2.4-GHz ZigBee Transmitter Using a Function-Reuse Class-F DCO-PA and an ADPLL Achieving 22.6% (14.5%) System Efficiency at 6-dBm (0-dBm) Pout," *IEEE J. Solid-State Circuits*, vol. 52, Jul. 2016.

If the functions of VCO and PA can be completely merged into one building block, which means we use only one block to realize the functions of VCO and PA simultaneously. Then the power consumption of the VCO and PA driver can be fully absorbed into the PA. Also, the voltage headroom limitation of the stacked VCO-PA can be avoided. Here, we propose a function-reuse DCO and PA to save power consumption while still achieve a large output power. Also, a single 6-port transformer is employed to combine the resonant tank of the DCO and matching network of the PA together which saves the chip area. The all-digital PLL is employed to withhold the DCO-PA frequency pulling due to the interferer picked up by the antenna as well as the load impedance variation. Also, the all-digital PLL can stabilize the DCO-PA frequency during the date transmission.

10

Class-F PA

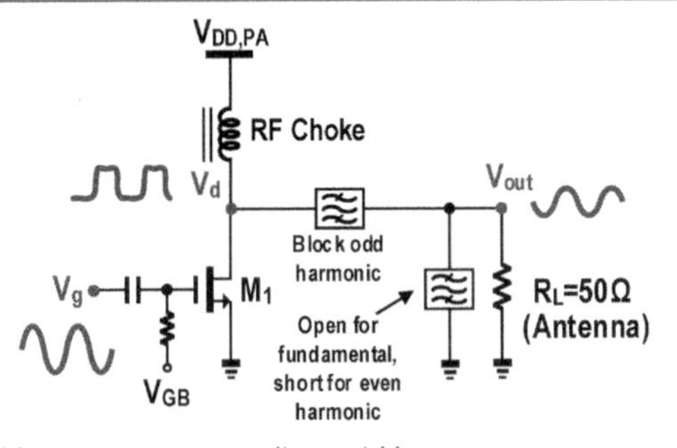

- Resemble a square-wave voltage at V_d
- Reduce the voltage magnitude across M_1 when it is conducting → reduce the power loss
- By simply blocking the 3rd harmonics → a maximum drain efficiency of ~88% can be achieved

S. Kee, "The class E/F family of harmonic-tuned switching power amplifiers," *Ph.D. dissertation, Dept. Elect. Eng., California Inst. Technol., Pasadena, CA, USA, 2002.*

The class-F PA can achieve a high-power efficiency by resembling a square-wave voltage at the drain node Vd by properly design the load to block odd harmonics and short the even harmonics. The sharp rising and falling edges of the square-wave voltage can reduce the voltage magnitude across the transistor M_1 when it is conducting and thus reduce the power loss. In practice, a maximum drain efficiency of ~88% can be achieved by simply blocking the third harmonic.

11

Concept of Class-F DCO-PA

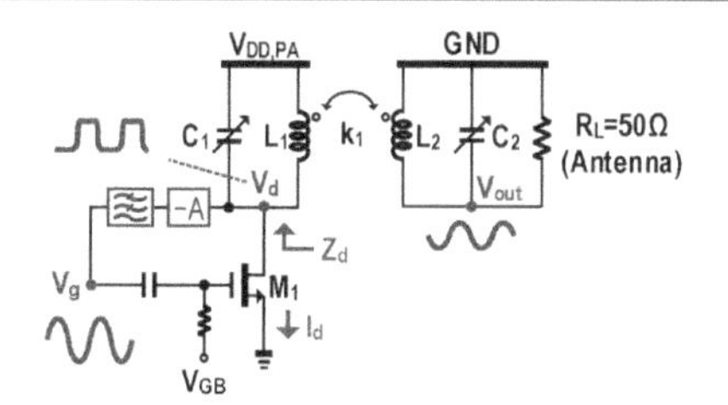

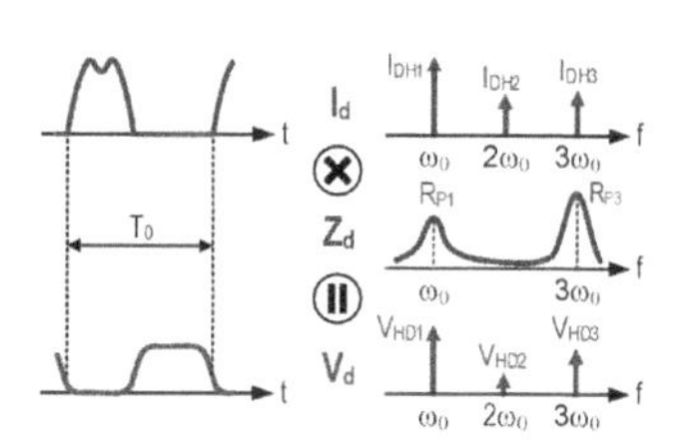

- Transformer tank
  - Generate 3rd-harmonic component at Vd
  - Impedance transformation from output (R_L) to drain (Z_d)
- Feedback from drain (V_d) to gate (V_g) for oscillation

To generate the square-wave voltage waveform at V_d, the high-order transformer tank can be employed to create another impedance peak located at $3\omega_0$. By multiplying with the 3rd harmonic component in the drain current I_d, a strong 3rd harmonic component at V_d can be generated. Here M_1 is biased at the class-C mode to increasing the 3rd harmonic component in I_d. The transformer tank also helps change the ratio between the output impedance R_L and the drain impedance Z_d according to the desired output power. To make the class-F PA self-oscillate, we can amplify V_d with a gain of -A and feedback it to the gate of M_1. A low-pass filter serves to remove the harmonics and restore a sinusoid waveform at V_g node.

Then the question is how to design the transformer tank (L_1, L_2, C_1, C_2 and k_1) to make the high frequency peak of Z_d locate at $3\omega_0$ while keep the loss introduced by the transformer tank as low as possible?

12

How to Design the Transformer Tank?

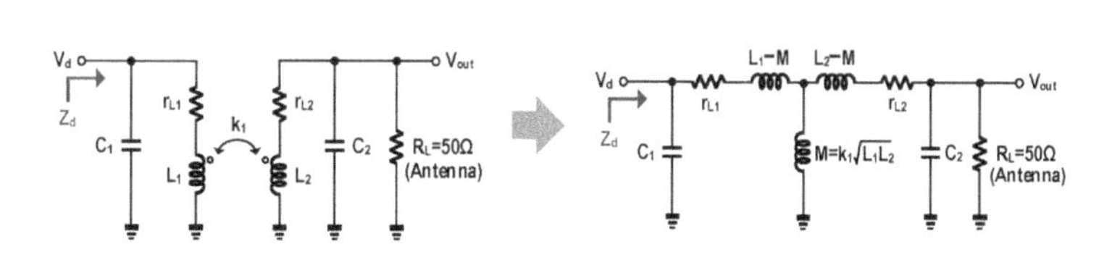

$$Z_d = \omega L_1 \cdot \frac{-(1-k_1^2) + jQ_{load}\omega_1^2[\omega_2^2 - \omega^2(1-k_1^2)]}{Q_{load}[\omega^4(1-k_1^2) - \omega^2(\omega_1^2+\omega_2^2) + \omega_1^2\omega_2^2] + j\omega_2^2[\omega_1^2 - \omega^2(1-k_1^2)]}$$

$$\omega_1 = 1/\sqrt{L_1C_1} \qquad \omega_2 = 1/\sqrt{L_2C_2} \qquad Q_{load} = R_L/\omega L_2$$

- Two resonate frequencies (ω_H, ω_L) can be achieved by letting $\angle Z_d$ to be 0

To calculate the impedance Zd looked from the drain node, the T-model for the transformer tank can be used. A loading impedance of 50 Ω is put here to represent the typical antenna impedance. By defining $\omega_1=1/\sqrt{L_1C_1}$, $\omega_2=1/\sqrt{L_2C_2}$ and $Q_{load}=R_L\omega L_2$, we can get the expression for Zd. Here ω_1 and ω_2 represent the resonant frequency of two separate tanks L_1, C_1 and L_2, C_2 when there is no magnetic coupling between them. Q_{load} is the quality factor of the secondary coil L_2 when loaded by the R_L.

By forcing the phase of Z_d to be zero, we can find the solutions for the two resonant frequencies, i.e. ω_H and ω_L, for this 4th-order tank.

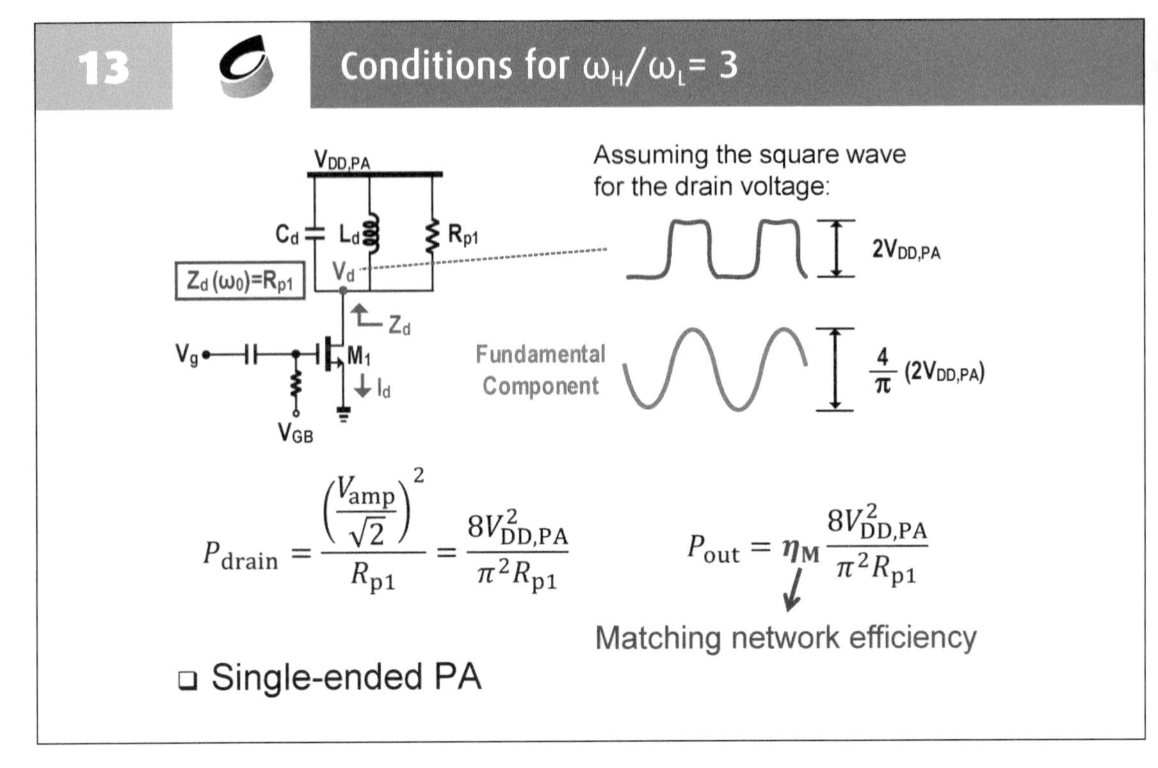

Since analytical solutions for ω_H and ω_L are quite complicated. We instead use the contour plot to study the conditions that guarantees ω_H/ω_L=3, which can provide more insights. Here the figure plots the contour curves of ω_H/ω_L=3 for different k_1 and X combinations where X=$(\omega_2\omega_1)^2$. We can see that if Q_{load} is quite large, e.g. $Q_{load}=\infty$, there are two solutions of X that can guarantee ω_H/ω_L=3 for the same k_1 as long as $k_1 < 0.8$. On the other hand, if k_1 is small, only one solution of X can be founded for certain k_1. Although we find some relationships between k_1 and X, there are still too many combinations. Thus, we need to use more constraints to further narrow the range of k_1 or X.

14  Determine the Impedance at the Drain [1]

Another constraint we can use is the impedance looked from the drain node. This impedance can be determined by the desired output power. Let us look at the single-ended PA topology at first. For the class-F operation, we can assume the square-wave voltage waveform for V_d. To achieve high efficiency, we want the swing of V_d to be as large as possible for certain V_{DD}. In our topology, the maximum voltage swing of V_d is twice of the supply voltage, i.e. $2V_{DD,PA}$. According to the Fourier Series, the fundamental component is a sinusoid waveform with a voltage swing of $8V_{DD,PA}/\pi$. By applying this sinusoid waveform on the drain resistance R_{p1}, we can find the power at the drain node at fundamental frequency P_{drain}. Since the matching network also introduces loss in practice, the output power at fundamental frequency P_{out} can be obtained by multiplying P_{drain} with the matching network efficiency η_M. By doing so we can find the relationship between the PA output power and drain impedance R_{p1}.

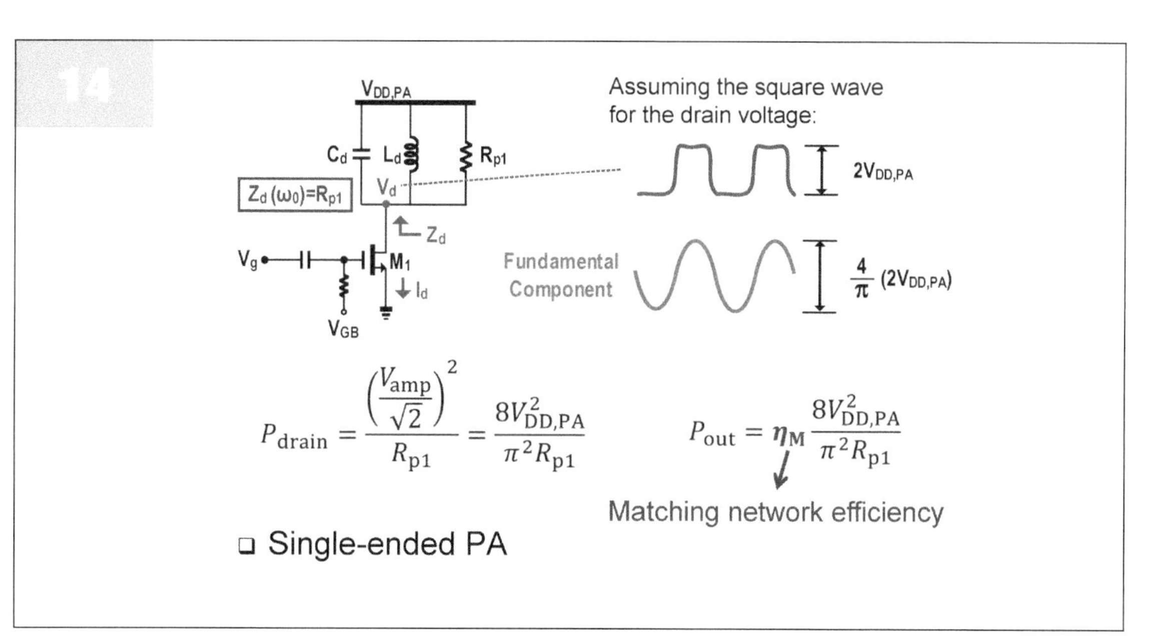

15

Determine the Impedance at the Drain [2]

$Z_d(\omega_0)=R_{p1}$

$V_{d+}-V_{d-}$: $4V_{DD,PA}$

Fundamental Component: $\frac{4}{\pi}(4V_{DD,PA})$

$$P_{drain}=\frac{\left(\frac{V_{amp}}{\sqrt{2}}\right)^2}{R_{p1}}=\frac{32V_{DD,PA}^2}{\pi^2 R_{p1}} \qquad P_{out}=\eta_M\frac{32V_{DD,PA}^2}{\pi^2 R_{p1}}$$

- Differential PA
- Example: η_M=55%, $V_{DD,PA}$=1V, P_{out}=7dBm ➔ R_{p1}=100Ω
  If $V_{DD,PA}$=0.5V ➔ P_{out}=1dBm

What will happen if we use the differential PA architecture? If we use the same tank and supply voltage here and just change the input from single-ended to differential, we expect that the differential output swing now equals to $V_{d+}-V_{d-}$ is doubled and so is the output swing at fundamental frequency. Thus, Pdrain will increase to four times of that in the single-ended PA. And Pout will also increase to four times if we assume η_M is unchanged.

As an example, if we assume η_M=55% and $V_{DD,PA}$=1 V, we need a Rp1 of 100 Ω to get a 7 dBm output power. If we reduce the supply voltage by half, the same R_{p1} will only result in an output power of 1 dBm. In this design, we will just set R_{p1}=100 Ω to give an output power of 1 dBm at $V_{DD,PA}$=0.5 V.

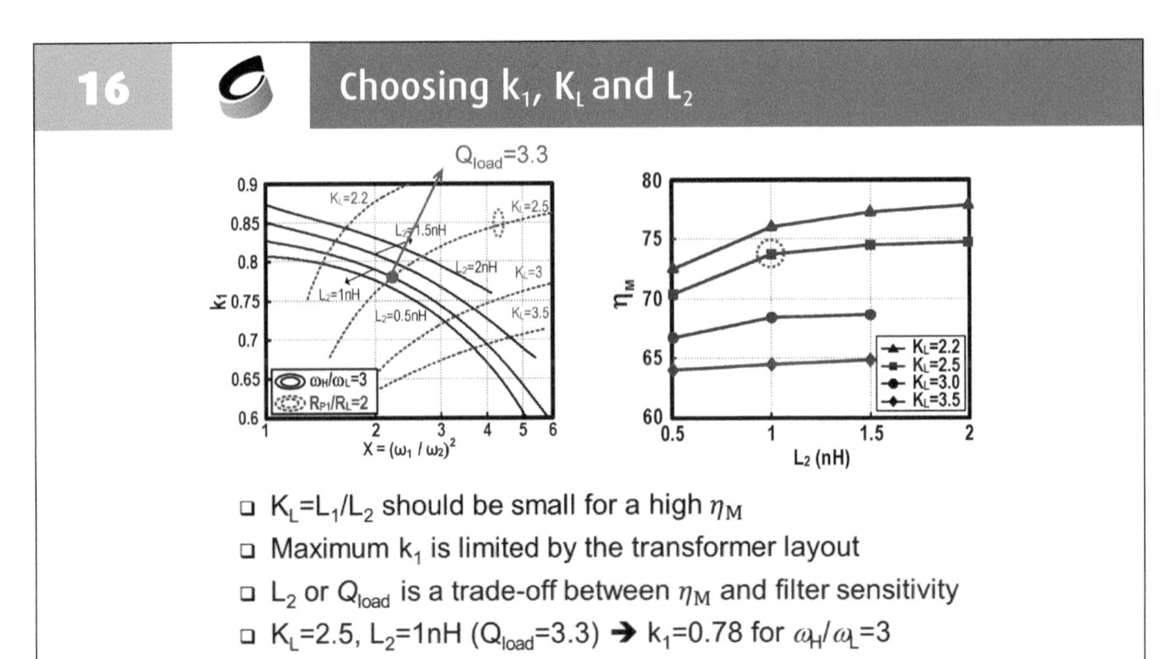

After determining the Rp_1, we can fix the impedance transfer ratio $Rp_1/R_L=2$. Then we come back to the contour plot of $\omega_H/\omega_L=3$ for different L_2 and add another set of contours of $R_{p1}/R_L=2$ for different K_L in the blue dashed line in the left figure. Here, $K_L=L_1/L_2$ is the inductance ratio between the primary and secondary coils of the transformer which is one important fact that determines the impedance transfer ratio. Compared with the figure in slide 14, we convert Q_{load} to L_2 here using the definition that $L_2=RL/\omega Q_{load}$.

Now the intersection points of the two contour plots give all the possible solutions of k_1 and X combinations that can satisfy the conditions for both $\omega_H/\omega_L=3$ and $R_{p1}/R_L=2$. By extracting these solutions, we can simulate the matching network efficiency η_M as shown in the right figure. First, we can see that a high η_M requires a small K_L. Then we come back to the left figure, a small K_L requires a large k_1 if X is kept the same. Generally, a large k_1 helps to improve the η_M. However, the maximum k_1 is limited by the physical layout of the transformer. Here, we choose $K_L=2.5$ that results in a practical k_1 in the range from 0.77 to 0.81. The transformer tank can also help filter the harmonic components at the output. The choice of L_2 is a trade-off between η_M and filter sensitivity since k_2 directly determines the Q_{load}. In this design, we choose L_2=1nH which can provide a reasonable attenuation for the harmonics with Q_{load}=3.3 while still keeping a high η_M of ~74%.

17  Proposed Class-F DCO-PA

This slide shows the complete schematic of the proposed Class-F DCO-PA. The differential PA topology is employed to suppress the even harmonics in the output after the differential to single-ended conversion realized by the transformer coils L_1 and L_2. To make the PA oscillate, V_d is feedback to V_g through another coil L_3 that is magnetically coupled to L_1, which can help boost the voltage swing at the gate and eliminate the bimodal oscillation. We will discuss the bimodal oscillation issue in detail in the next few slides. Besides, to stabilize the DCO frequency, the capacitor bank at drain is controlled by the all-digital PLL (ADPLL). This arrangement can achieve finer DCO resolution since the capacitor at drain node has less effect on the oscillation frequency than the capacitor at the gate node.

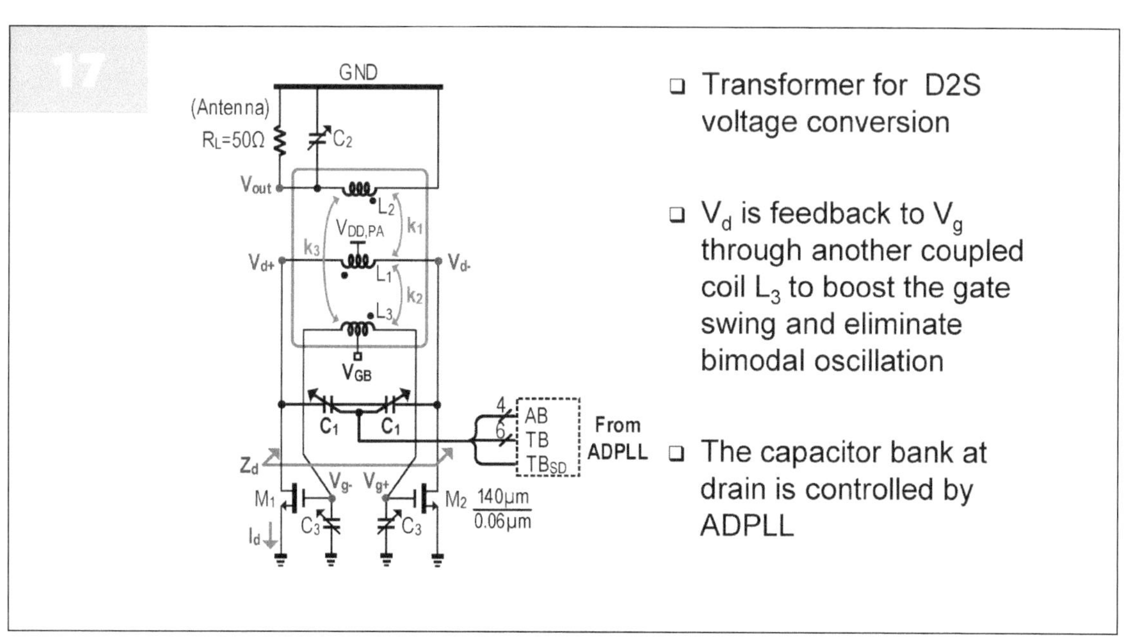

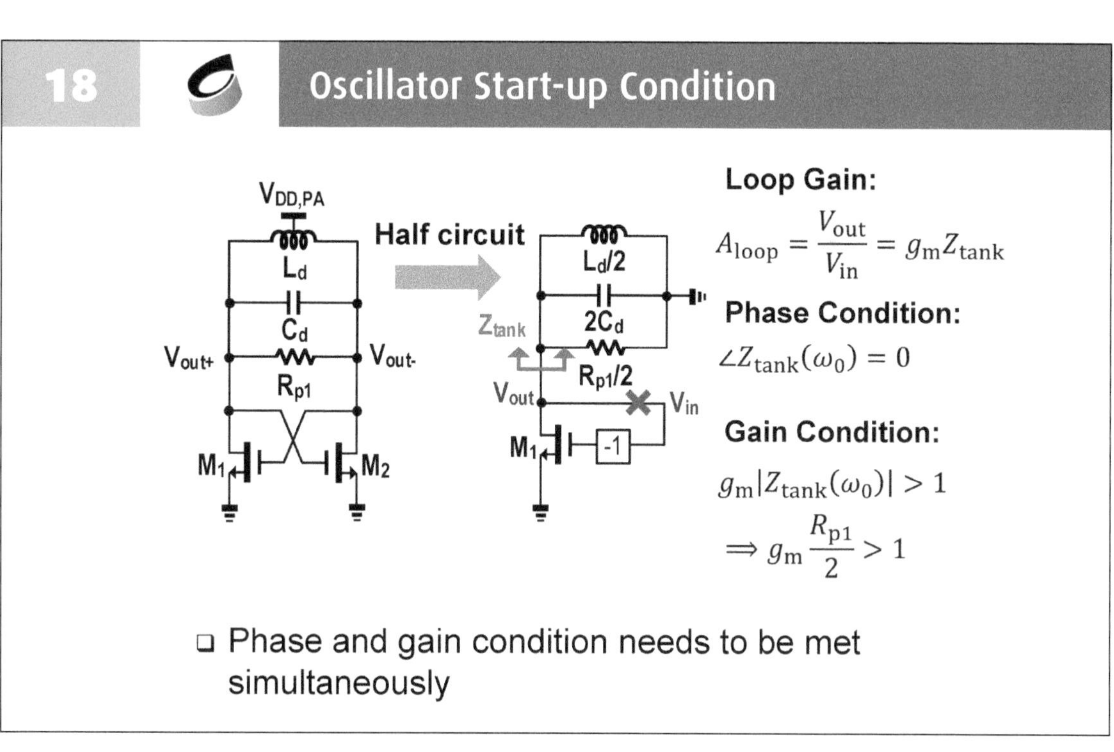

To explain the bimodal oscillation, let us go back to review the oscillator startup condition at first. To analyze the startup condition of the typical oscillator topology that using a cross-coupled pair to compensate the tank loss, we can use the half circuit as shown in the right figure. To find the loop gain, we can break the loop at any point. Here we break the loop at the point between Vout and the gain of −1. Then the loop gain would equal to gmZtank. Oscillation can start up at ω_0 only when both the phase and gain conditions are satisfied, i.e. $A_{loop}(\omega_0)$=0 and $A_{loop}(\omega_0)$>1. Since $|Z_{tank}(\omega_0)|$ just equals to $R_{p1}/2$ when $Z_{tank}(\omega_0)$=0. The transcondutance gm of M_1 or M_2 must be larger than $1/R_{p1}$ to satisfy the gain condition.

19

Then let us look at the frequency response of a simple 2nd-order LC tank. This tank only has a unique frequency ω_0 that satisfies the phase condition. Also, the tank impedance only has a unique impedance peak located at ω_0. If we use this simple LC tank, the oscillator will always oscillate at this unique frequency ω_0.

Frequency Response of a LC-Tank

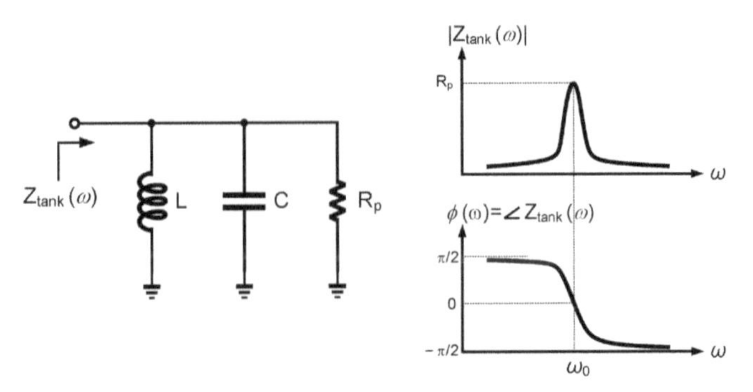

- $\angle Z_{tank} = 0$ at only ω_0 ➔ an unique oscillation frequency can be guaranteed

20

If we look at the frequency response of a 4th-order transformer tank, as shown in the left figure, the amplitude of tank impedance Z_{tank} when looked from the primary coil L_1 will have two peaks R_{p1} and R_{p2} located at ω_L and ω_H that both satisfy the phase conditions. Then the problem comes. If R_{p1} and R_{p2} are close to each other and the g_m of the cross-coupled transistors is larger than both $2/R_{p1}$ and $2/R_{p2}$, the oscillator may have a chance to oscillate at the frequency of either ω_L and ω_H when starting up. This phenomenon of ambiguous oscillation frequency is called bimodal oscillation, which should be avoided in the practical design. Then can we change the oscillator topology to avoid the bimodal oscillation?

Frequency Response of a Transformer-Tank

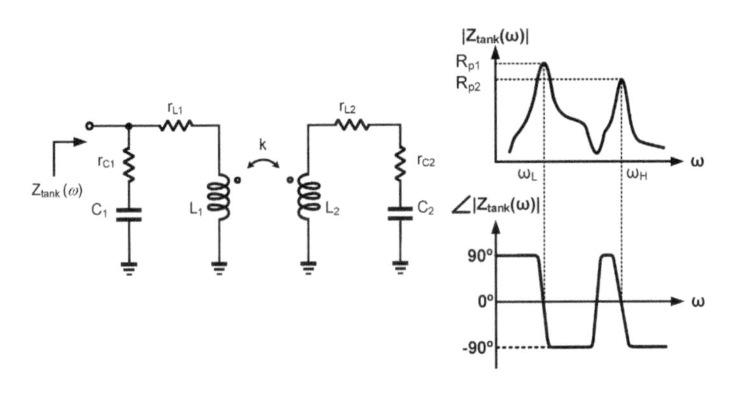

- One port configuration: $\angle Z_{tank} = 0$ at both ω_L and ω_H ➔ Bimodal oscillation

21

Two-Port Oscillator

$A_{loop} = g_m Z_{21}$

Phase Condition:

$\angle Z_{21} = 0$

Gain Condition:

$g_m |Z_{21}(\omega_L)| > 1$

- Two-port configuration: $\angle Z_{21} = 0$ at only ω_L ➔ unique oscillation frequency

If we re-arrange the connection of the transconductance to configure the oscillator as a two-port design, the phase and gain conditions will all related to the impedance Z_{21} instead of Z_{11}. For the amplitude of Z_{21} as shown in the right top figure, it still has two peaks located at ω_L and ω_H similar to that of Z_{11}. However, only the phase of Z_{21} at ω_L equals to zero and satisfy the phase condition. Thus, the two-port oscillator will only oscillate at ω_L which avoids the bimodal oscillation.

22

Design of k_2 and L_3 [1]

L_3 (nH)	10	8	2.8	0.7	0.55
k_2	0.25	0.35	0.45	0.55	0.65

- Different L_3 and k_2 combination to keep P_{out} = 7dBm
- Small k_2 ➔ Large gate amplitude

In our DCO-PA design, the oscillator is configured with a two-port topology through an additional coil L_3 coupled to coil L_1 with a coupling coefficient k_2. It is worth to note that k_1 is not necessary for the DCO-PA function. However, it is impossible to completely isolate L_2 and L_3 due to the practical layout limitation. Here, it is reasonable to assume k_3 is a fraction of k_2, i.e. $k_3 = 0.83k_2$. To design k_2 and L_3, first we find the different L_3 and k_2 combinations that keep a constant output power of 7 dBm, as shown in the table. Then we simulate the DCO-PA performance using these k_2 and L_3 values. The right bottom figure shows the simulated voltage swing of the gate voltage V_g for different values of k_2 at V_{GB} = 0.24 V. It can be seen that a small k_2 and a large L_3 result in a large swing of V_g.

23 

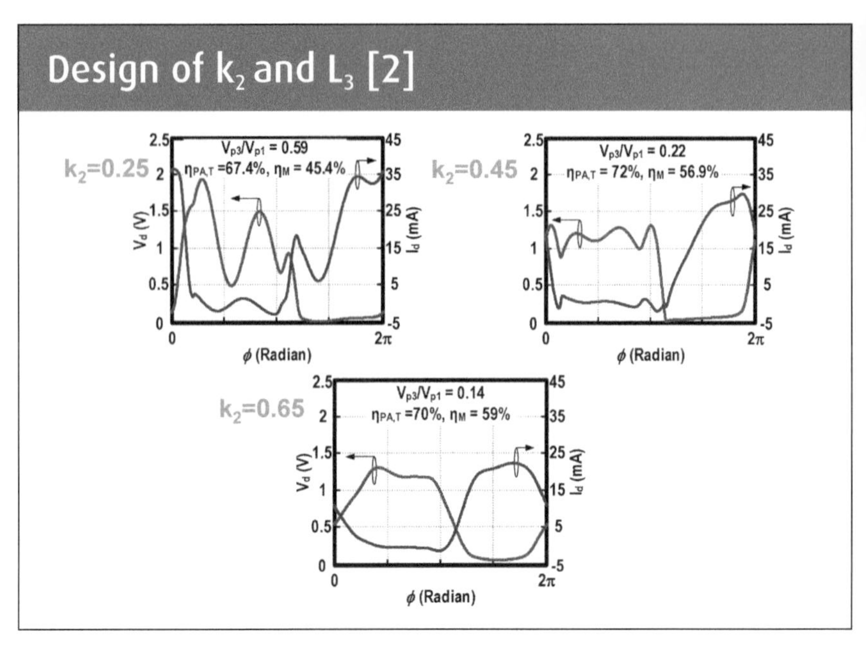

This slide shows the plots of drain voltage V_d and current I_d for different combinations of k_2 and L_3. The ratio of the third harmonic to the fundamental voltage V_{p3}/V_{p1} increases when k_2 goes down because the large swing of V_g boosts the third harmonic current by imposing M_1 and M_2 to operate like a switch. For $k_2 = 0.25$, a large V_{p3}/V_{p1} of 0.59 induces a deep valley in the voltage waveform forcing it to deviate significantly from a square wave. The large amplitude of the overlapped current and voltage raises the power loss on M_1 and M_2.

24 

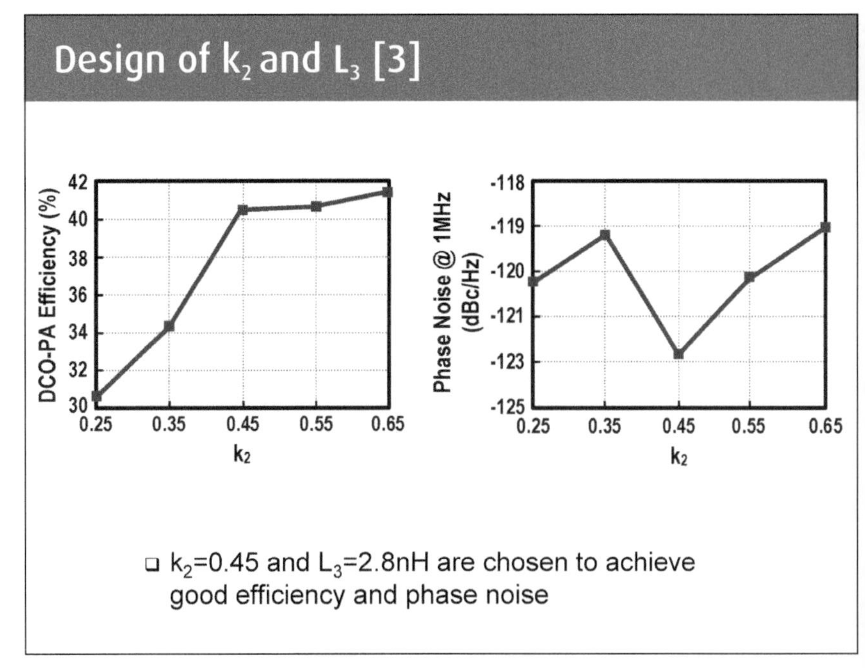

Here the left figure reveals that the DCO-PA efficiency grows with k_2, but saturates when $k_2 > 0.45$. Thus, a large k_2 is preferable to improve the power efficiency and reduce the voltage swing at the gate of M_1/M_2 for better reliability. According to the simulation results in the right figure, a k_2 of 0.45 would result in an optimized phase noise performance. Thus, we choose a moderate k_2 of 0.45 to balance the power efficiency and phase-noise performance.

25 

Phase Noise Analysis

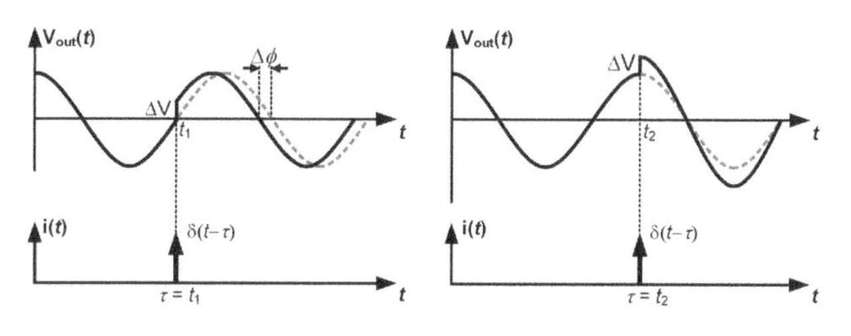

$$\Delta\phi = \underbrace{\Gamma(\omega_0\tau)}_{\text{ISF}}\frac{\Delta q}{q_{\max}} = \Gamma(\omega_0\tau)\frac{\Delta V}{V_0}$$

- Linear and Time-Variant (LTV) Model
- $\Gamma(\omega_0\tau)$: Impulse Sensitivity Function (ISF)

A. Hajimiri and T. H. Lee, "A general theory of phase noise in electrical oscillators," *IEEE J. Solid-State Circuits*, vol. 33, Feb. 1998.

Actually, the phase noise performance of an oscillator can be evaluated by the linear and time-variant model proposed by A. Hajimiri and T. H. Lee in 1998. The model is time variant since the phase perturbation **ΔΦ** depends on the time when the noise charge is injected to the oscillator. As shown in the left figure, if the noise charge is injected at the zero-crossing point of V_{out}, i.e. t_1, **ΔΦ** would reach its maximum. On the other hand, **ΔΦ** becomes zero if the noise charge is injected when the sinusoid waveform of V_{out} reaches the peak. Since the injected noise charge is usually much smaller than the signal charge swing, **ΔΦ** is proportional to the noise charge amplitude. Thus, the model is linear. The normalized phase perturbation is characterized by the dimensionless impulse sensitivity function (ISF), i.e. $\Gamma(\omega_0\tau)$...

26 

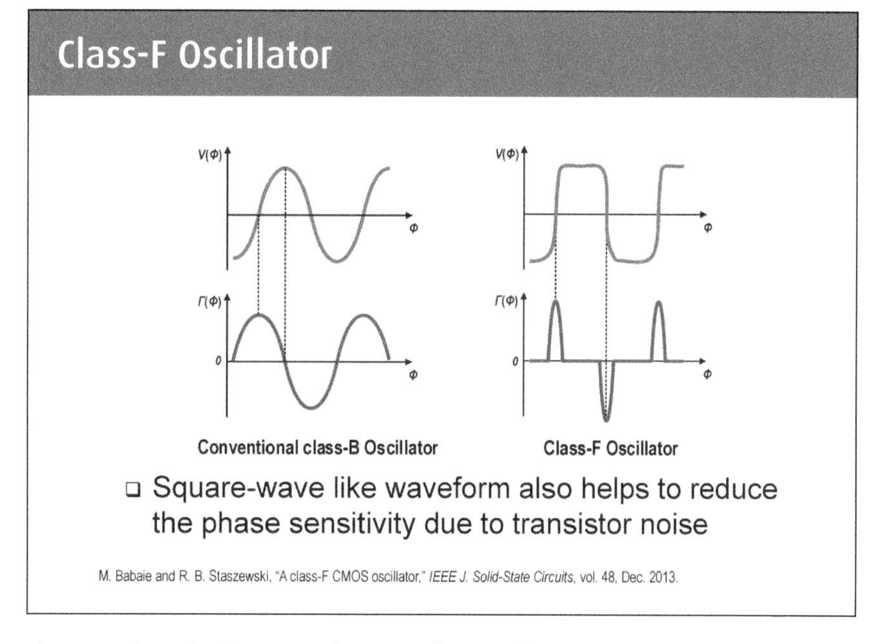

By comparing the ISFs for different oscillator topologies, we can tell which topology has a superior noise performance. The ISF of the conventional class-B oscillator with a sinusoid output waveform also has a sinusoid shape, as shown in the left figure. For the class-F oscillator that has a square-wave like output waveform, the shape of its ISF becomes narrow pulses as shown in the right figure. Since the rms value of the ISF for the class-F oscillator is smaller than that for the class-B oscillator, the class-F oscillator will have a reduced phase sensitivity due to the transistor noise.

27 

Layout of the Six-Port Transformer

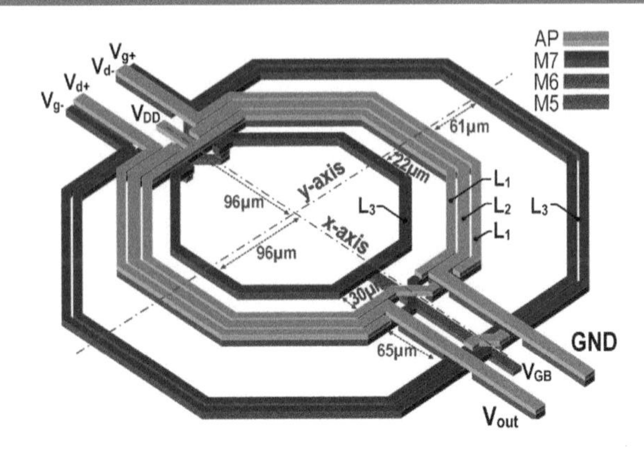

- L_1=2.46nH, L_2=0.92nH and L_3=5.62nH
- k_1=0.7, k_2=0.45 and k_3=0.37

The layout of the six-port transformer is non-trivial since it has to balance a number of parameters, $L_{1\text{-}3}$, and $k_{1\text{-}3}$. To achieve a large coupling coefficient k_1 between L_1 and L_2, we have to interleave the two coils in the layout with a minimum space of 2 μm between the adjunct metal traces. Then, we pick a turn ratio of 2:1 to realize a K_L of ~2.5 and a k_1 of ~0.7. To improve the η_M, we route in parallel both the 1.2 μm Alucap (AP) layer and the 0.9 μm top metal (M7) for coils L_1 and L_2 to enhance Q_{L1} and Q_{L2}. Due to the limited contact area between the AP layer and M7 allowed by the technology, we use only M7 for the coil L_3. To realize a large inductance of 5.6 nH and a moderate coupling coefficient k_2 of 0.45, we draw L_3 in three turns: placing one at the inner side of L_1 and L_2 and the other at the outer side of L_1 and L_2. Since k_2 mainly depends on the space between L_1 and the two large outer turns of L_3, the diameter of the size of L_3 represents a degree of freedom to obtain the size of L_3.

28 

EM Simulation for Transformer

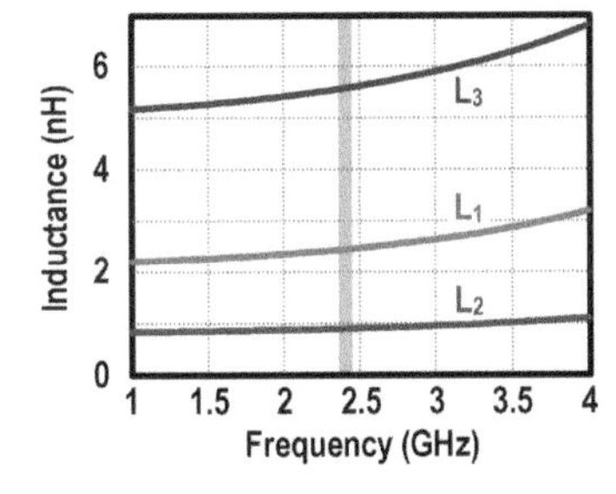

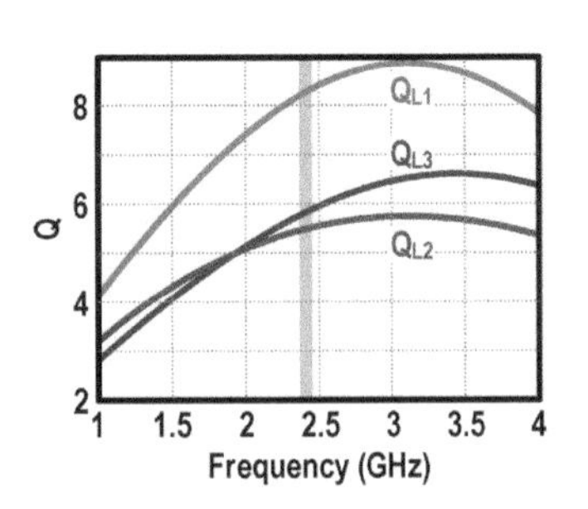

- Electromagnetic (EM) simulation
- Q_{L1}=8.2, Q_{L2}=5.6 and Q_{L3}=5.9

This slide shows the Electromagnetic simulation results for inductance and the quality factor Q. The Qs for L_1, L_2 and L_3 are 8.2, 5.6 and 5.9, respectively.

29 

All-Digital PLL (ADPLL) [1]

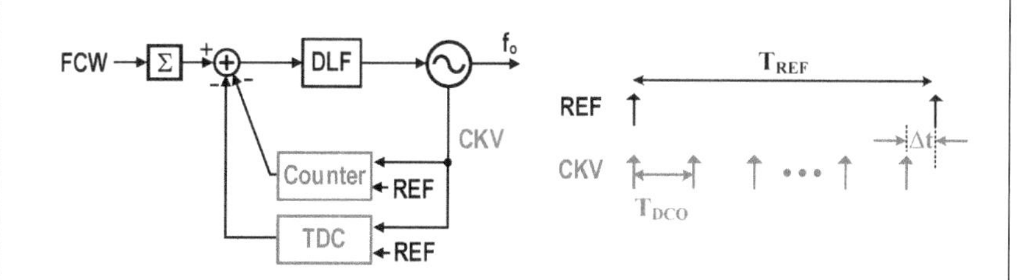

- ADPLL function
  - Detect the DCO phase (frequency)
  - Control the DCO frequency to $f_{DCO} = \text{FCW} \times f_{REF}$

R. B. Staszewski and P. T. Balsara, "All-Digital Frequency Synthesizer in Deep-Submicron CMOS," *John Wiley & Sons, Inc.*, 2016.

To stabilize the DCO-PA frequency, an all-digital PLL is employed in this design. Basically, the ADPLL has two tasks. One is to detect the DCO phase and compare it with an accurate phase provided by the reference clock REF. The other is to control and stabilize the DCO frequency based on the phase detection results. The digital counter and the time-to-digital converter are usually used for the phase detection. As shown in the timing diagram at the right figure, the counter can detect there are how many DCO cycles within one REF cycle while the TDC serves to detect the corresponding quantization error. Thus, the input range of the TDC is one DCO period T_{DCO}.

30 

All-Digital PLL (ADPLL) [2]

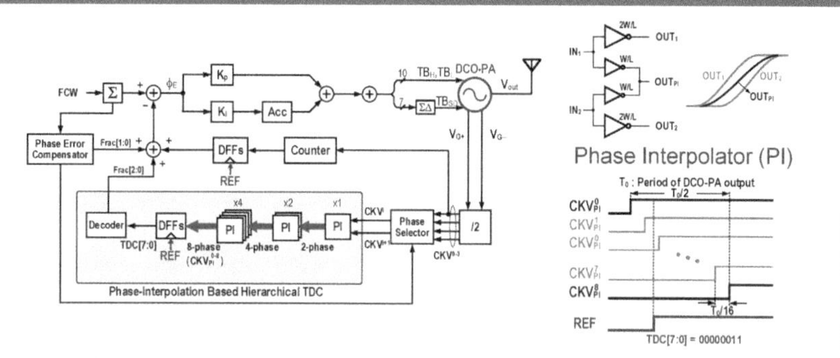

- Phase selection to reduce the input range of the TDC by 4x
- Phase-Interpolation to avoid the TDC gain normalization
- TDC resolution: 26 ps

J.-W. Lai et. al., "A 0.27mm² 13.5dBm 2.4GHz All-Digital Polar Transmitter Using 34%-Efficiency Class-D DPA in 40nm CMOS," *IEEE ISSCC*, Feb. 2013.

D. Miyashits, et. al., "A -104dBc/Hz In-Band Phase Noise 3GHz All Digital PLL with Phase Interpolation Based Hierarchical Time to Digital Convertor," *IEEE Symp. on VLSI Circuits*, Jun. 2011.

The left figure shows the fractional-N ADPLL architecture. Operating at half of the DCO-PA's output frequency, the power consumption of the counter and TDC can be reduced by half. Since the phase relationship between the REF and $CKV^{0\text{-}3}$ can be predicted by the accumulation of the fractional part of the frequency control word FCW, the CKV^{i} and CKV^{i+1}, with their rising edges closer to that of the REF, are selected as the inputs to the TDC. This phase selection scheme can reduce the input range of the time-to-digital converter (TDC) by 4x. To operate under a low supply voltage, we utilize a phase-interpolation-based hierarchical TDC with a time resolution of 26 ps to avoid the TDC gain normalization.

31 

To obtain the FM we use a two-point modulation scheme by controlling the same varactor bank for frequency tracking. We add the TX FM data to both the FCW and the control word of the DCO to achieve an all-pass characteristic from the TX FM Data to the output frequency.

Frequency Modulation

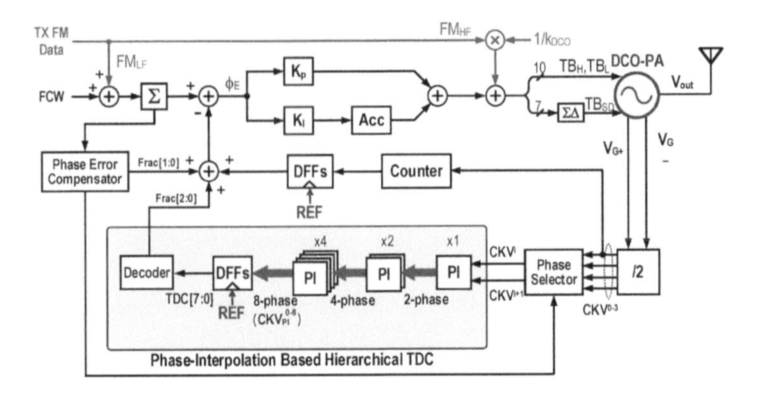

- Two-point modulation to achieve an all-pass characteristic from TX FM Data to output frequency

32 

The ultra-low power transmitter prototype using the DCO-PA is fabricated in ST 65nm CMOS technology without ultra-thick metal. It occupies a core area of 0.39 mm^2.

Chip Micrograph

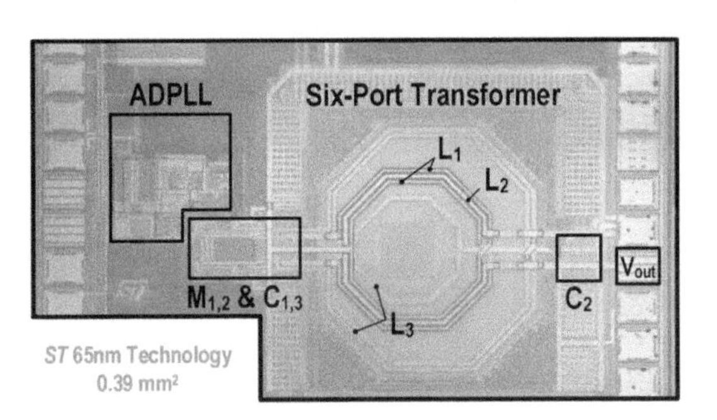

- ST 65nm CMOS Technology (No ultra-thick Metal)
- Core area: 0.39 mm^2

33

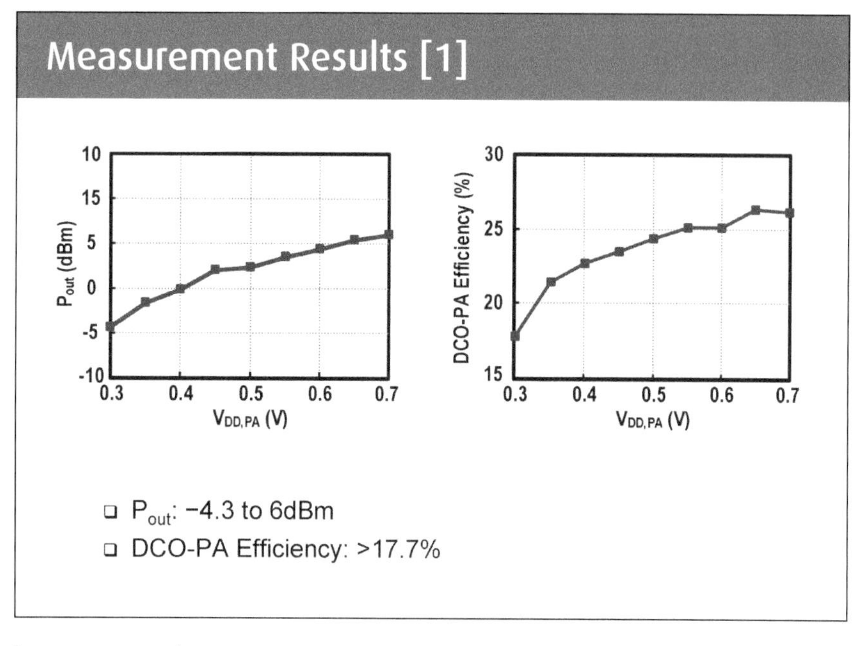

Here shows the measured P_{out} and DCO-PA efficiency versus $V_{DD,PA}$ under V_{GB} = 0.18 V. The cable and board losses of ~2dB is de-embedded when obtaining the output power. The system efficiency reaches 26.2% at 6-dBm P_{out}. A scalable $V_{DD,PA}$ for the DCO-PA can keep high the back-off efficiency (>17.7 %), while offering a wide P_{out} range from −4.3 to 6 dBm.

34

Measurement Results [2]

- DCO-PA phase noise at 2.4GHz: −128.5dBc/Hz@3.5MHz offset frequency
- ADPLL (1) reference: 32MHz, (2) power: 2.4mW (3) settling time: ~13 μs, (4) reference/fractional spur: ~ −72.5 dBc/−41.9 dBc

The measured DCO-PA phase noise at 2.4GHz is −128.5dBc/Hz@3.5MHz offset frequency which is well below −103dBc/Hz requirement of the Zigbee standard. With a 32MHz reference, the ADPLL consumes 2.4 mW and it takes 13 μs for its output to settle down to a frequency error of ±40 ppm, i.e. ±98kHz, of the desired 2.4GHz frequency. When operating at another channel frequency, such as 2.402GHz, the measured reference and fractional spurs are −72.5 and −41.9dBc, respectively, at FCW = 75.0625.

35

Measurement Results [3]

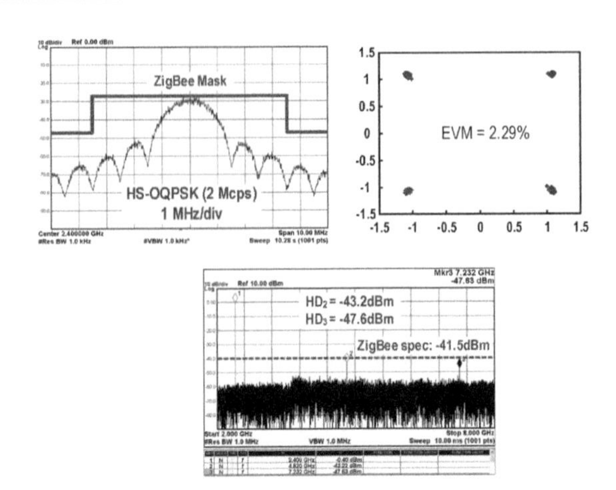

- Zigbee HS-OQPSK modulation spectrum and EVM
- Output spectrum at 0-dBm output

The top left figure shows the measured spectra for HS-OQPSK modulation, which complies with the ZigBee spectral mask with adequate margin. The achieved EVM of 2.29% is well below the specification of 35%. The bottom figure shows the measured output spectrum when delivering a single tone at 0 dBm including cable and PCB losses, the HD_2 and HD_3 are −43.2 and −47.6 dBm, respectively, after adding an external 0.5-pF capacitor for harmonic filtering.

36

Measurement Results [4]

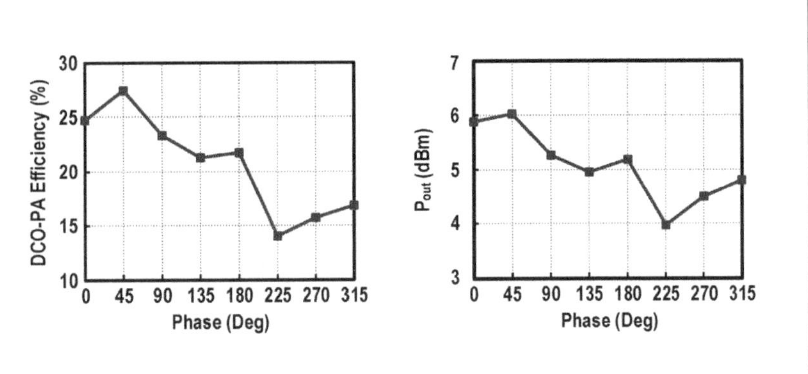

- DCO-PA efficiency and P_{out} under VSWR 1.5:1

Antenna impedance mismatch can potentially pull the oscillation frequency of the DCO-PA, but it should be a slow process induced by humans or other means. Thus, the ADPLL should be able to correct it swiftly if the variation of the oscillation frequency is within the frequency tuning range of the DCO-PA. Under a VSWR of 1.5:1 as a static impedance mismatch, the free-running oscillation frequency of the DCO-PA changes from 2.376 to 2.408 GHz. The ADPLL successfully locks the DCO-PA at all angles, while keeping reasonable DCO-PA efficiency of > 14 % and P_{out} of > 4dBm.

37 

Measurement Results [5]

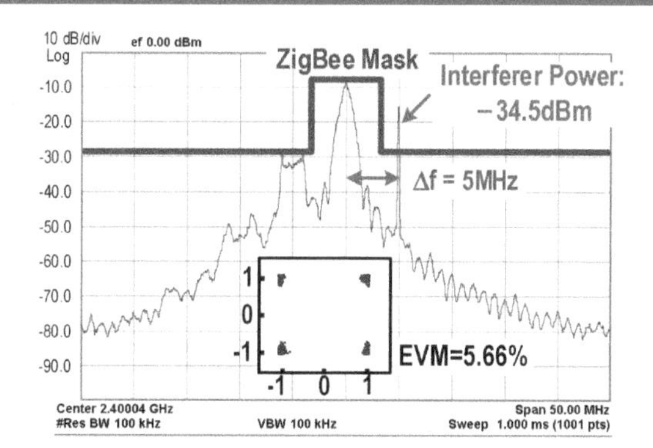

❑ Zigbee modulation spectrum and EVM in presence of a interferer

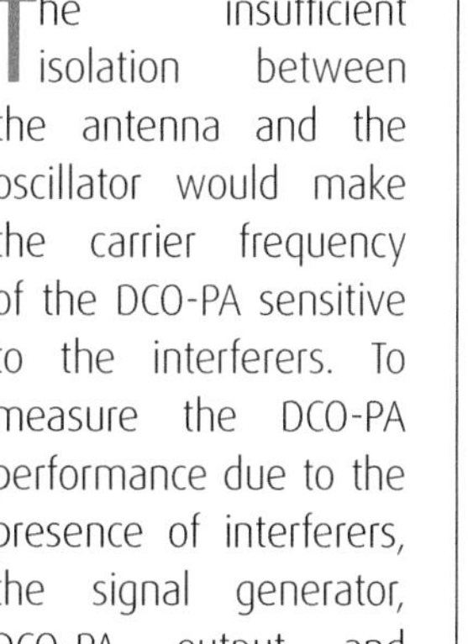

The insufficient isolation between the antenna and the oscillator would make the carrier frequency of the DCO-PA sensitive to the interferers. To measure the DCO-PA performance due to the presence of interferers, the signal generator, DCO-PA output and spectrum analyzer are connected to a circulator which allows the signal to travel in the direction of signal generator to DCO-PA output to spectrum analyzer. When the data is transmitted, the maximum interferer power at Δf = 5 MHz that still meets the ZigBee mask requirement is −34.5dBm and the corresponding EVM is degraded to 5.66%.

38 

Performance Summary

Parameters	This Work	M. Babaie et al. JSSC'16 [4]	C. Li et al. JSSC'16 [6]	J. Prummel et al. JSSC'15 [7]	Y. -H. Liu et al. ISSCC'15 [5]
Applications	2.4 GHz ZigBee	2.4 GHz BLE	2.4 GHz BLE	2.4 GHz BLE	2.4 GHz BLE/ ZigBee/ 2M Proprietary
Architectures	Function-Reuse Class-F DCO-PA + ADPLL	Class-E/F_2 PA + LC-DCO + ADPLL	PA-VCO+ Analog PLL	Class-D PA + LC-VCO + Analog PLL	Class-D PA + LC-DCO + ADPLL
On-chip inductors Or transformers	1 (Shared by DCO, PA and MN)	2 (1 for LC-VCO, 1 for MN)	1 (1 for LC-VCO)	3 (1 for LC-VCO, 2 for MN)	2 (1 for LC-DCO, 1 for MN)
On-Chip MN	Yes	Yes	No	Yes	Yes
Active Area (mm^2)	0.39	0.65	0.35*	0.53*	0.8*
Supply Voltage (V)	0.3 to 0.7 (DCO-PA) 0.7 (ADPLL)	0.5 (DCO) 1 (ADPLL+PA)	1.2	1.2	1
Power Consumption @ P_{out} (DCO/VCO + PA + PA Driver)	4.4 @ 0 dBm 15.3 @ 6 dBm	3.6 @ 0 dBm 5.5 @ 3 dBm	4.46 @ −1 dBm	6.8 @ 0 dBm	3.45 @ −2 dBm
Power Efficiency @ P_{out} (DCO/VCO + PA + PA Driver)	22.6% @ 0 dBm 26.1% @ 6 dBm	28% @ 0 dBm 36% @ 3 dBm	17.9% @ −1 dBm	14.7% @ 0 dBm	18.3% @ −2 dBm
TX Power Consumption (mW) @ P_{out}	6.8 @ 0 dBm 17.7 @ 6 dBm	4.4 @ 0 dBm 6.3 @ 3 dBm	4.56 @ −1 dBm	10.1 @ 0 dBm	4.2 @ −2 dBm
System Efficiency @ P_{out}	14.5% @ 0 dBm 22.6% @ 6 dBm	23% @ 0 dBm 32% @ 3 dBm	17.5% @ −1 dBm #	9.9% @ 0 dBm	15% @ −2 dBm
Settling Time (μS)	13	15	N/A	N/A	15
HD_2/HD_3 @ P_{out} (dBm)	−43.2 / −47.6 & @ 0 dBm P_{out}	−50 / −47 @ 0 dBm P_{out}	< −40 / −40 @ −1.6 dBm P_{out}	<−54 / −52 @ 0 dBm P_{out}	−50.9 / −54.5 @ −2 dBm P_{out}
VCO Phase Noise @ 3.5 MHz (dBc/Hz)	−128.5	−126.9 to −127.9 ^	−131.9 ^	−122.3 ^	N/A
Technology	65 nm CMOS	28 nm CMOS	130 nm CMOS	55 nm CMOS	40 nm CMOS

* Estimated from die photo   & With a 0.5 pF external capacitor   # PLL is off during transmission

^ Normalized from phase noise @ 1-MHz [4], @ 2-MHz [6] and @ 2.5-MHz [7] offset frequency.

This table compares the performance of the proposed DCO-PA and ULP TX with the state-of-the-art. In a comparison with the existing solutions using separated VCO and PA or current re-use VCO-PA, the proposed function-reuse Class-F DCO-PA achieves the smallest chip area of 0.39 mm^2 among all the TXs with on-chip matching networks as well as a competitive power efficiency of 22.6% at a 0-dBm P_{out}.

About the Editors

Yan Lu

Yan Lu received the BEng and MSc degrees in Microelectronics from South China University of Technology, Guangzhou, China, in 2006 and 2009, respectively; and the PhD degree in Electronic and Computer Engineering from the Hong Kong University of Science and Technology (HKUST), Hong Kong, in 2013.

In 2014, he joined the State Key Laboratory of Analog and Mixed-Signal VLSI of University of Macau as an Assistant Professor. His current research interests include, wireless power transfer circuits and systems, RF energy harvesting, highly-integrated power management solutions.

From 2013 to 2018, Yan has contributed 9 ISSCC papers which are distributed in three sub-areas of power management IC: wireless power transfer (3), low-dropout regulators (3), and switched-capacitor DC-DC converters (3). In addition, Yan has co-authored 40+ journal/conference papers,three book chapters, and one book entitled "CMOS Integrated Circuit Design for Wireless Power Transfer" published by Springer. Dr. Lu served as a member of Technical Program Committee of several IEEE conferences, including ISCAS, VLSI-DAT, APCCAS, etc. He also serves as a reviewer of many journals/conferences including IEEE JSSC, TCAS-I, TPEL, etc. Yan was a recipient of the Outstanding Postgraduate Student Award of Guangdong (Canton) Province 2008, the IEEE Solid-State Circuits Society Pre-Doctoral Achievement Award 2013-2014, and the IEEE CAS Society Outstanding Young Author Award 2017. He is a senior member of IEEE.

Chi-Seng Lam

Chi-Seng Lam received the B.Sc., M.Sc. and Ph.D. degrees in Electrical and Electronics Engineering from the University of Macau (UM), Macao, China, in 2003, 2006, and 2012 respectively. From 2006 to 2009, he was an Electrical and Mechanical Engineer in UM. In 2009, he simultaneously worked as a Laboratory Technician and started to pursue his Ph.D. degree, and completed his Ph.D. degree within 3 years. In 2013, he was a Postdoctoral Fellow with The Hong Kong Polytechnic University, Hong Kong, China. He is currently an Assistant Professor in the State Key Laboratory of Analog and Mixed-Signal VLSI, UM, Macao, China. He has co-authored 2 books: *Design and Control of Hybrid Active Power Filters* (Springer, 2014) and *Parallel Power Electronics Filters in Three-Phase Four-Wire Systems - Principle, Control and Design* (Springer, 2016), 2 US patent, 2 Chinese patents and more than 60 technical journals and conference papers. His research interests include integrated power electronics controllers, power management integrated circuits, power quality compensators, smart grid technology, renewable energy, etc.

Dr. Lam received the IEEE Power and Energy Society (PES) Chapter Outstanding Engineer Award in 2016, the Macao Science and Technology Invention Award (Third-Class) and R&D Award for Postgraduates (Ph.D.) in 2014 and 2012 respectively. He also received Macao Government Ph.D. Research Scholarship in 2009-2012, Macao Foundation Postgraduate Research Scholarship in 2003-2005, the 3rd RIUPEEEC Merit Paper Award in 2005, BNU Affinity Card Scholarship in 2001.

Dr. Lam is an IEEE Senior Member. He was the GOLD Officer, Student Branch Officer and a Secretary of IEEE Macau Section in 2007, 2008 and 2015. He is currently the Vice-Chair of IEEE Macau Section, the Chair of IEEE Macau CAS/COM Joint Chapter, and the Secretary of IEEE Macau PES/PELS Joint Chapter.He served as a member of Organizing Committee or Technical Program Committee of several international conferences, such as: IEEE TENCON 2015, ASP-DAC 2016, etc.

About the Authors

Wei-Sung Chang

Wei-Sung Chang was born in Taoyuan, Taiwan, in 1986. He received the B.S. degree and M.S. degree from National Tsing Hua University (NTHU), Hsinchu, Taiwan, in 2008 and 2010, respectively, and the Ph.D.E.E degree from National Taiwan University (NTU), Taipei, Taiwan in 2016.

In 2016, he joined the SerDes Group, MediaTek, Hsinchu, where he has been involved in high-speed wireline transceiver circuits. His current research interests include mixed-mode frequency synthesizer, SerDes transceivers, and ADC.

[Chapter 3]

Lin Cheng

Lin Cheng received the B. Eng degree from Hefei University of Technology, Hefei, China, in 2008, the M. Sc. degree from Fudan University, Shanghai, China, in 2011, and the Ph.D. degree from The Hong Kong University of Science and Technology (HKUST), Kowloon, Hong Kong, in 2016, respectively.

He is currently a Postdoctoral Research Associate with the Department of Electronic and Computer Engineering, HKUST. He was an Intern Analog Design Engineer with OmniVision, Shanghai, China, from Dec. 2010 to July 2011, and with Broadcom Limited, San Jose, USA, from Sept. 2015 to Mar. 2016. His research interests include power management circuits and systems, wireless power transfer circuits and systems, switched-inductor power converters and low dropout regulators.

Dr. Cheng was the recipient of the IEEE Solid-State Circuits Society Pre-Doctoral Achievement Award 2014–2015 and HKUST School of Engineering PhD Fellowship Award 2014-2015.

[Chapter 1]

Wing-Hung Ki

Wing-Hung Ki received the B.Sc. degree from the University of California, San Diego, CA, USA, in 1984, the M.Sc. degree from the California Institute of Technology, Pasadena, CA, USA, in 1985, and the Ph.D. degree from the University of California, Los Angeles, CA, USA, in 1995, all in electrical engineering.

In 1992, he joined Micro Linear Corporation, San Jose, CA, USA, as a Senior Design Engineer with the Department of Power and Battery Management, working on the design of power converter controllers. In 1995, he joined Hong Kong University of Science and Technology, Hong Kong, where he is now a Professor with the Department of Electronic and Computer Engineering. His research interests include power management circuits and systems, switched-inductor and switched-capacitor power converters, low dropout regulators, wireless power transfer for biomedical implants, and analog IC design methodologies.

Prof. Ki served as an Associate Editor of the *IEEE Transactions on Circuits and Systems II* from 2004 to 2005 and from 2012 to 2013, *IEEE Transactions on Power Electronics* since 2016, and the International Technical Program Committee (ITPC) of the *IEEE International Solid-State Circuits Conference* (ISSCC) from 2010 to 2014.

[Chapter 1]

Man-Kay Law

Man-Kay Law received the B.Sc. degree in Computer Engineering and the PhD degree in Electronic and Computer Engineering from Hong Kong University of Science and Technology (HKUST), in 2006 and 2011, respectively.

From February 2011, he joined HKUST as a Visiting Assistant Professor. He is currently an Associate Professor with the State Key Laboratory of Analog and Mixed-Signal VLSI, Faculty of Science and Technology, University of Macau, Macao. His research interests are on the development of ultra-low power sensing circuits and integrated energy harvesting techniques for wireless and biomedical applications.

He was a co-recipient of the ASQED Best Paper Award (2013), A-SSCC Distinguished Design Award (2015) and ASPDAC Best Design Award (2016). He also received the Macao Science and Technology Invention Award (2nd Class) by Macau Government - FDCT (2014). He is a senior member of the IEEE, and serves as a technical committee member in both the IEEE CAS committee on Sensory Systems as well as Biomedical Circuits and Systems.

[Chapter 6]

Tai-Cheng Lee

Tai-Cheng Lee was born in Taiwan in 1970. He received the B.S. degree from National Taiwan University in 1992, the M.S. degree from Stanford University in 1994, and the Ph.D. degree from the University of California, Los Angeles, in 2001, all in electrical engineering.

He worked for LSI Logic from 1994 to 1997 as a circuit design engineer. Since 2001, he has been with the Electrical Engineering Department and GIEE, National Taiwan University, where he is a Professor. His main research interests are in mixed-signal and analog circuit design, data converters, and PLL systems.

Prof. Lee is the TPC on ISSCC , where he is also Far East vice chair. He serves as associate editor (AE) of TCAS-1 and has served AE for TCAS-2 from 2012 to 2015.

[Chapter 3]

Yan Lu

Yan Lu received the BEng and MSc degrees in Microelectronics from South China University of Technology, Guangzhou, China, in 2006 and 2009, respectively; and the PhD degree in Electronic and Computer Engineering from the Hong Kong University of Science and Technology (HKUST), Hong Kong, in 2013.

In 2014, he joined the State Key Laboratory of Analog and Mixed-Signal VLSI of University of Macau as an Assistant Professor. His current research interests include, wireless power transfer circuits and systems, RF energy harvesting, highly-integrated power management solutions.

From 2013 to 2018, Yan has contributed 9 ISSCC papers which are distributed in three sub-areas of power management IC: wireless power transfer (3), low-dropout regulators (3), and switched-capacitor DC-DC converters (3). In addition, Yan has co-authored 40+ journal/conference papers,three book chapters, and one book entitled *CMOS Integrated Circuit Design for Wireless Power Transfer* published by Springer. Dr. Lu served as a member of Technical Program Committee of several IEEE conferences, including ISCAS, VLSI-DAT, APCCAS, etc. He also serves as a reviewer of many journals/conferences including IEEE JSSC, TCAS-I, TPEL, etc. Yan was a recipient of the Outstanding Postgraduate Student Award of Guangdong (Canton) Province 2008, the IEEE Solid-State Circuits Society Pre-Doctoral Achievement Award 2013-2014, and the IEEE CAS Society Outstanding Young Author Award 2017. He is a senior member of IEEE.

[Chapter 2]

Jun Ohta

Jun Ohta received the B.E., M.E., and Dr. Eng. degrees in applied physics, all from the University of Tokyo, Japan, in 1981, 1983, and 1992, respectively.

In 1983, he joined Mitsubishi Electric Corporation, Hyogo, Japan. From 1992 to 1993, he was a visiting researcher in Optoelectronics Computing Systems Center, University of Colorado at Boulder. In 1998, he joined Graduate School of Materials Science, Nara Institute of Science and Technology (NAIST), Nara, Japan as Associate Professor. He was appointed as Professor in 2004. His current research interests are smart CMOS image sensors, retinal prosthesis, and biomedical-photonic LSIs. He received several awards including "The National Commendation for Invention" in 2001, "The Izuo Hayashi Award, Japanese Society of Applied Physics" in 2009, etc.

He is a member of Japan Society of Applied Physics (Fellow), ITE Japan (Fellow), IEE Japan, IEICE Japan, IEEE (Senior Member), and OSA.

[Chapter 5]

Jun Yin

Jun Yin received the B.Sc. and the M.Sc. degrees in Microelectronics from Peking University, Beijing, China, in 2004 and 2007, respectively, and the Ph.D. degree in Electronic and Computer Engineering (ECE) from Hong Kong University of Science and Technology (HKUST), Hong Kong, China, in 2013.

He is currently an Assistant Professor at the State Key Laboratory of Analog and Mixed-Signal VLSI, University of Macau (UM), Macao, China. His research interests are on the low-power CMOS wireless transceivers for IoT application, analog/digital PLLs and integrated oscillators.

Dr. Yin serves as the associate editor of Integration, the VLSI Journal. He has co-authored one technical book (*Transformer-Based Design Techniques for Oscillators and Frequency Dividers*, Springer, 2015).

[Chapter 8]

Dixian Zhao

Dixian Zhao received the B.Sc. degree in microelectronics from Fudan University, Shanghai, China, in 2006, the M.Sc. degree in microelectronics from Delft University of Technology (TU Delft), the Netherlands, in 2009, and the Ph.D. degree in electrical engineering at University of Leuven (KU Leuven), Belgium, in 2015.

From 2008 to 2009, he was with Philips Research, Eindhoven, where he designed a 60-GHz beamforming transmitter for presence detection radar. From 2009 to 2010, he was a research assistant at TU Delft, working on the 94-GHz wideband receiver for imaging radar. From 2010 to 2015, he was a research associate at KU Leuven, where he developed several world-class 60-GHz and E-band transmitters and power amplifiers. Since April 2015, he has joined Southeast University, China, and he is now a Full Professor. His current research interests include RF and millimeter-wave integrated transceivers and power amplifiers for 5G MIMO, Satellite and Radar applications.

Prof. Zhao serves as a member of the Technical Program Committee (TPC) of several conferences, including ESSCIRC, ASICON and IWS. He was the recipient of the 1000-Young-Talent Award in 2016, the Innovative and Entrepreneurial Talent of Jiangsu Province in 2016, the IEEE Solid-State Circuits Society Predoctoral Achievement Award in 2014, the Chinese Government Award for Outstanding Students Abroad in 2013, the Top-Talent Scholarship from TU Delft in 2007 and 2008, and the Samsung Fellowship from Fudan University in 2005. He has authored and co-authored more than 30 IEEE journal and conference papers, one book (60-GHz and E-band Power Amplifiers and Transmitters, Springer Press, 2015), one book chapter (chapter 8, mm-Wave Silicon Power Amplifiers and Transmitters, Cambridge University Press, 2016), and has 3 US patents issued.

[Chapter 7]

Yan Zhu

Yan Zhu received the B.Sc. degree in electrical engineering and automation from Shanghai University, Shanghai, China, in 2006, and the M.Sc. and Ph.D. degrees in electrical and electronics engineering from the University of Macau Macao, China, in 2009 and 2011, respectively. She is now an assistant professor with the State Key Laboratory of Analog and Mixed-Signal VLSI, University of Macau, Macao, China.

She received Best Paper award in ESSCIRC 2014, the Student Design Contest award in A-SSCC 2011, the Chipidea Microelectronics Prize and Macao Scientific and Technological R&D Awards in 2012 and 2014 for outstanding Academic and Research achievements in Microelectronics. She has published more than 50 technical journals and conference papers in her field of interests, and holds 3 US patents. Her research interests include low-power and wideband high-speed Nyquist A/D converters as well as digitally assisted data converter designs.

[Chapter 4]